SPRINGER PROTOCOLS HANDBOOKS

For further volumes:
http://www.springer.com/series/8623

Human Embryonic and Induced Pluripotent Stem Cells

Lineage-Specific Differentiation Protocols

Edited by

Kaiming Ye and Sha Jin

College of Engineering, University of Arkansas, Fayetteville, AR, USA

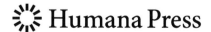

Editors
Kaiming Ye, Ph.D
College of Engineering
University of Arkansas
Fayetteville, AR, USA
kye@uark.edu

Sha Jin, Ph.D
College of Engineering
University of Arkansas
Fayetteville, AR, USA
sjin@uark.edu

ISSN 1949-2448 e-ISSN 1949-2456
ISBN 978-1-61779-266-3 e-ISBN 978-1-61779-267-0
DOI 10.1007/978-1-61779-267-0
Springer New York Dordrecht Heidelberg London

Library of Congress Control Number: 2011934970

© Springer Science+Business Media, LLC 2011
All rights reserved. This work may not be translated or copied in whole or in part without the written permission of the publisher (Humana Press, c/o Springer Science+Business Media, LLC, 233 Spring Street, New York, NY 10013, USA), except for brief excerpts in connection with reviews or scholarly analysis. Use in connection with any form of information storage and retrieval, electronic adaptation, computer software, or by similar or dissimilar methodology now known or hereafter developed is forbidden.
The use in this publication of trade names, trademarks, service marks, and similar terms, even if they are not identified as such, is not to be taken as an expression of opinion as to whether or not they are subject to proprietary rights.

Printed on acid-free paper

Humana Press is part of Springer Science+Business Media (www.springer.com)

Preface

Human embryonic stem (hES) cells, especially the newly developed human induced pluripotent stem (hiPS) cells, have become one of the most promising renewable cell sources for regenerative medicine and tissue engineering. Both hES and iPS cells are capable of self-renewal and differentiation. They can be cultured in vitro for extended periods of time. Such unique properties make these cells very promising for cell replacement therapy. Because of their huge potential in disease treatment and life quality improvement, enormous efforts have been made to develop new methodologies to translate lab discoveries in stem cell research into bedside clinical technologies.

While many new ideas and new approaches in embryonic stem cell research emerge every day, they are widely scattered among laboratories around the world. Thus, the systematic collection of these ideas and approaches will be helpful for researchers and students who want to explore these areas and transform their discoveries into the next generation of regenerative medicine and tissue engineering technologies. This book presents a comprehensive collection of protocols developed by leading scientists in the field. The topics covered include techniques used for maintenance of hES and hiPS cells in either small or large scale; techniques for directing hES and hiPS cell lineage specification; techniques for enhancing maturity of differentiated hES and hiPS cells within three-dimensional scaffolds; techniques for reprogramming adult cells into hiPS cells; techniques for generating patient-specific hiPS cells; and techniques for translating hES and hiPS cell research into new therapies. The book consists of 5 sections and 34 chapters. Chapter 1: Feeder-free growth of undifferentiated human embryonic stem cells, Chap. 2: Growth of human embryonic stem cells in long-term hypoxia, Chap. 3: Laboratory-scale purification of a recombinant E-cadherin-IgG Fc fusion protein that provides a cell surface matrix for extended culture and efficient subculture of human pluripotent stem cells, Chap. 4: Scale-up of single cell-inoculated suspension cultures of human embryonic stem cells, Chap. 5: Three-dimensional culture for expansion and differentiation of embryonic stem cells, Chap. 6: Expansion of pluripotent stem cells in defined xeno-free culture system, Chap. 7: Pluripotent stem cells in vitro from human primordial germ cells, Chap. 8: Cryopreservation of human embryonic stem cells and induced pluripotent stem cells, Chap. 9: Induced pluripotent stem cells from cord blood CD133⁺ cells using Oct4 and Sox2, Chap. 10: Generation, maintenance, and differentiation of hiPS cells from cord blood cells, Chap. 11: Generation of iPS cells from human umbilical vein endothelial cells by lentiviral transduction, and their differentiation to neuron lineage, Chap. 12: Generation of human induced pluripotent stem cells from endoderm origin cells, Chap. 13: Derivation of human induced pluripotent stem cells on autologous feeders, Chap. 14: Human mesenchymal stem cells and iPS cells (preparation methods), Chap. 15: Retroviral-vector-based approaches for the generation of human induced pluripotent stem cells from fibroblasts and keratinocytes, Chap. 16: Generation of nonviral integration-free induced pluripotent stem cells from plucked human hair follicles, Chap. 17: Generation of iPS cells from human skin biopsy, Chap. 18: Generation of human iPS cells from human primary amnion cells, Chap. 19: In Vitro two-dimensional endothelial differentiation of human embryonic stem cells, Chap. 20: A feeder-free culture method for the high efficient production of subcultural vascular endothelial cells from human embryonic stem cells, Chap. 21: hES cell feeder-free culture and feeder-independent maintenance of human embryonic stem cells and directed differentiation into endothelial cells under hypoxic condition, Chap. 22: Differentiation of endothelial cells from human embryonic stem cells and induced pluripotent stem cells, Chap. 23: Differentiation of human embryonic and induced pluripotent stem cells into blood cells in coculture with murine stromal cells, Chap. 24: Generation of multipotent $CD34^+$ $CD45^+$ hematopoietic progenitors from human induced pluripotent stem cells, Chap. 25: Adipogenic differentiation of human induced pluripotent stem cells, Chap. 26: Chondrogenic differentiation of hESC in micromass culture, Chap. 27: Deriving hepatic endoderm from pluripotent stem cells, Chap. 28: Multistage hepatic differentiation from human induced pluripotent stem cells, Chap. 29: Hepatic maturation of hES cells by using a murine MSC line derived from fetal livers, Chap. 30: Generation of lung epithelial-like tissue from hESC by air–liquid interface culture, Chap. 31: Directed differentiation of human embryonic stem cells into selective neurons on nanoscale ridge/groove pattern arrays, Chap. 32: Neural differentiation of human ES and iPS cells in three-dimensional collagen and Matrigel™ gels, Chap. 33: Single-cell transcript profiling of differentiating human embryonic stem cells: Lineage-specific differentiation protocols, and Chap. 34: Using endogenous mRNA expression patterns to visualize neural differentiation of human pluripotent stem Cells.

Finally, we would like to take this opportunity to acknowledge all of the contributors for their dedicated works and their kindness to share their wisdom and their expertise in the field. We are also grateful to Mr. Patrick J. Marton at Humana Press for his support and encouragement. Without his encouragement, the publishing of this book would not have been possible. In addition, we would like to thank Mr. David C. Casey for his editing and support on publishing this book. We hope this book will further stimulate the development of new technology for lineage-specific differentiation of hES and hiPS cells for generating a full spectrum of clinically relevant cell lineages for cell replacement therapies. We anticipate that this book will contribute to the development of new technologies for tissue engineering and regenerative medicine.

Fayetteville, AR *Kaiming Ye*
Fayetteville, AR *Sha Jin*

Contents

Preface... v
Contributors... xi

PART I MAINTENANCE AND CRYOPRESERVATION OF ES AND iPS CELLS

1 Feeder-Free Growth of Undifferentiated Human Embryonic Stem Cells.......... 3
   Dong-Youn Hwang
2 Growth of Human Embryonic Stem Cells in Long-Term Hypoxia............... 13
   Vladimir Zachar, Simon C. Weli, Mayuri S. Prasad, and Trine Fink
3 Laboratory-Scale Purification of a Recombinant E-Cadherin–IgG Fc
   Fusion Protein That Provides a Cell Surface Matrix for Extended Culture
   and Efficient Subculture of Human Pluripotent Stem Cells................... 25
   Masato Nagaoka and Stephen A. Duncan
4 Scale-Up of Single Cell–Inoculated Suspension
   Cultures of Human Embryonic Stem Cells................................. 37
   Harmeet Singh, Pamela Mok, and Robert Zweigerdt
5 Three-Dimensional Culture for Expansion
   and Differentiation of Embryonic Stem Cells.............................. 51
   Guang-wei Sun, Xiao-xi Xu, Nan Li, Ying Zhang, and Xiao-jun Ma
6 Expansion of Pluripotent Stem Cells in Defined, Xeno-Free Culture System..... 59
   Kristiina Rajala
7 Pluripotent Stem Cells In Vitro from Human Primordial Germ Cells........... 71
   Behrouz Aflatoonian and Harry D. Moore
8 Cryopreservation of Human Embryonic Stem Cells
   and Induced Pluripotent Stem Cells..................................... 85
   Frida Holm

PART II GENERATION OF HUMAN INDUCED PLURIPOTENT STEM CELLS:
         VIRAL AND NONVIRAL VECTOR-BASED NUCLEAR REPROGRAMMING

9 Induced Pluripotent Stem Cells (iPSC) from Cord
   Blood CD133⁺ Cells Using Oct4 and Sox2................................ 93
   Alessandra Giorgetti, Nuria Montserrat, and Juan Carlos Izpisua Belmonte
10 Generation, Maintenance, and Differentiation of Human iPS
   Cells from Cord Blood... 113
   Naoki Nishishita, Chiemi Takenaka, and Shin Kawamata

11 Generation of iPS Cells from Human Umbilical Vein Endothelial Cells
by Lentiviral Transduction and Their Differentiation to Neuronal Lineage 133
*Maria V. Shutova, Ilya V. Chestkov, Alexandra N. Bogomazova,
Maria A. Lagarkova, and Sergey L. Kiselev*

12 Generation of Human Induced Pluripotent Stem
Cells from Endoderm Origin Cells 151
Hua Liu, Su Mi Choi, and Yoon-Young Jang

13 Derivation of Human Induced Pluripotent Stem Cells on Autologous Feeders 161
Kazutoshi Takahashi

14 Human Mesenchymal Stem Cells and iPS Cells (Preparation Methods) 173
Hiroe Ohnishi, Yasuaki Oda, and Hajime Ohgushi

15 Retroviral Vector-Based Approaches for the Generation of
Human Induced Pluripotent Stem Cells from Fibroblasts and Keratinocytes 191
Athanasia D. Panopoulos, Sergio Ruiz, and Juan Carlos Izpisua Belmonte

16 Generation of Nonviral Integration-Free Induced Pluripotent
Stem Cells from Plucked Human Hair Follicles 203
Ann Peters and Elias T. Zambidis

PART III GENERATION OF PATIENT-SPECIFIC iPS CELLS
FOR CLINICAL APPLICATION

17 Generation of iPS Cells from Human Skin Biopsy 231
Katie Avery and Stuart Avery

18 Generation of Induced Pluripotent Stem Cells from Human Amnion Cells 249
*Masashi Toyoda, Shogo Nagata, Hatsune Makino, Hidenori Akutsu,
Takashi Tada, and Akihiro Umezawa*

PART IV LINEAGE-SPECIFIC DIFFERENTIATION OF hES AND iPS CELLS

19 In Vitro Two-Dimensional Endothelial Differentiation
of Human Embryonic Stem Cells 267
Xiaolong Lin, Hua Jiang, Zack Zhengyu Wang, and Tong Chen

20 Feeder-Free Culture for High Efficiency Production of Subculturable
Vascular Endothelial Cells from Human Embryonic Stem Cells 277
Kumiko Saeki

21 Feeder-Independent Maintenance of Human Embryonic Stem Cells
and Directed Differentiation into Endothelial Cells Under Hypoxic Condition 295
Xiuli Wang

22 Differentiation of Endothelial Cells from Human Embryonic
Stem Cells and Induced Pluripotent Stem Cells 311
Shijun Hu, Preston Lavinghousez, Zongjin Li, and Joseph C. Wu

23 Differentiation of Human Embryonic and Induced Pluripotent Stem
Cells into Blood Cells in Coculture with Murine Stromal Cells 321
*Feng Ma, Yanzheng Gu, Natsumi Nishihama, Wenyu Yang,
Ebihara Yasuhiro, and Kohichiro Tsuji*

24 Generation of Multipotent CD34⁺CD45⁺ Hematopoietic
   Progenitors from Human Induced Pluripotent Stem Cells   337
   Tea Soon Park, Paul W. Burridge, and Elias T. Zambidis

25 Adipogenic Differentiation of Human Induced Pluripotent Stem Cells   351
   *Michio Noguchi, Masakatsu Sone, Daisuke Taura, Ken Ebihara,
   Kiminori Hosoda, and Kazuwa Nakao*

26 Chondrogenic Differentiation of hESC in Micromass Culture.   359
   Deborah Ferrari, Guochun Gong, Robert A. Kosher, and Caroline N. Dealy

27 Deriving Metabolically Active Hepatic Endoderm
   from Pluripotent Stem Cells   369
   *Claire N. Medine, Zara Hannoun, Sebastian Greenhough,
   Catherine M. Payne, Judy Fletcher, and David C. Hay*

28 Multistage Hepatic Differentiation from Human Induced Pluripotent Stem Cells . .   387
   Su Mi Choi, Hua Liu, Yonghak Kim, and Yoon-Young Jang

29 Hepatic Maturation of hES Cells by Using a Murine
   Mesenchymal Cell Line Derived from Fetal Livers.   397
   Takamichi Ishii and Kentaro Yasuchika

30 Generation of Lung Epithelial-Like Tissue from hESC
   by Air–Liquid Interface Culture   405
   *Lindsey Van Haute, Gert De Block, Inge Liebaers, Karen Sermon,
   and Martine De Rycke*

31 Direct Differentiation of Human Embryonic Stem Cells
   into Selective Neurons on Nanoscale Ridge/Groove Pattern Arrays   413
   Kye-Seong Kim, Hosup Jung, and Keesung Kim

PART V DIRECTED DIFFERENTIATION OF HES AND IPS CELLS
        IN 3D ENVIRONMENTS

32 Neural Differentiation of Human ES and iPS Cells
   in Three-Dimensional Collagen and Matrigel™ Gels.   427
   Eric Derby, Dezhong Yin, Wei-Qiang Gao, and Wu Ma

PART VI QUALITATIVE AND QUANTITATIVE ANALYSIS OF DYNAMICS
         OF HES CELL DIFFERENTIATION

33 Single-Cell Transcript Profiling of Differentiating Embryonic Stem Cells   445
   *Jason D. Gibson, Caroline M. Jakuba, Craig E. Nelson,
   and Mark G. Carter*

34 Using Endogenous MicroRNA Expression Patterns
   to Visualize Neural Differentiation of Human Pluripotent Stem Cells   465
   Agnete Kirkeby, Malin Parmar, and Johan Jakobsson

Index. *481*

Contributors

BEHROUZ AFLATOONIAN • *IVF Unit, Madar Hospital, Yazd, Iran*
HIDENORI AKUTSU • *Department of Reproductive Biology, National Institute for Child Health and Development, Tokyo, Japan*
KATIE AVERY • *Institute of Medical Biology, A*STAR, Singapore*
STUART AVERY • *Institute of Medical Biology, A*STAR, Singapore*
JUAN CARLOS IZPISUA BELMONTE • *Gene Expression Laboratory, Salk Institute for Biological Studies, La Jolla, CA, USA; Center of Regenerative Medicine in Barcelona, Barcelona, Spain*
ALEXANDRA N. BOGOMAZOVA • *Vavilov Institute of General Genetics RAS, Moscow, Russia*
PAUL W. BURRIDGE • *Institute for Cell Engineering, Stem Cell Program, Johns Hopkins University School of Medicine, Baltimore, MD, USA*
MARK G. CARTER • *Center for Regenerative Biology, University of Connecticut, Storrs, Storrs, CT, USA; Animal Science Department, University of Connecticut, Storrs, Storrs, CT, USA*
TONG CHEN • *Department of Hematology, Huashan Hospital, Fudan University, Shanghai, China*
ILYA V. CHESTKOV • *Vavilov Institute of General Genetics RAS, Moscow, Russia*
SU MI CHOI • *Stem Cell Biology Laboratory, Johns Hopkins University School of Medicine, Baltimore, MD, USA*
GERT DE BLOCK • *Department of Embryology and Genetics, Vrije Universiteit Brussel, Brussels, Belgium*
MARTINE DE RYCKE • *Department of Embryology and Genetics, Vrije Universiteit Brussel, Brussels, Belgium; Centre for Medical Genetics, UZ Brussel, Brussels, Belgium*
CAROLINE N. DEALY • *Center for Regenerative Medicine and Skeletal Development, Department of Reconstructive Sciences, Department of Orthopaedic Surgery, University of Connecticut Health Center, Farmington, CT, USA*
ERIC DERBY • *BioDefense and Emerging Infections, American Type Culture Collection, Manassas, VA, USA*
STEPHEN A. DUNCAN • *Department of Cell Biology, Neurobiology and Anatomy, Medical College of Wisconsin, Milwaukee, WI, USA*
KEN EBIHARA • *Department of Medicine and Clinical Science, Kyoto University Graduate School of Medicine, Kyoto, Japan*
DEBORAH FERRARI • *Center for Regenerative Medicine and Skeletal Development, Department of Reconstructive Sciences, Department of Orthopaedic Surgery, University of Connecticut Health Center, Farmington, CT, USA*
TRINE FINK • *Laboratory for Stem Cell Research, Aalborg University, Aalborg, Denmark*
JUDY FLETCHER • *MRC Centre for Regenerative Medicine, Edinburgh, UK*

WEI-QIANG GAO • *Renji Stem Cell Research Center, Med-X Research Institute, Shanghai Jiao Tong University, Shanghai, China*

JASON D. GIBSON • *Molecular and Cell Biology Department, University of Connecticut, Storrs, CT, USA*

ALESSANDRA GIORGETTI • *Center of Regenerative Medicine in Barcelona, Barcelona, Spain*

GUOCHUN GONG • *Center for Regenerative Medicine and Skeletal Development, Department of Reconstructive Sciences, Department of Orthopaedic Surgery, University of Connecticut Health Center, Farmington, CT, USA*

SEBASTIAN GREENHOUGH • *MRC Centre for Regenerative Medicine, Edinburgh, UK*

YANZHENG GU • *Division of Stem Cell Processing, Centre for Stem Cell Biology and Regenerative Medicine, Institute of Medical Science, University of Tokyo, Tokyo, Japan; Jiangsu Provincial Stem Cell Key Laboratory, Suchoow University, Suzhou, China*

ZARA HANNOUN • *MRC Centre for Regenerative Medicine, Edinburgh, UK*

DAVID C. HAY • *MRC Centre for Regenerative Medicine, Edinburgh, UK*

FRIDA HOLM • *Division of Obstetrics and Gynecology, Department of Clinical Science, Intervention and Technology, Karolinska Institutet, K57, Karolinska University Hospital, Huddinge, Stockholm, Sweden*

KIMINORI HOSODA • *Department of Medicine and Clinical Science, Kyoto University Graduate School of Medicine, Kyoto, Japan*

SHIJUN HU • *Department of Medicine, Division of Cardiology, Stanford University School of Medicine, Stanford, CA, USA; Department of Radiology, Stanford University School of Medicine, Stanford, CA, USA*

DONG-YOUN HWANG • *CHA Stem Cell Institute, College of Medicine, CHA University, South Korea*

TAKAMICHI ISHII • *Department of Surgery, Graduate School of Medicine, Kyoto University, Kyoto, Japan*

JOHAN JAKOBSSON • *Department of Experimental Medical Sciences, Wallenberg Neuroscience Center, Lund University, Lund, Sweden*

CAROLINE M. JAKUBA • *Molecular and Cell Biology Department, University of Connecticut, Storrs, CT, USA*

YOON-YOUNG JANG • *Stem Cell Biology Laboratory, Johns Hopkins University School of Medicine, Baltimore, MD, USA*

HUA JIANG • *Gynecology and Obstetrics Hospital, Fudan University, Shanghai, China*

HOSUP JUNG • *School of Mechanical and Aerospace Engineering, Seoul National University, Seoul, South Korea*

SHIN KAWAMATA • *The Basic Research Group for Regenerative Medicine, Foundation for Biomedical Research and Innovation (FBRI), Kobe, Japan; Riken Center for Developmental Biology, Kobe, Japan*

KEESUNG KIM • *School of Mechanical and Aerospace Engineering, Seoul National University, Seoul, South Korea*

KYE-SEONG KIM • *Department of Anatomy and Cell Biology, College of Medicine, Hanyang University, Seoul, South Korea; Department of Biomedical Science, Hanyang University Graduate School of Biomedical Science and Engineering, Seoul, South Korea*

YONGHAK KIM • *Stem Cell Biology Laboratory, Johns Hopkins University School of Medicine, Baltimore, MD, USA*
AGNETE KIRKEBY • *Department of Experimental Medical Sciences, Wallenberg Neuroscience Center, Lund University, Lund, Sweden*
SERGEY L. KISELEV • *Vavilov Institute of General Genetics RAS, Moscow, Russia*
ROBERT A. KOSHER • *Center for Regenerative Medicine and Skeletal Development, Department of Reconstructive Sciences, Department of Orthopaedic Surgery, University of Connecticut Health Center, Farmington, CT, USA*
MARIA A. LAGARKOVA • *Vavilov Institute of General Genetics RAS, Moscow, Russia*
PRESTON LAVINGHOUSEZ • *Department of Medicine, Division of Cardiology, Stanford University School of Medicine, Stanford, CA, USA; Department of Radiology, Stanford University School of Medicine, Stanford, CA, USA*
NAN LI • *Laboratory of Biomedical Material Engineering, Dalian Institute of Chemical Physics, Chinese Academy of Sciences, Dalian, PR China*
ZONGJIN LI • *Department of Medicine, Division of Cardiology, Stanford University School of Medicine, Stanford, CA, USA; Department of Radiology, Stanford University School of Medicine, Stanford, CA, USA*
INGE LIEBAERS • *Department of Embryology and Genetics, Vrije Universiteit Brussel, Brussels, Belgium*
XIAOLONG LIN • *Gynecology and Obstetrics Hospital, Fudan University, Shanghai, China*
HUA LIU • *Stem Cell Biology Laboratory, Johns Hopkins University School of Medicine, Baltimore, MD, USA*
FENG MA • *Division of Stem Cell Processing, Centre for Stem Cell Biology and Regenerative Medicine, Institute of Medical Science, University of Tokyo, Tokyo, Japan; Institute of Blood Transfusion, Chinese Academy of Medical Sciences, Chengdu, China; State Key Laboratory of Experimental Hematology, Institute of Hematology, Chinese Academy of Medical Sciences, Tianjin, China; Jiangsu Provincial Stem Cell Key Laboratory, Suchoow University, Suzhou, China*
WU MA • *Stem Cell Center, American Type Culture Collection, Manassas, VA, USA; Ectycell, Romainville, France*
XIAO-JUN MA • *Laboratory of Biomedical Material Engineering, Dalian Institute of Chemical Physics, Chinese Academy of Sciences, Dalian, PR China*
HATSUNE MAKINO • *Department of Reproductive Biology, National Institute for Child Health and Development, Tokyo, Japan*
CLAIRE N. MEDINE • *MRC Centre for Regenerative Medicine, Chancellor's Building, Edinburgh, UK*
PAMELA MOK • *Institute of Medical Biology (IMB), Singapore; Cardiovascular Research Institute, University of California San Francisco, San Francisco, CA, USA*
NURIA MONTSERRAT • *Center of Regenerative Medicine in Barcelona, Barcelona, Spain*
HARRY D. MOORE • *Centre for Stem Cell Biology, Department of Biomedical Science, Sheffield University, Sheffield, UK*
MASATO NAGAOKA • *Department of Cell Biology, Neurobiology and Anatomy, Medical College of Wisconsin, Milwaukee, WI, USA*
SHOGO NAGATA • *Stem Cell Engineering, Institute for Frontier Medical Sciences, Kyoto University, Kyoto, Japan*

KAZUWA NAKAO • *Department of Medicine and Clinical Science, Kyoto University Graduate School of Medicine, Kyoto, Japan*
CRAIG E. NELSON • *Molecular and Cell Biology Department, University of Connecticut, Storrs, CT, USA*
NATSUMI NISHIHAMA • *Division of Stem Cell Processing, Centre for Stem Cell Biology and Regenerative Medicine, Institute of Medical Science, University of Tokyo, Tokyo, Japan*
NAOKI NISHISHITA • *The Basic Research Group for Regenerative Medicine, Foundation for Biomedical Research and Innovation (FBRI), Kobe, Japan;*
MICHIO NOGUCHI • *Department of Medicine and Clinical Science, Kyoto University Graduate School of Medicine, Kyoto, Japan*
YASUAKI ODA • *Tissue Engineering Research Group, Health Research Institute, National Institute of Advanced Industrial Science and Technology (AIST), Amagasaki, Hyogo, Japan*
HAJIME OHGUSHI • *Tissue Engineering Research Group, Health Research Institute, National Institute of Advanced Industrial Science and Technology (AIST), Amagasaki, Hyogo, Japan*
HIROE OHNISHI • *Tissue Engineering Research Group, Health Research Institute, National Institute of Advanced Industrial Science and Technology (AIST), Amagasaki, Hyogo, Japan*
ATHANASIA D. PANOPOULOS • *Gene Expression Laboratory, Salk Institute for Biological Studies, La Jolla, CA, USA*
TEA SOON PARK • *Institute for Cell Engineering, Stem Cell Program, Johns Hopkins University School of Medicine, Baltimore, MD, USA*
MALIN PARMAR • *Department of Experimental Medical Sciences, Wallenberg Neuroscience Center, Lund University, Lund, Sweden*
CATHERINE M. PAYNE • *MRC Centre for Regenerative Medicine, Edinburgh, UK*
ANN PETERS • *Institute for Cell Engineering, Stem Cell Program, Johns Hopkins University School of Medicine, Baltimore, MD, USA*
MAYURI S. PRASAD • *Laboratory for Stem Cell Research, Aalborg University, Aalborg, Denmark*
KRISTIINA RAJALA • *REGEA, Institute for Regenerative Medicine, University of Tampere, Tampere University Hospital, Tampere, Finland*
SERGIO RUIZ • *Gene Expression Laboratory, Salk Institute for Biological Studies, La Jolla, CA, USA*
KUMIKO SAEKI • *Department of Disease Control, Research Institute, National Center for Global Health and Medicine, Tokyo, Japan*
KAREN SERMON • *Department of Embryology and Genetics, Vrije Universiteit Brussel, Brussels, Belgium*
MARIA V. SHUTOVA • *Vavilov Institute of General Genetics RAS, Moscow, Russia*
HARMEET SINGH • *Institute of Medical Biology (IMB), Singapore*
MASAKATSU SONE • *Department of Medicine and Clinical Science, Kyoto University Graduate School of Medicine, Kyoto, Japan*
GUANG-WEI SUN • *Laboratory of Biomedical Material Engineering, Dalian Institute of Chemical Physics, Chinese Academy of Sciences, Dalian, PR China*

TAKASHI TADA • *Stem Cell Engineering, Institute for Frontier Medical Sciences, Kyoto University, Kyoto, Japan*
KAZUTOSHI TAKAHASHI • *Center for iPS Cell Research and Application, Institute for Integrated Cell-Material Sciences, Kyoto University, Kyoto, Japan*
CHIEMI TAKENAKA • *The Basic Research Group for Regenerative Medicine, Foundation for Biomedical Research and Innovation (FBRI), Kobe, Japan;*
DAISUKE TAURA • *Department of Medicine and Clinical Science, Kyoto University Graduate School of Medicine, Kyoto, Japan*
MASASHI TOYODA • *Department of Reproductive Biology, National Institute for Child Health and Development, Tokyo, Japan*
KOHICHIRO TSUJI • *Division of Stem Cell Processing, Centre for Stem Cell Biology and Regenerative Medicine, Institute of Medical Science, University of Tokyo, Tokyo, Japan*
AKIHIRO UMEZAWA • *Department of Reproductive Biology, National Institute for Child Health and Development, Tokyo, Japan*
LINDSEY VAN HAUTE • *Department of Embryology and Genetics, Vrije Universiteit Brussel, Brussels, Belgium*
XIULI WANG • *Department of Biomedical Engineering, Tufts University, Medford, MA, USA; Dalian Institute of Chemical and Physics, Chinese Academy of Sciences, Liaoning, China*
ZACK ZHENGYU WANG • *Center for Molecular Medicine, Maine Medical Center Research Institute, Scarborough, ME, USA*
SIMON C. WELI • *Section for Fish Health, National Veterinary Institute, Oslo, Norway*
JOSEPH C. WU • *Department of Medicine, Division of Cardiology, Stanford University School of Medicine, Stanford, CA, USA; Department of Radiology, Stanford University School of Medicine, Stanford, CA, USA; Institute of Stem Cell Biology and Regenerative Medicine, Stanford University School of Medicine, Stanford, CA, USA*
XIAO-XI XU • *Laboratory of Biomedical Material Engineering, Dalian Institute of Chemical Physics, Chinese Academy of Sciences, Dalian, PR China; Graduate School of the Chinese Academy, Becjing, PR China*
WENYU YANG • *Division of Stem Cell Processing, Centre for Stem Cell Biology and Regenerative Medicine, Institute of Medical Science, University of Tokyo, Tokyo, Japan; State Key Laboratory of Experimental Hematology, Institute of Hematology, Chinese Academy of Medical Sciences, Tianjin, China*
KENTARO YASUCHIKA • *Department of Surgery, Graduate School of Medicine, Kyoto University, Kyoto, Japan*
EBIHARA YASUHIRO • *Division of Stem Cell Processing, Centre for Stem Cell Biology and Regenerative Medicine, Institute of Medical Science, University of Tokyo, Tokyo, Japan*
DEZHONG YIN • *Stem Cell Center, American Type Culture Collection, Manassas, VA, USA*
VLADIMIR ZACHAR • *Laboratory for Stem Cell Research, Aalborg University, Aalborg, Denmark*

ELIAS T. ZAMBIDIS • *Institute for Cell Engineering, Stem Cell Program, Johns Hopkins University School of Medicine, Baltimore, MD, USA; Divisions of Pediatric Oncology and Cancer Biology, Sidney Kimmel Comprehensive Cancer Center at Johns Hopkins, Baltimore, MD, USA*

YING ZHANG • *Laboratory of Biomedical Material Engineering, Dalian Institute of Chemical Physics, Chinese Academy of Sciences, Dalian , PR China*

ROBERT ZWEIGERDT • *Institute of Medical Biology (IMB), Singapore; Hannover Medical School (MHH), Department of Cardiac, Thoracic, Transplantation and Vascular Surgery (HTTG), Leibniz Research Laboratories for Biotechnology and Artificial Organs (LEBAO), Hannover, Germany*

Part I

Maintenance and Cryopreservation of ES and iPS Cells

Chapter 1

Feeder-Free Growth of Undifferentiated Human Embryonic Stem Cells

Dong-Youn Hwang

Abstract

Conventionally, human embryonic stem cells (hESCs) are cultured on feeder cells. The most commonly used feeder cells are mouse embryonic fibroblasts. It is thought that the feeder cells provide an optimal microenvironment for the undifferentiated growth of hESCs by supplying currently unidentified extracellular matrix components and growth factors. Although these feeder cells are efficient, the feeder-dependent hESC culture system may not be suitable for drug screening or other research purposes that require the use of feeder-free culture methods. In this chapter, we describe in detail protocols for three widely used feeder-free systems for culturing hESCs: Matrigel-mTeSR1, CELLstart-STEMPRO® hESC SFM, and vitronectin-mTeSR1.

Key words: Human embryonic stem cells, Feeder-free culture, Matrigel, mTeSR1, CELLstart, STEMPRO® hESC SFM, Vitronectin

1. Introduction

Since the first embryonic stem cell culture was established (1), human embryonic stem cells (hESCs) have drawn widespread attention for their potential to treat many currently incurable diseases. Because of their ability to differentiate into most cell types found in the human body, hESCs provide valuable options not only for therapeutic advancements but also for basic and clinical research. Conventionally, hESCs are grown on feeder cells such as primary mouse embryonic fibroblasts (MEFs) or STO cells (mouse embryonic fibroblast cell line), which secrete extracellular matrix (ECM) components and growth factors required for the undifferentiated growth of hESCs. However, in some cases, hESCs must be cultured without feeder cells. For example, the effects of drug treatments or genetic modifications on the physiology and function of hESCs can be obscured by alterations in the feeder cell conditions. In addition, the use of feeder-free culture systems results in low levels of experimental variation and allows easy standardization of experimental conditions due to the low batch-to-batch variation of the ECMs. Large-scale cultivation of hESCs for future clinical applications would also require the use of feeder-free culture systems.

Feeder-free hESC culture was first reported in 2001 (2). In this method, hESCs are cultured either on Matrigel or laminin using MEF-conditioned medium. hESCs were maintained in an undifferentiated state for more than 130 passages under these conditions without losing pluripotency. Since then, several different feeder-free culture systems have been established. For example, in 2005, Ludwig et al. developed a serum-free defined culture medium, TeSR1, and they succeeded in culturing hESC lines for 2–6 months (3). In this case, Matrigel or a mixture of collagen IV, laminin, fibronectin, and vitronectin was shown to work efficiently with TeSR1 medium (3). Various efforts have been made by other groups to develop feeder-free hESC culture systems (4–8).

Several feeder-free hESC culture systems are commercially available. Among them, Matrigel-mTeSR1 and CELLstart-STEMPRO® hESC SFM have been used in many laboratories. A recent report has suggested that vitronectin, an abundant protein found in the ECM, works well with mTeSR1 (9). Recently, we extensively compared these three feeder-free systems for hESC culture and reported both the similarities and differences between these systems (10).

Intriguingly, the size and thickness of the hESC colonies vary depending on the culture system used (Fig. 1). For CHA6-hESCs, colonies cultured using CELLstart-STEMPRO® hESC SFM were much bigger, albeit thinner, than those grown using the other two systems. The number of cells in each colony was greater with the CELLstart-STEMPRO® hESC SFM culture system, and cell cycle analysis indicated that hESCs grew faster with the CELLstart-STEMPRO® hESC SFM culture system relative to the other two culture systems. The sizes of the CHA6-hESC colonies cultured using both Matrigel-mTeSR1 and vitronectin-mTeSR1 were very similar to the sizes of those cultured on MEF feeder cells (10).

Regardless of the differences in size and thickness among colonies grown using the different feeder-free culture systems, all three systems seemed to be able to efficiently support the undifferentiated growth of hESCs as demonstrated by the expression of hESC markers (10). Gene expression profiles of hESCs cultured in these three feeder-free systems were very similar to those under feeder-dependent conditions, indicating that the undifferentiated state of hESCs can be maintained using all three feeder-free culture systems (Fig. 2).

These results suggest that the three feeder-free culture methods tested (CELLstart-STEMPRO® hESC SFM, Matrigel-mTeSR1,

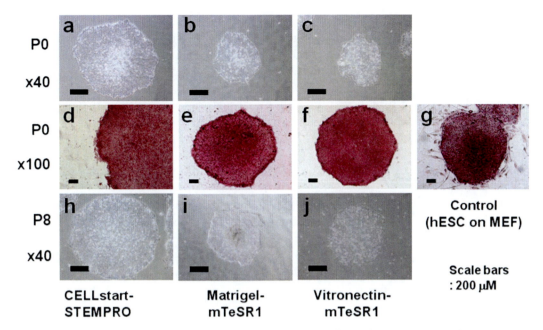

Fig. 1. Morphological comparison of hESC colonies cultured using three feeder-free systems and a feeder-dependent system. CHA6-hESCs (CHA Hospital, Seoul, Korea) were used for this characterization. Scale bars: (**a–c**), 500 μm; (**d–f**), 100 μm; and (**h–j**), 500 μm. (This figure was adapted from (10) with minor modifications).

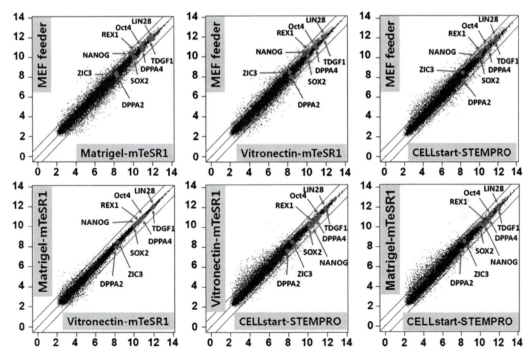

Fig. 2. Pairwise comparisons of gene expression profiles among hESCs cultured under feeder-free and feeder-dependent conditions. Affymetrix GeneChip Human Gene 1.0 ST arrays containing 28,869 gene-level probe sets were used. The scatter plots show that most of the genes were located close to the diagonal line of the plots, indicating similar gene expression patterns among hESCs grown using different culture systems. (This figure was adapted from (10) with minor modifications).

and vitronectin-mTeSR1) can be used to grow hESCs in the undifferentiated state without the need of feeder cells, thus providing unique opportunities for drug screening or for research involving genetic modifications of hESCs.

In this chapter, protocols for the three feeder-free hESC culture systems used in our laboratories are described in detail.

2. Materials

2.1. Reagents for the Matrigel-mTeSR1 Feeder-Free hESC Culture System

1. BD Matrigel (hESC-qualified, 5 ml) (BD Bioscience, Cat No. 354277). Stable for at least 3 months if stored at −20°C.
2. Dulbecco's modified Eagle's medium: Nutrient Mixture F-12 (DMEM-F12) (Invitrogen, Cat No. 11320-033). Store at 2–8°C. Protect from light.
3. mTeSR1 Basal Medium (Stem Cell Technology, Cat No. 05851). Store at 2–8°C. Protect from light. Stable for 1 year.
4. mTeSR1 Supplement (Stem Cell Technology, Cat No. 05852). Store at −20°C. Avoid repeated freezing and thawing. Stable for at least 1 year.

2.2. Reagents for the CELLstart-STEMPRO® hESC SFM Feeder-Free hESC Culture System

1. CELLstart (Invitrogen Cat No. A10142-01). Store at 2–8°C. Protect from light.
2. Dulbecco's modified Eagle's medium: Nutrient Mixture F-12 (DMEM-F12) with GlutaMAX (Invitrogen, Cat No. 10565-018). Store at 2–8°C. Protect from light.
3. STEMPRO® hESC Supplement (Invitrogen, Cat No. A10006-01). Store at −5°C to −20°C. Protect from light.
4. Bovine serum albumin (BSA), 25% (Invitrogen, Cat No. A10008-01). Store at 2–8°C. Protect from light.
5. 2-Mercaptoethanol (Invitrogen, Cat No. 21985-023). Store at room temperature.
6. Basic fibroblast growth factor (bFGF) (CHA Bio & Diostech, Korea, Cat No. FGF-080411). Store aliquots at −20°C. Avoid repeated freeze–thaw cycles.
7. A 0.2-μm syringe filter, hydrophilic (Sartorius Stedim Biotech, Cat No. 16534).

2.3. Reagents for the Vitronectin-mTeSR1 Feeder-Free hESC Culture System

1. Vitronectin (BD Bioscience, Cat No. 354238). Store at −5°C to −20°C.
2. mTeSR1 basal medium (Stem Cell Technology, Cat No. 05850). Store at 2–8°C. Protect from light.
3. mTeSR1 Supplement (Stem Cell Technology, Cat No. 05851). Store at −20°C. Avoid repeated freezing and thawing.

2.4. Reagents and Cultureware for General Use

1. Tissue culture supplies:
   4-well plate (NUNC, Cat No. 176740),
   24-well plate (NUNC, Cat No. 142475),
   12-well plate (NUNC, Cat No. 150628),
   6-well plate (NUNC, Cat No. 176740),
   35-mm dish (NUNC, Cat No. 153066),
   60-mm dish (NUNC, Cat No. 150288),
   100-mm dish (NUNC, Cat No. 150679).
2. Dulbecco's phosphate-buffered saline (DPBS; 1X) with calcium and magnesium (Cat. No. 14040-133, Invitrogen). Store at 4°C.

3. Methods

3.1. Culturing Human Pluripotent Stem Cells Using the Matrigel-mTeSR1 System

Matrigel is a rich source of basement membrane matrix derived from Engelbreth–Holm–Swarm (EHS) mouse sarcoma cells and is known to contain laminin, entactin, collagen IV, and heparin sulfate proteoglycan. The gelatinous protein mixture remains in a liquid state at 2–8°C and becomes a gel quickly at temperatures above 22°C. Therefore, all sterilized tips and Eppendorf tubes should be kept chilled before handling Matrigel.

3.1.1. Aliquoting Matrigel

1. Store vials of BD Matrigel at –20°C.
2. When ready to aliquot Matrigel, thaw the vials by incubating them at 4°C overnight.
3. The following day, aliquot 100 µl of liquefied Matrigel into 600-µl Eppendorf tubes and store at –20°C.

3.1.2. Coating Cell Culture Dishes with Matrigel

1. Take the required number of frozen aliquots out of the –20°C freezer and keep them at 4°C overnight. It would also be convenient to keep sterile pipette tips (yellow and blue tips) and serological pipettes at 4°C overnight.
2. The following day, dilute Matrigel 1–100 using cold DMEM-F12 (Invitrogen, Cat No. 11320-033), followed by thorough mixing.
3. Add Matrigel to the culture dishes at a volume of roughly 105 µl/cm^2 culture area (see Table 1).
4. Allow the Matrigel-containing cultureware to sit at room temperature for 1–2 h.
5. Just before use, remove the remaining Matrigel solution from the coated cultureware and use directly without washing.

Table 1
The volume of matrix needed to coat culture surfaces, and the volume of feeder-free hESC culture media used according to the type of cultureware

Culturewares	Culture area	Volume of matrices for coating	Volume of media
24-well plate	1.9 cm^2 per well	200 µl per well	1 ml per well
12-well plate	3.5 cm^2 per well	370 µl per well	2 ml per well
6-well plate	9.6 cm^2 per well	1 ml per well	3 ml per well
35-mm dish	8.8 cm^2	925 µl	3 ml
60-mm dish	21.5 cm^2	2.25 ml	5 ml
100-mm dish	56.7 cm^2	6 ml	12.5 ml

3.1.3. Culturing Human Embryonic Stem Cells

1. Thaw a bottle of frozen mTeSR1 5X Supplement, and add 100 ml of the supplement to 400 ml of mTeSR1 Basal Medium. Store the complete mTeSR1 medium at 4°C for up to 2 weeks.
2. Add an appropriate amount of complete mTeSR1 medium to each culture container (see Table 1).
3. Seed mechanically dissected hESC fragments into each culture dish. The exact number required will depend on the purpose of the experiment. For routine maintenance, it may be reasonable to seed 40–100 hESC fragments of 1–2 mm^2 per 60-mm dish.
4. Incubate the cells under standard incubation conditions (37°C, 5% CO_2) for 48 h without changing the medium.
5. From the third day after seeding, change the medium daily.
6. Usually, at 5–7 days after seeding, the hESCs will form fairly large colonies and will need to be passaged.

3.2. Culturing Human Embryonic Stem Cells Using the CELLstart-STEMPRO® hESC SFM System

CELLstart is an ECM commercially available from Invitrogen (GIBCO). STEMPRO® hESC SFM, also produced by Invitrogen, is a defined medium recommended for use in combination with CELLstart. Notably, CELLstart is known to contain only human-derived components (xenogen-free), although the detailed compositions of CELLstart and STEMPRO® hESC SFM have not been disclosed.

Interestingly, STEMPRO® hESC SFM also works well in combination with Matrigel in our hands, while CELLstart does not seem to be compatible with mTeSR1 in supporting the undifferentiated growth of hESCs (unpublished data).

3.2.1. Coating Cell Cultureware with CELLstart

1. Store vials of CELLstart (Invitrogen, 2 ml per vial) at 4°C (CELLstart can be stored for up to 12 months).

2. Before coating the cultureware, dilute CELLstart 1:50 in 1X DPBS, and store at 4°C.
3. Add the diluted CELLstart solution to the cultureware roughly at 105 μl/cm² culture area (see Table 1).
4. Incubate the CELLstart-containing culture at 37°C and 5% CO_2 for 2 h.
5. Just before use, remove the remaining CELLstart solution from the coated cultureware, and use directly without washing.

3.2.2. Culturing Human Embryonic Stem Cells Using the CELLstart-STEMPRO® hESC SFM System

1. Thaw a bottle of STEMPRO® hESC Supplement (Invitrogen, 10 ml per bottle), transfer 1 ml aliquots to Eppendorf tubes, and store at −20°C until use.
2. Prepare a stock solution of bFGF (20 ng/μl in H_2O), filter the solution (0.2 μm sterile syringe filter), make aliquots, and store them at −20°C until use.
3. To prepare a 50 ml complete STEMPRO® hESC SFM, mix the three reagents as follows:
   - DMEM-F12 with GlutaMAX (Invitrogen, Cat No. 10565–018): 45.4 ml
   - STEMPRO® hESC Supplement (Invitrogen, Cat No. A10006-01): 1 ml
   - Bovine serum albumin 25% (Invitrogen, Cat No. A10008-01): 3.6 ml
4. Filter the mixture using 0.2-μm sterile syringe filter attached to 50-ml disposable syringe.
5. Finally, add 18.2 μl of 2-mercaptoethanol (Invitrogen, Cat No. 21985-023) and 20 μl of bFGF stock solution (20 ng/μl) to the filtrate to generate complete STEMPRO® hESC SFM.
6. Add an appropriate volume of complete STEMPRO® hESC SFM to the cultureware (see Table 1).
7. Seed mechanically dissected hESC fragments into each culture dish. The exact number required will depend on the purpose of the experiment. For routine maintenance, it may be reasonable to seed 40–80 hESC fragments of 1–2 mm² in size per 60-mm dish.
8. Incubate the cells under standard incubation conditions (37°C and 5% CO_2) for 48 h without changing the medium.
9. Change the medium daily starting on the third day after seeding.
10. Usually, hESCs need to be passaged every 5–6 days when they are cultured using the CELLstart-STEMPRO® hESC SFM feeder-free system. Please note that hESCs grow faster using this culture system than when using the other two feeder-free culture systems (10).

3.3. Culturing Human Embryonic Stem Cells Using the Vitronectin-mTeSR1 System

Vitronectin, a 75-kDa glycoprotein, is known to exist abundantly in ECM and in the blood. This protein has roles in many processes, including cell adhesion, migration, proliferation, and fibrinolysis. It interacts with certain macromolecules such as heparin. It has been shown that hESCs attach to vitronectin through integrin $\alpha V\beta 5$ (9). Vitronectin can support the undifferentiated growth of hESCs not only in MEF-conditioned medium but also in defined media such as mTeSR1 (9). A recent report demonstrated that the N-terminal 54 amino acids of vitronectin, which contain somatomedin B and Arg-Gly-Asp (RGD) sequences, are sufficient to mediate its efficient binding to hESCs (11).

3.3.1. Coating the Cell Culture Containers with Vitronectin

1. Store vials of lyophilized vitronectin at 4°C until use. Lyophilized vitronectin is known to be stable for at least 3 months at 2–8°C.
2. To make a working solution, bring the whole vial of vitronectin (BD Biosciences, 0.25 mg) to room temperature, and dissolve 0.25 mg of vitronectin in 50 ml of 1X DPBS. Aliquot 5 ml of dissolved vitronectin (5 µg/ml) into 15-ml Corning tubes, and then store at −20°C.
3. To coat the culture dishes, thaw the appropriate amount of vitronectin, and add to the cultureware (see Table 1).
4. Allow the vitronectin-containing cultureware to sit at room temperature for 1–2 h.
5. Remove the remaining solution, and use the coated dishes directly without washing.
6. If not being used immediately, the coated culture containers can be sealed with Parafilm and stored at 4°C for up to 1 week.

3.3.2. Culturing Human Embryonic Stem Cells Using the Vitronectin-mTeSR1 Feeder-Free System

1. Thaw a bottle of frozen mTeSR1 5X Supplement, and add 100 ml of the supplement to 400 ml of mTeSR1 Basal Medium. Store the complete mTeSR1 medium at 4°C for up to 2 weeks.
2. Add an appropriate amount of mTeSR1 medium to each cultureware (see Table 1).
3. Seed mechanically dissected hESC fragments into each culture dish. The exact number required will depend on the purpose of the experiment. For routine maintenance, it may be reasonable to seed 40–100 hESC fragments of 1–2 mm^2 in size per 60-mm dish.
4. Incubate the cells under standard incubation conditions (37°C, 5% CO_2) for 48 h before changing the medium.
5. Feed the cells every day with fresh medium starting on the third day after seeding.
6. Usually, hESCs need to be cultured for 5–7 days between passages when using the vitronectin-mTeSR1 feeder-free system.

4. Notes

4.1. Matrigel-mTeSR1 Culture System

1. Make sure Matrigel must be kept on ice during the whole aliquoting and coating procedures because it solidifies above 10°C. In addition, all the pipette tips, serological pipettes, and tubes should be prechilled before use.
2. Matrigel aliquots are stable for up to 6 months if stored at −70°C.
3. Matrigel-coated dishes can be stored for up to 1 week at 4°C. In this case, wrap the culture containers with Parafilm to avoid drying.
4. Complete mTeSR1 medium is stable at 4°C for up to 2 weeks. It can also be aliquoted and stored frozen at −20°C for up to 6 months. Before use, thaw the aliquots at 4°C overnight.
5. It would be better not to disturb the cells until they stably attach to the Matrigel and begin dividing (~48 h). If the need arises to open the incubator during this time period, open and close the incubator door very gently.
6. Start changing the medium from the third day after seeding. From this time point, replace mTeSR1 medium daily.

4.2. CELLstart-STEMPRO® hESC SFM Condition

1. There is no need to aliquot CELLstart solution.
2. Just before use, dilute CELLstart in 1X DPBS containing calcium and magnesium (Cat. No. 14040-133, Invitrogen).
3. Although the standard dilution rate of CELLstart is 1:50, this ratio can be adjusted according to hESC lines used.
4. It would be better if CELLstart-coated dishes are prepared on the day of use. However, the coated dishes can be stored for a couple of days at 4°C. In this case, wrap the culture containers with Parafilm to avoid drying.
5. STEMPRO® hESC Supplement should be stored at −20°C. Do not store the Supplement in frost-free freezer.
6. When you prepare complete STEMPRO® hESC SFM, try to filter the mixture containing DMEM-F12 with GlutaMAX, STEMPRO® hESC Supplement, and BSA (just before adding 2-mercaptoethanol and bFGF), since debris derived from BSA may interfere with attachment and growth of hESCs.
7. Complete STEMPRO® hESC SFM can be stored at 4°C for up to 1 week.
8. It would be better not to disturb the cells until they stably attach to the CELLstart and begin dividing (~48 h). If the need arises to open the incubator during this time period, open and close the incubator door gently.
9. Start changing the medium daily from the third day after seeding.

4.3. Vitronectin-mTeSR1 Feeder-Free Condition

1. Vitronectin aliquots must be stored at −20°C, and is recommended to use within 1 month. Do not store in frost-free freezer and avoid multiple freeze–thaw cycles.
2. The vitronectin-coated culturewares can be sealed with Parafilm and stored at 4°C for up to 1 week.
3. Do not culture hESCs on vitronectin longer than 1 week, since the matrix starts to be detached from the bottom of the cultureware.
4. Complete mTeSR1 medium is stable at 4°C for up to 2 weeks. The complete medium can be also aliquoted and stored frozen at −20°C for up to 6 months. Before use, thaw the aliquotes at 4°C overnight.
5. It would be better not to disturb the cells until they stably attach to the vitronectin and begin dividing (~48 h). If the need arises to open the incubator during this time period, open and close the incubator door very gently.
6. Start changing the medium from the third day daily after seeding.

References

1. Thomson J.A., Itskovitz-Eldor J., Shapiro S.S., Waknitz, M.A., Swiergiel, J.J., Marshall, V.S. Jones J.M. (1998) Embryonic stem cell lines derived from human blastocysts. 282, 1145–47.
2. Xu C., Inokuma M.S., Denham J., Golds K., Kundu P., Gold J.D., Carpenter M.K. (2001) Feeder-free growth of undifferentiated human embryonic stem cells. Nature Biotechnology 19, 971–4.
3. Ludwig T.E., Levenstein M.E., Jones J.M., Berggren W.T., Mitchen E.R., Frane J.L., Crandall L.J., Daigh C.A., Conard K.R., Piekarczyk M.S., Llanas R.A., Thomson J.A. (2006) Derivation of human embryonic stem cells in defined conditions. Nature Biotechnology 24, 185–7.
4. Li Y., Powell S., Brunette E., Lebkowski J., Mandalam R. (2005) Expansion of human embryonic stem cells in defined serum-free medium devoid of animal-derived products. Biotechnology and Bioengineering 91, 688–98.
5. Hakala H., Rajala K., Ojala M., Panula S., Areva S., Kellomäki M., Suuronen R., Skottman H. (2009) Comparison of biomaterials and extracellular matrices as a culture platform for multiple independently derived human embryonic stem cell lines. Tissue Eng Part A 15, 1–12.
6. Fletcher J., Ferrier P., Gardner J., Harkness L., Dhanjal S., Serhal P., Harper J., Delhanty J., Brownstein D.G., Prasad Y.R., Lebkowski J., Mandalam R., Wilmut I., De Sousa P.A. (2006) Variations in humanized and defined culture conditions supporting derivation of new human embryonic stem cell lines. Cloning Stem Cells 8, 319–34.
7. Skottman H., Dilber M.S., Hovatta O. (2006) The derivation of clinical-grade human embryonic stem cell lines. FEBS Letters 580, 2875–78.
8. Skottman H., Hovatta O. (2006) Culture conditions for human embryonic stem cells Reproduction. 132, 691–8.
9. Braam S.R., Zeinstra L., Litjens S., Ward-van Oostwaard D., van den Brink S., van Laake L., Lebrin F., Kats P., Hochstenbach R., Passier R., Sonnenberg A., Mummery C.L. (2008) Recombinant vitronectin is a functionally defined substrate that supports human embryonic stem cell self renewal via {alpha}V{beta}5 integrin. Stem Cells 26, 2257–65.
10. Yoon T.M., Chang B., Kim H.T., Jee J.H., Kim D.W., Hwang D.Y. (2010) Human Embryonic Stem Cells (hESCs) Cultured Under Distinctive Feeder-Free Culture Conditions Display Global Gene Expression Patterns Similar to hESCs from Feeder-Dependent Culture Conditions. Stem Cell Rev 26, 2257–65.
11. Prowse A.B., Doran M.R., Cooper-White J.J., Chong F., Munro T.P., Fitzpatrick J., Chung T.L., Haylock D.N., Gray P.P., Wolvetang E.J. (2010) Long term culture of human embryonic stem cells on recombinant vitronectin in ascorbate free media. Biomaterials 31, 8281–8.

Chapter 2

Growth of Human Embryonic Stem Cells in Long-Term Hypoxia

Vladimir Zachar, Simon C. Weli, Mayuri S. Prasad, and Trine Fink

Abstract

Human embryonic stem cells (hESCs) hold a great promise for regenerative medicine and tissue engineering. In order to obtain uniform hESC cultures without spontaneous differentiation, which is of interest for basic investigations as well as the development of future therapeutic protocols, it is important that specific culture conditions are adhered to. Here, we describe in detail a procedure for propagation of hESCs that by virtue of exposure of the cultures to low atmospheric oxygen (5%) enables the maintenance of their undifferentiated phenotype in long term. The critical steps and impact of possible modifications on the final outcome are discussed, and useful hints are provided to streamline the troubleshooting.

Key words: Human embryonic stem cells, Long-term culture, Hypoxia, Pluripotency, Self-renewal, Manual microdissection

1. Introduction

The embryonic stem cells (ESCs) stand out in that they have practically unlimited life span and have a potential to differentiate into all body tissues, except for placental trophoblasts (1). These properties render ESCs a very valuable source for prospective therapeutic approaches based on cell regeneration and replacement and tissue-engineered spare parts. ESCs are considered especially useful in applications where the lineages follow the endo- and neuroectodermal differentiation pathways, since there is a lack of suitable somatic stem cell associated with these particular specifications. Recently, a promising progress has been done with in vitro production of liver, lung, neural, or insulin-secreting cells (2–5).

The key prerequisites for in vitro propagation of human ESCs (hESCs) are currently well defined; nevertheless, there are a number of alternative protocols. They differ in the use of different growth factors, media, gaseous environment, and the mode of passaging (6, 7). Attention has to be given to all these factors, since their compound effect constitutes a microenvironmental setting that determines the behavior and long-term properties of a given hESC line. In this regard, differences have been observed between different hESC lines with respect to their biological responses to, for example, trypsinization or dependence on feeder cells (8). It is plausible that such specific features reflect implementation of unique conditions during isolation and growth.

A frequently encountered drawback during hESC culture is the spontaneous differentiation (9, 10). The exact mechanism underlying this phenomenon is not clearly understood, although there is a plethora of evidence linking developmentally important pathways, such as Notch, sonic hedgehog, Wnt, and bone morphogenetic protein-associated signaling, to the control of ESC self-renewal and differentiation (11–14). Recently, it has been demonstrated that mild hypoxic conditions have the capacity to enhance the maintenance of hESC pluripotency by inhibiting the spontaneous differentiation in short as well as long term (15–17), and at least some of the above-mentioned factors appear to be involved (18).

Since the use of hypoxia is a very straightforward and simple approach, it may be of interest to all basic and translational biomedical research as well as industrial applications, where the access to uniform ESC cultures is necessary. It is generally more feasible to expose cultures to short-term hypoxia during the regular passaging intervals. The procedure becomes, however, more difficult when the hypoxic culturing is carried out over extended periods of time and the harmful reoxygenation is to be avoided. In the following sections, we will describe in detail a procedure for long-term maintenance and characterization of hESCs in hypoxic atmosphere.

2. Materials

2.1. Growth and Irradiation of Human Foreskin Fibroblasts (HFFs)

1. HFFs (American Type Culture Collection, CRL-2429).
2. Iscove's Modified Dulbecco's Modified Eagle Medium (IDMEM) (GIBCO/Invitrogen, Carlsbad, CA) is stored at 4°C.
3. Fetal bovine serum (FBS) (GIBCO/Invitrogen) is stored at −80°C and is used to supplement IDMEM in final concentration of 10%.
4. Penicillin/streptomycin 1,000 U/mL/1 mg/mL stock solution mixture (GIBCO/Invitrogen) is stored in aliquots of 10 mL at −20°C.

5. The feeder growth medium, which is fully supplemented IDMEM, is stored at 4°C and is stable for 2 weeks. It is prepared using following volumes to obtain 500 mL:

   448 mL IDMEM

   50 mL FBS

   2 mL penicillin/streptomycin (see Note 1)

6. Phosphate-buffered saline (PBS) (GIBCO/Invitrogen).

7. Trypsin/EDTA blend. Trypsin (GIBCO/Invitrogen) is obtained as a solution of 2.5% in PBS. It is diluted 10 times with PBS and sterile filtered. Aliquots of 10 mL are stored at –20°C. EDTA (anhydrous, crystalline, cell culture tested) (Sigma-Aldrich, Brøndby, Denmark) is dissolved in PBS at a concentration of 0.02%, sterile filtered, and stored in 10 mL aliquots at 4°C. To prepare the blend, trypsin and EDTA solutions are mixed 1:1. The blend is stored at 4°C and is stable for up to a week.

8. Trypan blue solution 0.4%, cell culture tested and sterile filtered (Sigma-Aldrich).

9. Hemocytometer.

10. Tissue culture flasks T-175.

11. Centrifuge tubes (50 mL).

12. 35-mm tissue culture dishes (Cell Bind; Corning, Amsterdam, the Netherlands).

13. Gamma cell irradiator (Gammacell 2000; Mølsgaard Medical, Ganløse, Denmark) (see Note 2).

2.2. Propagation of hESCs

1. Knockout Dulbecco's Modified Eagle Medium (KDMEM) (GIBCO/Invitrogen) is stored at 4°C.

2. Knockout Serum Replacer (GIBCO/Invitrogen) is stored in aliquots of 50 mL at –20°C and is used to supplement KDMEM in final concentration of 20%.

3. L-glutamine 200 mM stock solution (GIBCO/Invitrogen) is stored in aliquots of 10 mL at –20°C and is used to supplement KDMEM in a final 2 mM concentration.

4. Beta-mercaptoethanol 50 mM stock solution (GIBCO/Invitrogen) is stored in aliquots of 200 µL at –20°C and is used to supplement KDMEM in a final 0.1 mM concentration.

5. Nonessential amino acid 10 mM (100×) stock solution (GIBCO/Invitrogen) is stored at 4°C and is used to supplement KDMEM to a final 0.1 mM concentration.

6. Recombinant human bFGF (BioSource Europe, Nivelles, Belgium) is stored in a concentration of 25 µg/µL in aliquots of 50 µL at –80°C and is used to supplement KDMEM in the final concentration of 4 ng/mL.

7. The hESC growth medium is prepared without bFGF and is stored at 4°C for up to 2 weeks. Just prior to use, 0.16 μL of bFGF is added per mL media. The medium is prepared using following volumes to obtain 500 mL:

   387 mL KDMEM

   100 mL Knockout Serum Replacer

   5 mL L-glutamine

   1 mL Beta-mercaptoethanol

   5 mL Nonessential amino acids

   2 mL Penicillin/streptomycin (same stock solution as in Sect. 2.1)

8. Surgical disposable scalpels no. 15 (Aesculap, Tuttlingen, Germany).

9. Stereo microscope in the hypoxia workstation (see Note 3).

10. Xvivo System integrated workbench-incubator glovebox with controllable atmosphere (BioSpherix, Ltd., Redfield, NY) (see Note 4).

2.3. Immunofluorescence Staining

1. Hoechst 33342 (Gibco/Invitrogen) is obtained as a solution at 10 mg/mL. The working solution is made ready by diluting 1:100 in PBS to a final 100 μg/mL. The working solution is stable for several months if stored in the dark at 4°C.

2. Phosphate-buffered 4% paraformaldehyde.

3. Bovine serum albumin (BSA) (Standard grade; Europa Bioproducts, Cambridge, United Kingdom) is stored at 4°C. It is used at working concentration of 1%, 2%, or 4%, depending on the assay, by diluting in PBS.

4. Triton X-100 (Sigma-Aldrich) working solution is made freshly by diluting to 0.2% final concentration in 4% BSA.

5. SSEA-1-specific mouse monoclonal antibody (sc-21702; Santa Cruz Biotechnology, Santa Cruz, CA) is supplied at 200 μg/mL and stored at 4°C. Prior to use, it is diluted 200-fold in 1% BSA. It should be used within a couple of hours.

6. Cy-5 goat anti-mouse conjugate (AP130S; Millipore, Bedford, MA) is mixed upon receipt with an equal volume of glycerol and stored at −20°C. Prior to use, the antibody is further diluted 100-fold (total dilution is 200-fold) in 1% BSA and should be used within a couple of hours.

7. Oct4-specific rabbit polyclonal antibody (ab19857; Abcam, Cambridge, United Kingdom) is mixed upon receipt with an equal volume of glycerol and stored at −20°C. Prior to use, it is further diluted 100-fold (total dilution is 200-fold) in 1% BSA and should be used within a couple of hours.

8. FITC-conjugated goat anti-rabbit polyclonal antibody (ab6717, Abcam) is mixed upon receipt with an equal volume of glycerol and stored at −20°C. Prior to use, the antibody is further diluted 150-fold (total dilution is 300-fold) in 1% BSA and should be used within a couple of hours.

9. Wide-field fluorescence system Axio Observer Z1 integrated with AxioCam MRm camera and controlled by the AxioVision software package (all Carl Zeiss, Göttingen, Germany).

2.4. Real-Time PCR

1. Aurum Total RNA Mini Kit (Bio-Rad Laboratories, Hercules, CA).
2. iScript cDNA synthesis kit (Bio-Rad Laboratories).
3. iQ SYBR Green Supermix (Bio-Rad Laboratories).
4. Diethylpyrocarbonate (DEPC)-water (Sigma-Aldrich) is stored at 4°C.
5. PCR primers for pluripotency markers Oct4 and Nanog and internal reference 18 S rRNA:

   Oct4 forward primer: 5′-CTG GTT CGC TTT CTC-3′

   Oct4 reverse primer: 5′-GGG GGT TCT ATT TGG-3′

   Nanog forward primer: 5′-AGG AAG AGT AGA GGC-3′

   Nanog reverse primer: 5′-CAA CTG GCC GAA GAA-3′

   18 S rRNA forward primer: 5′-AGG ACC GCG GTT CTA TTT TGT TGG-3′

   18 S rRNA reverse primer: 5′-CCC CCG GCC GTC CCT CTT A-3′

   Oct4 and Nanog primers are diluted to a working concentration of 10 pmol/μL; 18 S rRNA primers are diluted to 5 pmol/μL. All primers are stored at −20°C.

6. 96-well translucent plates (iCycler iQ PCR plates; Bio-Rad Laboratories).
7. ND-1000 UV-Vis spectrophotometer (NanoDrop Technologies, Wilmington, DE).
8. MyIQ single-color real-time PCR detection system (Bio-Rad Laboratories).

3. Methods

Good laboratory practice should strictly be observed, especially in all steps requiring sterile handling. All media prepared in-house should be sterile filtered using a 0.2-μm exclusion cell culture grade filters (TPP, Trasadingen, Switzerland).

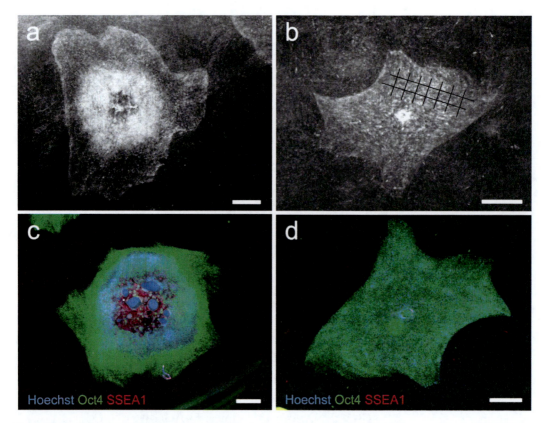

Fig. 1. Effect of hypoxia on the self-renewal of hESCs during long-term cultures. The CLS1 (**a** and **b**) and CLS2 (**c** and **d**) lines were cultured for 18 months in ambient air conditions (**a** and **c**) or hypoxic atmosphere of 5% oxygen (**b** and **d**). Four weeks after passaging, the spontaneous differentiation in the central parts of the colonies from normoxic conditions is discernible in the dark field images (**a** and **b**) and after immunofluorescence imaging of gene markers (**c** and **d**). The grid in (**b**) indicates the microdissection pattern. The scale bars indicate 1 mm.

3.1. Long-Term Maintenance of hESCs in Hypoxia

With regard to the continuous propagation in hypoxic atmosphere, the greatest challenge is to maintain stable gaseous conditions during the subculture steps. This means that the manual microdissection of hESC colonies has to be performed within enclosure with controllable environment, and feeder culture and media have to be properly pre-equilibrated to these conditions. Xvivo System workstation appears especially well suited for this purpose. Different oxygen tensions can be used, but 5% offers the best compromise between the proliferation rate and suppression of spontaneous differentiation. In our hands, the different hESC lines display undiminished replication rate and other properties of ESCs for over 18 months (approx. 25 passages) (Fig. 1).

3.1.1. Preparation of Feeder Cells

A large stock of feeder cells can be prepared and kept frozen at $-140°C$ (see Note 5). One confluent T175 flask typically yields $5-6 \times 10^6$ cells, which is sufficient for eight dishes. In the following sections, we will describe the preparation of approximately 40 dishes of feeder cells.

1. An aliquot of 2×10^6 HFFs is thawed and seeded in a T175 flask in 20 mL of feeder growth medium. The cells will reach confluency after approximately 4 days.

2. When the cells reach confluency, the medium is removed, and the cells are washed twice with 10 mL of PBS to prepare for trypsinization.

3. 4 mL of Trypsin/EDTA blend is added and incubated at 37°C.

4. The detachment is assessed visually under the microscope at 3-min intervals and can be aided by manually tapping the flask.

5. When cells have detached, the trypsinization is stopped by adding 12 mL of feeder growth medium.

6. The cells are divided into eight T175 flasks, and the growth medium is added to achieve a total of 20 mL per flask. After approximately a week, the cells will reach confluency and may be subcultured, as described above.

7. Cells from one flask are typically used to continue the propagation of feeder cells and are divided into eight new T175 flasks. For additional expansion of cells, repeat steps 2–7 (see Note 6).

8. Cells from the remaining seven flasks are pooled.

9. An aliquot of the cell suspension is mixed with an equal volume of trypan blue, and the yield of live cells is determined using a hemocytometer.

10. The cells are divided into four 50-mL centrifuge tubes, and the concentration is adjusted with feeder growth medium to 2.5×105 cells/mL.

11. The cells are irradiated with 35 Gray, and 2.5 mL aliquots are then seeded into 35-mm tissue culture dishes yielding 625,000 cells/dish.

12. The cultures are grown overnight in CO_2 incubator in standard normoxic conditions, after which the medium is changed to 3 mL of hESC growth medium. The cultures are then transferred to incubation chambers with preselected hypoxic atmosphere, and after 2 h, they are ready for transfer of hESCs. The dishes with feeder cells may be used for seeding with hESC for up to 1 week (see Note 7).

3.1.2. Passaging of hESCs

1. Every 3–4 weeks, incisions are made in the hESC colonies using a disposable scalpel under the guidance of stereo microscope (50-fold magnification) to demarcate several rectangular subsections of up to 0.3×0.3 mm in size. With the tip of the scalpel, the hESC subsections are gently released from the culture dish (Fig. 1b) (see Note 8).

2. Each of the subsections is transferred with a micropipette in 8 μL volume and deposited on the feeder monolayer.

3. Approximately six subsections are seeded in a single dish. After seeding, the dishes are left on the work space for 2 h to allow the subsections to settle onto the feeders, and the dishes are transferred to the incubation chambers. Caution is taken to avoid swirling, which can result in irregular distribution of the colonies.

4. The hESC growth medium is completely replaced twice a week (see Note 9).

3.2. Determination of hESC Pluripotency

The pluripotency status can be assessed by identification of cell surface or intracellular markers by immunostaining or real-time RT-PCR. There are several hESC markers, such as Oct4, Nanog, SSEA3 and 4; in addition, the differentiation marker SSEA1 is routinely used to counterstain for the cells that departed from the hESC pool. The methods in the following text will be exemplified using immunofluorescence staining for Oct4 and SSEA1 and RT-PCR for Oct4 and Nanog. Functional assessment normally also involves differentiation assays that are not dealt within this chapter.

3.2.1. Immunofluorescence Staining

1. After predetermined period of growth, a portion of 35 μL of Hoechst 33342 reagent is added to the media of the cultures in hypoxia to achieve final concentration of 10 μg/mL. Gently swirl the culture dishes 3–4 times to aid proper mixing, and incubate them for 30 min.

2. The cultures are removed from hypoxia, the medium is discarded, and the cells are washed with 1 mL of PBS for 5 min at room temperature (RT). All subsequent incubation steps are done in the dark, i.e., with aluminum foil cover over the dishes.

3. The cells are fixed with buffered 4% formaldehyde for 20 min at RT.

4. The cells are washed twice with 1 mL of PBS for 3 min, followed by blocking with 2% BSA for 10 min at RT.

5. The blocking solution is replaced with 0.5 mL of SSEA1 antibody per dish and incubated for 1 h at RT.

6. The primary antibody is removed by washing twice with 1 mL of PBS for 3 min, and the cultures are incubated with 0.5 mL of Cy-5 goat anti-mouse conjugate for 30 min at RT.

7. To detect nucleus-localized Oct4, cell permeabilization is necessary. To accomplish this, wash the cells twice with 1 mL of PBS for 3 min and subsequently incubate them with 1 mL of 0.2% Triton X-100 in 4% BSA for 1 h at 37°C.

8. The detergent is removed by washing twice with 1 mL of PBS for 3 min at RT, and 0.5 mL of Oct4 antibody is added and incubated for 1 h at RT.

9. The unbound antibody is removed by washing twice with 1 mL of PBS for 3 min, and the specific immune complexes are revealed by adding 0.5 mL of FITC-conjugated goat anti-rabbit polyclonal antibody and incubating for 30 min at RT.

10. The dishes are washed twice with 1 mL of PBS for 3 min at RT.

11. After the final washing step, the preparations are preserved in PBS. They can be analyzed immediately or stored at 4°C for at least 2 weeks.

12. The immunofluorescence imaging microscopy is done with the aid of Axio Observer Z1 wide-field fluorescence system. The MosaiX module enables tiling of images to capture whole colony area using three-channel full resolution based on 2.5–10-fold objective magnification.

3.2.2. Semiquantitative Real-Time PCR

Isolation of RNA and cDNA Synthesis

RNA isolation and cDNA synthesis is performed with Aurum Total RNA Mini and iScript cDNA synthesis kits, respectively, according to manufacturer's instructions. Steps that do not involve the commercial kits are detailed below. For RNA isolation, we have found that five to six colonies will yield sufficient RNA materials for the assay.

1. A 1.5-mL microcentrifuge tube with 1 mL of ice-cold PBS is placed on ice in the hypoxia workstation.

2. The hESC colonies are released gently by using the scalpel. The stereo microscope in the hypoxia workstation is used to control that the embryonic cells are fully separated from underlying layer of feeder cells (see Note 10).

3. The floating cell clumps are recovered in 8 µL of medium using a micropipette and transferred to the ice-cold PBS (see Note 11).

4. The cells are centrifuged at $300 \times g$ for 1 min, and the supernatant is removed. The pellets are resuspended in 350 µL of lysis solution from the Aurum RNA Mini Kit supplemented with 1% β-mercaptoethanol, and thoroughly repetitive pipetting is applied until the complete lysis is achieved. The rest of the procedure is carried out as instructed by the manufacturer.

5. The concentration of eluted total RNA is determined spectrophotometrically, and the sample is used immediately for the synthesis of cDNA (see Note 12).

6. To produce cDNA, a reaction mix is prepared containing 100 ng total RNA, 4 µL 5× iScript reaction mix, 1 µL iScript reverse transcriptase, and DEPC-nuclease-free water to a final volume of 20 µL. The rest of the procedure is performed as suggested by the manufacturer.

Amplification Assay

1. Each reaction is performed in duplicate and is set up using 8 µL of cDNA, 13 µL of iQ SYBR Green Supermix reaction components, 0.03 pmol of 18 S primers or 0.192 pmol of gene-specific primers, and water to make the total volume of 25 µL.
2. The amplification is performed using two-temperature cycling consisting of a single annealing/extension step of 30 s at 60°C and a denaturation step of 15 s at 95°C, for a total number of 40 cycles.
3. To confirm the quality of each run, the occurrence of primer dimers is monitored by invoking a melting curve function of the program.
4. The relative transcriptional levels are assessed by extrapolation from amplification of a fourfold dilution series of pooled cDNA.
5. For each sample, the levels of Oct4 and Nanog are normalized to the levels of 18 S rRNA.

4. Notes

1. The final concentration of penicillin and streptomycin is 40 U/mL and 40 µg/mL, respectively, which is 2.5-fold lower than the standard concentration used for cell culture.
2. If a gamma irradiator is not in-house, blood banks often have one. We have prepared the feeder cells in the centrifuge tubes, transported to blood bank, irradiated, and transported back in about 90 min without noticeable cell death.
3. SteREO Lumar.V12 is due to its motorization especially suitable to be incorporated in the hypoxia workstation. It enables contrasting stereomicroscopy as well as fluorescence imaging.
4. The workstation enables incubation and handling in an uninterrupted hypoxic environment buffered with 5% CO_2 and nitrogen. Several humidified incubation chambers with variable oxygen concentrations can be used simultaneously.
5. When the cells are received from ATCC, they are thawed and expanded in a T175 tissue culture flask, and designated passage (P) 0. Upon confluency, the cells are trypsinized and divided into 10 T175 tissue culture flasks (P1). When P1 cultures are confluent, the cells are frozen in aliquots of 2×10^6 cells. These aliquots can be thawed and expanded to replenish the stock or support passaging of hESCs.
6. We use cells up to P10, after which a new aliquot of frozen cells is thawed and propagated.

7. Culturing the feeders on Cell Bind (Corning) culture dishes enhances integrity of the monolayer. Nevertheless, when the feeder cells are cultured after irradiation for more than 3–4 weeks, there is a tendency for detachment.

8. The number of sections available from a single colony varies depending on its size, but approximately 10–15 can be obtained. The center, which represents the original seed, is usually not selected, neither are sections from the outermost area of the colony. It is not important whether the cuts penetrate deeper in the feeder layer, such that some feeder cells are transferred together with the hESCs.

9. For pre-equilibration, the medium already with bFGF is placed in hypoxia. A maximum of 25 mL is used per T175 cell culture flask for 2 h. The flask is placed horizontally to allow for rapid release of oxygen from the medium.

10. It is quite critical for the quality of RNA that the scrapping of hESCs is achieved without contamination with feeder cells. By placing the scalpel perpendicularly to the colony and very gently scraping while confirming the separation from feeders microscopically, the colonies can be released. The same type of scalpel is used as for passaging of hESCs. This procedure requires some practice, so we recommend including several additional dishes with hESC colonies for training purposes for the first couple of experiments.

11. We try to recover as many pieces of stem cells as possible per 8 µL media. We repeat the recovery of stem cells until all the material has been harvested and transferred to the ice-cold PBS.

12. Alternatively, Bioanalyzer (Agilent Technologies, Naerum, Denmark) can be used to measure concentration as well as determine integrity of the RNA.

References

1. Thomson, J. A., Itskovitz-Eldor, J., Shapiro, S. S., Waknitz, M. A., Swiergiel, J. J., Marshall, V. S., and Jones, J. M. (1998) Embryonic stem cell lines derived from human blastocysts *Science* **282**, 1145–7.

2. Roelandt, P., Pauwelyn, K. A., Sancho-Bru, P., Subramanian, K., Ordovas, L., Vanuytsel, K., Geraerts, M., Firpo, M., De Vos, R., Fevery, J., Nevens, F., Hu, W. S., and Verfaillie, C. M. Human embryonic and rat adult stem cells with primitive endoderm-like phenotype can be fated to definitive endoderm, and finally hepatocyte-like cells *PLoS One* **5**, e12101.

3. Van Haute, L., De Block, G., Liebaers, I., Sermon, K., and De Rycke, M. (2009) Generation of lung epithelial-like tissue from human embryonic stem cells *Respir Res* **10**, 105.

4. Zhou, J., Su, P., Li, D., Tsang, S., Duan, E., and Wang, F. High-Efficiency Induction of Neural Conversion in hESCs and hiPSCs with a Single Chemical Inhibitor of TGF-beta Superfamily Receptors *Stem Cells*.

5. Shirasawa, S., Yoshie, S., Yokoyama, T., Tomotsune, D., Yue, F., and Sasaki, K. A novel stepwise differentiation of functional pancreatic exocrine cells from embryonic stem cells *Stem Cells Dev*.

6. Skottman, H., Narkilahti, S., and Hovatta, O. (2007) Challenges and approaches to the culture of pluripotent human embryonic stem cells *Regen Med* **2**, 265–73.

7. Silvan, U., Diez-Torre, A., Arluzea, J., Andrade, R., Silio, M., and Arechaga, J. (2009) Hypoxia and pluripotency in embryonic and embryonal carcinoma stem cell biology *Differentiation* **78**, 159–68.

8. Akopian, V., Andrews, P. W., Beil, S., Benvenisty, N., Brehm, J., Christie, M., Ford, A., Fox, V., Gokhale, P. J., Healy, L., Holm, F., Hovatta, O., Knowles, B. B., Ludwig, T. E., McKay, R. D., Miyazaki, T., Nakatsuji, N., Oh, S. K., Pera, M. F., Rossant, J., Stacey, G. N., and Suemori, H. Comparison of defined culture systems for feeder cell free propagation of human embryonic stem cells *In Vitro Cell Dev Biol Anim* **46**, 247–58.

9. Lysdahl, H., Gabrielsen, A., Minger, S. L., Patel, M. J., Fink, T., Petersen, K., Ebbesen, P., and Zachar, V. (2006) Derivation and characterization of four new human embryonic stem cell lines: the Danish experience *Reprod Biomed Online* **12**, 119–26.

10. Xu, R. H., Peck, R. M., Li, D. S., Feng, X., Ludwig, T., and Thomson, J. A. (2005) Basic FGF and suppression of BMP signaling sustain undifferentiated proliferation of human ES cells *Nat Methods* **2**, 185–90.

11. Schroeder, T., Meier-Stiegen, F., Schwanbeck, R., Eilken, H., Nishikawa, S., Hasler, R., Schreiber, S., Bornkamm, G. W., and Just, U. (2006) Activated Notch1 alters differentiation of embryonic stem cells into mesodermal cell lineages at multiple stages of development *Mech Dev* **123**, 570–9.

12. Mfopou, J. K., De Groote, V., Xu, X., Heimberg, H., and Bouwens, L. (2007) Sonic hedgehog and other soluble factors from differentiating embryoid bodies inhibit pancreas development *Stem Cells* **25**, 1156–65.

13. Davidson, K. C., Jamshidi, P., Daly, R., Hearn, M. T., Pera, M. F., and Dottori, M. (2007) Wnt3a regulates survival, expansion, and maintenance of neural progenitors derived from human embryonic stem cells *Mol Cell Neurosci* **36**, 408–15.

14. Toh, W. S., Yang, Z., Liu, H., Heng, B. C., Lee, E. H., and Cao, T. (2007) Effects of culture conditions and bone morphogenetic protein 2 on extent of chondrogenesis from human embryonic stem cells *Stem Cells* **25**, 950–60

15. Ezashi, T., Das, P., and Roberts, R. M. (2005) Low O2 tensions and the prevention of differentiation of hES cells *Proc Natl Acad Sci USA* **102**, 4783–8.

16. Prasad, S. M., Czepiel, M., Cetinkaya, C., Smigielska, K., Weli, S. C., Lysdahl, H., Gabrielsen, A., Petersen, K., Ehlers, N., Fink, T., Minger, S. L., and Zachar, V. (2009) Continuous hypoxic culturing maintains activation of Notch and allows long-term propagation of human embryonic stem cells without spontaneous differentiation *Cell Prolif* **42**, 63–74.

17. Zachar, V., Prasad, S. M., Weli, S. C., Gabrielsen, A., Petersen, K., Petersen, M. B., and Fink, T. The effect of human embryonic stem cells (hESCs) long-term normoxic and hypoxic cultures on the maintenance of pluripotency *In Vitro Cell Dev Biol Anim* **46**, 276–83.

18. Weli, S. C., Fink, T., Cetinkaya, C., Prasad, S. M., and Zachar, V. (2010) Notch and Hedgehog signaling cooperate to maintain self-renewal of human embryonic stem cells exposed to low oxygen concentration *Int J Stem Cells* **3**(2), 129–37.

Chapter 3

Laboratory-Scale Purification of a Recombinant E-Cadherin-IgG Fc Fusion Protein That Provides a Cell Surface Matrix for Extended Culture and Efficient Subculture of Human Pluripotent Stem Cells

Masato Nagaoka and Stephen A. Duncan

Abstract

The culture of human pluripotent stem cells under defined conditions most commonly relies on the use of Matrigel as a substrate upon which the cells remain in an undifferentiated state. Matrigel is a complex mixture of extracellular matrices and growth factors derived from mouse Engelbreth–Holm–Swarm sarcoma cells. The complexity and lot-to-lot variation of Matrigel preparations has prompted the search for more defined substrates that can support pluripotent stem cell culture. A recombinant human E-cadherin-IgG Fc domain fusion protein (E-cad-Fc) has recently been shown to be extremely efficient in facilitating the culture of human embryonic stem (ES) cells and induced pluripotent stem (iPS) cells under completely defined conditions. This fusion protein is particularly appealing because binding requires the cellular expression of E-cadherin, which is one hallmark of pluripotent stem cells, and so the substrate is selective of an undifferentiated cell state. In addition, cells can be removed from the substrate using gentle enzyme-free dissociation buffers containing chelating reagents, which ensures maintenance of cell surface epitopes and high cell viability during subculture. Here, we provide a detailed protocol for the purification of the E-cad-Fc substrate.

Key words: Human pluripotent stem cells, Feeder-free, Defined culture conditions, Recombinant E-cadherin surface

1. Introduction

Human pluripotent stem cells have the potential to be used for cell therapy as well as the study of human disease and development (1). Human embryonic stem (hES) cells were first developed in 1998 (2), and recently human-induced pluripotent stem (hiPS) cells were generated from differentiated somatic cells (3, 4). Initially, the establishment and sustained culture of human pluripotent stem cells required growth on a layer of mouse embryonic fibroblasts (MEF), which presumably provide factors that sustain cell pluripotency and viability (2). Unfortunately, for many purposes, including the possible production of pluripotent stem cells for therapeutic use and drug discovery, a reliance on feeder fibroblasts is far from ideal. Recently, a variety of cell culture media have been described that can circumvent the need for feeder cells as long as the pluripotent stem cells are cultured on an appropriate substratum, which is usually Matrigel (5, 6).

Matrigel is a secreted product of Engelbreth–Holm–Swarm (EHS) sarcoma cells that consists of a mixture of extracellular matrix proteins as well as a variety of associated growth factors and other secreted polypeptides (7–9). There have been considerable efforts devoted to the generation of highly defined culture conditions to support pluripotent stem cell growth that eliminate xenogeneic components such as serum, feeder cells, and ill-defined matrices. These efforts have resulted in the production of a number of culture systems that rely upon defined media and recombinant protein substrates, including laminin-511 (10, 11), collagen (12), vitronectin (13), as well as cell-adhesive recombinant peptides (14, 15) and synthetic polymers (16). Such materials can eliminate the risk of contamination of cells with pathogens; moreover, the use of highly defined culture conditions is likely to be necessary to facilitate the therapeutic use of human pluripotent stem cells by ensuring GMP level standardization of the cultures.

We have recently described the use of an alternative matrix that efficiently supports pluripotency of human iPS and ES cells in defined medium over an extended period of culture (>70 passages) (17). E-cadherin, a Ca^{2+}-dependent cell–cell adhesion molecule (18, 19), is essential for intercellular adhesion and colony formation of mouse embryonic stem cells (20, 21). Several reports suggest that E-cadherin levels in hES cells decrease during differentiation (22, 23), and high expression of E-cadherin is characteristic of undifferentiated pluripotent stem cells. Based on this, we reasoned that coating plastic cell culture dishes with a fusion protein consisting of the human E-cadherin extracellular domain and the IgG Fc domain (hE-cad-Fc) could provide a substrate that would facilitate cell attachment and enrich for pluripotency through selective adhesion of E-cadherin-expressing cells. This was indeed successful, and we found that the hE-cad-Fc substrate could support human stem cell pluripotency for >70 passages. The attachment through E-cadherin has the added benefit that cells can be easily passaged using enzyme-free dissociation buffers containing EDTA, which maintain cell surface epitopes and result in high levels of cell viability. Although a similar matrix has recently become commercially available (StemAdhere™, Primorigen Biosciences, Madison, WI, http://www.primorigen.com; Celagix, Kanagawa, Japan), here we describe a detailed protocol for laboratory-scale production and purification of hE-cad-Fc.

2. Materials

2.1. Reagents and Supplies for Production of hE-cad-Fc

1. HEK293 clone 1C8 cells expressing recombinant hE-cad-Fc (17).
2. Complete DMEM: DMEM (Millipore, Billerica, MA) supplemented with 10% fetal bovine serum (FBS), 2 mM glutamine (Millipore), 1 mM sodium pyruvate (Millipore), nonessential amino acids (Millipore), and penicillin (100 units/ml)/streptomycin (100 μg/ml) (Millipore); store at 4°C.
3. Serum-free medium (CHO-S-SFM II, Invitrogen, Carlsbad, CA); store at 4°C.
4. Trypsin (0.25%)/EDTA (0.02%) (Sigma-Aldrich, Saint Louis, MO); store at −20°C.
5. Phosphate-buffered saline (PBS) (Sigma-Aldrich).
6. Tissue culture-treated dishes, 100 mm (#430167, Corning, Lowell, MA).
7. Bottle-top filter (pore size: 0.22 μm) (Corning).
8. Spinner flask (BellCo Glass, Vineland, NJ).
9. Variomag Biosystem stirrer (Thermo Scientific, Waltham, MA).

2.2. Reagents for Purification of Recombinant hE-cad-Fc

1. Wash buffer: 20 mM phosphate buffer (pH 7.0): Adjust pH to 7.0 by mixing 20 mM Na_2HPO_4 solution with 20 mM NaH_2PO_4 solution; then filter through a 0.22-μm filter and store at 4°C.
2. Elution buffer: 0.1 M sodium citrate (pH 2.7): Adjust pH to 2.7 by mixing 0.1 M citric acid monohydrate solution with 0.1 M sodium citrate dihydrate solution; then filter through a 0.22-μm filter and store at 4°C.
3. 1.0 M Tris–HCl (pH 9.0): Store at 4°C.
4. Phosphate-buffered saline+ (PBS+): PBS supplemented with 0.9 mM $CaCl_2$ and 0.9 mM $MgCl_2$.
5. Protein A column (HiTrap rProtein A FF, 1 ml or 5 ml, GE Healthcare, Pittsburgh, PA).
6. Ultrapure water.
7. 20% Ethanol/water.
8. Dialysis membrane (Slide-A-Lyzer G2, 20 kDa, Thermo Scientific).
9. Syringe filter (pore size: 0.22 μm) (Corning).
10. BCA protein assay kit (Thermo Scientific).
11. Peristaltic pump (P-1, GE Healthcare).

2.3. Immunoblotting/ SDS-PAGE

1. 2 × Laemmli sample buffer: 126 mM Tri-HCl, 4% SDS, 20% glycerol, 12% 2-mercaptoethanol, bromophenol blue
2. 30% Acrylamide/bis-acrylamide (29:1) solution
3. 1.5 M Tris–HCl, pH 8.8
4. 0.5 M Tris–HCl, pH 6.8
5. 10% Sodium dodecyl sulfate (SDS)
6. N,N,N',N',-Tetramethylethylenediamine (TEMED)
7. 10% Ammonium persulfate (APS)
8. Running buffer: 2.5 mM Tris base, 19.2 mM glycine, 0.01% SDS
9. SimplyBlue™ SafeStain (Invitrogen)
10. Transfer buffer: 2.5 mM Tris base, 19.2 mM glycine, 20% methanol
11. Immun-Blot PVDF membrane (Bio-Rad, Hercules, CA)
12. Blot absorbent filter paper (Bio-Rad)
13. 0.1% Tween 20/PBS (PBS-T)
14. Blocking solution: 5% nonfat skim milk in PBS-T
15. Antibody diluting solution: 1% nonfat skim milk in PBS-T
16. Horseradish peroxidase (HRP)-conjugated anti-mouse IgG antibody (Bio-Rad)
17. SuperSignal West Pico Chemiluminescent Substrate (Thermo Scientific)
18. X-ray film

2.4. Materials for hES Cell Culture

1. Non-treated polystyrene (bacteriological grade) plates or dishes (e.g., #351147, BD Biosciences, San Jose, CA; Corning; Nalge Nunc, Rochester, NY).
2. mTeSR1 medium (Stem Cell Technologies, Vancouver, Canada); dispense supplement into aliquots and store at −20°C. Complete medium should be used within 2 weeks.
3. Cell Dissociation Buffer, PBS-based (Invitrogen), Versene/ EDTA 0.02% (Lonza), or PBS (Sigma-Aldrich).
4. Accutase (Millipore).
5. PBS (Sigma-Aldrich).
6. 50-ml and 15-ml conical tubes (Corning).

3. Methods

3.1. Production of hE-cad-Fc

1. Thaw one tube of 1C8 cells (293 cells that secrete hE-cad-Fc) and dilute in 10 ml of complete DMEM; collect cells by centrifugation at $200 \times g$ for 5 min.

2. After removing supernatant, resuspend cells in 10 ml of fresh medium, plate in a 100-mm tissue culture-treated dish, and culture at 37°C, 5% CO_2.
3. As cells approach confluence, wash with PBS, harvest using Trypsin/EDTA, and collect by centrifugation.
4. Divide cells between twenty 100-mm tissue culture-treated dishes and culture until confluent using a 50:50 mixture of complete DMEM and CHO-S-SFM II media.
5. Collect cells by incubating with Trypsin/EDTA followed by centrifugation, and determine viable cell numbers by counting.
6. Transfer cells to a spinner flask at a density of $3-5 \times 10^5$ cells/ml in 25:75 mixture of complete DMEM and CHO-S-SFM II media with an impeller speed set at 100 rpm.
7. When cells become confluent, collect cells by centrifugation at $200 \times g$ for 5 min and resuspend in fresh CHO-S-SFM II medium.
8. Conditioned medium can be collected weekly following steps (9 and 10).
9. Harvest half of the culture and remove cells and debris by centrifugation at $500 \times g$ for 5 min.
10. Filter supernatant using a 0.22-μm filter (see Note 1) and store conditioned medium at −20°C until ready to purify the hE-cad-Fc.
11. Repeat steps 9 and 10 as necessary.
12. Expression of hE-cad-Fc can be monitored by Western blotting (see Sect. 3.3). If levels decline, initiate a new culture of 1C8 cells from seed stocks.

3.2. Purification of hE-cad-Fc

The working flow rate is 1 ml/min for a 1-ml column or 5 ml/min for a 5-ml column, and all purification steps should be performed at 4°C. Keep aliquots of all fractions at each step for analysis by SDS-PAGE and immunoblotting.

1. Thaw conditioned medium and filter again using a 0.22-μm filter to remove precipitates.
2. Wash a HiTrap rProtein A FF column with ten column volumes of wash buffer and control flow rate using a peristaltic pump.
3. (Optional) wash column with five column volumes of elution buffer followed by ten column volumes of wash buffer.
4. Apply conditioned medium at 4°C from step 11 of Sect. 3.1.
5. Wash column with ten column volumes of wash buffer.
6. Elute with five column volumes of elution buffer, collecting the eluate in a conical tube containing 1 column volume of 1 M Tris–HCl (pH 9.0) (see Note 2).

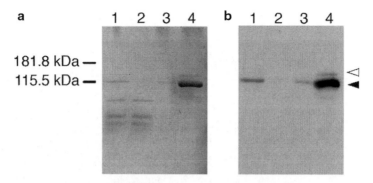

Fig. 1. Analysis of the hE-cad-Fc purification by SDS-PAGE and immunoblot analysis. (**a**) The processed form of hE-cad-Fc (120 kDa; closed arrowhead) was isolated with high purity using a rProtein A column and could be identified by SDS-PAGE after staining the gel with SimplyBlue™ SafeStain. The faint upper band is an unprocessed form of the hE-cad-Fc protein, which contains a pro-domain (open arrowhead). (**b**) The identity of the 120-kDa protein was confirmed to be hE-cad-Fc by immunoblot using an anti-mouse IgG antibody. Lane 1: conditioned medium before purification, lane 2: flow through fraction, lane 3: washed fraction, lane 4: eluted protein.

7. Mix eluate gently and dialyze against 1 l of PBS+ for 3 days at 4°C with stirring. Replace buffer with fresh PBS daily.

8. Sterilize the purified protein solution by passing through a 0.22-μm syringe filter and dispense into aliquots.

9. Store aliquots at −20°C to −80°C (aliquots can be stored for at least 6 months).

10. Measure the concentration of hE-cad-Fc using a BCA protein assay kit following the manufacturer's instruction and analyze the purity of the recombinant protein by SDS-PAGE and immunoblot following standard procedures (Sect. 3.3). A predominant band of ~120 kDa should be observed on a SimplyBlue™ stained 7.5% gel (Fig. 1a). The identity of the protein can be confirmed by immunoblot analysis (Fig. 1b).

3.3. SDS-PAGE and Immunoblotting

1. Mix each fraction (pre-purification, flow through, wash fraction, and eluate) with 2 × Laemmli sample buffer and boil at 95°C for 5 min.

2. Quick chill on ice.

3. Prepare a 7.5% SDS-polyacrylamide gel.

4. Load samples onto the individual wells and run the gel at 200 V.

5. Stain the gel with SimplyBlue™ SafeStain following the manufacturer's instructions (Fig. 1a) (see Note 3).

6. For immunoblot analysis, transfer a duplicate gel to PVDF membrane.

7. Wash the membrane once with PBS-T and incubate the membrane with blocking solution at room temperature for 1 h.

8. Prepare antibody solution by diluting HRP-conjugated anti-mouse IgG antibody with antibody diluting solution at a ratio of 1/2,000:1/10,000.

9. Discard the blocking solution and incubate the membrane with antibody solution overnight at 4°C.

10. Wash the membrane three times by shaking in the PBS-T for 5 min at room temperature.

11. Incubate the membrane with SuperSignal West Pico Chemiluminescent Substrate for 5 min at room temperature, and expose to X-ray film for 5 s to 2 min (Fig. 1b).

3.4. Coating of Non-treated Polystyrene Dishes with hE-cad-Fc

1. Remove an aliquot of E-cadherin solution from the freezer 1 day before use and thaw the solution overnight at 4°C (see Note 4).
2. Dilute in PBS to a concentration of 20 μg/ml (see Note 5).
3. Add diluted solution to each well or plate and tap the plate to make sure that the surface is completely covered by the solution (Table 1) (see Note 6).
4. Incubate at 37°C for 1 h.
5. Wash the surface with culture medium once before plating the human pluripotent stem cells (see Note 7).

3.5. Adaptation of Human Pluripotent Stem Cells to hE-cad-Fc Following Culture on Matrigel-Coated Dishes or on an MEF Feeder Layer

1. Wash cells once with PBS.
2. Add appropriate amount of Accutase (Table 1).
3. Incubate at room temperature for no longer than 5 min (see Note 8).
4. Tap the plate, and if cells are ready to detach, add twice the volume of medium.
5. Gently mix cells by pipetting and collect the detached cells into a conical tube.
6. Collect cells by centrifugation at $200 \times g$ for 5 min, then remove supernatant.
7. Resuspend cells in mTeSR1 medium (or any other medium that supports the culture of human pluripotent stem cells including feeder cell-conditioned medium) and seed onto the hE-cad-Fc-coated tissue culture dish (see Note 9).

3.6. Passage of Human Pluripotent Stem Cells from a hE-cad-Fc-Treated Surface

1. Wash cells once with PBS.
2. Add appropriate amount of Cell Dissociation Buffer or Versene/EDTA (Table 1).
3. Incubate at room temperature for 30–60 s (see Note 10).
4. Tap the plate, and if cells are ready to detach, immediately add twice the volume of culture medium (Fig. 2b).

Table 1
Volumes of hE-cad-Fc and Accutase/Cell Dissociation Buffer

	96-well plate	24-well plate	6-well plate	60-mm dish
Volume of hE-cad-Fc	50 μl/well	300 μl/well	1 ml/well	2 ml
Volume of Accutase/Cell Dissociation Buffer	–	250 μl/well	500 μl/well	1 ml

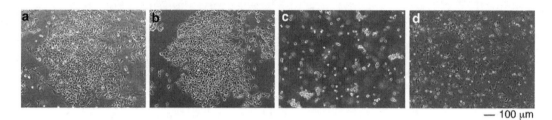

Fig. 2. Morphology of H9 human ES cells cultured on a hE-cad-Fc-coated surface. (**a**) Cells are almost confluent and ready for passage. (**b**) After treatment with Cell Dissociation Buffer for 2 min, the loss of cell–cell contact can be observed. (**c**) Following dissociation of cells by gentle pipetting, they tend to predominantly form small clusters. (**d**) Cells were observed attached to hE-cad-Fc-coated surface after 4 h.

5. Gently mix cells by pipetting and collect detached cells into a conical tube.
6. Collect cells by centrifugation at $200 \times g$ for 5 min and remove supernatant.
7. Resuspend the cell pellet in medium that is compatible with human pluripotent stem cell culture (e.g., mTeSR1) and transfer cells onto a new hE-cad-Fc-coated plate at an appropriate dilution.
8. Change the medium daily and passage the cells before they reach confluence (Fig. 2a).
9. Confirm that cells are undifferentiated and pluripotent following standard procedures (optional).

4. Notes

1. It is recommended that a master stock of cells are frozen at 1×10^7 cells/ml in a mixture of 92.5% CHO-S-SFM II (50% conditioned SFM + 50% of fresh SFM)/7.5% DMSO. Cells that have been adapted to serum-free medium can be recovered at a density of 5×10^5 cells/ml in CHO-S-SFM II medium by directly seeding into a spinner flask and continuing with Sect. 3.1, step 7.

2. After use, the HiTrap rProtein A FF column can be washed according to the manufacturer's instruction and stored at 4°C.

3. If the concentration of unprocessed protein (>120 kDa) is equal to or greater than the mature protein, adhesion of human pluripotent stem cells to the coated polystyrene surface will be inhibited (Fig. 1a). We have found that unlike culture using CHO-S-SFM II, some commercially available serum-free media dramatically increase the relative abundance of unprocessed form of hE-cad-Fc.

4. To avoid freeze and thaw cycles, hE-cad-Fc can be stored at 4°C for several months.

5. Optimal concentration of coating is best determined empirically. However, we have found that adsorption of optimally prepared hE-cad-Fc onto the polystyrene surfaces is saturated at 10–20 μg/ml (17).

6. It is absolutely crucial to use non-treated polystyrene culture dishes, which we believe facilitate the directional binding of the hE-cad-Fc through hydrophobic interactions between the IgG Fc domain and the polystyrene surface.

7. If the culture dish surface is coated with hE-cad-Fc protein, it becomes hydrophilic due to the presence of the E-cadherin domain.

8. Accutase treatment results in the degradation of the E-cadherin molecule on the cell surface (17). If cells are incubated too long with Accutase or other proteolytic enzymes, they will be incapable of attaching to the hE-cad-Fc-coated surface. It is, therefore, crucial to avoid excess degradation of E-cadherin by ensuring that Accutase treatment is kept as short as possible, optimally ≤1 min. If collecting the cells using Accutase proves to be problematic, it is possible to substitute Cell Dissociation Buffer or Versene/EDTA buffer to retrieve cells from Matrigel. If the pluripotent stem cells are cultured on feeders, they can be alternatively transferred using manual subculture procedures.

9. If cells are collected using Accutase, the initial plating efficiency can be low due to loss of E-cadherin from the cell surface. If this is the case, the number of cells plated should be increased for initial seeding onto the hE-cad-Fc-coated surface. Once cells are established on the hE-cad-Fc surface, they can easily be passaged with >90% viability as described in Sect. 3.6.

10. It is crucial to keep the incubation of the human pluripotent stem cells with Cell Dissociation Buffer or Versene/EDTA to a minimum to ensure high efficiency of cell attachment through E-cadherin interactions. If incubation in Ca^{2+}-free buffers is extensive, the stem cell plating efficiency is markedly reduced.

References

1. Lerou, P.H., Daley, G.Q. (2005) Therapeutic potential of embryonic stem cells. Blood Rev 19, 321–31.
2. Thomson, J.A., Itskovitz-Eldor, J., Shapiro, S.S., Waknitz, M.A., Swiergiel, J.J., Marshall, V.S., Jones, J.M. (1998) Embryonic stem cell lines derived from human blastocysts. Science 282, 1145–47.
3. Takahashi, K., Tanabe, K., Ohnuki, M., Narita, M., Ichisaka, T., Tomoda, K., Yamanaka, S. (2007) Induction of pluripotent stem cells from adult human fibroblasts by defined factors. Cell 131, 861–72.
4. Yu, J., Vodyanik, M.A., Smuga-Otto, K., Antosiewicz-Bourget, J., Frane, J.L., Tian, S., Nie, J., Jonsdottir, G.A., Ruotti, V., Stewart, R., Slukvin, I.I., Thomson, J.A. (2007) Induced pluripotent stem cell lines derived from human somatic cells. Science 318, 1917–20.
5. Stewart, M.H., Bendall, S.C., Bhatia, M. (2008) Deconstructing human embryonic stem cell cultures: niche regulation of self-renewal and pluripotency. J Mol Med 86, 875–86.
6. Xu, C., Inokuma, M.S., Denham, J., Golds, K., Kundu, P., Gold, J.D., Carpenter, M.K. (2001) Feeder-free growth of undifferentiated human embryonic stem cells. Nat Biotechnol 19, 971–74.
7. Kleinman, H.K., McGarvey, M.L., Liotta, L.A., Robey, P.G., Tryggvason, K., Martin, G.R. (1982) Isolation and characterization of type IV procollagen, laminin, and heparin sulfate proteoglycan from the EHS sarcoma. Biochemistry 21, 6188–93.
8. Kleinman, H.K., McGarvey, M.L., Hassell, J.R., Star, V.L., Cannon, F.B., Laurie, G.W., Martin, G.R. (1986) Basement membrane complexes with biological activity. Biochemistry 25, 312–18.
9. Vukicevic, S., Kleinman, H.K., Luyten, F.P., Roberts, A.B., Roche, N.S., Reddi, A.H. (1992) Identification of multiple active growth factors in basement membrane Matrigel suggests caution in interpretation of cellular activity related to extracellular matrix components. Exp Cell Res 202, 1–8.
10. Miyazaki, T., Futaki, S., Hasegawa, K., Kawasaki, M., Sanzen, N., Hayashi, M., Kawase, E., Sekiguchi, K., Nakatsuji, N., Suemori, H. (2008) Recombinant human laminin isoforms can support the undifferentiated growth of human embryonic stem cells. Biochem Biophys Res Commun 375, 27–32.
11. Rodin, S., Domogatskaya, A., Ström, S., Hansson, E.M., Chien, K.R., Inzunza, J., Hovatta, O., Tryggvason, K. (2010) Long-term self-renewal of human pluripotent stem cells on human recombinant laminin-511. Nat Biotechnol 28, 611–15.
12. Furue, M.K., Na, J., Jackson, J.P., Okamoto, T., Jones, M., Baker, D., Hata, R., Moore, H.D., Sato, J.D., Andrews, P.W. (2008) Heparin promotes the growth of human embryonic stem cells in a defined serum-free medium. Proc Natl Acad Sci USA 105, 13409–14.
13. Braam, S.R., Zeinstra, L., Litjens, S., Ward-van Oostwaard, D., van den Brink, S., van Laake, L., Lebrin, F., Kats, P., Hochstenbach, R., Passier, R., Sonnenberg, A., Mummery, C.L. (2008) Recombinant vitronectin is a functionally defined substrate that supports human embryonic stem cell self-renewal via $\alpha v\beta 5$ integrin. Stem Cells 26, 2257–65.
14. Melkoumian, Z., Weber, J.L., Weber, D.M., Fadeev, A.G., Zhou, Y., Dolley-Sonneville, P., Yang, J., Qiu, L., Priest, C.A., Shogbon, C., Martin, A.W., Nelson, J., West, P., Beltzer, J.P., Pal, S., Brandenberger, R. (2010) Synthetic peptide-acrylate surfaces for long-term self-renewal and cardiomyocyte differentiation of human embryonic stem cells. Nat Biotechnol 28, 606–10.
15. Derda, R., Musah, S., Orner, B.P., Klim, J.R., Li, L., Kiessling, L.L. (2010) High-throughput discovery of synthetic surfaces that support proliferation of pluripotent cells. J Am Chem Soc 132, 1289–95.
16. Villa-Diaz, L.G., Nandivada, H., Ding, J., Nogueira-de-Souza, N.C., Krebsbach, P.H., O'Shea, K.S., Lahann, J., Smith, G.D. (2010) Synthetic polymer coatings for long-term growth of human embryonic stem cells. Nat Biotechnol 28, 581–83.
17. Nagaoka, M., Si-Tayeb, K., Akaike, T., Duncan, S.A. (2010) Culture of human pluripotent stem cells using completely defined conditions on a recombinant E-cadherin substratum. BMC Dev Biol 10, 60.
18. Takeichi, M. (1994) Morphogenetic roles of classic cadherins. Curr Opin Cell Biol 7, 619–27.
19. Gumbiner, B.M. (2005) Regulation of cadherin-mediated adhesion in morphogenesis. Nat Rev Mol Cell Biol 6, 622–34.
20. Larue, L., Antos, C., Butz, S., Huber, O., Delmas, V., Dominis, M., Kemler, R. (1996) A role for cadherins in tissue formation. Development 122, 3185–94.
21. Dang, S.M., Gerecht-Nir, S., Chen, J., Itskovitz-Eldor, J., Zandstra, P.W. (2004) Controlled, scalable embryonic stem cell differentiation culture. Stem Cells 22, 275–82.
22. Eastham, A.M., Spencer, H., Soncin, F., Ritson, S., Merry, C.L., Stern, P.L., Ward, C.M. (2007) Epithelial-mesenchymal transition events during

human embryonic stem cell differentiation. Cancer Res 67, 11254–62.
23. Ullmann, U., In't Veld, P., Gilles, C., Sermon, K., De Rycke, M., Van de Velde, H., Van Steirteghem, A., Liebaers, I. (2007) Epithelial-mesenchymal transition process in human embryonic stem cells cultured in feeder-free conditions. Mol Hum Reprod 13, 21–32.

Chapter 4

Scale-Up of Single Cell–Inoculated Suspension Cultures of Human Embryonic Stem Cells

Harmeet Singh, Pamela Mok, and Robert Zweigerdt

Abstract

We have developed a simple yet highly reproducible technique of upscaling single cell–inoculated human embryonic stem cells (hESCs) in suspension cultures, using defined, nonconditioned media. Mass expansion of hESC was readily achieved by serially upscaling 2-ml static cultures to 50-ml spinner flasks. Use of the Rho kinase (ROCK) inhibitor Y-27632 (Ri) was found to enable cell survival in suspension by promoting hESC reaggregation. This treatment resulted in 44% (vs. 5–10% in untreated controls) of inoculated cells being rescued after 24 h. We further optimized a heat shock treatment in combination with Ri, which significantly increased the percentage of surviving cells from 44% to 60%. Interestingly, our data suggest that E-cadherin plays a role in hESC aggregation. The dissociation and reaggregation upon passaging may function as a "purification" step toward an E-cadherin and pluripotency marker-enriched population. In addition to the use of Ri and heat shock treatment, a media comparison revealed that mTeSR was superior to the knockout medium in supporting cell proliferation and inhibition of cell differentiation in our expansion protocol. Our upscaling strategy proved to be highly robust and practical, with significant potential to provide pluripotent cells on a clinically relevant scale. Nevertheless, our data highlight a significant line-to-line variability with respect to culture conditions and the need for a critical assessment of novel methods with numerous relevant cell lines.

Key words: Human embryonic stem cells, Bioreactor, Suspension culture, Heat shock, E-cadherin

1. Introduction

The clinical applications of hESC and their derivatives require mass production in reproducible and scalable suspension cultures. Processing of anchorage-dependent cells such as ES cells is facilitated by dissociation into single cells for passaging. This is crucial for controlled scale-up process and automation, where bioreactors should be seeded with reproducible numbers of evenly distributed cells. Experience with murine embryonic stem cell (mESC) cultures over the past three decades has recently resulted in the development of relatively large-scale processes of up to 2 l in stirred bioreactors (1–3). mESCs are routinely passaged as single cells generated by enzymatic treatment and remain vital when seeded in suspension.

Three-dimensional (3D) suspension culture of ES cells results in the formation of aggregates. This process is influenced by several important parameters, including cell concentration at inoculation, bioreactor and impeller design, stirring speed (1), and the culture medium composition. The addition of leukemia inhibitory factor (LIF) that activates the Jak/Stat pathway largely prevents mESC differentiation in both conventional plastic-adherent two-dimensional (2D) culture as well as in aggregates in 3D suspension culture (4). Unfortunately, LIF fails to maintain self-renewal of human ESC (5). In recent studies, investigators have focused on adapting 2D hESC culture to 3D suspension by microcarriers (MCs). MCs are particles which allow the translation of anchorage-dependent cells into suspension culture by providing an enlarged attachment surface. In a first study using coated polystyrene MCs (6), cell-to-carrier attachment and initial cell expansion were achieved, but cell growth ceased over successive passages. Continuous long-term propagation of hESC on cellulose MCs was established by Oh and coworkers (7); however, cell attachment and propagation relied on MC coating with Matrigel, a poorly defined basement membrane. Historically, MC technology has been applied to scale up suspension culture of conventional cell lines for the production of recombinant proteins, vaccines, or antibodies (8), whereby the final product is ultimately separated from the cells. In contrast, MCs might impose technical and regulatory hurdles with respect to stem cell generation for clinical purposes, as cells in such situations comprise the final therapeutic treatment.

This chapter describes an approach that can be used to propagate and expand hESCs in reproducible, scalable, stirred suspension cultures which are independent of additional substrates or MCs (9). By using an inhibitor of Rho kinase (Y-27632; Ri) in combination with heat shock treatment, we have systematically developed a method of forming controlled aggregates in suspension from single hESC inoculations in defined, nonconditioned culture media (KO and mTeSR). Three different hESC lines of early to late passages were used, and results were verified by performing a high number of experimental repeats. Experimental conditions were systematically optimized in 2-ml static cultures and progressively upscaled to 10-ml and finally 50-ml stirred suspension cultures (Fig. 1). Expression of pluripotent markers and the ability to form teratomas that persisted over at least five serial passages in 50-ml bioreactors were validated as well. Interestingly, these features were strongly dependent on cell line and possibly passage number.

It has been reported that Ri supports survival of hESC dissociated into single cells (10) and the formation of aggregates in

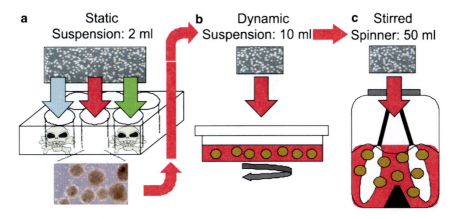

Fig. 1. Schematic representation and strategy used to optimize and scale up single cell-inoculated hESC suspension cultures (9).

suspension culture (11). To assess the potential application of Ri in 3D hESC expansion, we added Y-27632 (10 µM) in hESC culture medium (KO) post single cell dissociation and analyzed aggregate formation and cell survival in suspension culture 24 h later using Trypan Blue staining. Pretreatment with Y-27632 before hESC dissociation was not more advantageous as compared to direct addition to suspended cells.

Heat shock can induce the expression of genes that promote cell survival. To test if heat shock could increase vital cell recovery after cell dissociation, dissociated cells were subjected to a 30-min heat shock treatment at different times post cell dissociation. We found that heat shock treatment alone was most effective when applied 2 h after cell dissociation.

A potential synergistic effect of Ri and heat shock treatment was then investigated. A further increase in surviving cells (41–44% to 52–60%) was induced by combining Ri with heat shock treatment. Dissociated cells in Ri-containing culture medium were subjected to 30 min of heat shock, carried out 2 h after cell dissociation. Varying heat shock treatment times (at 0, 4, or 8 h after cell dissociation) in combination with Ri did not outperform the Ri with heat shock at 2 h (S) protocol (termed Ri + S = RiS), which was applied throughout the study.

Twenty-four hours post single cell inoculation consisted of both cell aggregates and un-aggregated single cells. Aggregates were then separated from the remaining single cells, and both populations were analyzed. While the vitality of single cells was rather low (<10%), essentially all cells that had formed clusters appeared vital.

The mechanism of cell aggregation was investigated by analyzing cell surface expression of E- and N-cadherin (E-cad, N-cad) by flow cytometry. Both transmembrane proteins are known to mediate Ca^{2+}-dependent cell–cell adhesion and are expressed in hESC (12, 13). For all three cell lines tested, there was an apparent enrichment for an E-cad-high cell population in aggregates. This suggests that cells with a higher E-cad level have a higher likelihood to agglomerate and subsequently survive in suspension.

Our expansion strategy was upscaled from a Petri dish platform to a bioreactor level (14). RiS-treated cells were dissociated into single-cell suspensions and seeded at a density of 1×10^6 cells/ml in 50 ml mTeSR medium in stirred spinner flasks (5×10^7 cells/flask). Optimization of the stirring speed resulted in extensive cell aggregation at 30 rpm. Increasing the stirring speed to 40 rpm reproducibly resulted in the formation of relatively uniform aggregates for all three cell lines. As with the Petri dish platform, increasing the stirring speed was detrimental to the cells. Under these conditions, an average cluster size of approximately 15,000–25,000 cells per aggregate was observed on day 7, being approximately tenfold larger as compared to aggregates generated within agitated Petri dishes or static suspension.

Serial passaging by aggregate dissociation every 7 days consistently resulted in the re-formation of aggregates. Cell expansion per passage was approximately twofold, yielding 1×10^8 to 1.2×10^8 cells per spinner flask, irrespective of the hESC line. This was a substantial improvement compared to the agitated Petri dish platform. Importantly, the cell yield remained extremely stable over five passages, highlighting the robustness and high reproducibility of the culture system. Karyotype analysis was performed to assess the genomic integrity of cells after five passages in spinner flasks, and no abnormalities were detected in any of the three lines.

In order to be of use in a clinical setting, the pluripotency of hESCs must be maintained. Analysis of pluripotency markers revealed persistent (at least over five serial passages) expression of pluripotency markers for two out of three cell lines (hES2 and hES3), which were used at higher passage numbers (>86) (Fig. 2). In contrast, rapid downregulation of Oct4, Tra-1-60, and SSEA4 was observed for ESI049, used at lower (20–36) passages. For the hES2 and hES3 lines, injection of passage 5/day 7 cells into SCID mice resulted in teratoma formation in 5 out of 5 mice, while medium-receiving controls did not cause teratoma formation (Fig. 3). Successful in vivo differentiation into derivatives of all three germ layers strongly underscored the pluripotency of our suspension-derived cells. Furthermore, when a directed differentiation protocol

---

Fig. 2. Growth kinetics, flow cytometry, and immunohistochemical analyses of cell aggregates harvested from 50-ml spinner flasks. (**a**) Depicted are cell yields obtained at the end of passages 1–5. (**b**) While cell yields remained stable over five passages, significant reductions in Oct4, SSEA4, and Tra-1-60 expression were observed for ESI049 cells. In contrast, all markers remained highly expressed in hES2 and hES3 cells. (**c**) Micrography of aggregates at day 7 passages 3 and 5, respectively. *Red scale bars* represent 250 μm. (**d**) ESI049 aggregates contained much larger cystic cavities, and only a small cluster of positive stained cells were observed, unlike hES2 and hES3 aggregates. *White bars* represent 500 μm; nuclear Hoechst stain is depicted in *blue* (9).

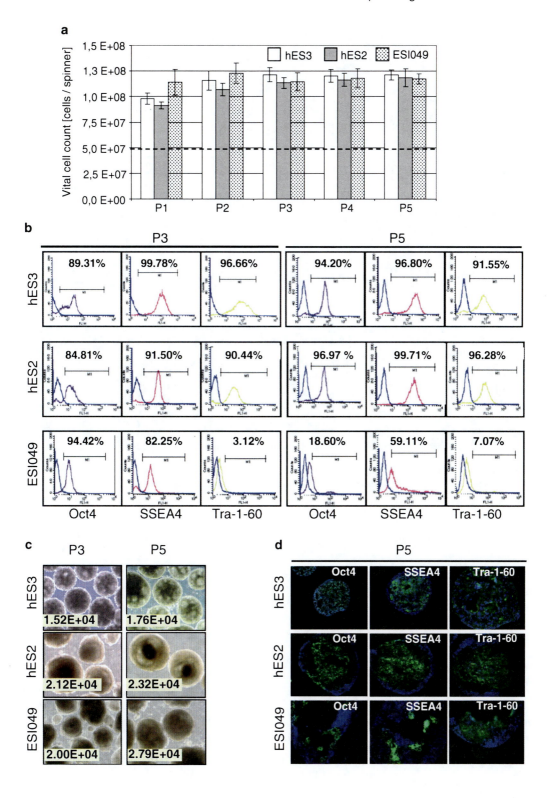

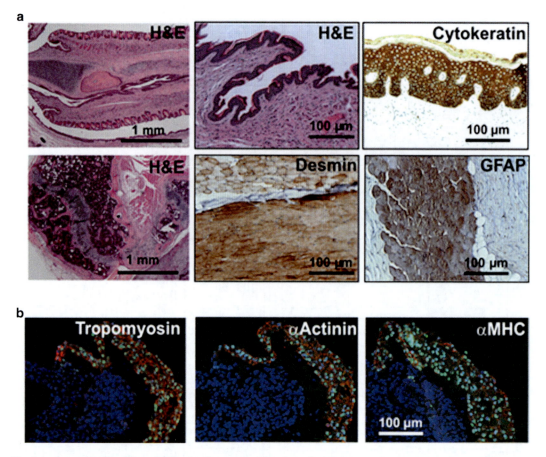

Fig. 3. In vivo and in vitro differentiation of cells generated in spinner flasks at day 7 of passage 5. (**a**) Teratoma formation at 5 weeks after intramuscular cell injections. (**b**) Spinner flask-derived aggregates were differentiated toward the cardiomyogenic lineage and stained with tropomyosin, α-MHC (α-myosin heavy chain), and α-actinin (*red*). Nuclear anti-Nkx2.5-specific staining is shown in green nuclei, while cell nuclei were stained with Hoechst (*blue*) (9).

toward cardiomyogenesis was applied to our cell aggregates, the cell aggregates differentiated into cardiomyocytes to a similar extent as hESC grown in conventional 2D culture.

2. Materials

2.1. Cell Lines and Culture Media

1. hESC lines: hES3 (Passage number 138–166), hES2 (Passage number 83–123), and ESI049 (Passage number 20–36; ES Cell International, Singapore).

2. Feeder cells: γ-irradiated human foreskin fibroblasts (Passage number 11; Ortec International).

3. hESC culture media: (a) mTeSR (Stemcell Technologies); (b) Knockout DMEM (KO) medium supplemented with 20% Knockout Serum Replacement, 50 ng/ml basic FGF, 2 mM L-glutamine, and 1% MEM nonessential amino acids (all Invitrogen). Store at 4°C.

4. Feeder culture medium: low glucose DMEM supplemented with 10% fetal bovine serum and 2 mM L-glutamine (all Invitrogen). Store at 4°C.

2.2. Cell Dissociation and Inoculation

1. Phosphate-buffered saline (PBS) solutions (Invitrogen): PBS+ (with Ca^{2+} and Mg^{2+}) and PBS− (Ca^{2+}, Mg^{2+} free).
2. TrypLE Select™ (Invitrogen). Store at 4°C.
3. 1 mg/ml Collagenase IV (Invitrogen). Store aliquots frozen at −20°C.
4. 40-µm cell strainer (BD Falcon).
5. 10 µM Y-27632 (Merck) or ROCK inhibitor. Reconstituted in sterile DI water. Store aliquots frozen in the dark at −20°C.
6. Trypan Blue (Sigma-Aldrich).

2.3. Cell Culture Plastics and Equipment

1. 6-well plates: Ultra-low attachment plates (Corning).
2. 10-cm Petri dishes. Non-tissue culture-treated plates (BD Falcon).
3. 100-ml vessel incorporating a single stirring pendulum (CELLSPIN, IBS Integra Biosciences).
4. Orbital shaker (Labotron Infors 11 T).
5. 15- or 50-ml polystyrene tubes (BD Falcon).

2.4. Cardiomyocyte Differentiation

1. Differentiation medium (15): 97.68% high glucose DMEM, 1% L-glutamine, 1% nonessential amino acids, 0.18% β-mercaptoethanol, 0.138% human transferrin, and 0.002% 100 µg/ml selenite (all Invitrogen). Store at 4°C.
2. SB203580 or p38 MAPK inhibitor, reconstituted in DMSO (Sigma-Aldrich). Store aliquots frozen in the dark at −20°C.

2.5. Flow Cytometry

1. Cytofix/Cytoperm™ Permeabilization Kit (BD).
2. Primary antibodies: mouse antihuman (a) Oct4 (Santa Cruz Biotechnology), (b) SSEA4, (c) Tra-1-60 (both Millipore), (d) E-cadherin (Santa Cruz Biotechnology), and (e) N-cadherin (Sigma-Aldrich). Dilutions used were 1:10, 1:50, 1:40, 1:50, and 1:50, respectively, per 1×10^6 cells.
3. Secondary antibody: goat anti-mouse FITC-conjugated IgG and IgM (Invitrogen). Dilution of 1:100 used.
4. Flow cytometry (FACS) buffer: PBS− with 5% FBS.
5. Blocking solution: PBS− with 10% goat serum (Zymed).
6. BD FACSCalibur.

2.6. Teratoma Studies

1. Male SCID mice aged 8–12 weeks.
2. Sterile 1-ml syringe with a 23-G needle for intramuscular injection (BD).
3. Bouin's solution (Sigma-Aldrich).

2.7. Immuno-histochemistry

(a) Cell aggregates

1. OCT freezing medium (Leica).
2. Cryostat model CM3050S (Leica).
3. 0.1% Triton X-100 in PBS.
4. Blocking buffer: PBS and 5% goat serum (Zymed, Invitrogen).
5. Primary antibodies: mouse antihuman (a) Oct4, (b) SSEA4, (c) Tra-1-60 (all as used in flow cytometry), (d) MF-20 (Developmental Studies Hybridoma Bank), (e) α-actinin (Sigma-Aldrich), and (f) Nkx2.5 (Santa Cruz Biotechnology). Concentrations for MF-20, α-actinin, and Nkx2.5 were 1:200, 1:800, and 1:500, respectively.
6. Secondary antibody: goat anti-mouse FITC-conjugated IgG and IgM (Invitrogen). Dilution of 1:100 used.
7. Hoechst 33242 (Invitrogen).
8. Fluorsave (Calbiochem).

(b) Teratomas

9. Primary antibodies: staining for cytokeratin (endoderm), desmin (mesoderm), and GFAP (ectoderm), respectively, were: (a) polyclonal rabbit anti-cytokeratin wide spectrum screening, (b) desmin clone D33, and (c) rabbit anti-GFAP (all from Dako). Dilutions used were 1:250, 1:100, and 1:250, respectively.
10. Secondary antibodies: staining for cytokeratin, desmin, and GFAP were: (a) EnVision + HRP anti-Rabbit, (b) EnVision + HRP anti-Mouse, and (c) ENVision + HRP anti-Rabbit (all from Dako), respectively, undiluted.
11. Protein blocking solution: 10% goat serum (Zymed, Invitrogen) in Tween buffer (PBS with 0.05% Tween, or PBS-T).
12. 3% H_2O_2 in methanol for blocking endogenous peroxidase.
13. Chromogen DAB (Dako).
14. Hematoxylin (Dako).

3. Methods

While the processes of inoculation with Ri, heat shock treatment, teratoma injections, flow cytometry, and immunohistochemistry analyses were all critical and will be described in detail in this section, the aggregate-dissociation protocol that we optimized is the key in determining the success of this strategy. This dissociation protocol enables cells to be counted as single cells, and consequently reinoculated at specific numbers. Furthermore, this methodology resulted in excellent cell viability.

3.1. Cell Culture and Inoculation

1. Rinse hESC cultures with PBS– (Ca^{2+}, Mg^{2+} free) and dissociate from feeder layers by incubating cells with 3 ml TrypLE Select™ per 10-cm culture dish, at 37°C for 10 min.
2. Gently tap side of culture dish. Feeder layer should lift off as a "sheet" upon tapping. Neutralize immediately with 4 ml KO medium.
3. Swirl feeder layer (sheet) around plate using a 1-ml pipette tip to "release" attached hESC (see Note 2).
4. Remove feeder layer, leaving the hESC suspension behind.
5. Strain hESC suspension through a 40-µm cell strainer placed in a 50-ml polystyrene tube.
6. Rinse culture dish with additional 3 ml of PBS– and repeat step 5 to "retrieve" any residual cells.
7. Centrifuge cell suspension at 1,200 rpm for 4 min and aspirate supernatant.
8. Resuspend cell pellet in 10 ml (or more if necessary) of either KO or mTeSR media (see Note 3).
9. Count cells using a hemocytometer and Trypan Blue solution.
10. Remove required cell numbers (see step 12) and resuspend in an appropriate volume of mTeSR medium to achieve desired cell concentration.
11. Add 1 µl Ri to every ml of cell suspension. Note that Ri is only used at this point – Ri should not be used in subsequent daily medium exchanges.
12. Inoculate cells into a desired plate format: (a) 6-well tissue culture plate (2.5×10^5 cells/ml), (b) 10-cm non-tissue culture-treated Petri dishes (1×10^6 cells/ml), and (c) spinner flasks (1×10^6 cells/ml). Transfer cells to a CO_2 incubator at 37°C (see Note 4).
13. For 6-well tissue culture plate, DO NOT agitate. For 10-cm non-tissue culture-treated Petri dishes, agitate on an orbital shaker at *35 rpm* placed within a CO_2 incubator at 37°C

(see Note 3). For spinner flasks, place entire culture platform in a CO_2 incubator at 37°C and set stirrer speed to *40 rpm* (see Note 5).

14. Where required, apply a 30-min heat shock treatment by incubating cell suspension at 43°C. Use a separate incubator for heat shock. Transfer cells back to a 37°C incubator after 30 min.

3.2. Cell Dissociation and Serial Passaging

1. Harvest cell aggregates from 6-well plates or 10-cm Petri dishes and transfer to 15- or 50-ml (depending on quantity) polystyrene tubes. Harvest cell aggregates from spinner flasks and transfer into two 50-ml polystyrene tubes. Leave cell aggregates to settle for 10–15 min (see Note 6).
2. Aspirate medium and add 1 (for 6-well plates), 3 (for 10-cm plates), or 10 ml (for spinner flask cultures) of collagenase IV (1 mg/ml).
3. Transfer cells to a CO_2 incubator for 45 min and gently agitate on an orbital shaker/rotator at 35 rpm to gently enhance aggregate dissociation.
4. Remove aggregates and leave to stand for 5–10 min, allowing aggregates to settle.
5. Aspirate collagenase IV and gently rinse aggregates in PBS−. Allow aggregates to settle.
6. Aspirate PBS− and add TrypLE Select™ to the aggregates (similar volumes to collagenase IV above) and place polystyrene tubes on the orbital shaker within a CO_2 incubator for 10 min (6-well plates and Petri dish aggregates) or 20 min (spinner flask-derived aggregates). Note that ESI049 hESC aggregates from spinner flasks required 30 min for near complete dissociation, being relatively differentiated (see Note 7).
7. Remove aggregates from the incubator and neutralize with cell culture medium. Gently dissociate aggregates into single cells with a 5- or 10-ml pipette by pipetting up and down.
8. Count the single-cell suspension. Note that viability was generally >95%.
9. For serial passaging, cells were reinoculated as single cells.

3.3. Cardiomyocyte Differentiation

1. Incubate aggregates overnight in differentiation medium.
2. Change medium the next day and add SB203580 (5 mM).
3. Change medium every 3 days.
4. Beating cardiomyocytes are seen from days 10 to 12 and thereafter.

3.4. Flow Cytometry

1. Fix 1×10^6 cells for 10 min with 1 ml of Cytofix/Cytoperm.
2. Centrifuge cells at $300 \times g$ and aspirate Cytofix/Cytoperm.

3. Rinse cell pellet with FACS buffer and centrifuge again at $300 \times g$. Repeat once.
4. Resuspend fixed cells in blocking buffer and block for 1 h.
5. Rinse cells and resuspend in 50 µl of FACS buffer.
6. Add primary antibodies at desired concentration to cells and incubate at room temperature (RT) for 1 h.
7. Repeat step 3 three times.
8. Add secondary antibodies at desired concentration to cells in 50 µl of buffer and incubate in the dark for 1 h at RT.
9. Repeat step 3 three times.
10. Resuspend cells in 500 µl of FACS buffer and analyze with BD FACSCalibur.

3.5. Teratoma Studies

1. Clean right hind limb of the mouse with a sterile alcohol swab.
2. Guide a sterile 1-ml syringe with a 23-G needle containing either (a) hESC suspension (100 µl, 5×10^4 cells/µl) or (b) medium (100 µl) into the right hind leg quadriceps of the mouse along its long axis and toward the muscle center.
3. Assess mice visually for teratoma formation every 2–3 days using a grading system (16).
4. Upon appearance of grade three teratomas (teratomas that impede locomotion), sacrifice mice by CO_2 asphyxiation within a ducted chamber.
5. Carefully excise teratomas from the surrounding muscle tissue and fix in Bouin's solution for immunohistochemical analysis.

3.6. Immunohistochemistry

(a) Cell aggregates

1. Fix aggregates in 4% paraformaldehyde for 15 min at RT and preserve in 25% sucrose solution at 4°C overnight.
2. Snap-freeze aggregates in OCT medium by immersing in liquid nitrogen.
3. Section frozen aggregates at 6 µm and gently place on glass slides.
4. Permeate sections with 0.1% Triton X-100 solution for 10 min at RT.
5. Incubate sections in blocking buffer for 60 min at RT.
6. Add primary antibodies and incubate for 1 h at RT.
7. Wash sections three times with PBS– for 10 min per rinse.
8. Add secondary antibodies and incubate at RT for 1 h in the dark (as used for flow cytometry).
9. Repeat step 7.
10. Stain nuclei with Hoechst 33242 (diluted 1:10,000) for 30 s.

11. Wash samples twice with PBS– in the dark.
12. Mount sections with Fluorsave to reduce bleaching and place coverslip over stained sections.
13. View samples under a standard fluorescence microscope.

(b) Teratomas

14. Embed excised teratomas in paraffin.
15. Use a microtome to cut 8-µm sections and gently transfer sections to glass slides.
16. De-wax paraffin sections via a standard xylene and ethanol series.
17. Wash sections in deionized water (dH_2O).
18. Block endogenous peroxidase with 3% H_2O_2 in methanol for 15 min at RT.
19. Wash samples in dH_2O and block with 10% goat serum in Tween buffer for 20 min at RT.
20. Add primary antibodies for endoderm (cytokeratin), mesoderm (desmin), and ectoderm (GFAP) at desired concentration, and incubate samples overnight at 4°C.
21. Wash samples under running dH_2O for 5 min.
22. Protein block with 10% goat serum in Tween buffer for 20 min at RT.
23. Repeat step 17.
24. Add secondary antibodies at desired concentration and incubate for 30 min at RT.
25. Repeat step 21.
26. Add chromogen DAB to samples for 3–4 min.
27. Repeat step 21.
28. Stain nuclei with hematoxylin for 1 min.
29. Repeat step 21.
30. Mount sections with standard mounting medium and place coverslip over stained sections.
31. View samples with a standard phase-contrast microscope.

4. Notes

1. It may help to tilt the plate while swirling the feeder layer and occasionally dip the feeder layer into the media to release cells more effectively.
2. Prior trials involving various seeding densities showed that higher densities would (e.g., 5×10^5 or 1×10^6 cells/ml) result in large, irregular chain-like aggregate formations.

3. First gently dissociate cell pellet by tapping the bottom of the polystyrene tube.

4. Results revealed the formation of larger, irregular aggregates that occurred at 20 rpm especially, and to a lesser degree at 30 rpm. This was due to a percentage of cell aggregates being unable to stay suspended continuously, therefore sinking to the base of the culture dish. Rotational speeds of 40 and 50 rpm were found to result in lower cell counts, likely due to the negative effects of excessive shear.

5. Previous spinner flask culture trials performed at 20, 30, and 50 rpm revealed that speeds below 40 rpm resulted in aggregates eventually sinking to the bottom of the vessel, thereby forming clumps of tissue. While trials performed at 50 rpm resulted in uniform cell aggregates, cell counts were noted to be lower as compared to 40 rpm.

6. Rather than subjecting the delicate hESC aggregates to the additional stresses of centrifugation, they were instead allowed to gently settle due to gravitational influence.

7. While trypsin-EDTA was initially used to dissociate hESC, it was found to be excessively harsh and, furthermore, did not favor reaggregation of cells, resulting in considerably lower cell counts. This harsh treatment was also found to adversely affect cell counts, while also potentially affecting the accuracy of flow cytometry results. The ESI049 aggregates encountered were partially differentiated and were especially "clumpy." Additional time was required to dissociate these cell aggregates into single cells.

References

1. Schroeder, M., Niebruegge, S., Werner, A., et al. (2005) Differentiation and lineage selection of mouse embryonic stem cells in a stirred bench scale bioreactor with automated process control, *Biotechnol Bioeng 92*, 920–933.

2. zur Nieden, N. I., Cormier, J. T., Rancourt, D. E., et al. (2007) Embryonic stem cells remain highly pluripotent following long term expansion as aggregates in suspension bioreactors, *J Biotechnol 129*, 421–432.

3. Niebruegge, S., Nehring, A., Bar, H., et al. (2008) Cardiomyocyte production in mass suspension culture: embryonic stem cells as a source for great amounts of functional cardiomyocytes, *Tissue Eng Part A 14*, 1591–1601.

4. Taiani, J., Krawetz, R. J., Nieden, N. Z., et al. (2009) Reduced Differentiation Efficiency of Murine Embryonic Stem Cells in Stirred Suspension Bioreactors, *Stem Cells Dev*.

5. Daheron, L., Opitz, S. L., Zaehres, H., et al. (2004) LIF/STAT3 signaling fails to maintain self-renewal of human embryonic stem cells, *Stem Cells 22*, 770–778.

6. Phillips, B. W., Horne, R., Lay, T. S., et al. (2008) Attachment and growth of human embryonic stem cells on microcarriers, *J Biotechnol 138*, 24–32.

7. Oh, S. K., Chen, A. K., Mok, Y., et al. (2009) Long-term microcarrier suspension cultures of human embryonic stem cells, *Stem Cell Res*.

8. Zweigerdt, R. (2009) Large Scale Production of Stem Cells and Their Derivatives, *Adv Biochem Eng Biotechnol*.

9. Singh, H., Mok, P., Balakrishnan, T., et al. (2010) Up-scaling single cell-inoculated suspension culture of human embryonic stem cells, *Stem Cell Res 4*, 165–179.

10. Watanabe, K., Ueno, M., Kamiya, D., et al. (2007) A ROCK inhibitor permits survival of

11. Li, X., Krawetz, R., Liu, S., *et al.* (2009) ROCK inhibitor improves survival of cryopreserved serum/feeder-free single human embryonic stem cells, *Hum Reprod 24*, 580–589.
12. Eastham, A. M., Spencer, H., Soncin, F., *et al.* (2007) Epithelial-mesenchymal transition events during human embryonic stem cell differentiation, *Cancer Res 67*, 11254–11262.
13. Avery, K., Avery, S., Shepherd, J., *et al.* (2008) Sphingosine-1-phosphate mediates transcriptional regulation of key targets associated with survival, proliferation, and pluripotency in human embryonic stem cells, *Stem Cells Dev 17*, 1195–1205.
14. Dang, S. M., Gerect-Nir, S., Chen, J., *et al.* (2004) Controlled, scalable embryonic stem cell differentiation culture, *Stem Cells 22*, 275–282.
15. Xu, X. Q., Graichen, R., Soo, S. Y., *et al.* (2008) Chemically defined medium supporting cardiomyocyte differentiation of human embryonic stem cells, *Differentiation 76*, 958–970.
16. Hentze, H., Soong, P. L., Wang, S. T., *et al.* (2009) Teratoma formation by human embryonic stem cells: Evaluation of essential parameters for future safety studies, *Stem Cell Res*.

Reference 10 (partial): dissociated human embryonic stem cells, *Nat Biotechnol 25*, 681–686.

Chapter 5

Three-Dimensional Culture for Expansion and Differentiation of Embryonic Stem Cells

Guang-wei Sun, Xiao-xi Xu, Nan Li, Ying Zhang, and Xiao-jun Ma

Abstract

Although embryonic stem (ES) cells have the regenerative capability of producing any tissue in our body, the development of efficient methods to generate a large number of ES cells with high purity has still remained a technical challenge before they can be used routinely in clinical trials. In this chapter, we describe a simple procedure to establish an ES cell 3D culture system through alginate-poly-L-lysine-alginate (APA) microencapsulation. The APA microcapsule provides an environment that promotes better cell growth and differentiation without requiring feeder layers.

Key words: APA microcapsule, 3D culture, ES cells, Expansion, Differentiation

1. Introduction

Embryonic stem (ES) cells have the regenerative compatibility to potentially produce any tissue in the body (1–3). However, the successful use of ES cells for clinical application requires the development of efficient methods to generate a large number of ES cells with high purity while maintaining their pluripotency. Unfortunately, in the course of in vitro passaging, the cells often spontaneously differentiate (4). A number of studies have been carried out to optimize and standardize the growth conditions (5, 6). Extensive efforts have also been made to optimize growth media, the use of growth factors (7, 8), and the passaging procedures (9). Furthermore, the replacement of mouse embryonic fibroblast (MEF) feeder layers with extracellular matrix (ECM) or their derivative matrices, such as Matrigel or an ECM derived from MEFs, has also been explored (10, 11). However, these stationary two-dimensional (2D) systems foster the formation of gradients of dissolved oxygen, nutrients, and metabolites, thus resulting in heterogeneous environments that can compromise optimal cell growth and viability (12). In addition, the above culture systems also make the process more labor-intensive and time-consuming when scale-up is necessary.

Advance in studies of ES cell interactions with their environment has led to the development of many new material-based approaches to control stem cell proliferation and differentiation. Biomaterial-based scaffolds that have been widely used in stem cell expansion include hydrogel (6, 13), microcarriers (14–17), porous or fibrous scaffolds (18–20), and microcapsules (21–27). A body of evidence suggests that 3D cultures can potentially mimic the in vivo microenvironment, thus promoting better ES cell growth, differentiation, maturation, and even better protein secretion without using feeder layers (21–25). Biomaterials provide cues based on their chemistry, mechanics, and structure, which regulate ES cell fate decisions.

In this chapter, we describe a simple procedure to establish ES cell 3D culture system by using alginate-poly-L-lysine-alginate (APA) microcapsule. APA microencapsulation is one of the most well-studied encapsulation technologies, including entrapment of cells in alginate gel beads, the formation of alginate-poly-L-lysine membrane, and the liquefying of the alginate gel core to leave the cell floating in the center of microcapsules (28). The semipermeable membrane of microcapsules allows free diffusion of nutrients, oxygen, and toxic metabolites to support cell growth, but prevents direct contact with immunoglobulins, immunological moieties, and immune cells (28). Due to biocompatibility and biodegradability of APA capsules, cell microencapsulation technology has been widely used for in vitro cell culture and in vivo cell therapy (29, 30). Encapsulated ES cells exhibit a distinct growth pattern (Fig. 1), and the expression of typical marker genes of the undifferentiated ES cells can last over 2 weeks in vitro. In contrast to the in vitro culture, encapsulated ES cells grow much faster in vivo and reduce the expression of stemness markers (Fig. 2) (21). Thus, the APA-ES cell system may provide an optimal model to maintain ES cells in vitro and differentiate them in vivo. In addition, the APA microencapsulation can also be an alternative technology for large-scale ES cell cultures since the microcapsulation membrane can protect cells from mechanical damage of shear forces associated with agitation and aeration (31). Although the protocol shown here uses mouse ES cells as a model system, the technique can be readily augmented to handle human ES cells.

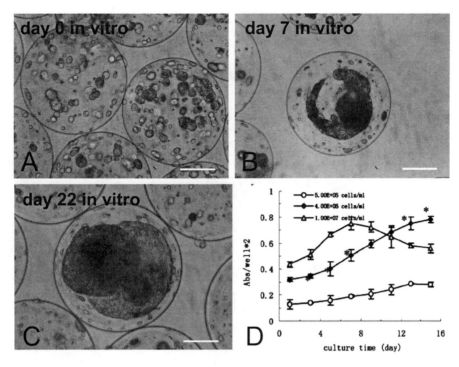

Fig. 1. A typical growth profile of mouse embryonic stem (mES) cells enclosed in liquefied APA microcapsules (**a–c**) and their expansion rate in vitro (**d**). Scale bar: 100 μm (Reproduced from (21) with permission from (Ma X)).

2. Materials

2.1. Culture Medium of Undifferentiated ES Cells

1. Dulbecco's Modified Eagle's Medium (DMEM) with high glucose concentration (H-DMEM, Gibco/BRL, Grand Island, NY); store at 4°C
2. 15% Fetal calf serum (FCS, Gibco/BRL, Grand Island, NY); store at −20°C
3. 0.1 mM β-Mercaptoethanol (β-Me, Sigma-Aldrich, St. Louis, MO)
4. 1% Nonessential amino acids (Sigma-Aldrich, St. Louis, MO); store at 4°C
5. 2 mM L-Glutamine (Sigma-Aldrich, St. Louis, MO); store at −20°C
6. 10^3 U/ml LIF (Gibco/BRL, Grand Island, NY); store at 4°C

2.2. Preparation of APA Microcapsules

1. 1.5% (w/v) and 0.15% (w/v) Sterile sodium alginate (0.22-μm syringe filters, Sigma-Aldrich, St. Louis, MO); stable for up to 1 month at 4°C
2. 100 mM Sterile calcium chloride solution

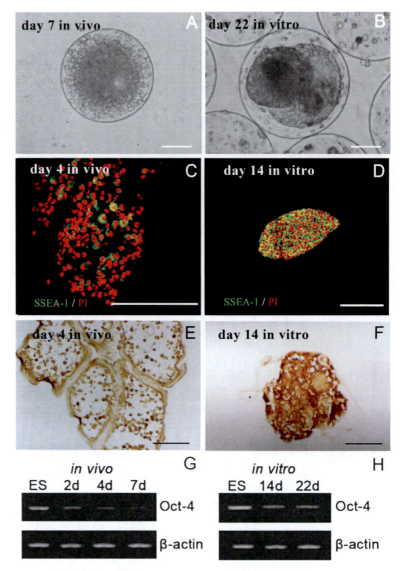

Fig. 2. Micrograph images of the growth profile of ES cells enclosed in liquefied APA microcapsules cultured in vivo (**a**) and in vitro (**b**). Compared with in vitro cultures, ES cells grow much faster in vivo and nearly fill up all the space of microcapsules at day 7. (Scale bar: 100 μm.) The sections of the encapsulated ES cells are labeled with antibodies against SSEA-1 (**c** and **d**) and alkaline phosphatase (ALP) (**e** and **f**). The expression of SSEA-1 (**c**) and ALP (**e**) in vivo is decreased with the increase in the cell passage numbers. This is different from the slow differentiation process of encapsulated ES cells in vitro (**d** and **f**). RT-PCR assay of the expression of transcription factor gene Oct-4 in encapsulated ES cells in vivo and in vitro is given in (**g**) and (**h**), respectively. The housekeeping gene, β-actin, serves as an internal control (Reproduced from (21) with permission from (Ma X)).

3. 0.05% (w/v) Sterile poly-L-lysine solution (29 kDa, 0.22-μm syringe filter sterilized, Sigma-Aldrich, St. Louis, MO); store at 4°C

4. 55 mM Sterile sodium citrate solution

5. 0.9% (w/v) Sterile sodium chloride
6. An electrostatic droplet generator
7. A syringe pump

3. Methods

Alginate is an anionic polysaccharide distributed widely in brown algae. It can easily form hydrate gel in the presence of divalent cations such as Ca^{2+}. The poly-L-lysine can chemically cross-link calcium alginate gel beads at the surface to form a membrane that is permselective and immunoprotective. The methodology described here has been used to encapsulate many different types of cells, including mammalian cells, bacteria, and so on. It is a simple and quick procedure to gain large quantity of microcapsules, which can provide a mild microenvironment to maintain better cell activity.

3.1. Cell Microencapsulation

The preparation procedure is largely based on the previously described method (28, 32) (see Note 1). It requires an electrostatic droplet generator, which produces smaller (diameter of less than 300 μm) and more uniform microcapsules (Fig. 3).

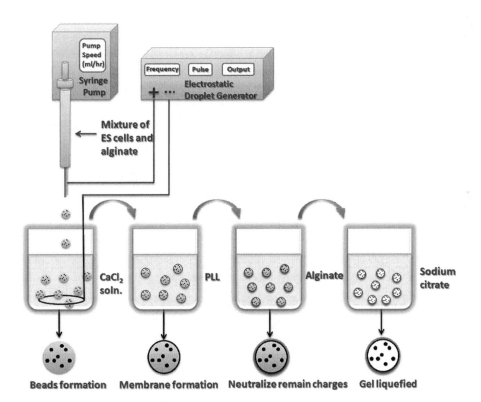

Fig. 3. Microencapsulation of ES cells by alginate-poly-L-lysine-alginate microcapsules.

1. Harvest ES cells using 0.25% trypsin–0.04% EDTA and count.
2. Resuspend cells in 1.5% (w/v) sodium alginate (see Note 2).
3. Extrude the ES cell/alginate suspension using a syringe pump into 100 mM calcium chloride solution under an electrostatic field to form calcium alginate gel beads (see Fig. 3) (see Notes 3 and 4).
4. Add 0.05% poly-L-lysine solution to the calcium alginate gel beads to form a semipermeable membrane (see Note 5).
5. Discard the supernatant and add 0.9% (w/v) sodium chloride to wash the spheres thoroughly (see Note 6).
6. Resuspend the spheres in 0.15% (w/v) alginate solution for 5 min (see Note 7).
7. Discard the supernatant and add 0.9% (w/v) sodium chloride to wash the spheres thoroughly.
8. Resuspend the membrane-enclosed gel beads in 55 mM sodium citrate to liquefy the gel core.
9. Wash the microcapsules thoroughly with 0.9% (w/v) sodium chloride and culture medium.

3.2. Culture of Encapsulated ES Cells In Vitro

Microencapsulated ES cells (without feeder layers) are cultured as 10% (packed volume of capsules/volume of medium) suspension in H-DMEM medium containing 15% FCS (without LIF), and incubated at 37°C in a humidified 5% CO_2. Half the medium needs to be refreshed every other day.

3.3. Implantation of Encapsulated ES Cells

Microcapsules are separated into aliquots of 800 capsules, washed in normal saline for 10 min in order to remove any trace of impurity (operated under sterile conditions), and then implanted intraperitoneally using a sterilized 18-gauge catheter (see Notes 8 and 9).

3.4. Explantation of Encapsulated ES Cells

The mice are sacrificed by CO_2 vapor, and the skin and muscle layer are cut open with scissors to reveal the peritoneum. Free-floating capsules are retrieved by washing/flushing the peritoneal cavity with PBS and obtained for further investigation (see Note 10).

4. Notes

1. All procedures of cell microencapsulation can be operated under sterile conditions at room temperature.
2. The optimal cell density is $5–6 \times 10^6$/ml sodium alginate.

3. Decide the diameters of the gel beads by the parameters of an electrostatic droplet generator, such as voltage, frequency, velocity of pump, and so on.

4. The gelation is maintained for at least 20 min.

5. The porosity of membrane is increased with the increase of the poly-L-lysine solution volume or reacting time, or vice versa.

6. Gel beads or microcapsules can be separated by gravity in solution.

7. Use 0.15% (w/v) sodium alginate solution to neutralize the remaining cationic charges on the surface of the spheres. This step is necessary to improve capsular durability and biocompatibility.

8. For implantation, the quantity of microcapsules can be adjusted to the age or species of murine.

9. Using 18-gauge catheter is to ensure the microcapsules' intactness.

10. Microcapsules retrieved from peritoneal explantation can continue to be cultured in vitro if operated under sterile conditions. They can also be preserved at any condition for later use.

References

1. Odorico, J.S., Kaufman, D.S., Thomson, J.A. (2001) Multilineage Differentiation from Human Embryonic Stem Cell Lines. Stem Cells **19**, 193–204.

2. Itskovitz-Eldor, J., Schuldiner, M., Karsenti, D., Eden, A., Yanuka, O., Amit, M., Soreq, H., Benvenisty, N. (2000) Differentiation of human embryonic stem cells into embryoid bodies compromising the three embryonic germ layers. Mol Med **6**, 88.

3. Keller, G.M. (1995) In vitro differentiation of embryonic stem cells. Curr Opin Cell Biol **7**, 862–869.

4. Xu, R.-H., Peck, R.M., Li, D.S., Feng, X., Ludwig, T., Thomson, J.A. (2005) Basic FGF and suppression of BMP signaling sustain undifferentiated proliferation of human ES cells. Nat Meth **2**, 185–190.

5. Bigdeli, N., Andersson, M., Strehl, R., Emanuelsson, K., Kilmare, E., Hyllner, J., Lindahl, A. (2008) Adaptation of human embryonic stem cells to feeder-free and matrix-free culture conditions directly on plastic surfaces. J Biotechnol **133**, 146–153.

6. Gerecht, S., Burdick, J., Ferreira, L., Townsend, S., Langer, R., Vunjak-Novakovic, G. (2007) Hyaluronic acid hydrogel for controlled self-renewal and differentiation of human embryonic stem cells. Proc Natl Acad Sci **104**, 11298.

7. Chen, H.F., Kuo, H.C., Chien, C.L., Shun, C.T., Yao, Y.L., Ip, P.L., Chuang, C.Y., Wang, C.C., Yang, Y.S., Ho, H.N. (2007) Derivation, characterization and differentiation of human embryonic stem cells: comparing serum-containing versus serum-free media and evidence of germ cell differentiation. Hum Reprod **22**, 567–577.

8. Wang, L., Schulz, T.C., Sherrer, E.S., Dauphin, D.S., Shin, S., Nelson, A.M., Ware, C.B., Zhan, M., Song, C.-Z., Chen, X., Brimble, S.N., McLean, A., Galeano, M.J., Uhl, E.W., D'Amour, K.A., Chesnut, J.D., Rao, M.S., Blau, C.A., Robins, A.J. (2007) Self-renewal of human embryonic stem cells requires insulin-like growth factor-1 receptor and ERBB2 receptor signaling. Blood **110**, 4111–4119.

9. Ellerström, C., Strehl, R., Noaksson, K., Hyllner, J., Semb, H. (2007) Facilitated Expansion of Human Embryonic Stem Cells by Single-Cell Enzymatic Dissociation. Stem Cells **25**, 1690–1696.

10. Mallon, B.S., Park, K.-Y., Chen, K.G., Hamilton, R.S., McKay, R.D.G. (2006) Toward xeno-free culture of human embryonic stem cells. Int J Biochem Cell Biol **38**, 1063–1075.

11. Skottman, H., Hovatta, O. (2006) Culture conditions for human embryonic stem cells. Reproduction **132**, 691–698.
12. Hassell, T., Gleave, S., Butler, M. (1991) Growth inhibition in animal cell culture. Appl Biochem Biotechnol **30**, 29–41.
13. Ferreira, L.S., Gerecht, S., Fuller, J., Shieh, H.F., Vunjak-Novakovic, G., Langer, R. (2007) Bioactive hydrogel scaffolds for controllable vascular differentiation of human embryonic stem cells. Biomaterials **28**, 2706–2717.
14. Tielens, S., Declercq, H., Gorski, T., Lippens, E., Schacht, E., Cornelissen, M. (2007) Gelatin-Based Microcarriers as Embryonic Stem Cell Delivery System in Bone Tissue Engineering: An in-Vitro Study. Biomacromolecules **8**, 825–832.
15. Nie, Y., Bergendahl, V., Hei, D.J., Jones, J.M., Palecek, S.P. (2009) Scalable culture and cryopreservation of human embryonic stem cells on microcarriers. Biotechnol Prog **25**, 20–31.
16. Hwang, Y.S., Cho, J., Tay, F., Heng, J.Y.Y., Ho, R., Kazarian, S.G., Williams, D.R., Boccaccini, A.R., Polak, J.M., Mantalaris, A. (2009) The use of murine embryonic stem cells, alginate encapsulation, and rotary microgravity bioreactor in bone tissue engineering. Biomaterials **30**, 499–507.
17. Lock, L.T., Tzanakakis, E.S. (2009) Expansion and Differentiation of Human Embryonic Stem Cells to Endoderm Progeny in a Microcarrier Stirred-Suspension Culture. Tissue Eng Part A **15**, 2051–2063.
18. Levenberg, S., Huang, N., Lavik, E., Rogers, A., Itskovitz-Eldor, J., Langer, R. (2003) Differentiation of human embryonic stem cells on three-dimensional polymer scaffolds. Proc Natl Acad Sci USA **100**, 12741.
19. Mao, G.-h., Chen, G.-a., Bai, H.-y., Song, T.-r., Wang, Y.-x. (2009) The reversal of hyperglycaemia in diabetic mice using PLGA scaffolds seeded with islet-like cells derived from human embryonic stem cells. Biomaterials **30**, 1706–1714.
20. Gerecht-Nir, S., Cohen, S., Ziskind, A., Itskovitz-Eldor, J. (2004) Three-dimensional porous alginate scaffolds provide a conducive environment for generation of well-vascularized embryoid bodies from human embryonic stem cells. Biotechnol Bioeng **88**, 313–320.
21. Wang, X., Wang, W., Ma, J., Guo, X., Yu, X., Ma, X. (2006) Proliferation and Differentiation of Mouse Embryonic Stem Cells in APA Microcapsule: A Model for Studying the Interaction between Stem Cells and Their Niche. Biotechnol Prog **22**, 791–800.
22. Maguire, T., Novik, E., Schloss, R., Yarmush, M. (2006) Alginate-PLL microencapsulation: Effect on the differentiation of embryonic stem cells into hepatocytes. Biotechnol Bioeng **93**, 581–591.
23. Dang, S.M., Gerecht-Nir, S., Chen, J., Itskovitz-Eldor, J., Zandstra, P.W. (2004) Controlled, Scalable Embryonic Stem Cell Differentiation Culture. Stem Cells **22**, 275–282.
24. Dean, S.K., Yulyana, Y., Williams, G., Sidhu, K.S., Tuch, B.E. (2006) Differentiation of Encapsulated Embryonic Stem Cells After Transplantation. Transplantation **82**, 1175–1184.
25. Siti-Ismail, N., Bishop, A.E., Polak, J.M., Mantalaris, A. (2008) The benefit of human embryonic stem cell encapsulation for prolonged feeder-free maintenance. Biomaterials **29**, 3946–3952.
26. Batorsky, A., Liao, J., Lund, A.W., Plopper, G.E., Stegemann, J.P. (2005) Encapsulation of adult human mesenchymal stem cells within collagen-agarose microenvironments. Biotechnol Bioeng **92**, 492–500.
27. Li, X., Liu, T., Song, K., Yao, L., Ge, D., Bao, C., Ma, X., Cui, Z. (2006) Culture of Neural Stem Cells in Calcium Alginate Beads. Biotechnol Prog **22**, 1683–1689.
28. Ma, X., Vacek, I., Sun, A. (1994) Generation of alginate-poly-l-lysine-alginate (APA) biomicrocapsules: the relationship between the membrane strength and the reaction conditions. Artif Cells Blood Substit Biotechnol **22**, 43–69.
29. Peirone, M., Ross, C.J.D., Hortelano, G., Brash, J.L., Chang, P.L. (1998) Encapsulation of various recombinant mammalian cell types in different alginate microcapsules. J Biomed Mater Res **42**, 587–596.
30. Orive, G., Hernández, R., Gascón, A., Calafiore, R., Chang, T., De Vos, P., Hortelano, G., Hunkeler, D., Lacík, I., Shapiro, A., Pedraz, J. (2003) Cell encapsulation: promise and progress. Nat Med **9**, 104–107.
31. Li, H., Jiang, H., Wang, C., Duan, C., Ye, Y., Su, X., Kong, Q., Wu, J., Guo, X. (2006) Comparison of two types of alginate microcapsules on stability and biocompatibility in vitro and in vivo. Biomedical Materials **1**, 42.
32. Lim, F., Sun, A.M. (1980) Microencapsulated islets as bioartificial endocrine pancreas. Science **210**, 908–910.

Chapter 6

Expansion of Pluripotent Stem Cells in Defined, Xeno-Free Culture System

Kristiina Rajala

Abstract

The hallmark of pluripotent stem cells is their nearly unlimited self-renewal capacity, and their potential to differentiate into a diverse range of specialized cell types. These unique properties make stem cells important research tools, in vitro models for pharmaceutical testing, and an attractive source of various cell types for regenerative therapies. For stem cell technology to be fully exploited, however, culture systems must be improved to enable large-scale production, and safety ensured. Most stem cell culture systems developed to date utilize undefined, xenogeneic products that pose a risk of a severe immune response and the transmission of infections. In this chapter, we describe a robust method for the expansion of pluripotent stem cells in defined and xeno-free culture conditions. Both mechanical and single-cell enzymatic passaging can be applied with this method. This procedure can be adopted for both basic research purposes and clinical applications.

Key words: Pluripotent stem cell culture, Defined, xeno-free culture system, Stem cell proliferation, Stem cell self-renewal, hES cell culture

1. Introduction

Pluripotent stem cells tend to maintain tight contacts with their neighbors and grow in colonies in culture. In general, pluripotent stem cells are challenging to culture as they tend to follow their natural cell fate and differentiate spontaneously in vitro. Most culture conditions result in some level of unwanted spontaneous differentiation of pluripotent stem cells. Differentiation is a result of many complex interactions with intrinsic and extrinsic factors, including growth factors, ECM molecules and components, environmental stressors, and direct cell-to-cell interactions (1). While some spontaneously differentiated cells usually appear at the margin and at the center of pluripotent stem cell colonies, an ideal culture condition provides growth support with minimal amounts

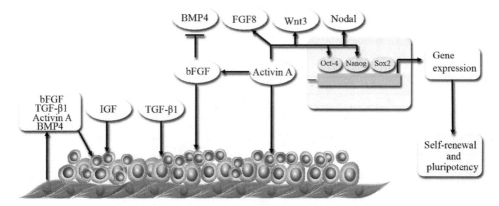

Fig. 1. Molecules regulating pluripotent stem cell renewal. Growth factors such as bFGF, IGF, TGF-β1, and Activin A stimulate pluripotent stem cell growth via direct and indirect mechanisms, whereas BMP signaling induces premature differentiation.

of differentiated cells. Since the first establishment of permanent human embryonic stem cell (hESC) line in 1998 (2), various culture conditions have been described for the establishment and expansion of pluripotent stem cells. Over the last few years, a number of molecular factors and signaling pathways that play a major role in maintaining self-renewal have been identified (Fig. 1).

The diverse culture conditions utilized for the in vitro expansion and differentiation of stem cells influence the gene expression profiles of stem cells and, hence, probably many of the cell properties (3). Most stem cell lines established to date have been directly or indirectly exposed to xenogeneic products during their derivation, expansion, or differentiation in vitro. The exposure of stem cells to xenogeneic products increases the risk of graft rejection and severe immune response in the recipient (4, 5). Xenogeneic immunogen N-glycolylneuraminic acid (Neu5Gc) for which a preformed antibody exists in humans has been identified from stem cells cultured with xenogeneic products (6, 7). More recently, Sakamoto and coworkers reported the identification of another predominant immunogen apoB-100 that was acquired by stem cells from xenogeneic products in the culture environment (8, 9). Other potential risks to the recipient include viral or bacterial infections, prions, and as yet unidentified zoonoses (10). The use of completely defined conditions will allow for a better understanding of stem cell regulations and provide more reproducible results. Therefore, for research purposes, as well as for the clinical application of stem cells, the development of a completely defined, xeno-free, and standardized culture conditions is highly desirable.

In this chapter, we describe a robust method for the expansion of pluripotent stem cells in defined and xeno-free culture conditions. Human ESC lines can be successfully derived using the described culture method. The culture method has been developed to be used with feeder cells and mechanical passaging of cells and is designed to meet the special requirements of stem cell culture and the quality requirements for clinical use of stem cells (11). However, to enable upscaling and large-scale cultivation, a single-cell enzymatic dissociation (SCED) method has been successfully used with the defined and xeno-free culture system (11, 12). The method has been shown to maintain the normal karyotype and differentiation potential of stem cells both in vitro and in vivo after long-term culture. Pluripotent stem cells can also be frozen and thawed using the xeno-free medium (11). Importantly, the culture system we describe is widely applicable and should therefore be of general use to facilitate reproducible and safe cultivation of pluripotent stem cells for various applications.

2. Materials

2.1. Preparation of Defined, Xeno-Free RegES Medium

1. Glycine (Sigma-Aldrich), store at room temperature
2. L-Histidine (Sigma-Aldrich), store at room temperature
3. L-Isoleucine (Sigma-Aldrich), store at room temperature
4. L-Methionine (Sigma-Aldrich), store at room temperature
5. L-Phenylalanine (Sigma-Aldrich), store at room temperature
6. L-Proline (Sigma-Aldrich), store at room temperature
7. L-Hydroxyproline (Sigma-Aldrich), store at room temperature
8. L-Serine (Sigma-Aldrich), store at room temperature
9. L-Threonine (Sigma-Aldrich), store at room temperature
10. L-Tryptophan (Sigma-Aldrich), store at room temperature
11. L-Tyrosine (Sigma-Aldrich), store at room temperature
12. L-Valine (Sigma-Aldrich), store at room temperature
13. Thiamine hydrochloride (Sigma-Aldrich), store at room temperature
14. Glutathione (reduced, Sigma-Aldrich), store at room temperature
15. L-Ascorbic acid 2-phosphate magnesium salt (Sigma-Aldrich), store at room temperature
16. Human serum albumin (Sigma-Aldrich), store at 4°C (see Note 1)
17. Linoleic acid (Cayman Chemicals), store at −20°C
18. Arachidonic acid (Cayman Chemicals), store at −20°C
19. Oleic acid (Sigma-Aldrich), store at −20°C
20. Retinol solution (Sigma-Aldrich), store at 4°C, air- and daylight sensitive, keep in the dark
21. Recombinant human transferrin (Sigma-Aldrich), store at 4°C

22. Recombinant human insulin (Invitrogen), store at −20°C
23. Recombinant human Activin A (R&D Systems), store at −20°C
24. Recombinant human basic fibroblast growth factor (bFGF, R&D Systems), store at −20°C
25. Trace elements A, B, and C (Mediatech/Cellgro), store at 4°C
26. MEM nonessential amino acid (NEAA) solution (Lonza), store at 4°C
27. GlutaMAX (Invitrogen), store at −20°C
28. β-Mercaptoethanol (Invitrogen), store at 4°C
29. Knockout-Dulbecco's minimal essential medium (KO-DMEM) or DMEM high glucose 4.5 g/L (Invitrogen), store at 4°C
30. Cell culture grade MilliQ water (Millipore)
31. Filter manifold or disposable filter flasks, 0.22-μm pore size (Millipore)
32. Osmometer
33. pH meter
34. 50 mL Sterile tubes (Nunc)

2.2. Preparation of Feeder Cells

1. hFF cells (CRL-2429, ATCC), store at liquid nitrogen
2. Iscove's modified Dulbecco's medium (IMDM) with L-glutamine and 25 mM HEPES (Invitrogen), store at 4°C
3. Penicillin-streptomycin (Lonza), store at −20°C
4. MEM nonessential amino acids (Lonza), store at 4°C
5. Human serum (Sigma-Aldrich) or fetal bovine serum (FBS, Invitrogen), store at −20°C (see Note 2)
6. PBS (Lonza), store at 4°C
7. Trypsin-EDTA (Invitrogen), store at −20°C
8. 50 mL Centrifuge tubes (Nunc)
9. 10-cm Cell culture plates (Nunc)
10. 3.5-cm Cell culture plates (Becton Dickinson)

2.3. Expansion of Pluripotent Stem Cells in RegES Medium

1. RegES medium

2.4. Mechanical Passaging of Pluripotent Stem Cell Cultures

1. RegES medium
2. Mitotically inactivated feeder cell plates
3. Scalpel

2.5. Single-Cell Enzymatic Passaging of Pluripotent Stem Cell Cultures

1. PBS (Lonza), store at room temperature
2. TrypLE Select (Invitrogen), store at 4°C
3. RegES medium
4. Mitotically inactivated feeder cell plates

3. Methods

3.1. Preparation of Defined, Xeno-Free RegES Medium (See Note 3)

1. Weigh the following amino acids:

Reagent	Solubility (mg/mL) in 1 M HCl	Final concentration (mg/L) in medium
L-Isoleucine	50	615
L-Methionine	50	44
L-Phenylalanine	50	336
L-Tryptophan	10	82
L-Tyrosine	25	84

Dissolve the amino acids in 1 M HCl.

2. Weigh the following amino acids:

Reagent	Solubility (mg/mL) in H_2O	Final concentration (mg/L) in medium
Glycine	100	53
L-Histidine	50	183
L-Hydroxyproline	50	15
L-Proline	50	600
L-Serine	50	162
L-Threonine	50	425
L-Valine	25	454

Dissolve the amino acids in cell culture grade water.

3. Weigh 9 mg thiamine and dissolve in cell culture grade water.
4. Weigh 1.5 mg reduced glutathione and dissolve in cell culture grade water.
5. Weigh 50 mg ascorbic acid and dissolve in cell culture grade water.
6. Weigh 10 g human serum albumin and dissolve in 100 mL cell culture grade water.
7. Next day/after finishing steps 1–6, weigh 8 mg transferrin and dissolve in cell culture grade water (see Note 4).

8. Mix amino acid solutions, reduced glutathione, and ascorbic acid.
9. Adjust the pH of the solution to 7.0–7.4 using 5 M NaOH.
10. Add 1 mg of linoleic, 1 mg of oleic, and 1 mg of arachidonic acid (diluted in 5 mg/mL in absolute ethanol) to human serum albumin solution.
11. Add human serum albumin/lipid solution, transferrin, and thiamine.
12. Add 5 mL of trace elements A, B, and C.
13. Add cell culture grade water to give approximately 180 mL volume.
14. Measure and adjust pH, if needed, to 7.4.
15. Add cell culture grade water to give the 200 mL of volume.
16. Add 10 mL NEAA.
17. Add 10 mL GlutaMAX.
18. Add 1.96 mL β-mercaptoethanol.
19. Add 2.3 mL retinol solution.
20. Add 2.5 mL insulin solution.
21. Dissolve bFGF to PBS containing 0.1% human serum albumin as a carrier protein. Add 8 μg bFGF.
22. Dissolve Activin A to PBS containing 0.1% human serum albumin as a carrier protein. Add 5 μg Activin A.
23. Adjust osmolarity with 5 M NaCl to 320–330 mOsm/kg.
24. Add KO-DMEM basal medium to give the desired volume of 1 L and filter sterilize the final medium.
25. Aliquot the medium to appropriate batches. The medium can be stored at −20°C for up to 3 months and at 4°C for 1 week. Store the medium in the dark and avoid unnecessary contact with light.

3.2. Preparation of Feeder Cells

Feeder cells are required to support the growth of undifferentiated hESC and iPS cells. Human foreskin fibroblast cells (hFFs, CRL-2429, ATCC, Manassas, USA) are recommended to be used as feeder cells with RegES medium (see Note 5). Confluent monolayers of hFFs should be mitotically inactivated by irradiation (40 Gy) before use.

3.2.1. Culture and Passaging of hFF Feeder Cells

1. Culture hFFs in Iscove's modified Dulbecco's medium (IMDM) with L-glutamine and 25 mM HEPES supplemented with 1% penicillin-streptomycin, 0.1 mM NEAA, and 10% human serum or 10% fetal bovine serum (FBS). Change the media once a week.
2. Cells should be split when they reach confluence. A split based on a seed density 5×10^3 cells/cm^2 is recommended.

3. Discard the medium and wash the cells twice with PBS.
4. Aspirate PBS and add 2 mL of 0.25% trypsin-EDTA solution for a 10-cm culture dish.
5. Incubate at 37°C for 5 min. Check microscopically whether the cells have detached.
6. Add 5 mL hFF medium, break up the cell clumps by gently pipetting up and down several times, and collect the cells in 50 mL centrifuge tube.
7. Centrifuge at 1,300 rpm for 7 min.
8. Discard the supernatant and resuspend the pellet in 1 mL hFF medium.
9. Count the number of cells, plate cells at $5 \times 10^3/cm^2$, and incubate at 37°C with 5% CO_2.

3.2.2. Irradiation and Plating of hFF Cells (see Note 6)

1. At confluence, hFF cells are mitotically inactivated by irradiation.
2. Discard the medium and wash the cells twice with PBS.
3. Aspirate PBS and add 2 mL 0.25% trypsin-EDTA solution for a 10-cm culture dish.
4. Incubate at 37°C for 5 min. Check microscopically whether the cells have detached.
5. Add 5 mL hFF medium, break up the cell clumps by gently pipetting up and down several times, and collect the cells in 50 mL centrifuge tube.
6. Centrifuge at 1,300 rpm for 7 min.
7. Discard the supernatant and resuspend the pellet in 1 mL of hFF medium.
8. Irradiate the cells (40 Gy).
9. Count the number of cells, plate cells at $3.8 \times 10^4/cm^2$ in hFF medium, and incubate at 37°C with 5% CO_2.

3.3. Expansion of Pluripotent Stem Cells in RegES Medium

It is recommended to adapt stem cells to RegES medium by using 50/50 mix of RegES medium and conventional culture medium at first passage. If the 100% culture performs poorly, a more cautious adaptation should be performed. For example, use a 20/80 mix of RegES medium and conventional culture medium at first passage, a 50/50 mix at second passage, a 80/20 mix at third passage, followed by 100% RegES medium at the fourth passage and thereafter. Culture of pluripotent stem cells in RegES medium should result in even colonies with defined borders (Fig. 2) (see Note 7).

1. Thaw the frozen aliquoted RegES medium in a 37°C water bath. Minimize dwell time. The RegES medium can be stored at 4°C in the dark for 1 week.
2. Pre-equilibrate RegES medium to temperature and gases before use.

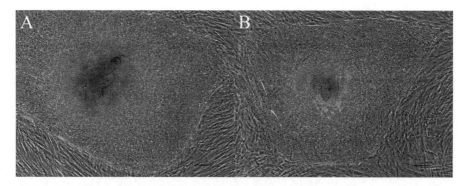

Fig. 2. Representative images of hESC lines derived and cultured in RegES medium: (**a**) hESC line Regea 07/046 at passage 50, (**b**) hESC line Regea 08/013 at passage 38.

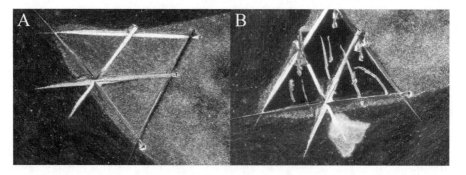

Fig. 3. Consecutive steps in the mechanical passaging of pluripotent stem cells: (**a**) Select and cut only undifferentiated areas of the colonies. (**b**) Detach the pieces with a needle or pipette tip and transfer the pieces to new feeder cell plates.

3. Incubate cells at 37°C with 5% CO_2 in air atmosphere until the cells reach 70–80% confluence.

4. Change the medium every day or when pH decreases.

5. Passage the growing cell colonies to new feeder cell plates at 5–7-day intervals (see Note 8).

6. Overtly differentiated areas should be cut out during passaging intervals as they tend to induce differentiation in other colonies.

3.4. Mechanical Passaging of Pluripotent Stem Cell Cultures

1. Pre-equilibrate RegES medium to temperature and gases before use.

2. Change RegES medium to mitotically inactivated feeder cell plates at least 30 min before use.

3. For mechanical passaging, cut the undifferentiated parts of the colonies into 5–10 pieces and transfer on new feeder cell plates (20–30 pieces/plate) (see Fig. 3).

4. Right after plating cells, gently swirl the plate back-and-forth and side-to-side and incubate at 37°C with 5% CO_2 (see Note 9).

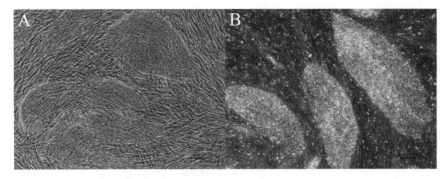

Fig. 4. Representative images of hESC colonies passaged with SCED method. The resulting colonies are smaller and thinner when compared to colonies resulting from the mechanical passaging.

3.5. Single-Cell Enzymatic Passaging of Pluripotent Stem Cell Cultures

Facilitated expansion of human embryonic stem cells by single-cell enzymatic dissociation using TrypLE Select (Invitrogen) can be used with RegES medium (12). The method enables large-scale culture of pluripotent stem cells.

1. Wash cells once with PBS.
2. Aspirate PBS and add 0.5 mL of 1× TrypLE Select to each cell culture plate.
3. Incubate at 37°C for 3–5 min. Check microscopically when the cells start to detach. When the hESC colonies start to round up from the feeder layer, break the cell sheet apart to a single-cell suspension by mixing with a pipette.
4. Centrifuge at 400 g for 5 min, discard the supernatant, and resuspend the hESC pellet in RegES medium.
5. Plate the single-cell suspension onto cell culture plates containing mitotically inactivated hFF layer. For the initial passage, use split ratios of 1:2–1:8. Routinely, split ratios between 1:4 and 1:40 can be used (see Note 9).
6. Depending on the growth rate of the individual pluripotent stem cell line, the cells should be passaged every 6–12 days (Fig. 4).

4. Notes

1. It is advisable to test several batches of human serum albumin as different batches from the same manufacturer can differ substantially in quality.
2. As with human serum albumin, different batches of human serum and FBS can substantially differ in quality and should be tested prior to use.

3. In steps 1–6, the solubilization may require several hours.
4. Transferrin has a limited shelf life of 1 week at 4°C.
5. Other types of hFF feeder cells should be tested for their applicability to maintain the undifferentiated growth of pluripotent stem cells prior to use with RegES medium. Mouse embryonic fibroblast feeder cells (MEFs) can also be used with RegES medium.
6. hFF feeder cells can be safely used until they reach passage 15 (13, 14).
7. Culture under 5% hypoxia may increase the number of undifferentiated colonies (15, 16).
8. Do not passage the cells too early as they will plate poorly and differentiate. The cultures should be maintained at a high density. If the density of colonies drops too low, the culture will tend to deteriorate showing more differentiated cells.
9. Some cell death at passaging is normal.

References

1. Peerani, R., Rao, B.M., Bauwens, C., Yin, T., Wood, G.A., Nagy, A., Kumacheva, E., Zandstra, P.W. (2007) Niche-mediated control of human embryonic stem cell self-renewal and differentiation. EMBO J **26**, 4744–4755.
2. Thomson, J., Itskovitz-Eldor, J., Shapiro, S., Waknitz, M., Swiergiel, J., Marshall, V., Jones, J. (1998) Embryonic stem cell lines derived from human blastocysts Science **282**, 1145–7.
3. Skottman, H., Stromberg, A.M., Matilainen, E., Inzunza, J., Hovatta, O., Lahesmaa, R. (2006) Unique gene expression signature by human embryonic stem cells cultured under serum-free conditions correlates with their enhanced and prolonged growth in an undifferentiated stage. Stem Cells **24**, 151–67.
4. Bradley, J.A., Bolton, E.M., Pedersen, R.A. (2002) Stem cell medicine encounters the immune system. Nat Rev Immunol **2**, 859–71.
5. Selvaggi, T.A., Walker, R.E., Fleisher, T.A. (1997) Development of antibodies to fetal calf serum with arthus-like reactions in human immunodeficiency virus-infected patients given syngeneic lymphocyte infusions. Blood **89**, 776–9.
6. Martin, M., Muotri, A., Gage, F., Varki, A. (2005) Human embryonic stem cells express an immunogenic nonhuman sialic acid. Nat med **11**, 228–232.
7. Heiskanen, A., Satomaa, T., Tiitinen, S., Laitinen, A., Mannelin, S., Impola, U., Mikkola, M., Olsson, C., Miller-Podraza, H., Blomqvist, M., Olonen, A., Salo, H., Lehenkari, P., Tuuri, T., Otonkoski, T., Natunen, J., Saarinen, J., Laine, J. (2007) N-glycolylneuraminic acid xenoantigen contamination of human embryonic and mesenchymal stem cells is substantially reversible. Stem Cells **25**, 197–202.
8. Sakamoto, N., Tsuji, K., Muul, L.M., Lawler, A.M., Petricoin, E.F., Candotti, F., Metcalf, J.A., Tavel, J.A., Lane, H.C., Urba, W.J., Fox, B.A., Varki, A., Lunney, J.K., Rosenberg, A.S. (2007) Bovine apolipoprotein B-100 is a dominant immunogen in therapeutic cell populations cultured in fetal calf serum in mice and humans. Blood **110**, 501–508.
9. Hisamatsu-Sakamoto, M., Sakamoto, N., Rosenberg, A.S. (2008) Embryonic stem cells cultured in serum-free medium acquire bovine apolipoprotein B-100 from feeder cell layers and serum replacement medium. Stem Cells **26**, 72–8.
10. Cobo F, Stacey G, Hunt C, Cabrera C, Nieto A, Montes R, Cortes J, Catalina P, Barnie A, Concha A (2005): Microbiological control in stem cell banks: approaches to standardisation. Appl Microbiol Biotechnol **68**, 456–66.
11. Rajala, K., Lindroos, B., Hussein, S.M., Lappalainen, R.S., Pekkanen-Mattila, M., Inzunza, J., Rozell, B., Miettinen, S., Narkilahti, S., Kerkelä, E., Aalto-Setälä, K., Otonkoski, T., Suuronen, R., Hovatta, O., Skottman, H. (2010) A Defined and Xeno-free Culture Method Enabling the Establishment of Clinical-grade Human Embryonic, Induced Pluripotent and Adipose Stem Cells. Plos One **5**, e10246.

12. Ellerström, C., Strehl, R., Noaksson, K., Hyllner, J., Semb, H. (2007) Facilitated expansion of human embryonic stem cells by single-cell enzymatic dissociation. Stem Cells **25**, 1690–6.
13. Hovatta, O., Mikkola, M., Gertow, M., Stromberg, A., Inzunza, J., Hreinsson, J., Rozell, B., Blennow, E., Andang, M., Ahrlund-Richter, L.(2003) A culture system using human foreskin fibroblasts as feeder cells allows production of human embryonic stem cells. Hum Reprod **18**, 1404–1409.
14. Amit, M., Margulets, V., Segev, H., Shariki, K., Laevsky, I., Coleman, R., Itskovitz-Eldor, J. (2003) Human feeder layers for human embryonic stem cells. Biol Reprod **68**, 2150–2156.
15. Prasad, S., Czepiel, M., Cetinkaya, C., Smigielska, K., Weli, S., Lysdahl, H., Gabrielsen, A., Petersen, K., Ehlers, N., Fink, T., Minger, S., Zachar, V. (2009) Continuous hypoxic culturing maintains activation of Notch and allows long-term propagation of human embryonic stem cells without spontaneous differentiation. Cell Prolif **1**, 63–74.
16. Westfall, S., Sachdev, S., Das, P., Hearne, L., Hannink, M., Roberts, R., Ezashi, T. (2008) Identification of oxygen-sensitive transcriptional programs in human embryonic stem cells. Stem cells dev **5**, 869–881.

Chapter 7

Pluripotent Stem Cells In Vitro from Human Primordial Germ Cells

Behrouz Aflatoonian and Harry D. Moore

Abstract

In the mouse, embryonic germ (EG) cell lines can be generated from primordial germ cells (PGCs) recovered from the genital ridge of the developing fetus. These EG cells have the capacity of self-renewal, are pluripotent, and have the potential to differentiate to other cell types of the body similar to embryonic stem cells (ESCs). In this chapter, we describe a simple procedure to derive and keep in culture putative human EG (hEG) cells. Fetal material (5–8 weeks' gestation) was obtained following full local ethical approval and patient consent. Fetal gonads were recovered by dissection and treated with collagenase type IV. Cells were washed by centrifugation and cocultured with inactivated mouse embryonic fibroblasts (MEFs) feeder layer in different conditions. These cell lines can be characterized with specific markers. It was not possible to maintain an undifferentiated hEG cell line for more than 30 cell passages.

Key words: Germ cells, Human embryonic germ cells, Human embryonic stem cells, Primordial germ cells, Stem cells

1. Introduction

Reproduction is one of the most critical procedures in life with the aim of transferring genetic information from one generation to the next. The germ cell is determined early in fetal life to perform this task and produce gametes, spermatozoa, and oocytes depending on the sex of the individual. The earliest stage of the germ cells is referred to as a primordial germ cell (PGC). These cells are undifferentiated precursors of either gamete (sperm or oocyte) and originate at the beginning of gastrulation in the epiblast region adjacent to the extraembryonic ectoderm (1, 2). In vivo, PGCs migrate to the genital ridges where they are surrounded by gonadal interstitial cells to commit to specified germ cell development to finally give rise to gametes in adulthood, depending on the development of a testis or ovary (3). In vitro, PGCs only survive for a short period in culture but can be transformed to EG cells (4, 5), which, in the mouse, have been demonstrated to have pluripotent capacity, and can form chimeric offspring when injected into blastocysts and implanted into foster females as well as differentiated in culture to various cell phenotypes (6, 7). EG cells of the mouse show very extensive proliferation and differentiation, and with care can be passaged indefinitely in culture. Apart from the mouse, the other species where isolated EG cells have shown indications of chimera development is in the pig (8), where a single chimeric piglet was produced (from 186 embryos). However, the cell lines all failed by passage 29 or 6 months of culture.

In 1998, Shamblott and coworkers (9) derived putative human EG cell lines from the genital ridge of first trimester fetuses obtained after termination of pregnancy. These cell lines demonstrated proliferative and pluripotent potential in vitro, although, for obvious ethical reasons, true germ line pluripotency could not be verified. Since then, several other groups have derived putative hEG cell lines which show pluripotent potential in culture, although the claim that these cells are definitive EG cell lines has also been questioned (11–13).

Although hEG cells (9) were first derived around the same time as their pluripotent hESC counterparts (13), there has been much less attention focused on the potential use of hEG cells for applications in regenerative medicine than on the use of hESC. The cause of this difference is mainly twofold. First, although both stem cells are of controversial origin (i.e., embryo and fetus) and their derivation necessitates comprehensive ethical and, in some countries, governmental approval, the recovery of PGCs from early human fetal tissue presents additional practical issues. The time of collection of the donated tissue is important (as tissue is recovered after termination of pregnancy) that can limit access to suitable samples to a greater extent than for obtaining preimplantation embryos for hESC derivation. Second, although the initial generation of hEG cells is a relatively simple procedure (9), the maintenance of well-defined cell lines through extended passage in culture has proved to be difficult to date (10, 11). The reason for this is not clear, but it has had restricted the wide distribution of well-characterized hEG cell lines for exploring potential cell therapies. Despite these problems, hEG cells still represent an important alternative to hESCs because they potentially have a different epigenetic status that might ultimately prove to be significant for cell therapy applications. Moreover, EG cells are important tools for studying factors involved in PGC survival, proliferation, and regulation that have clinical relevance such as in testicular cancer (14).

Previous studies on PGC development and culture in mouse have provided an indication that the mammalian PGC niche in vitro can be applied to modify and improve culture systems of hEG cells.

For instance, it has been shown that stem cell factor (SCF) is necessary for the survival and proliferation of mouse PGCs (6, 15, 16). Also, leukemia inhibitory factor (LIF) has been shown to have an effect on PGC survival in culture (6, 17). Together these factors promote proliferation of PGCs for a limited period of culture. With the addition of bFGF and the presence of membrane-associated SCF and LIF, murine PGCs, at least, show extensive proliferation in vitro (3). Bone morphogenic protein-4 (BMP4; (18)) and retinoic acid (RA; (19)) are other factors that influence survival and generation of PGCs and germ cells (20).

Knowing this valuable information, an in vitro culture condition was devised for enhancing proliferation and survival of hEG cells.

2. Materials

1. Trypsin/EDTA.

    0.25% trypsin (w/v) (Gibco: Cat. No. 15090-046) and 5 mM EDTA (Sigma: Cat. No. E-5134).

2. DMEM/FCS.

    Dulbecco's MEM (Gibco: Cat. No. 41965-039) supplemented with 10% fetal calf serum (FCS; preferably of US origin), 2 mM L-glutamine (Gibco: Cat. No. 21051-016), and antibiotic Gentamicin (20 μg/ml; Gibco: Cat. No. 15750-037).

3. Mito-c-DMEM/FCS.

    Add 2 mg Mitomycin-c (Sigma: Cat. No. M-4287) to 200 ml DMEM/FCS and sterilize by passing through a 0.2-μm cellulose acetate filter.

    Store at 4°C and use within 4 weeks.

2.1. Preparation of hES Medium

For 200 ml hES medium (21):

(a) Add 10 ml PBS (w/o Ca^{2+}, Mg^{2+}) to 0.146 g L-glutamine in a 15 ml tube.

(b) Add 7 μl of β-mercaptoethanol (14.3 M β-mercaptoethanol; Sigma M-7154) to the L-glutamine/PBS and mix well.

(c) Add the following cell culture medium into a 225-ml 0.2-μm cellulose acetate filtering unit:

- 160 ml Knockout DMEM (Gibco: Cat. No. 10829-018)
- 40 ml Knockout Serum Replacement (KOSR) (Gibco: Cat. No. 10828-028)
- 2 ml L-Glutamine/β-mercaptoethanol solution
- 2 ml 100× Nonessential amino acid solution (Gibco: Cat. No. 11140-035)
- 400 μl of 2 μg/ml bFGF stock

Human bFGF (Gibco: Cat. No.13256-029): 10 μg human bFGF dissolved in 5 ml of 0.1% BSA in PBS (w/o Ca^{2+}, Mg^{2+})

- Aliquot into 400 μl and store at –20°C
- Antibiotics (optional, e.g., 120 μl Gentamicin)

(d) Filter.

(e) Store at 4°C and use within 2 weeks.

2.1.1. Final Concentration

80% Knockout DMEM

20% Gibco knockout serum replacement

1% Nonessential amino acid solution

1 mM L-Glutamine

0.1 mM β-Mercaptoethanol

4 ng/ml Human bFGF

2.2. Collagenase IV Solution

Add 1 mg/ml collagenase type IV (Gibco: Cat. No. 17104-019) to DMEM/F12.

Sterilize with a 0.2-μm cellulose acetate filter.

Store at 4°C and use within 2 weeks.

2.3. Conditioned Medium from Neonatal Mouse Testis

The procedure was based on the method of (22). Neonatal mouse testes (NMT) from MF1 strain of mice were obtained from the Animal Facility (Sheffield University) and cultured in DMEM medium supplemented with 10% FCS and 20 μg/ml gentamicin (Fig. 1). Cells (both Sertoli, primary spermatogonia and interstitial) from NMT were cultured, passaged, frozen, and thawed by following the same procedures as for MEFs (Sects. 3.1 and 3.2). Medium was collected every 5 days as conditioned medium and stored at –20°C. This conditioned medium (35 ml) was added to a novel EG culture medium to make 100 ml, filtered, and used as NMT-conditioned medium.

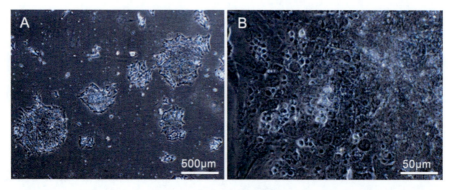

Fig. 1. Cells from neonatal mouse testis were cultured with DMEM + 10% FCS with gentamicin. (**a**) Colonies of NMT cells look similar in culture. (**b**) Different cell types of cells were observed within the colonies at higher magnification.

3. Methods

3.1. Derivation of Mouse Embryonic Fibroblasts (MEFs)

1. Perform a schedule 1 cervical dislocation on a 13.5-day pregnant mouse (strains MF-1 or CF-1 are suitable).
2. Liberally cover abdomen with 70% alcohol and cut through skin and peritoneum to expose the uterine horns. Remove the uterine horns and place in a Petri dish containing PBS (w/o Ca^{2+}, Mg^{2+}).
3. Remove the embryos from the embryonic sac and then dissect out and discard the placenta and membranes. Decapitate and eviscerate the embryos and wash the remaining carcasses 3 times with PBS (w/o Ca^{2+}, Mg^{2+}).
4. Place the carcasses in a clean Petri dish and then mince with a scalpel blade for as long as you can manage.
5. Add 2 ml trypsin/EDTA and incubate at 37°C for 10–20 min.
6. Add 5 ml DMEM/FCS, transfer to a 15-ml centrifuge tube, and aspirate vigorously.
7. Allow large chunks to settle by gravity and then transfer supernatant to a T75 flask, and add a further 15 ml DMEM/FCS.
8. Place flask in the 37°C incubator overnight.
9. The next day, replace medium to remove floating cellular debris. Allow to grow until around 90% confluent and then freeze.

3.2. Passage of MEFs

1. Aspirate medium from flask and wash cells once with PBS (w/o Ca^{2+}, Mg^{2+}).
2. Add 1 ml trypsin/EDTA each to T75 flasks and incubate at 37°C for 5 min.
3. Tap flask to dislodge cells and break up clumps. Add 9 ml DMEM/FCS and aspirate to form a single cell solution and then transfer to a 15-ml centrifuge tube.
4. Spin down at 200 g for 5 min.
5. Remove supernatant and gently flick the bottom of the tube to disperse pellet.
6. Add 10 ml of DMEM/FCS and then distribute equally between fresh T75 flasks. Place it in a 37°C and 5% CO_2 incubator.

3.3. Mitomycin-c Inactivation of MEFs

Caution: Mitomycin-c is harmful. Use gloves and other appropriate protection when handling Mitomycin-c. Dispose of Mitomycin-c solutions as hazardous waste.

1. Remove medium from a flask of MEFs and replace with enough Mito-c-DMEM/FCS to cover cells in a layer a few millimeters deep.
2. Place in a 37°C and 5% CO_2 incubator for 2–3 h.
3. While the MEFs are incubating, pretreat tissue culture plastic (6-well plates or T25 flasks) by covering with a 0.1% gelatin solution for at least 1–2 h.
4. Aspirate Mito-c-DMEM/FCS.
5. Wash 3 times with PBS.
6. Add 4 ml trypsin/EDTA each to T75 flasks (1 ml for T25 flask) and incubate at 37°C for 5 min. Tap the flask to dislodge and detach MEFs.
7. Terminate trypsinization using 9 ml DMEM/FCS, gently aspirate to disperse clumps, transfer to a 15-ml centrifuge tube, and pellet by centrifugation at $200 \times g$ for 5 min.
8. Aspirate supernatant and resuspend pellet in 10 ml DMEM/FCS if cell counting is necessary.
9. Remove the gelatin solution from the tissue culture plates or flasks.
10. Count MEFs with a hemocytometer and seed out at 1.5×10^5 cells for every T25 flask.
11. The MEFs can be used after about 5–6 h but are best left to settle overnight.
12. Use within 1 week.

3.4. SCF Transfected MEFs as Feeders

Sl⁴-m220 MEFs, which express only the membrane-associated stem cell factor (SCF; steel factor; kit Ligand; (3)), were obtained from Professor Peter Andrews (University of Sheffield) and cultured, passaged, frozen, thawed, and inactivated to form feeder layer the same as for normal MEFs.

3.5. Fetal Gonad Collection, Dissection, and Disaggregation

1. Permission was granted from the local research ethics committee to obtain fetal tissue (first trimester) from women undergoing therapeutic termination of pregnancy. Women were counseled by an independent research nurse at first presentation to clinic and provided fully informed consent. Fetal age was initially estimated from the time of the last menstrual period (gestation age + 2 weeks). The developmental age of the fetus was determined by taking a micrograph of the forehand and comparing digit development with average developmental morphology (23). This was accurate to about 0.5 week.
2. Following abortion, fetal tissue was placed in pots of cold DMEM medium (supplemented with antibiotic) and transferred to the laboratory within 1 h.

3. The fetus was extracted from any extraembryonic membranes and washed clear of blood before being placed in a 13-cm Petri dish in DMEM medium. Under a stereoscopic dissecting microscope in a class II laminar flow hood, each gonadal ridge and mesonephros were recovered using the cutting edges of two 19-gauge needles (with 1 ml syringes attached as handles).

4. These were transferred to a smaller Petri dish, and the gonadal ridges recovered by further needle dissection. The gonadal ridges were minced into small pieces, and the cells then disaggregated by incubation in 100 µl drops of 0.1% collagenase type IV in DMEM medium at 37°C in 5% CO_2 in air for 1 h. The cells were then recovered by aspiration and washed by transfer through five drops of fresh DMEM medium before culture.

3.6. Generation of hEG Cells

The method of Shamblott and coworkers (9) was used initially with inactivated MEFs rather than STO fibroblasts.

1. A novel EG medium consisted of 65 ml standard hES medium (Sect. 2.3) supplemented with 10 ng/ml BMP4 (R& D system), 1000 U/ml of human recombinant leukemia inhibitory factor (hrLIF; Chemicon), 10 µM forskolin (Sigma), 2 µM RA (Sigma), and 35 ml NMT-conditioned medium (Sect. 2.5). The medium was filtered using a 225-ml 0.2-µm filter, stored at 4°C, and used within 2 weeks.

2. Mitotically inactivated MEFs were first established in 4-well culture plates (Nunc) or small culture Petri dishes (Falcon; 3.5 cm) as a feeder layer. Fibroblast medium was removed 24 h before the derivation procedure and replaced with EG medium. Isolated cells from the gonadal ridge of a fetus were normally seeded into 5–10 wells of feeders depending on the developmental stage (tissue from 12 weeks' gestation provided much more tissue than from 8 week). Cultures were grown at 37°C in 5% CO_2 in air (95% humidity).

3. Following initial isolation of putative hEG cell lines with the original standard published method, investigations were made with:
   - Cells cultured with DMEM/FBS without MEF feeder layer (Fig. 2)
   - Cells cultured with standard hES medium with inactivated MEF feeder layer (Fig. 3)
   - Cells cultured with a modified EG medium and a SCF transfected inactivated MEF feeder layer (Fig. 4)

3.7. Passage of hEG Cells

Putative hEG colonies established usually within 10 days of first genital ridge disaggregation and colonies were then passaged after 4–7 days (depending on the proliferation rate).

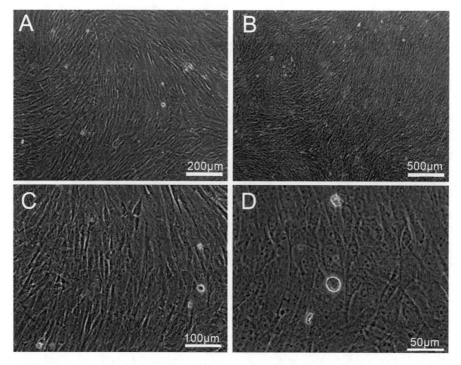

Fig. 2. (**a, b**) Culture of human fetal gonadal tissues after enzymatic treatment in standard hEG without feeder layer. After several passages, cells similar to fibroblasts became proliferated. These cells were used initially for the derivation of Shef7 (hESC) line (24). (**c, d**) Higher magnification of micrograph of (**a**) and (**b**).

1. Aspirate medium from flask.
2. Add 2 ml of collagenase each to T25 flasks and incubate at a 37°C and 5% CO_2 incubator for 15–20 min.
3. Gently scrape with the tip of a glass Pasteur pipette rounded off with a flame.
4. Add 5 ml of EG medium and gently aspirate. Transfer to a 10-ml centrifuge tube.
5. Spin at 800 rpm for 3 min at 4°C.
6. While the cells are spinning, remove the medium from the fresh flasks of MEF feeders and wash once with PBS.
7. Aspirate supernatant, leaving hEG cell pellet.
8. Remove PBS from MEFs.
9. Gently flick tube to disperse hEG pellet.
10. Gently resuspend hEG cell pellet in an appropriate volume of hES medium (e.g., 4 ml for a 1:4 split into 4 × T25 flasks (at 1 ml/T25)) and distribute between flasks of feeders. Add 4 ml EG medium per T25 flask.
11. Carefully place in a CO_2 incubator, maintaining an even distribution of cells across the flask.

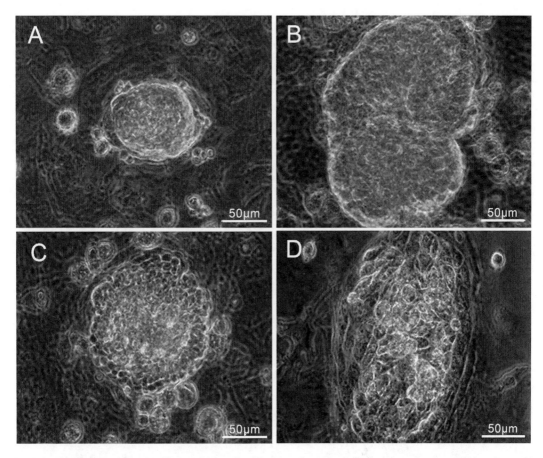

Fig. 3. (**a–d**) Different colonies in different stages of proliferation of putative human embryonic germ cells in culture with standard hES medium on feeder layer of mitotic inactivated MEFs.

3.8. Indirect Immunofluorescent Localization of Specific Markers for Characterization of hEG Cells

1. Human EG cells were washed with PBS containing 1% fetal calf serum before fixing in 4% paraformaldehyde (PFA) in PBS (pH 7.4) for 15 min.

2. Samples were washed twice for 5 min in PBS containing 1% fetal calf serum followed by incubation in 0.1% Triton X in PBS for 5 min.

3. Samples were washed twice for 5 min in PBS containing 1% fetal calf serum and incubated overnight at 4°C with specific primary antibodies against hEG cells (markers highly expressed in hEG cells; (9)). Antibodies were diluted in PBS (1/5 SSEA1; 1/5, TRA-1-81; 1/5, TRA-1-60; 1/5, TRA-2-49; 1/5, TRA-2-54) (see (22)).

4. Colonies were then washed twice in PBS and incubated with appropriate secondary antibodies for 1 h in 37°C.

5. Preparations were covered with mounting medium (Vectashield; Vector Laboratories, USA) or PBS and examined by microscopy using phase contrast and UV excitation optics (Figs. 5 and 6).

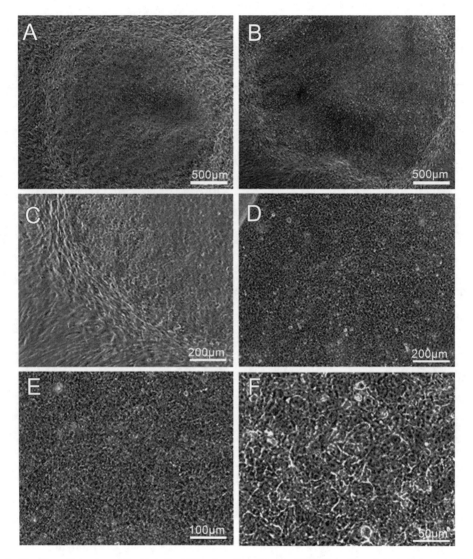

Fig. 4. Proliferating colonies (**a** and **b**) of putative hEG cells cultured in a modified EG medium. Cells in higher magnifications (**c–f**) show very similar morphology to ES and EC cells.

4. Notes

1. All instruments and solutions used should be sterile.
2. Mouse fetuses at more than 10.5 days' gestation are considered to have ethical status by the Home Office and must be humanely killed immediately.
3. Try to keep the MEFs actively dividing prior to mitotic inactivation, ideally by passaging the day before use.
4. Do not use MEFs after passage 4–5.

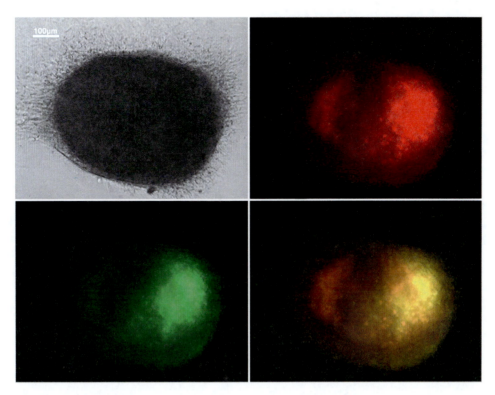

Fig. 5. Characterization of large hEG colony (cultured with hES medium) with specific markers SSEA1 (*red*) and TRA-2-49 (*green*) and merged (*orange*). Scale bar = 100 μm.

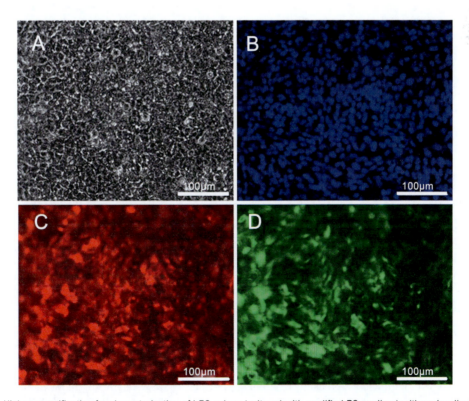

Fig. 6. Higher magnification for characterization of hEG colony (cultured with modified EG medium) with co-localization of specific markers. (**a**) Phase contrast, (**b**) DAPI (*blue*), (**c**) TRA-1-60 (*red*), and (**d**) TRA-2-49 (*green*).

5. Mito-c-DMEM/FCS: Add 2 mg Mitomycin-c (Sigma: Cat. No. M-4287) to 200 ml DMEM/FCS and sterilize by passing through a 0.2-μm cellulose acetate filter. Store at 4°C and use within 4 weeks.

6. DMEM/FCS must be used to seed feeders.

7. Human EG cells require partially feeding every 2 days with fresh EG medium.

8. Areas of spontaneous differentiation can be removed from cultures by picking with a sterile pipette. Alternatively, undifferentiated hEG colonies can be picked, either by (a) picking colonies with a pipette and transferring to fresh feeder layers or (b) performing a longer collagenase treatment, which should preferentially loosen/detach the undifferentiated hEG colonies. These can then be removed by gently drizzling medium over the surface of the culture followed by transferal to fresh feeder layers. It may be necessary to mechanically break up the colonies a little by trituration.

9. Disaggregation of hEG colonies with 0.05% trypsin/EDTA solution was inefficient and often resulted in few subsequent colonies.

References

1. Ginsburg M., Snow M.H., McLaren A. (1990) Primordial germ cells in the mouse embryo during gastrulation. Development 110, 521–528
2. Geijsen N., Horoschak M., Kim K., Gribnau J. et al. (2004) Derivation of embryonic germ cells and male gametes from embryonic stem cells. Nature 427, 106–107
3. Kucia M., Machalinski B., Ratajczak M.Z. (2006) The developmental deposition of epiblast/germ cell-line derived cells in various organs as a hypothetical explanation of stem cell plasticity? Acta Neurobiol Exp (War) 66, 331–341
4. Matsui Y., Zsebo K., Hogan B.L. (1992) Derivation of pluripotential embryonic stem cells from murine primordial germ cells in culture. Cell 70, 841–847
5. Resnick J.L., Bixler L.S., Cheng L. et al. (1992). Long-term proliferation of mouse primordial germ cells in culture. Nature 359, 550–551
6. Matsui Y., Toksoz D., Nishikawa S. et al. (1991) Effect of Steel factor and leukaemia inhibitory factor on murine primordial germ cells in culture. Nature 353, 750–752
7. Labosky P.A., Barlow D.P., Hogan B.L. (1994) Embryonic germ cell lines and their derivation from mouse primordial germ cells. Ciba Found Symp 182, 157–168
8. Shim H., Gutierrez-Adan A., Chen L.R. et al. (1997). Isolation of pluripotent stem cells from cultured porcine primordial germ cells. Biol Reprod 57, 1089–1095
9. Shamblott M.J., Axelman J., Wang S. et al. (1998) Derivation of pluripotent stem cells from cultured human primordial germ cells. Proc Natl Acad Sci 95, 13726–13731
10. Turnpenny L., Brickwood S., Spalluto C.M. et al. (2003) Derivation of human embryonic germ cells: an alternative source of pluripotent stem cells. Stem Cells 21, 598–609
11. Liu S., Liu H., Pan Y. et al. (2004) Human embryonic Germ cells isolation from early stages of post-implantation embryos. Cell Tissue Res 318, 525–531
12. Aflatoonian B., Moore H. (2005) Human primordial germ cells and embryonic germ cells, and their use in cell therapy. Curr Opin Biotechnol 16, 530–535
13. Thomson J.A., Itskovitz-Eldor J., Shapiro S.S. et al. (1998) Embryonic stem cell lines derived from human blastocysts. Science 282, 1145–1147
14. Rajpert-De Meyts E., Bartkova J., Samson M. et al. (2003) The emerging phenotype of the

testicular carcinoma in situ germ cell. APMIS 111, 267–278
15. Dolci S., Williams D.E., Ernst M.K. et al. (1991) Requirement for mast cell growth factor for primordial germ cell survival in culture. Nature 352, 809–811
16. Godin I., Deed R., Cooke J. et al. (1991) Effects of the steel gene product on mouse primordial germ cells in culture. Nature 352, 807–809
17. De Felici M., Dolci S. (1991) Leukaemia inhibitory factor sustains the survival of mouse primordial germ cells cultured on TM4 feeder layers. Dev Biol 147, 281–284
18. Lawson K.A., Dunn N.R., Roelen B.A. et al. (1999). Bmp4 is required for the generation of primordial germ cells in the mouse embryo. Genes & Dev 13, 424–436
19. Baleato R.M., Aitken R.J., Roman S.D. (2005) Vitamin A regulation of BMP4 expression in the male germ line. Dev. Biol 286, 78–90
20. Aflatoonian B., Ruban L., Jones M. et al. (2009) In vitro post-meiotic germ cell development from human embryonic stem cells. Hum Reprod 24, 3150–3159
21. Amit M., Itskovitz-Eldor J. (2006) Maintenance of human embryonic stem cells in animal serum- and feeder layer-free culture conditions. Methods Mol Biol 331, 105–113
22. Lacham-Kaplan O., Chy H., Trounson A. (2006) Testicular cell conditioned medium supports differentiation of embryonic stem cells into ovarian structures containing oocytes. Stem Cells 24, 266–273
23. Moore K.L., Persaud T.V.N. (2000) Colour atlas of clinical embryology. 2nd edition. Saunders, London
24. Aflatoonian B., Ruban I., Shamsuddin S. et al. (2010) Generation of Sheffield (Shef) human embryonic stem cell lines using a microdrop culture system. In Vitro Cell Del.-Animal 46, 236–241

Chapter 8

Cryopreservation of Human Embryonic Stem Cells and Induced Pluripotent Stem Cells

Frida Holm

Abstract

Until now, it has been problematic to obtain high survival rates after thawing of human embryonic stem cells (hESCs) and later also induced pluripotent stem cells (iPSCs). Already in 1994, Freshner and coworkers established the conventional slow freezing/rapid thawing methods by using a culture medium supplemented with 10% dimethylsulfoxide (DMSO). The cell survival using this method was low. It has been improved since then in various ways, for example, using a ROCK inhibitor to freeze dissociated hESCs. Later, Reubinoff and coworkers published a completely new method called vitrification, wherein an open pulled straw was used. It showed good results by using vitrification in some cell types including hESCs; yet it has been problematic and difficult to handle. Since it is mostly suitable for small amount of cells, this is inconvenient for researchers because they often need large amount of cells. A novel, chemically defined effective xeno-free cryopreservation system has recently given excellent results for the cryostorage of both hESCs and iPSCs, frozen in cell aggregates or as single cells. It is based on an optimal constitution of cryoprotectants, 10% DMSO, glucose, and a high polymer described in the Japanese Pharmacopoeia (JP), plus NaCl, KCl, Na_2HPO_4, and $NaHCO_3$ as pH adjustors for maintenance of the cell function, all dissolved in phosphate-buffered saline (PBS). The procedure is slow freezing without any extra equipment. Over 90% survival has been achieved.

Key words: Human embryonic stem cells, Defined, Cryopreservation, Survival, Differentiation

1. Introduction

Together with optimized clinical grade cultures, it is just as important to obtain an efficient clinical grade cryopreservation system for hESCs and iPSCs, due to the great potential in regenerative medicine, which has been described for the last decade (1–3), as well as for research banks and transport between research centers worldwide.

Since the start of research in stem cell biology in the late 1990s, the established freezing protocol has given constantly a low viability of the surviving cells which resulted in high differentiation rates.

It has not been improved much despite the fact that the research field itself has gone forward. This fact has resulted in an acute need of effective procedures for cryo-storage, which so far has been lacking (4, 5).

Cell cryopreservation has traditionally used serum and/or culture medium containing 10% DMSO as the only cryo protectant (permeable) to cryo-store large amount of cells in one vial (6, 7). While this method has been improved using various solutions (8, 9), it has not yielded too much improved result.

Vitrification in open pulled straws has been effectively used to cryo-store small number of cells (10). However, this method is difficult to handle. The requirement of keeping sterile during the process reduced the wide use of this method.

As stated above, both hESCs and iPSCs are highly sensitive to cryo-injury caused by cryopreservation. This has led to extensive studies in recent years (9–15). Already in 1994, it was described that hESCs could be cryopreserved using slow freezing and rapid thawing, which became one of the most commonly used cryopreservation methods in the field. However, the results have not been of satisfaction, since it is only possible to recover 70% of the clumps after thawing, and only 16% of clumps can develop into undersized hESC colonies with a high level of differentiation (10). A few years later, Reubinoff published a new method, the open pulled straw vitrification method, which has been used to cryopreserve embryos from bovine species (16). This method enabled to cryopreserve early stocks of hESCs. It has been problematic to establish in different laboratories since it is difficult to handle and hard to keep completely sterile. Thus, it is not an optimal method for future clinical usage. Martin Ibañez and coworkers reported, for the first time, successful cryopreservation of hESCs as single-cell suspension in the presence of Rho-associated kinase (ROCK) inhibitor Y-27632 (9, 17, 18). Their approach resulted in a 50% survival rate of hESCs compared with only 35% survival rate in the absence of the ROCK inhibitor.

Use of combinations of cryoprotectants in other cell types, such as embryos (19, 20), human oocytes (21), and red blood cells (7), have been used for many years, but in hESC banking such combinations were only reported in as late as 2010. This novel procedure can be performed without dedicated instrumentation which facilitates the usage. The medium used for cryopreservation is completely serum and animal substance free. It contains DMSO, anhydrous dextrose, and, a polymer as cryo protectant instead of using only one cryo protectant (DMSO). STEM-CELLBANKER is a chemically defined freezing medium and CELLOTION is a wash solution after thawing. For optimal dehydration and minimal ice crystal formation, the medium contains both permeating and non-permeating cryo protectants. The cells can be frozen directly at −80°C without using a programmed freezer.

This system using a defined freezing–thawing system offers an effective and simple option for banking of hESCs and iPSCs, allowing large cell numbers to be frozen in one vial, resulting in outstanding cell survival after freezing. While comparing this novel method with the standard method (10% DMSO in culture medium), the cells frozen in STEM-CELLBANKER™ will have a high viability (90–96%) without any impact on proliferation and differentiation, as compared to the standard freezing procedure where viability is usually much lower (49%) (22).

2. Materials

2.1. Maintenance and Culture of hESCs and iPSCs

Human embryonic stem cell lines, derived at Karolinska Institutet, Stockholm (23)

Induced pluripotent stem cell line established in collaboration with University of Geneva by transducing human skin fibroblasts with Oct-4, Nanog, Sox2, and Lin28 (24)

2.2. Feeder Layer

Human foreskin fibroblasts: (CRL-2429; ATCC, Manassas, VA, USA) mitotically inactivated by irradiation (40 Gy)

2.3. Culture Medium

1. Knockout Dulbecco's Modified Eagle's Medium (DMEM)
2. 20% Knockout SR (Gibco Invitrogen Corporation, Paisley, UK)
3. 0.5% Penicillin–streptomycin (Gibco Invitrogen Corporation, Paisley, UK)
4. 1% Nonessential amino acids (Gibco Invitrogen Corporation, Paisley, UK)
5. 0.5 mM 2-mercaptoethanol (Sigma-Aldrich Co, St Louis, USA)
6. 8 ng/ml basic fibroblast growth factor (R&D Systems, Oxon, UK)

2.4. Freezing Medium

Ready-to-use STEM-CELLBANKER™.

2.5. Thawing Medium

Ready-to-use CELLOTION™.

All compounds included in STEM-CELLBANKER™ and CELLOTION™ are of the Pharmacopoeia of the United States of America, European Pharmacopoeia, or JP grade.

3. Methods

3.1. Maintenance and Culture of hESCs and iPSCs

hESCs and iPSCs can be cultured on human foreskin fibroblast (CRL-2429, ATCC, Manassas, VA) (25) or feeder free on mTeSR1 or human recombinant laminin (26, 27). Before using these cells as supporting cells for hESCs and iPSCs, they need to be inactivated mitotically by using mitomycin C or gamma irradiation in 40 Gy. Their culture medium is Iscoves medium (Gibco, Invitrogen Corp, Paisley, UK) supplemented with 10% fetal calf serum (FCS; SDS, Sweden). The day after irradiation, the medium needs to be exchanged with a medium that is supplemented with serum replacement. hESCs and iPSCs can either be passaged mechanically by using a surgical scalpel or enzymatically using TrypLE Select and a Rho-associated kinase (ROCK) inhibitor Y-27632 (Merck Chemicals Ltd, Nottingham, UK) (9, 22, 28). They are passaged in a 5–7-day interval and transferred to freshly coated human foreskin fibroblasts or extra cellular matrix (ECM).

3.2. Cryopreservation

Undifferentiated cell aggregates or single cells from hESCs or iPSCs can be transferred to a cryo vial. When freezing cell aggregates, transfer them immediately to the cryo vial and remove all surplus medium with a micropipette before adding cold (4–8°C) STEM-CELLBANKER™. Single cells need to be spinned down and resuspended with the freezing medium before adding cell suspension to the cryo vial. Immediately store the vials at −80°C overnight. A programmed freezer is not needed. Transfer cryo vials for long-term storage to liquid nitrogen tanks the day after treatment. Traditionally, hESCs and later iPSCs are frozen using 10% DMSO in a cold culture medium, which for the time being is the most often used slow freezing system for these cells. The method using STEM-CELLBANKER™ instead is identical to other methods, but gives a higher cell survival and faster recovery.

3.3. Thawing

Prepare fresh culture cells 30 min prior to thawing. Vials that are removed from liquid nitrogen should immediately be placed in a 37°C water bath. When only a small ice burger remains in the cryo vial, wipe the vial using 70% ethanol to avoid contamination from the water bath, then add cold (4–8°C) CELLOTION™ to the vials to dilute the existing STEM-CELLBANKER™. Transfer the contents to a sterile centrifuge tube, and then further dilute sevenfold using CELLOTION™ before centrifugation ($50 \times g$ for 7 min). Suspend the cell pellet with preheated culture medium and seed out to the fresh culture plates.

3.4. Assessment of Viability After Cryopreservation

Both hESCs and iPSCs can undergo small changes during the maintenance; thus, it is important to determine their status after each procedure. The ratio between surviving cells and the originally frozen cells should be analyzed by counting the cells using a hemocytometer chamber using the tryptan blue exclusion method. Monitor the cell number for at least three passages to get a reliable survival rate; after that the cultures should be regarded as stable with normal growth and expansion. Cell aggregates originally frozen should also be compared to the number of surviving colonies after thawing.

Using the calcein-esterase-based live/dead assay (Molecular Probes, Eugene, USA), according to the manufacturers' procedure, gives a more reliable quantitative picture. To determine the level of differentiation based on the morphological appearance, by inverted phase contrast microscopy, the undifferentiated colonies should have a large nucleus with an abundant cytoplasm.

4. Notes

It is of high importance that the freezing medium, STEM-CELLBANKER™, as well as thawing medium, CELLOTION™, are used cold (4–8°C). As described above, the survival of hESCs and iPSCs has been poor in the past using the traditionally slow freezing/rapid thawing methods. The method described in this chapter is a xeno-free and chemically defined system. It gives excellent survival rates. When using STEM-CELLBANKER™ and the recovery solution CELLOTION™, 90% survival rate can be expected for hESCs and iPSCs, whereas 50% survival rate can be achieved using the traditional method, which is based on permeating cryo protectants only. Even the morphological differences in these cells differ drastically from each other; cells recovered using this method have a normal and typical stem cell–like morphology. Cells frozen with STEM-CELLBANKER™ also express the immunocytochemical surface markers SSEA-4, TRA-1-81, as well as the nuclear marker Nanog. Also, mRNA expression of Oct-4 and Nanog was determined by real-time quantitative PCR, which confirms that the cells stay undifferentiated after being frozen and thawed with the novel method. These cells also showed a setting of normal chromosomes when analyzing their karyotype, which suggests that there has not been any impact of the chromosomes when they are stored for long term in liquid nitrogen, which is of high importance for future clinical application where a safe bank of stem cells will be needed without any impact on their quality.

References

1. Thomson JA, Itskovitz-Eldor J, Shapiro SS, et al. (1998) Embryonic stem cell lines derived from human blastocysts. Science;**282**: 1145–7.
2. Trounson A, Pera M. (2001) Human embryonic stem cells. Fertil Steril;**76**:660–1.
3. Klimanskaya I, Rosenthal N, Lanza R. (2008) Derive and conquer: sourcing and differentiating stem cells for therapeutic applications. Nat Rev Drug Discov;7:131–42.
4. Gearhart J. (1998) New potential for human embryonic stem cells. Science;**282**:1061–2.
5. Pera MF, Trounson AO. (2004) Human embryonic stem cells: prospects for development. Development;**131**:5515–25.
6. Grout B, Morris J, McLellan M. (1990) Cryopreservation and the maintenance of cell lines. Trends Biotechnol;**8**:293–7.
7. Meryman HT. (2007) Cryopreservation of living cells: principles and practice. Transfusion;**47**:935–45.
8. Li T, Zhou C, Liu C, Mai Q, Zhuang G. (2008) Bulk vitrification of human embryonic stem cells. Hum Reprod;**23**:358–64.

9. Martin-Ibanez R, Unger C, Stromberg A, Baker D, Canals JM, Hovatta O. (2008) Novel cryopreservation method for dissociated human embryonic stem cells in the presence of a ROCK inhibitor. Hum Reprod;23:2744–54.
10. Reubinoff BE, Pera MF, Vajta G, Trounson AO. (2001) Effective cryopreservation of human embryonic stem cells by the open pulled straw vitrification method. Hum Reprod;16:2187–94.
11. Richards M, Fong CY, Tan S, Chan WK, Bongso A. (2004) An efficient and safe xeno-free cryopreservation method for the storage of human embryonic stem cells. Stem Cells;22:779–89.
12. Zhou CQ, Mai QY, Li T, Zhuang GL. (2004) Cryopreservation of human embryonic stem cells by vitrification. Chin Med J (Engl);117:1050–5.
13. Ha SY, Jee BC, Suh CS, et al. (2005) Cryopreservation of human embryonic stem cells without the use of a programmable freezer. Hum Reprod;20:1779–85.
14. Heng BC, Ye CP, Liu H, Toh WS, Rufaihah AJ, Cao T. (2006) Kinetics of cell death of frozen-thawed human embryonic stem cell colonies is reversibly slowed down by exposure to low temperature. Zygote;14:341–8.
15. Yang PF, Hua TC, Wu J, Chang ZH, Tsung HC, Cao YL. (2006) Cryopreservation of human embryonic stem cells: a protocol by programmed cooling. Cryo Letters;27:361–8.
16. Vajta G, Holm P, Greve T, Callesen H. (1997) Vitrification of porcine embryos using the Open Pulled Straw (OPS) method. Acta Vet Scand;38:349–52.
17. Ishizaki T, Uehata M, Tamechika I, et al. (2000) Pharmacological properties of Y-27632, a specific inhibitor of rho-associated kinases. Mol Pharmacol;57:976–83.
18. Hu E, Lee D. (2005) Rho kinase as potential therapeutic target for cardiovascular diseases: opportunities and challenges. Expert Opin Ther Targets;9:715–36.
19. Skottman H, Stromberg AM, Matilainen E, Inzunza J, Hovatta O, Lahesmaa R. (2006) Unique gene expression signature by human embryonic stem cells cultured under serum-free conditions correlates with their enhanced and prolonged growth in an undifferentiated stage. Stem Cells;24:151–67.
20. Kuleshova LL, Shaw JM, Trounson AO. (2001) Studies on replacing most of the penetrating cryoprotectant by polymers for embryo cryopreservation. Cryobiology;43:21–31.
21. Liebermann J, Dietl J, Vanderzwalmen P, Tucker MJ. (2003) Recent developments in human oocyte, embryo and blastocyst vitrification: where are we now? Reprod Biomed Online;7:623–33.
22. Holm F, Strom S, Inzunza J, et al. (2010) An effective serum- and xeno-free chemically defined freezing procedure for human embryonic and induced pluripotent stem cells. Hum Reprod;25:1271–9.
23. Strom S, Inzunza J, Grinnemo KH, et al. (2007) Mechanical isolation of the inner cell mass is effective in derivation of new human embryonic stem cell lines. Hum Reprod;22:3051–8.
24. Unger C, Gao S, Cohen M, et al. (2009) Immortalized human skin fibroblast feeder cells support growth and maintenance of both human embryonic and induced pluripotent stem cells. Hum Reprod.
25. Hovatta O, Mikkola M, Gertow K, et al. (2003) A culture system using human foreskin fibroblasts as feeder cells allows production of human embryonic stem cells. Hum Reprod;18:1404–9.
26. Domogatskaya A, Rodin S, Boutaud A, Tryggvason K. (2008) Laminin-511 but not -332, -111, or -411 enables mouse embryonic stem cell self-renewal in vitro. Stem Cells;26:2800–9.
27. Rodin S, Domogatskaya A, Strom S, et al. (2010) Long-term self-renewal of human pluripotent stem cells on human recombinant laminin-511. Nat Biotechnol;28:611–5.
28. Watanabe K, Ueno M, Kamiya D, et al. (2007) A ROCK inhibitor permits survival of dissociated human embryonic stem cells. Nat Biotechnol;25:681–6.

Part II

Generation of Human Induced Pluripotent Stem Cells: Viral and Nonviral Vector-Based Nuclear Reprogramming

Chapter 9

Induced Pluripotent Stem Cells (iPSC) from Cord Blood CD133⁺ Cells Using Oct4 and Sox2

Alessandra Giorgetti, Nuria Montserrat,
and Juan Carlos Izpisua Belmonte

Abstract

Induced pluripotent stem cells (iPSCs) provide an invaluable resource for regenerative medicine to repair tissues damaged through disease or injury. Although human iPSCs have been generated using different type of somatic cells, such as skin fibroblasts and keratinocytes, neural stem cells, hepatocytes, and blood cells, it is still under debate on what is the best source for generating iPSCs. In this chapter, we discuss how to generate human iPSCs from cord blood (CB) CD133⁺ cells using only two transcription factors OCT4 and SOX2. The methods for establishment and maintenance of CB-derived iPS (CB-iPS) cells are similar, with some modifications, to those for human iPSCs derived from fibroblasts. In particular, this protocol includes a detailed procedure to isolate CB CD133⁺ cells from both fresh and frozen CB and a procedure to optimize retroviral infection to generate iPSCs from these cells. In addition, we describe methods for the characterization of CB-iPS cells, such as embryoid bodies formation and teratoma differentiation, to validate their pluripotent phenotype.

Key words: Cord blood, Stem cells, iPS cells, CD133+ cells, Pluripotency, Reprogramming

1. Introduction

Less than a decade ago, the prospect of reprogramming a human somatic cell looked bleak at best. However, in 2006, Yamanaka et al. have shown that it was apparently simple to revert the phenotype of a differentiated cell to pluripotency by overexpression of four transcription factors using murine fibroblasts (1). These cells were known as induced pluripotent stem cells (iPSCs) and appeared to be highly similar to hESC. Later, the same strategy was used to generate human iPSCs, first from skin fibroblasts and keratinocytes and then from neural stem cells, hepatocytes, as well as blood cells (2–5).

The possibility of reprogramming mature somatic cells allows for the production of pluripotent cells that carry the specific genome of individuals, providing an unprecedented experimental platform to model human diseases. Patient-specific iPSCs could help to establish in vitro disease models and might lead to the discovery of drugs for treating patients. Therefore, patient-specific iPSCs have been hailed as an enormous development for regenerative medicine since transplantation of individual iPSCs should not be subject to immune rejection. However, in many instances, a ready-to-use approach could be desirable, such as for cell therapy of acute conditions or when the patient's somatic cells are altered as a consequence of a chronic disease or aging. An option could be the generation of healthy iPSC lines with a wide genetic variety rather than patient-specific iPSCs, which will enable broader immune histocompatibility.

Cord Blood (CB) stem cells, currently widely used as a source of hematopoietic stem cells (HSC) for transplantation, appear ideally suited for the generation of clinically sound iPSCs in an allogenic setting. The main practical advantages of using CB stem cells are the relative ease of procurement, the absence of risks for the donors, and the ability to store fully tested and HLA-typed samples in public cord blood banks, available for immediate use. Since the first successful CB therapy (6), the basic science behind clinical application has advanced at a rapid rate. To date, more than 400,000 immunologically characterized CB units are available worldwide through a network of public banks, facilitating a rapid search for compatible donors for iPSCs generation (7, 8). In addition, CB cells are young, expected to carry minimal somatic mutations, and more immature than HSCs and HPCs derived from adult bone marrow (BM). A key observation in this regard is the evidence that the more immature the adult starting cell population, the easier and more efficient it may be to generate iPSCs (4).

We have recently reported that the overexpression of only two transcription factors, OCT4 and SOX2, is sufficient to reprogram CB CD133⁺ cells faster than fibroblasts and keratinocytes (9). Cord blood CD133⁺ cells express a subset of pluripotency-associated genes (OCT4, SOX2, NANOG, and CRIPTO), albeit at much lower levels than hESC. On the other hand, the endogenous levels of c-MYC and KLF4 are higher in CB CD133⁺ cells compared to fibroblasts and keratinocytes (10). From these data, it can be hypothesized that the combination of low levels of pluripotency markers with the high levels of c-MYC and KLF4 may allow for enhanced reprogramming of CB CD133⁺ cells.

The first part of this chapter is focused on the isolation of the CB CD133⁺ subpopulation from fresh and cryopreserved CB units. CD133 antigen is known as a stem cell marker for hematopoietic stem and progenitor cells and is a valid substitute for CD34, commonly used for HSCs enrichment. Moreover, since CD133 is not expressed in late hematopoietic progenitors, this allows for selecting a more homogeneous population enriched in HSCs.

CD133⁺ cells, which represent a small fraction of total nucleated cells in fresh CB (0.1–0.7%), are positively selected using an immunomagnetic separation system.

Usually the CB units (UBCs) are stored in freezing compact bags designed with two sealed compartments that allow for their separation before thawing. The UBCs are thawed using an automated system, the Sepax, following the standardized procedure implemented in the "Banc de Sang i Teixits" of Barcelona.

Another important point discussed here is the transduction of CB CD133⁺ cells using a retroviral approach. The standard protocol used for fibroblasts reprogramming needs to be modified for the transduction of hematopoietic stem cells, where the infection efficiency is usually low. The inefficient gene transfer can most likely be attributed to characteristics from both the target cell population and the retroviral vectors as well as to the interaction between these two. One factor of importance is that the human HSCs are quiescent and nondividing and that cell division is required for retroviral vectors to integrate into the genome of target cells (11). To overcome this limitation, it is important to perform a prestimulation of CD133⁺ cells for 24 h, in the presence of stem cell factor (SCF), thrombopoietin (TPO), interleukin 6 (IL-6), and Flt3-ligand. In addition, to increase the transduction efficiency, CD133⁺ cells are infected in the presence of RetroNectin, a fragment of fibronectin that binds both hematopoietic cells and vector particles, thereby enhancing gene transfer by colocalization of the cells with the vectors (12). Finally, CD133⁺ cells are infected three times, every 12 h; on day 3, post infection cells are transferred to a 6 well plate containing irradiated human fibroblasts (HFF) and hES medium and cultured until iPSC colonies appear.

Usually, as early as 9 days post transduction, small colonies start to appear. They develop into colonies with a typical hES morphology in about 15 days. It is important to carry out an exhaustive characterization of iPSCs in order to unequivocally demonstrate their pluripotency. Like hESCs, iPSCs should show a compact morphology and should be maintained for a long period of time in culture. They have to express markers of pluripotency (SSEA3, SSEA4, TRA 1-60, and TRA 1-81) as well as transcription factors considered fundamental to maintaining an undifferentiated state (OCT4, SOX2, and NANOG). In addition, they have to be able to differentiate in vitro and in vivo into cells of the three embryonic germ layers. For in vitro differentiation, the most common strategy includes embryoid bodies (EBs) formation through the culture of iPSC colonies in suspension. EBs cultured over different substrates and in the presence of specific media can differentiate toward the cell lineages of interest. Finally the in vivo pluripotency test consists of inducing the spontaneous differentiation of iPSCs once injected in severe immunodeficient (SCID) mice. The injected cells proliferate and differentiate, generating teratomas that contain structures and tissues derived from the three embryonic germ layers.

2. Materials

2.1. Umbilical Cord Blood Thawing

1. Sepax automated washing system (Biosafe)
2. Sepax cell separation kit CS-600 (Biosafe)
3. Human albumin 200 g/L (Instituto Grifols SA, cat. no.670612)
4. Rheomacrodex (Dextran 40) 100 g/L
5. 300 mL Lifecell bag
6. 60 mL syringe
7. Transfer bag (Baxter)

2.2. Isolation of Umbilical CB CD133+ Cells

1. Phosphate-buffered saline (PBS) without calcium and magnesium (Invitrogen, cat. no. 2531)
2. EDTA disodium 0.5 M (Sigma-Aldrich, cat no. E7889)
3. PBS containing 2 nM EDTA; store at 4°C
4. Lympholyte-H (Cederlane, CL5016); toxic by skin contact; wear gloves when handling
5. CD133 MicroBead Kit (Miltenyi Biotec, cat. no. 130-050-801); store at 4°C
6. Trypan Blue stain (Invitrogene, cat. no. 15250-061)
7. Mini-Macs separator (Miltenyi Biotec, cat. no. 130-090-312)
8. MS column (Miltenyi Biotec, cat. no. 130-041-301)
9. Preseparation filters (Miltenyi Biotec, cat. no. 130-041-407)
10. Syringe, 2.5 mL (PentaFerte, cat. no. 08L01)
11. Cell counter or hemocytometer
12. Conical tubes, 15 and 50 mL
13. Storage bottle, 500 mL (Corning, cat. no. 430282)
14. Tissue culture plates, 6 well
15. DMEM (Invitrogen, cat. no. 11965-092); store at 4°C
16. Fetal bovine serum (FBS, Invitrogen, cat. no. 10270-106); store at 4°C
17. GlutaMAX (Invitrogen, cat. no. 35050–038)
18. Penicillin/streptomycin (Invitrogen, cat. no. 15140-122)
19. DMEM complete medium, 500 mL: high glucose 440 mL DMEM, 50 mL FBS, 5 mL GlutaMAX (1 mM), and 5 mL penicillin-streptomycin (100 U mL^{-1} penicillin and 100 µg mL^{-1} streptomycin); store at 4°C and use in 2–3 weeks
20. Recombinant human stem cell factor (SCF, PeproTech, cat. no. 300-07) (see Note 1)
21. Recombinant human Flt3-ligand (Flt3, PeproTech, cat. no. 300-19) (see Note 1)

22. Recombinant human interleukin 6 (IL-6, PeproTech, cat. no. 200-06) (see Note 1)
23. Recombinant human thrombopoietin (TPO, PeproTech, cat. no. 300-18) (see Note 1)
24. Mouse IgG1 antihuman CD133-PE (Miltenyi Biotec, cat. no. 130-080-801); store at 4°C
25. Mouse IgG1 and k antihuman CD45-APC (Becton Dickinson, cat. no. 555485); store at 4°C

2.3. Retrovirus Production and CD133+ Transduction

1. pMSCV-based retroviral vectors expressing the reprogramming transgene OCT4 and SOX2 (Addgene, 20072 and 20073)
2. Phoenix amphotropic cells (ATCC, cat. no. SD 3443)
3. Opti-MEM® (Invitrogen, cat. no. 31985-062); store at 4°C
4. DMEM complete medium (see Sect. 2.2)
5. FuGENE 6 transfection reagent (Roche Applied Science, cat. no. 1181509001); store at 4°C
6. Polybrene (10 mg/mL) (Chemicon, cat. no. TR-1003-6); store at −20°C
7. 100-mm tissue culture dish
8. 15 mL-Conical tubes
9. RetroNectin (Takara, Otsu, Japan www.takara-bio.com, cat. no. T100A); store at −20°C
10. 24 well tissue culture plates
11. Filter, Millec-HV PVDF 0.45 μm (Millipore, cat. no. SLHV033RS)

2.4. Culturing Transduced CD133+ Cells

1. Mitotically inactivated human foreskin fibroblasts (ATCC, cat. no. CRL-2429)
2. Gelatin 0.1% solution (Millipore, cat. no. ES-006-B)
3. Knockout (KO-DMEM; Invitrogen, cat. no. 10829-018); store at 4°C
4. KO serum replacement (KOSR; Invitrogen, cat. no. 10828-028); store at −20°C
5. Nonessential amino acid solution (Invitrogen, cat. no. 11140-050); store at 4°C
6. 50 mM 2-mercaptoethanol (Invitrogen, cat. no. 31350-010); store at 4°C; toxic by inhalation and skin contact
7. Recombinant human fibroblast growth factor-basic (bFGF, PeproTech, cat. no. 100-18B); store aliquots at −20°C
8. 6 well tissue culture plates
9. DMEM complete medium (see Sect. 2.3)
10. hES medium, 500 mL: 387.5 mL KO-DMEM, 100 mL KO serum, 5 mL GlutaMAX (1 mM), 5 mL penicillin-streptomycin

(100 U mL^{-1} penicillin and 100 μg mL^{-1} streptomycin), 500 μL 2-mercaptoethanol, 5 mL nonessential amino acid solution, and 50 μL of bFGF (10 ng/mL); store at 4°C and use maximum in 1 week

11. Bottle-top filter system 0.22 μm, 500 mL (Millipore, cat. no. SCGPU05RE)

2.5. Culturing and Expanding iPSC Colonies

1. Mitotically inactivated human foreskin fibroblasts (ATCC, cat. no. CRL-2429)
2. Gelatin 0.1% solution (Millipore, cat. no. ES-006-B)
3. hES medium (see Sect. 2.4)
4. 6 well tissue culture plates
5. Stripper micropipette (Mid Atlantic, cat. no. MXL3-STR)
6. Stripper tips, 150 μm

2.6. iPS Cells Characterization

1. 60 mm cell cultures plates (Corning/Cultek, cat. no. 15430166)
2. Gelatin 0.1% (Chemicon, cat. no. ES-006-B)
3. L-Ascorbic acid (Sigma, cat. no. A4544-25g)
4. N2 supplement (Invitrogen, cat. no. 17502048)
5. B27 supplement (Invitrogen, cat. no. 17502044)
6. Knockout KO-DMEM (see Sect. 2.5)
7. EB medium: KO-DMEM, GlutaMAX 1×, NEEA 1X, 2-mercaptoethanol 0.1×, 10% FBS, and Pen-strepto 1×; store at 4°C
8. Neurobasal medium (Invitrogen, cat. no. 21103-049)
9. DMEM/F12 (1×) (Invitrogen, cat. no. 21331-046)
10. N2/B27 medium: 50% neural basal media, 50% DMEM/F12 media, N2 0.5%, B27 1%, GlutaMAX 1×, and Pen-strepto 1×; store at 4°C
11. hES media; store 4°C (see Sect. 2.4)
12. MEM alpha media, Ribonucleic acid, Desoxyribonucleic acid, 10% FBS, GlutaMAX 1×, and Pen-strepto 1×; store at 4°C
13. MEM NEEA 100× (Cambrex, cat. no. 13-114); store at 4°C
14. Cell line RCB1127: MCT3-G2/PA6 RIKEN BioResource Center
15. Ultra low attachment plates 60 mm (Corning/Cultek, cat. no. 153261)
16. Slide flask (Nunc cat. no. 170920)
17. Paraformaldehyde (PFA) 4% (Sigma, cat. no. P6148-500G); store at −20°C

2.7. Freezing iPS Cells

1. Dimethyl sulfoxide (DMSO) (Sigma, cat. no. D4540), toxic by inhalation and skin contact
2. Fetal bovine serum (FBS; Invitrogen, cat. no. 10270-106); store at 4°C
3. Freezing medium: mix 10% DMSO (vol/vol) and 90% FBS (vol/vol); store at 4°C use immediately
4. Cryovials (Sigma, cat. no. V7634-500EA)
5. Cryo 1°C Freezing container, "Mr. Frosty" (Nalgene, cat. no. 5100-0001)
6. Stripper micropipette and Stripper tips (see Sect. 2.5)

3. Methods

CB CD133$^+$ cells represent a small fraction of total nucleated cells (0.1–0.7%) in one CB unit. The quality of the CB unit is a critical step in obtaining good recovery and high purity of CD133$^+$ cells, after immunoselection. In particular, use samples not older than 16 h, with a minimum blood volume of 65 mL and containing not less than 2.5×10^6 CD34$^+$ cells.

3.1. Umbilical Cord Blood Thawing

This procedure is the standard method used at the Tissue Bank Umbilical Cord Blood of Barcelona for thawing and washing the cord blood units with a volume range of 25–105 mL.

1. Prepare a washing solution of 7.5% of rheomacrodex and 5% of human albumin (see Table 1); store at 4°C.
2. Remove the unit from the liquid nitrogen (LN2) tank.
3. Transfer the bag into another plastic bag and immerse the unit in the thermostatic bath at 37°C.
4. To speed up the thawing process, shake the bag in the bath and finger massage it.
5. As soon as the bag is thawed, remove from the bath and check the bag weight to start the thawing process.

Table 1
The composition of washing solution

Unit volume (mL)	Buffer volume (mL)	Rheomacrodex (mL)	Human albumin
25	300	225	75
50	400	300	100
>75	500	375	125

3.2. Isolation of Umbilical CB CD133⁺

6. Prepare the Sepax CS-600 washing kit in a laminar flow cabinet and connect with the cord blood bag. Proceed with the washing of the bag following the manufacturer's instructions, Sepax system by Biosafe.
7. Continue as described in Sect. 3.2.1.

Parents' informed consent must be obtained before the delivery, and all hospital ethical requirements must be met. When working with fresh CB units, the samples should be kept at 4°C during the transport. For optimum results and to prevent blood hemolysis or coagulum formation, the process should take place in ≤16 h. Umbilical cord blood is collected into a plastic blood collection bag by venipucture (see Fig. 1a).

3.2.1. Isolation of Umbilical Cord Blood Mononuclear Cells Using Lympholyte-H Density Gradient

1. From now on, all subsequent steps should be carried out in a tissue culture hood. Dilute the cord blood 1:3 with sterile PBS/EDTA in a sterile 500-mL bottle and layer 30 mL of diluted blood slowly on 15 mL of Lympholyte-H layer (ratio 3:1) using a 50-mL tube (Fig. 1b) (see Note 2).
2. Centrifuge the tubes for 25 min at 400×g at room temperature (see Note 3).
3. Collect the interphase cells (white layer) using a plastic pipette in a 50-mL tube (Fig. 1c). Fill up with sterile PBS/EDTA (ratio 1:1) and centrifuge for 5 min at 300×g at room temperature.
4. Remove the supernatant and discard without disturbing the pellet. Next, pool together the pellets, resuspending in 50 mL of PBS/EDTA.
5. Count the total mononuclear cells using hematocytometer and asses the viability using Trypan Blue exclusion method. Briefly, place 10 μL of mononuclear cells suspension in an appropriate tube containing 90 μL of Trypan Blue and mix gently. Place 10 μL of stained cells in a hemocytometer and count the number of alive (unstained) and dead cells stained blue. Calculate the average number of unstained cells in at least three quadrants and multiply by 10^4 to find the number of cells per ml. The percentage of viable cells is the number of viable cells divided by the number of total cells. Calculate the average number of unstained cells in at least three quadrants and multiply by 10^4 to find the number of cells per ml. The percentage of viable cells is the number of viable cells divided by the number of total cells.
6. Centrifuge the sample for 5 min at 200×g at room temperature to eliminate platelets.
7. Pipette off the supernatant and resuspend the cell pellet in a final volume of 300 μL of PBS/EDTA, per 10^8 total cells. When less than 10^8 total cells, use 300 μL of PBS/EDTA.

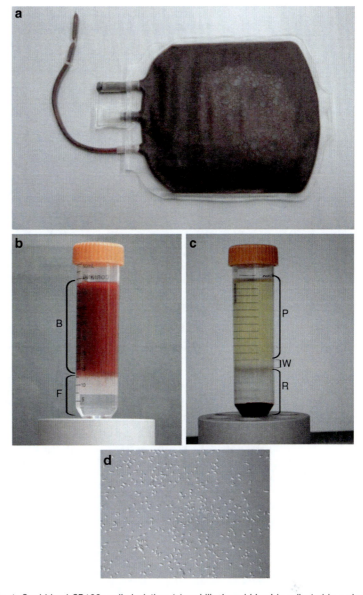

Fig. 1. Cord blood CD133+ cells isolation: (**a**) umbilical cord blood is collected in a plastic blood collection bag by venipucture; (**b**) diluted cord blood (B) is stratified on Lympholyte-H (F) for the isolation of mononuclear cells; (**c**) after centrifugation, there will be a defined lymphocyte layer (W) at the interface between plasma with platelets (P) and red blood cells on the bottom below the Lympholyte-H; and (**d**) cord blood CD133+ after immunoselection.

3.2.2. Isolation of CD133+ Cells

1. Add 100 µL FcR Blocking Reagent to 10^8 total cells resuspended in 300 µL of PBS/EDTA. Add 100 µL CD133 MicroBeads to 10^8 total cells (see Note 4).

2. Mix well and incubate cells for 30 min at 4–8°C.

3. Wash cells by adding PBS/EDTA up to 50 mL and centrifuge for 5 min at $300 \times g$ at room temperature.

4. Pipette off the supernatant and resuspend cell pellet in 500 μL PBS/EDTA to 10^8 total cells. Proceed to magnetic separation.
5. Use one MS column for up to 1×10^7 magnetically labeled cells or up to 2×10^8 total cells.
6. Place the column in the magnetic field of MACS Separator and place a 15-mL tube under the column. Rinse the column with 500 μL PBS/EDTA.
7. Apply 500 μL cell suspension onto the column and allow the negative cells to pass through.
8. Wash the column twice with 500 μL PBS/EDTA, then remove the column from separator and place the column on a new 15-ml collection Falcon tube.
9. Pipette 1 mL of PBS/EDTA onto the column and firmly flush out fraction with magnetically labeled cells using the plunger.
10. To obtain a highly purified population, repeat magnetic separation steps 6–9, applying the eluted cells to a new prefilled column.
11. Count the cells using hematocytometer and calculate the viability using Trypan Blue exclusion method (as described in Sect. 3.2.1).
12. In order to estimate the purity of the isolated CD133 positive cells, take 5×10^4 cells and perform the flow cytometry (FACS). Briefly, add 10 μL of CD133-PE (Miltenyi Biotec) and 20 μL of CD45APC (BD) in cell suspension (50,000 in 500 μL of PBS), mix well and incubate for 15 min in the dark at room temperature. Wash the cells with 2 mL of HSA2%-PBS and centrifuge at $600 \times g$ for 5 min. Resuspend cell pellet in 500 μL of HSA 2%-PBS containing propidium iodide (final concentration 10 μg/mL) to detect dead cells.
13. At this point, plate the CD133$^+$ cells in a concentration of 5,000/mL of complete DMEM supplemented of SCF (50 ng/mL), Flt3-ligand (50 ng/mL), IL-6 (10 ng/mL), and TPO (10/ng mL) in a 6 well plate. Incubate at 37°C, 5% CO_2 for 24 h. (see Fig. 1d) (see Note 5).

3.3. Retrovirus Production

1. Thaw a vial of Phoenix amphotropic 293 cells and plate approximately 2×10^6 cells in complete DMEM medium in 100-mm tissue culture dish and incubate at 37°C, 5% CO_2 for 2 days.
2. When the cells reach 80% confluence, aspirate medium, wash gently with PBS, and aspirate, and add 2 mL 0.05% Trypsin/EDTA. Incubate for 1 min at 37°C. Gently tap the tissue culture plate ensuring all cells are in suspension. Add 10 mL complete DMEM medium to the plate, collect and transfer cell suspension to a 50-mL tube.
3. Centrifuge at $200 \times g$ for 5 min at room temperature.

4. Resuspend pellet in 10 mL complete DMEM medium and count cells (as described in 3.2.1).

5. Plate 4×10^6 cells in 10 mL final volume in 100-mm culture dishes and place in a 37°C, 5% CO_2 incubator.

6. On the next day, cotransfect Phoenix amphotropic 293 cells with FuGENE:DNA complex according to the manufacturer's instructions. Briefly, place 0.873 mL of opti-MEM into separate 1.5-mL tube (one for each of the plasmids to be transfected) and add 27 µL of FuGENE 6 transfection reagent by gently tapping to mix. Incubate at room temperature for 5 min. Add 9 µg of pMSCV plasmids DNA dropwise into separate solutions of FuGENE 6/opti-MEM and mix by tapping with a finger. Incubate at room temperature for 15 min.

7. Add the FuGENE 6:DNA solution dropwise onto plate and return to a 37°C, 5% CO_2 incubator overnight.

8. After 24 h, gently change the media (10 mL complete DMEM medium/plate) and incubate in a 32°C, 5% CO_2 incubator overnight. At this time point, near 100% confluence of cells should be transfected. Monitor transfection efficiency using GFP reporter plasmid (see Fig. 2a, b).

9. Collect 5 mL viral supernatant from every plate 48 h after transfection using plastic pipettes and filter the supernatant through a 0.45 µm PVDF filter to remove any residual cells (see Note 6).

10. Add 5 mL of fresh complete DMEM to every plate and place in a 37°C, 5% CO_2 incubator.

11. Add 5 µL of polybrene for each ml viral supernatant needed.

12. Every 12 h, repeat steps 10–12 in order to collect more viral supernatant.

3.4. Retroviral Transduction of CD133⁺ Cells

3.4.1. Infection 1st Day

1. Dispense an appropriate volume of RetroNectin solution (15 µg/cm²) into each well of 24 well plate and incubate for 2 h at room temperature.

2. Remove RetroNectin solution and add 500 µL of PBS containing 2% human albumin (HSA) into each well for blocking and perform an incubation of 30 min at room temperature.

3. Remove the PBS/HSA solution and wash once with PBS.

4. Preload the viral supernatant derived from Phoenix amphotropic cells that have been filtered onto RetroNectin coated plate. Use equal amounts of each transcription factor to reach a total volume of 1 mL (500 µL of OCT4 and 500 µL of Sox2).

5. Set the plate into a centrifuge prewarmed at 32°C and centrifuge for 1 h at $2,000 \times g$ at 32°C.

6. During the centrifugation, collect the CD133⁺ cell from step 13 in a 15-mL Falcon tube and count the number of living cells

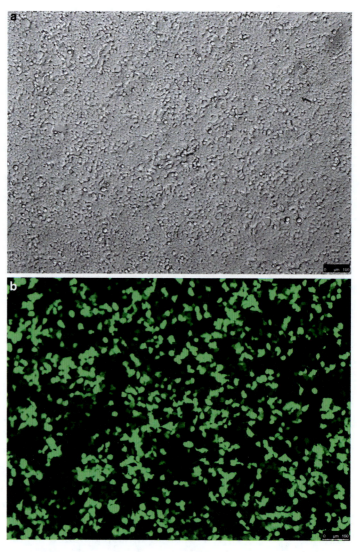

Fig. 2. Phoenix amphotropic 293 cells transfected using GFP reporter plasmid: (**a**) Contrast phase image of phoenix cells 48 h after tranfection (scale bar, 100 μm); (**b**) Transfection efficiency is monitored using GFP reporter plasmid.

(as described in Sect. 3.2.1). Centrifuge for 5 min at $200 \times g$ at room temperature.

7. Suspend the pellet cells in complete DMEM medium supplemented with cytokines (SCF, Flt3-ligand, IL-6, and TPO) at a concentration of 8×10^4 cells/mL.

8. Pipette off the viral supernatant from each well, taking care that the virus bound to the RetroNectin should not dry, and wash with 1 mL of PBS each well.

9. Pipette off PBS and add immediately 1 mL of cell suspension into each well. Incubate for 12 h 37°C, 5% CO_2 incubator.

3.4.2. Infection 2nd Day

1. Repeat steps 10–12 of Sect. 3.3.
2. Remove 500 μL from each well of 24 well plate containing CD133⁺ cells and add 500 μL of fresh viral supernatant to infect the cells a second time (see Note 7).
3. Incubate for 12 h in a 37°C, 5% CO_2 incubator.
4. Repeat steps 2–3 to infect the cells a third time and incubate for 24 h in a 37°C, 5% CO_2 incubator (see Note 8).

3.5. Culturing Transduced CD133⁺ Cells

It is important to transfer CD133⁺ cells into ES culture conditions no later than 48 h post transduction to induce the reprogramming process as soon as possible. In addition, keeping CD133⁺ cells for long periods in the presence of hematopoietic growth factors could induce differentiation through mature hematopoietic progenitors.

1. Plate human irradiated fibroblast (HFF) in a gelatin-treated 6 well plate.
2. 1 day after the last infection, collect the cells in a 15-mL Falcon tube and centrifuge for 5 min at 200×*g* at room temperature.
3. Resuspend in 1 mL hES medium and plate the infected CD133⁺ cells onto 6 well plate containing HFF feeder and 1 mL of hES medium.
4. After 2 days, change the medium daily and maintain in culture until the emergence of iPS cell colonies (see Note 9).
5. After 9 days, small colonies start to appear. At 15 days of culture, the colonies exhibit typical hESC morphology.

3.6. Picking and Expanding iPS Cell Colonies

3.6.1 1st Day

1. Identify colonies of iPS-like morphology and mark them on the bottom of the dish under the inverted microscope.
2. Prepare the required number of 6 well plates with irradiated HFF feeder layers, calculating 1 well per colony to be picked.

3.6.2 2nd Day

1. Aspirate the medium and add 2 mL of fresh hES cell medium in each well of reprogramming dishes and 6 well feeder plates.
2. Place the plate for 20 min in a 37°C, 5% CO_2 incubator.
3. Working on the steromicroscope placed inside in a tissue culture hood, manually pick up the single colonies from each well using a stripper micropipette. Transfer colony fragments into a well with HFF feeder. Incubate in a 37°C, 5% CO_2 incubator for 48 h.

3.6.3. 4th Day

1. Change the medium of each well. Many colony fragments should be attached to the feeder.
2. Change the medium daily. Attached colony fragments will form iPS-like colonies, ready to be passaged by day 7–8.

3.7. iPS Cells Characterization

3.7.1. Immunocytochemistry for Pluripotent Cell Markers

Detection of the expression of alkaline phosphatase activity is the first step to test the pluripotency of iPS cells (see Figs. 3a and 4a). However, it is important to also analyze the expression of pluripotent stem cell markers such as SSEA3, SSEA4, TRA 1-60, TRA1-81, OCT4, NANOG, and SOX2 by immunocytochemistry. In this section we will not describe in detail these procedures as they are standard practices in most laboratories (see Figs. 3b and 4b).

3.7.2. In Vitro Differentiation of iPS Cells by EB Formation

For in vitro differentiation, iPS cell lines are cultured as usual in hES cell media. Once the colonies reach a big size (about 7 days of culture), they are entirely lifted mechanically with a fine bore glass pasteur pipette. The colonies in suspension are cultured 24–48 h in ultra low attachment plates in hES cell media to induce the formation of EBs (see Fig. 4c).

Differentiation to endoderm:

1. After 24 h of EBs formation, hES media is exchanged to EB media and the EBs are cultured 2–3 more days in suspension.
2. Seed the EBs in slide flasks in EB media (with a previous coating of gelatine 0.1% for a minimum of 60 min in the incubator).
3. The medium will be changed every 2–3 days until 15–20 days before processing the samples.

Differentiation toward mesoderm:

1. After 24 h of the EBs formation, the hES media is changed by EB media supplemented with 0.5 mM of ascorbic acid and the EBs are cultured 2–3 more days in suspension.
2. Seed the EBs in slide flask with a previous coating of gelatine in EB media supplemented with 0.5 mM of ascorbic acid.
3. The medium will be changed every 2–3 days until 15–20 days before processing the samples.

Differentiation toward ectoderm:

1. After 24 h of the EBs formation, the hES cell media is changed by N2/B27 media and the EBs are cultured 3–4 more days in suspension.
2. Seed the EBs in slide flask in N2/B27 media over a monolayer of confluent PA6 cells.
3. The medium will be changed every 2–3 days until 14–16 days before processing the samples.

Fig. 3. Images related to iPS cells characterization derived using a fresh cord blood unit: (**a**) Images of established iPS cells before and after AP staining; (**b**) images of immunocytochemistry for pluripotency markers such as OCT4, SOX2, NANOG, TRA 1-81, TRA 1-60, SSEA-3, and SSEA-4. Blue indicates nuclei stained with DAPI (scale bar, 250 μm); (**c**) Generation of EBs using a fire finely bore glass pasteur pipette; (**d**) in vitro differentiation of iPS cells into the three primary germ layers (ectoderm TUJ-1 (green), endoderm AFP (green) and FOXA2 (red), and mesoderm-ASA (green) and SMA (red); and (**e**) Immunofluorescence analysis of teratoma sections after 60 days after intratesticular injection in SCID mice showing ectoderm (TUJ-1 and GFAP positive), endoderm (AFP and FOXA2 positive), and mesoderm (ASA and SMA positive) structures.

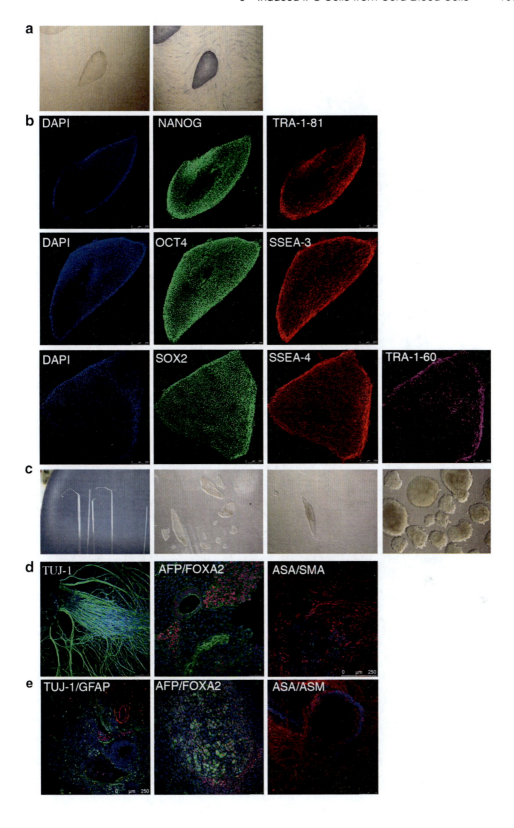

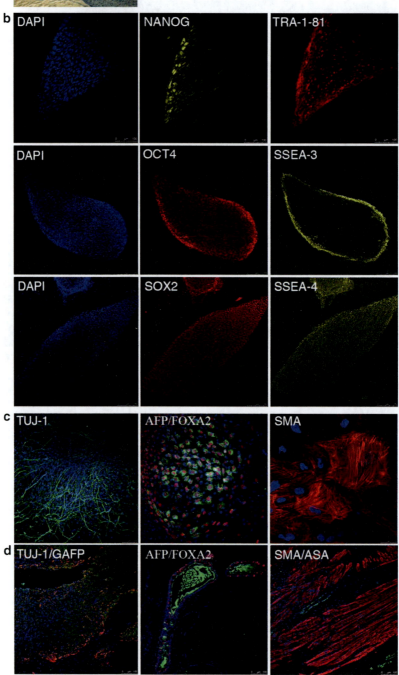

At the end of the culture, examine the cells with the inverted microscope if you have differentiated EBs and mark with a permanent marker the best area in order to make localization easier during the imaging step. Fix each slide flask for 20 min with 2 mL of 4% paraformaldehyde (RFA) at room temperature. Remove the PFA and wash twice for 5 min each with PBS and leave the flasks with 3 mL of PBS and leave at 4°C until use for immunofluorescence (see Figs. 3d and 4c).

3.7.3. In Vivo Differentiation of iPS Cells by Teratoma Assay

The last step, to evaluate the pluripotency of human iPSC lines, is the in vivo differentiation into derivates of the three embryonic germ layers upon injection into immunocompromised SCID beige mice.

1. Pick up the colonies mechanically and resuspend them in a final volume of 0.5 mL of hES cell medium and place in 1.5-mL Eppendorf tube. (Pick 3–4 confluent wells of 6 well-plate).
2. Centrifuge the cells at $200 \times g$ 2 min at room temperature.
3. Load a syringe of insulin, connected with a 25 G needle, with the supernatant (about 0.5 mL of hES cell media) leaving the pellet in the tube (40 µL approximately).
4. Resuspend the pellet lightly and aspirate it with the syringe.
5. Then, inject this 40 µL of pellet in the desired place of a SCID mouse.

After 8–12 weeks, the animals will be sacrificed and the teratoma will be collected. Process the samples for conventional histology or by immunohistochemistry techniques (see Figs. 3e and 4d).

3.8. Freezing iPS Cells

Established iPSCs cultures should be frozen at early passages to maintain the stock. When the cells reach confluence in the 6 well plate, it is time to make cryostocks.

1. Aspirate the medium and add 2 mL of fresh hES cell medium in each well.
2. Select about 20 colonies from each well and transfer the cell suspension to a 15-mL tube containing 2 mL prewarmed hES medium. Centrifuge for 1 min at $200 \times g$ at room temperature and aspirate the supernatant.

Fig. 4. Images related to iPS cells characterization derived using a frozen and thawed cord blood bag: (**a**) Images of established iPS cells AP positive; (**b**) Images of immunocytochemistry for pluripotency markers such as OCT4, SOX2, NANOG, TRA 1-81, SSEA-3, and SSEA-4. *Blue* indicates nuclei stained with DAPI (scale bar, 100 µm); (**c**) Images of immunocytochemistry for in vitro differentiated iPS cells into derivates of ectoderm (TUJ-1 positive), endoderm (AFP and FOXA2 positive),and mesoderm (SMA positive); and (**d**) Images of teratoma sections showing ectoderm (TUJ-1 and GFAP positive), endoderm (AFP and FOXA2 positive), and mesoderm (ASA and SMA positive) structures.

3. Add drop by drop 1 mL of freezing medium and transfer the cells into a freezing vial.
4. Keep the vials in a cell freezing container overnight at −80°C and then transfer them into a liquid nitrogen tank the next day.

4. Notes

1. Dilute all the growth factors in appropriate aliquots at the concentration of 5 ng/μL, according to the manufacturer's recommendations; store at −20°C.
2. Do not mix the tube and make sure that the layer is not disturbed.
3. Acceleration should be high and brake should be medium in order to cause minimal disruption to the layers.
4. Work fast and keep the cells cold using cold solutions to avoid nonspecific cell labeling. When working with lower cell numbers, use same volumes. When working with higher cell numbers, scale up all reagent volumes (e.g., for 2×10^8 total cells, use twice the volume of all indicated reagents).
5. Since the integration and expression of retroviral constructs requires mitotic division of the target cells, it is important to perform a prestimulation step 24 h in advance to induce the quiescent CD133+ cells to enter a proliferative state.
6. Use low protein binding filters to avoid trapping of virus and reduction of titer. Virus can be snap frozen in cryovials using liquid nitrogen and stored at −80°C or in liquid nitrogen for several months. Some loss in infectivity is observed upon freezing.
7. Add cytokine cocktail (half amount) to the fresh viral supernatant to keep the cells alive and in a proliferative state.
8. The infection efficiency, monitored by a constitutive GFP reporter retrovirus, could be variable (10–40%).
9. During the first week of culture, when CD133+ cells still grow in suspension, it is important to aspirate the medium gently using a 1 mL pipette and to add, dropwise, fresh medium in order to avoid detaching the few cells that have attached to the feeder layer.

References

1. Takahashi, K., and Yamanaka, S. (2006) Induction of pluripotent stem cells from mouse embryonic and adult fibroblast cultures by defined factors, *Cell 126*, 663–676.
2. Takahashi, K., Tanabe, K., Ohnuki, M., Narita, M., Ichisaka, T., Tomoda, K., and Yamanaka, S. (2007) Induction of pluripotent stem cells from adult human fibroblasts by defined factors, *Cell 131*, 861–872.
3. Aasen, T., Raya, A., Barrero, M. J., Garreta, E., Consiglio, A., Gonzalez, F., Vassena, R., Bilic, J., Pekarik, V., Tiscornia, G., Edel, M., Boue, S., and Belmonte, J. C. (2008) Efficient and rapid generation of induced pluripotent stem cells from human keratinocytes, *Nature biotechnology 26*, 1276–1284.
4. Kim, J. B., Greber, B., Arauzo-Bravo, M. J., Meyer, J., Park, K. I., Zaehres, H., and Scholer, H. R. (2009) Direct reprogramming of human neural stem cells by OCT4, *Nature 461*, 649–653.
5. Loh, Y. H., Agarwal, S., Park, I. H., Urbach, A., Huo, H., Heffner, G. C., Kim, K., Miller, J. D., Ng, K., and Daley, G. Q. (2009) Generation of induced pluripotent stem cells from human blood, *Blood 113*, 5476–5479.
6. Gluckman, E., Broxmeyer, H. A., Auerbach, A. D., Friedman, H. S., Douglas, G. W., Devergie, A., Esperou, H., Thierry, D., Socie, G., Lehn, P., and et al. (1989) Hematopoietic reconstitution in a patient with Fanconi's anemia by means of umbilical-cord blood from an HLA-identical sibling, *The New England journal of medicine 321*, 1174–1178.
7. Gluckman, E., and Rocha, V. (2009) Cord blood transplantation: state of the art, *Haematologica 94*, 451–454.
8. Rocha, V., Labopin, M., Sanz, G., Arcese, W., Schwerdtfeger, R., Bosi, A., Jacobsen, N., Ruutu, T., de Lima, M., Finke, J., Frassoni, F., and Gluckman, E. (2004) Transplants of umbilical-cord blood or bone marrow from unrelated donors in adults with acute leukemia, *The New England journal of medicine 351*, 2276–2285.
9. Giorgetti, A., Montserrat, N., Aasen, T., Gonzalez, F., Rodriguez-Piza, I., Vassena, R., Raya, A., Boue, S., Barrero, M. J., Corbella, B. A., Torrabadella, M., Veiga, A., and Izpisua Belmonte, J. C. (2009) Generation of induced pluripotent stem cells from human cord blood using OCT4 and SOX2, *Cell stem cell 5*, 353–357.
10. Giorgetti, A., Montserrat, N., Rodriguez-Piza, I., Azqueta, C., Veiga, A., and Izpisua Belmonte, J. C. Generation of induced pluripotent stem cells from human cord blood cells with only two factors: Oct4 and Sox2, *Nature protocols 5*, 811–820.
11. Miller, D. G., Adam, M. A., and Miller, A. D. (1990) Gene transfer by retrovirus vectors occurs only in cells that are actively replicating at the time of infection, *Molecular and cellular biology 10*, 4239–4242.
12. Hanenberg, H., Xiao, X. L., Dilloo, D., Hashino, K., Kato, I., and Williams, D. A. (1996) Colocalization of retrovirus and target cells on specific fibronectin fragments increases genetic transduction of mammalian cells, *Nature medicine 2*, 876–882.

Chapter 10

Generation, Maintenance, and Differentiation of Human iPS Cells from Cord Blood

Naoki Nishishita, Chiemi Takenaka, and Shin Kawamata

Abstract

This chapter describes a robust method for the generation of iPS cells from non-cultured human cord blood cells. We describe the preparation of the CD34+ fraction from cord blood mononuclear cells, the protocols to determine the pluripotency of the reprogrammed cells, the culture conditions for serial passages of iPS cells, and the protocol for cryopreservation of established iPS cells. As the efficiency of gene transfer to suspension cells is relatively low compared with adherent cells such as fibroblasts or mesenchymal stem cells, improvements in the efficiency of gene transfer to cord blood cells is a key issue in using them to generate iPS cells. Here, iPS cells are generated by introducing Yamanaka's four factors (Oct3/4, Sox2, Klf4, and c-Myc) with retroviruses, and, in addition, a short-hairpin RNA (shRNA) sequence with lentivirus that increases the rate of CD34+ cord blood cell proliferation. Enhanced green fluorescent protein (GFP) is also introduced to monitor the infection efficiency and the silence of exogenous genes. Infected cells are cultured in hematopoietic medium (X-VIVO 10 containing 50 ng/mL IL-6, 50 ng/mL soluble IL-6 R, 50 ng/mL SCF, 10 ng/mL TPO, and 20 ng/mL Flt-3-ligand) for 5 days after viral infection. Then, cells are transferred onto SNL76/7 feeder cells in human embryonic stem cell medium (DMEM/F-12 containing 20% KSR, 2 mM L-glutamine, 1% NEAA, 0.1 mM 2-ME, and 4 ng/mL bFGF). The ES cell-like colonies emerge in the following 4 weeks and are picked up to check their capacity for serial passage. The picked up colonies are stained to detect ALP activities and pluripotency-related molecules such as SSEA-4, TRA-1-60, and TRA-1-81 by immunochemistry. Further induction of endogenous pluripotency-related genes and silence of exogenous genes are determined by RT-PCR. The silencing of the introduced p53 knockdown lentiviral construct is checked by QRT-PCR. Finally, the differentiation potential of cells from ES cell-like colonies is determined by in vitro culture and teratoma formation assay with SCID mice. We explain each protocol in a practical way, focusing on problems we have encountered as stated in the Notes. We also utilize figures and photographs to facilitate comprehension of the protocols.

Key words: Induced pluripotent stem (iPS) cells, Cord blood cells, CD34+ cells, p53, Hematology

1. Introduction

Since the first report of mouse induced pluripotent stem (miPS) cells by retroviral vectors (1) in 2007, other exogenous gene delivery systems such as adenoviruses (2), sendai viruses (3), conventional plasmids (4), the piggyBac transposition system (5, 6), Cre-excisable viruses (7), and oriP/EBNA1-episomal vectors (8) have been reported. Further, protein delivery methods (9, 10) and small molecule delivery methods (11, 12) have also been reported to explore "safer" iPS cell generation. Importantly, the efficiency of iPS cell generation differs greatly depending on the reagents, methods, and cell sources used.

As for the cell source for generating iPS cells, a variety of cell types or tissues have been reported, including dermal fibroblasts (1), keratinocytes (13), hepatocytes (14), $CD34^+$ peripheral blood cells (15), thymocytes (16), and $CD34^+$ cord blood cells (17, 18). However, among these somatic cells, cord blood cells have a distinct advantage for iPS cell generation.

First, unlike the cells obtained by biopsy from a variety of tissues at different ages, freshly isolated (non-cultured) $CD34^+$ cord blood cells have distinct genetic and epigenetic profiles as hematopoietic stem cells and progenitors. Further, this fraction is the youngest stem cell population available following birth and free from postnatal individual genomic or epigenetic alterations caused by chemical and/or UV irradiation and also from genetic deletions related to rearrangements of the T/B cell receptor. Thus, generating iPS cells from this fraction facilitates our understanding of the reprogramming process by knowing the genetic profile of the cell source ($CD34^+$ cord blood cell), and the reprogrammed cell (iPS cells).

Second, by using cord blood cells as an iPS cell source, we are able to collaborate with existing public cord banks which preserve freshly isolated cord blood cells in a clinical setting with individual HLA typing information (19). Thus, the use of cord blood as a cell source for generating iPS cells has several advantages for basic research, drug discovery, and allogenic cell transplantation.

In this chapter, we describe an easy and robust method for generating iPS cells from $CD34^+$ cord blood cells to readily obtain a number of bona fide iPS cell clones from 2×10^4 virus-infected cells.

2. Materials

1. Pipetman (2, 10, 20, 200, and 1,000 μL; GILSON, Middleton, WI, USA)
2. Incubator, humidified, 37°C, 5% CO_2
3. Sterile biosafety cabinet
4. Liquid disposal system for aspiration
5. Centrifuge
6. Sampling tubes (1.5 mL; Eppendorf, Tokyo, Japan)

7. Sterile serological pipets (5, 10, and 25 mL; Becton Dickinson, Tokyo, Japan)
8. Sterile conical tubes (15 and 50 mL; Becton Dickinson)
9. Cell culture dishes (100- and 60-mm; Becton Dickinson)
10. Cell culture plates (6-well, 12-well, 24-well, and 96-well; Becton Dickinson)
11. Tips (0.1–10 µL, 1–200 µL, 100–1,000 µL; Sorenson BioScience, West Salt Lake City, UT, USA)
12. Cell scraper (9,000-220; ASAHI TECHNO GLASS, Tokyo, Japan)
13. Freezing stock vials (Cryogenic Vials; 2742-002; ASAHI TECHNO GLASS)
14. Cellulose acetate filters, 0.22-µm pore size (Millipore, Billerica, MA, USA)

2.1. Preparation for Retroviral Transduction: Reagents

1. Dulbecco's modified Eagle medium (DMEM; Invitrogen, Carlsbad, CA, USA).
2. Phosphate buffered saline without calcium and magnesium (PBS; Invitrogen).
3. Penicillin/streptomycin (Pen/Strep; ×100, 10,000 units/mL and 10,000 µg/mL; Invitrogen).
4. L-Glutamine, 200 mM (Invitrogen)
5. Fetal bovine serum (FBS; 2917354 CELLect; MP Biomedicals, Solon, OH, USA).
6. 0.05% Trypsin/5.3 mM EDTA solution (Invitrogen).
7. FuGene 6 transfection reagent (1814443; Roche, Basel, Switzerland).
8. Polybrene (10 mg/mL; Millipore).
9. 293T medium, 500 mL. Mix 440 mL DMEM, 50 mL FBS, 5 mL 200 mM L-glutamine, and 5 mL Pen/Strep. Store at 4°C.
10. 293T cells (Open Biosystems, Huntsville, AL, USA).
11. pMX-Oct3/4, -Sox2, -Klf4, -c-Myc, GFP, pCMV-VSV-G, pMDLg/p.PRE: retroviral constructs (Addgene, Cambridge, MA, USA).
12. Shp53 pLKO.1 puro, pMD2.G-VSV-G, psPAX2: lentivirus constructs (Addgene).
13. Cellulose acetate filters, 0.45-µm pore size (Millipore).

2.2. Reprogramming CD34+ Cord Blood Cells: Reagents

1. Dulbecco's modified Eagle medium: Nutrient Mixture F-12 HAM 1:1 (DMEM/F12; Invitrogen).
2. Knockout serum replacement (KSR; Invitrogen).
3. L-Glutamine, 200 mM (see Sect. 2.1).

4. Non-essential amino acid solution, ×100 (NEAA; Invitrogen).
5. 2-Mercaptoethanol, 55 mM (2-ME; Invitrogen).
6. Pen/Strep (see Sect. 2.1).
7. Recombinant basic fibroblast growth factor, human (bFGF; Wako, Osaka, Japan), 10 μg/mL. Reconstitute 50 μg of bFGF in 5 mL of 0.22 μm filtered water. Aliquot and store at −20°C.
8. Human ES medium, 550 mL. Mix 423 mL DMEM/F12, 110 mL KSR, 5.5 mL 200 mM L-glutamine, 5.5 mL NEAA, 1 mL 2-ME, and 5 mL Pen/Strep. Add 5 ng/mL bFGF into the medium before use. Store at 4°C up to 1 week.
9. PBS (see Sect. 2.1).
10. FBS (see Sect. 2.1).
11. DMEM (see Sect. 2.1).
12. SNL medium, 500 mL. Mix 440 mL DMEM, 50 mL FBS, 5 mL 200 mM L-glutamine, and 5 mL Pen/Strep. Store at 4°C.
13. 0.05% Trypsin/5.3 mM EDTA solution (Invitrogen).
14. Gelatin (G1890; Sigma, Tokyo, Japan). To make 0.1% gelatin solution, dissolve 1.0 g of gelatin powder in 1,000 mL of distilled water, autoclave, and store at 4°C.
15. Gelatin-coated culture dishes. To coat culture dishes, add enough volume of 0.1% gelatin solution to cover the entire area of the dish bottom. For example, 3 or 5 mL of gelatin solution is used for a 60- or 100-mm dish, respectively. Incubate the dish for at least 15 min at room temperature. Before use in culture, aspirate excess gelatin solution.
16. X-VIVO 10 (Lonza, Basel, Switzerland).
17. Interleukin-6 (IL-6; Peprotech, London, UK).
18. Soluble interleukin-6 receptor (IL-6 R; Peprotech).
19. Stem cell factor (SCF; Peprotech).
20. Thrombopoietin (TPO; Peprotech).
21. Flt-3-ligand (R&D Systems, Minneapolis, MN, USA).
22. Hematopoietic culture (HC) medium (for $CD34^+$ cells). Supplement X-VIVO 10 with 50 ng/mL IL-6, 50 ng/mL IL-6 R, 50 ng/mL SCF, 10 ng/mL TPO, and 20 ng/mL Flt-3-ligand. Store at 4°C up to a week.
23. SNL76/7 (SIM strain embryonic fibroblasts; European Collection of Cell Culture, Salisbury, UK).
24. Mitomycin C solution, 1 mg/mL (Nacalai Tesque, Kyoto, Japan).
25. Cell Banker 1(JUJI FIELD INC, Tokyo, Japan).
26. The $CD34^+$ fraction of cord blood cells (Lonza or Riken BRC, Ibaraki, Japan).

27. Direct CD34 progenitor Cell Isolation Kit (Miltenyi Biotech GmbH, Bergisch Gladbach Germany).
28. RNeasy Mini kit (QIAGEN, Tokyo, Japan).
29. PrimeScript RT reagent kit (Takara, Shiga, Japan).
30. ExTaq (Takara).

2.3. Maintenance of ES Cell-Like Colonies or iPS Cell Colonies on Feeder Cells: Reagents

1. Human ES medium (see Sect. 2.2).
2. SNL medium (see Sect. 2.2).
3. Gelatin-coated culture dishes (see Sect. 2.2).
4. Trypsin, 2.5% (Invitrogen).
5. Collagenase IV (Invitrogen), 10 mg/mL. Reconstitute 100 mg of collagenase IV in 10 mL of distilled water, and then sterilize with a 0.22-μm pore filter. Aliquot and store at −20°C.
6. $CaCl_2$ (Sigma), 1 M. Dissolve 1.1 g of $CaCl_2$ in 10 mL of distilled water, and then sterilize with a 0.22-μm pore filter. Store at 4°C.
7. CTK solution, 100 mL. Mix 10 mL 10 mg/mL collagenase IV, 10 mL 2.5% trypsin, 0.1 mL of 1 M $CaCl_2$, 20 mL of KSR, and 60 mL of PBS. Store at −20°C. Avoid repeated freezing and thawing.
8. Y-27632 (Rock inhibitor: 253-00513; Wako), 10 mM. Reconstitute 5 mg Y-27632 in 1.48 mL of distilled water. Aliquot and store at −20°C.
9. 0.05% Trypsin/5.3 mM EDTA solution (see Sect. 2.2).
10. Acetamide (Wako). Dissolve 5.9 g acetamide in 10 mL of distilled water. Store at room temperature.
11. Propylene glycol (Wako).
12. Dimethyl sulfoxide (DMSO; D2650, Sigma).
13. DAP213 solution, 10 mL. Add 1.43 mL of DMSO, 1 mL of 10 M acetamide, and 2.2 mL of propylene glycol to 5.37 mL of human ES medium. Store at −80°C.
14. SNL76/7 (see Sect. 2.2).

2.4. Embryoid Body-Mediated Differentiation of Reprogrammed Cells: Reagents

1. Ultra-low attachment plates (Corning, Tokyo, Japan).
2. PA6 medium, 550 mL. Mix 493 mL GMEM, 55 mL KSR, 5.5 mL NEAA, 1 mL 2-ME, and 5.5 mL Pen/Strep. Store at 4°C.
3. Alkaline Phosphatase kit (Vector, Burlingame, CA, USA).
4. 4% Paraformaldehyde (PFA; Wako).
5. Anti-Oct3/4 antibody (1:100 sc-5279; Santa Cruz Biotechnology, Santa Cruz, CA, USA).
6. Anti-TRA-1-81 antibody (1:200 MAB4381; Millipore).

7. Anti-TRA-1-60 antibody (1:200 MAB4360; Millipore).
8. Anti-SSEA-4 antibody (1:200 MAB4304; Millipore).
9. Anti-Nanog antibody (1:1,000 RCAB0003P; ReproCELL, Kanagawa, Japan).
10. Anti-E-cadherin antibody (1:100 610181; BD Bioscience, CA, USA).
11. Anti-α-fetoprotein antibody (AFP; 1:100 MAB1368; R&D Systems).
12. Anti-vimentin antibody (1:100 sc-5565; Santa Cruz Biotechnology).
13. Anti-α-smooth muscle actin antibody (α-SMA; 1:400 A-2547; Sigma).
14. Anti-βIII-tubulin antibody (1:200 T4026; Sigma).
15. Anti-Glial fibrillary acidic protein antibody (GFAP; 1:50 sc-6170; Santa Cruz Biotechnology).
16. Tyrosine hydroxylase (1:100 AB152; Millipore).
17. Alexa Fluor 488 goat anti-mouse (1:1,000; Invitrogen).
18. Alexa Fluor 594 rabbit anti-mouse (1:1,000; Invitrogen).
19. Alexa Fluor 594 goat anti-rabbit (1:1,000; Invitrogen).
20. Nuclei are stained with DAPI (1:1,000; Sigma).
21. Inverted microscope (BX51, IX71; Olympus, Tokyo, Japan).
22. Microscope (CKX31, Olympus).

2.5. Preparation of Teratoma Formation Assay with Reprogrammed Cells

Required supplies (equivalent supplies can be used if necessary):

1. Sterile syringes (1 mL Tuberculin syringes; Terumo, Kyoto, Japan)
2. Sterile needles ("28 G × 1/2 1", "26 G × 1/2 1"; Terumo)
3. Sterile conical tubes (15 mL, 50 mL; Becton Dickinson)
4. Immunocompromised mice: FOX CHASE SCID® C.B-17/lcr-scid/scidJcl (CLEA Japan, Tokyo, Japan) (see Note 1)
5. HBSS buffer (Invitrogen)
6. CTK solution (see Sect. 2.3)
7. 4% PFA (see Sect. 2.4)
8. CO_2 lab grade gas (located in animal housing area)
9. Tissue-Tek O.C.T. Compound (14-373-65, Andwin Scientific, Woodland Hills, CA, USA)
10. Nembutal solution (Dainippon Sumitomo Pharma, Osaka, Japan)

3. Methods

3.1. Retroviral Transduction

Retroviral constructs and lentiviral constructs can be obtained from Addgene. Produce viruses as described previously (17). Plate 293T cells at 3.6×10^6 cells per 100-mm dish and incubate overnight. The next day, transfect cells with 5 μg each of pMX-c-Myc, -Klf4, -Oct3/4, -Sox2, -GFP, and 2.5 μg each of pCMV-VSV-G and pMDLg/p.PRE (retroviral construct), or 6 μg shp53 pLKO.1 puro and 3 μg each of pMD2.G-VSV-G and psPAX2 (lentiviral construct) using FuGene6. Forty-eight hours after transfection, collect the supernatant from transfected cells and filter with a 0.45-μm-pore size cellulose acetate filter.

Day 0, in the evening:

1. Seed 293T cells, 3.6–4.0×10^6/10 mL, in a 100-mm dish. Prepare six dishes (see Note 2).

Day 1, in the morning:

2. After 16–18 h, aspirate the medium and replace with 3 mL of fresh serum-free DMEM 30 min prior to infection.
3. Dispense 600 μL DMEM (without serum) to a 1.5-mL tube (prepare six tubes for six different constructs).
4. Add 15 μL FuGene6 to the tube.
5. Tap the tube gently to mix (do not Vortex).
6. Add 5 μg of pMX-retroviral vector that integrates one reprogramming factor, 2.5 μg of pCMV-VSV-G, and 2.5 μg of pMDLg/p.PRE to the DMEM-FuGene6 mixture.
7. Add 6 μg of lentiviral construct shp53 pLKO.1 puro, 3 μg of pMD2.G-VSV-G, and 3 μg of psPAX2 to the DMEM-FuGene6 mixture.
8. Tap the tube gently and keep the tube for 30 min at room temperature.
9. Add the DNA-FuGene6 conjugate to the cells.
10. Swirl the plate gently, and then incubate the cells at 37°C, 5% CO_2.
11. After 2 h, add 3 mL of DMEM with 20% FBS (final 10% FBS).
12. Incubate the cells for 24 h at 37°C, 5% CO_2.

Day 2, in the morning:

13. After 24 h, aspirate the medium and add 6 mL of fresh DMEM with 10% FBS.
14. Then, incubate the cells for 24 h at 37°C, 5% CO_2 (see Fig. 1a left panel: phase contrast, right panel: fluorescent microscopic observation) (see Note 3).

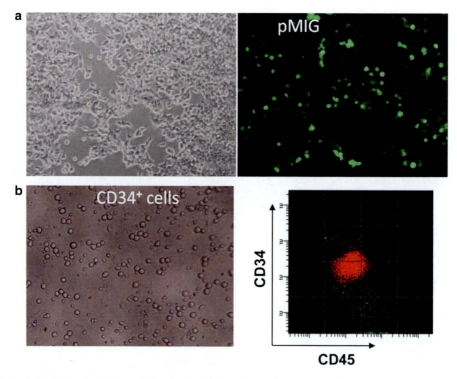

Fig. 1. Retroviral infection to CD34+ cord blood cells. (**a**) Retrovirus infection to 293T cells to generate viral sup. Phase contrast (*left panel*) and fluorescent image (*right panel*) of GFP-integrated pMX (pMIG) retrovirus construct infected 293T cells. (**b**) CD34+ fraction of cord blood cells used for reprogramming. Phase-contrast image (*left*) and flow cytometric analysis of CD34+ purified cord blood cells (*right panel*).

Day 3, in the morning:

15. Harvest the first virus-containing medium (viral supernatant) 48 h after infection and filter through a 0.45-μm filter (use this fresh supernatant for the gene transduction described in Sect. 3.2.2).
16. Place 6 mL of DMEM with 10% FBS per well.

Day 4, in the morning:

17. Harvest the second viral-containing medium (viral supernatant) 72 h after infection and filter through a 0.45-μm filter (use this fresh supernatant for gene transduction stated in Sect. 3.2.2).

3.2. Reprogramming CD34+ Cord Blood Cells

Human CD34+ cord blood cells can be obtained from Lonza or Riken BRC Japan. CD34+ cells can also be purified from freshly collected, whole cord blood in two steps. First, harvest mononuclear cells by the Ficoll gradient separation method (20) followed by sorting CD34+ cells with Direct CD34 progenitor Cell Isolation Kit (autoMACS separator or MACS® Technology). Alternatively, the HES method (21) can be substituted for gradient fractionation. Purity of the CD34+ cells should be determined by flow cytometry

(FACS Calibur or equivalent) (Fig. 1b left panel: phase contrast; right panel: purity of CD34$^+$ cells by flow cytometric analysis).

3.2.1. Preparation of Mitomycin C-Treated SNL (MMC-SNL) Feeder

1. Remove a frozen vial of SNL cells from the liquid nitrogen tank and put the vial into a 37°C water bath until most (but not all) of the liquid has thawed.
2. Transfer the cell suspension to a 15-mL tube and add 5 mL SNL medium to mix the cell suspension well.
3. Centrifuge at 170×g for 5 min at 4°C, and then aspirate the supernatant. Resuspend the cell pellet in 10 mL fresh SNL medium.
4. Transfer the cell suspension to a 100-mm dish (7.5×10^5 cells per dish) and incubate the cells in a 37°C, 5% CO_2 incubator for several days until the cells reach 80–90% confluency (see Note 4).
5. When the SNL cells reach 80–90% confluency in the 100-mm dish, aspirate the medium and wash once with PBS. After aspiration of PBS, add 2 mL of 0.05% trypsin/0.53 mM EDTA, and incubate the dish at 37°C for 5 min. Neutralize the trypsin/EDTA solution by the addition of 10 mL fresh SNL medium, and transfer the cell suspension to a 50-mL tube. Count the harvested cells and seed them in 10–15 dishes (1×10^6 cells per 100-mm dish) (see Note 5).
6. When the cells become 80–90% confluent after passage, add 100 µL of mitomycin C solution (1 mg/mL) to each 100-mm dish and incubate for 3 h at 37°C, 5% CO_2.
7. Aspirate the medium and wash the adherent cells three times with PBS. Then, add 8 mL of fresh SNL medium and incubate MMC-SNL cells for 24 h at 37°C, 5% CO_2.
8. Aspirate the medium and wash the cells with 10 mL of PBS. Remove the PBS thoroughly, and add 1 mL of 0.05% trypsin/0.53 mM EDTA solution and incubate at 37°C for 5 min. Add 10 mL of SNL medium to neutralize the trypsinization, and then transfer the cell suspension to a 15-mL conical tube.
9. Centrifuge at 170×g for 5 min at 4°C, and then discard the supernatant.
10. Aspirate the medium and resuspend the cell pellet in 500 µL of Cell Banker 1 per stock vial (approx. 5.0×10^6 cells per vial). Place the tubes in a −80°C freezer for 1 day, and then transfer to liquid nitrogen tank for frozen stock.
11. Take out a vial of frozen MMC-SNL cells from the liquid nitrogen tank and place the vial into a 37°C water bath until most (but not all) cells are thawed (see Note 6).

3.2.2. Generation of hES Cell-Like Colonies from CD34+ Cord Blood Cells

1. When using frozen CD34+ cells, culture them (4×10^4 cells) for 1 day in four wells of a 24-well plate with 500 µL of HC medium per well prior to virus infection. Skip this step when using freshly isolated cells. Suspend all the CD34+ cells in 18 mL of viral supernatant (3 mL from each of the six different constructs: Oct3/4, Sox2, Klf4, c-Myc, shTP53, and GFP) in 50-mL conical tubes in the presence of 4 µg/mL polybrene and transduce the viral constructs by spinoculation by centrifuging the tube at $1,750 \times g$ (3,000 rpm) for 3.5 h at 35°C (22).
2. Recover cells after spinoculation, and then culture in two wells of a 6-well plate with 2 mL of HC medium per well for 24 h at 37°C, 5% CO_2.
3. Second spinoculation (repeat step 1 followed by step 2).
4. Collect infected cells and grow them in two wells of a 6-well plate with 2 mL of HC medium per well for 5 days. Then, harvest all the cells, and resuspend in 6 mL of human ES medium and dispense in two 60-mm dishes (approximately 2×10^4 infected cord blood cells per 60-mm dish) pre-coated with MMC-SNL (culture the cells in 3 mL of human ES medium per dish) (see Note 7).
5. ES cell-like colonies emerge between 3 and 4 weeks after viral infection, and colonies are picked for colony selection (Fig. 2a, b).
6. Pick up ES cell-like colonies mechanically under the microscope using a Pipetman set at 180 µL. Transfer the cells into a 24-well plate pre-seeded with MMC-SNL to check for their ability to undergo serial passage. Pick up 10–15 colonies for further characterization of emergent colonies (Fig. 2b, c) (see Note 8).
7. Incubate the cells until the size of the colonies reaches around 500 µm in diameter (P1) (see Note 9).
8. Five–seven days after culturing on the 24-well plate, transfer the emergent colonies to 6-well plates pre-seeded with MMC-SNL (P2).
9. Select ES cell-like colonies based upon their morphology (round rim, uniform small cell size colony) (Fig. 3a, see Method 3.3) and seed in duplicate or triplicate for ALP staining (Fig. 2d and 10.3c, see Method 3.2.3), for immunohistochemical staining (Method 3.2.4), for RNA extraction (see Method 3.2.5), and for teratoma formation assay (see Method 3.5).

3.2.3. ALP Staining

1. Fix cells with 4% PFA in permeabilizing solution (0.2% Triton X-100 in PBS) for 20 min at 4°C.
2. Rinse the cells with PBS and treat with blocking buffer (PBS containing 0.1% BSA) for 15 min at room temperature.

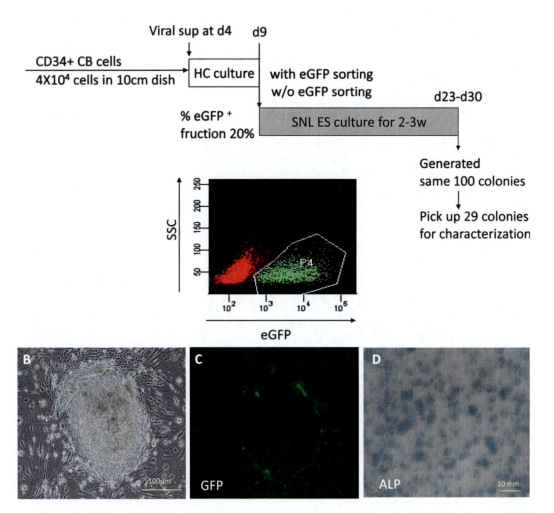

Fig. 2. Generation of human ES cell-like colonies. (a) CD34+ cord blood cells (4 × 10⁴) were cultured for 3 days in hematopoietic culture (HC) media prior to infection with retroviruses containing the Klf4, c-Myc, Oct3/4, Sox2, and eGFP genes and a lentivirus containing an shTP53 RNA construct, and cultured with hematopoietic cell culture for another 5 days. Infected cells (2 × 10⁴) were seeded on SNL feeder cells with or without GFP sorting (20% positive for eGFP) in human ES culture conditions. ES cell-like colonies were counted after 2 weeks and picked up in the following weeks. Phase contrast (b), GFP expression (c), and ALP staining (d) images of a typical colony. Note that GFP expression was limited to the rim of the colony.

3. Assess the alkaline phosphatase activity of cultured cells with a Vector Blue® Alkaline Phosphatase Substrate kit according to the manufacturer's instructions.

4. Examine the stained cells using an inverted phase-contrast microscope.

3.2.4. Immunostaining

1. Fix cells with 4% PFA solution overnight at 4°C.

2. Rinse the cells with PBS, followed by treatment with blocking buffer for 15 min at room temperature.

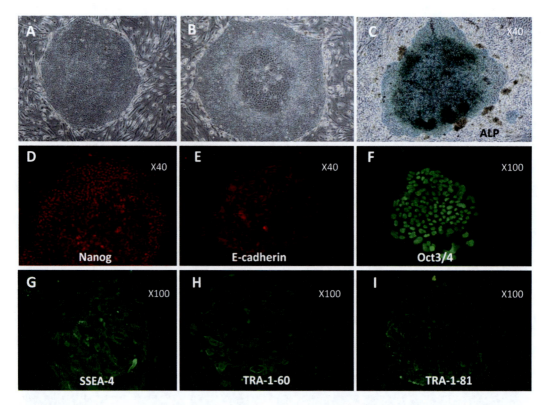

Fig. 3. Characterization of ES cell-like colonies. Those in good condition (**a**) contain small homogenous cells, while those in bad condition are overgrown with signs of differentiation in the center of the colony (**b**). ALP staining (**c**) and immunohistochemical staining of ES cell-like (good) colonies stained by antibodies against Nanog(**d**), E-cadherin(**e**), Oct3/4 (**f**), SSEA-4(**g**), TRA-1–60(**h**), and TRA-1–81(**i**) are shown.

3. Immunostain the ES cell-like colonies, using a set of antibodies listed in Materials Sect. 2.4, by following the manufacturer's instructions.

3.2.5. RT-PCR

1. Extract total RNA using the RNeasy Mini kit, according to the manufacturer's instructions.
2. Determine the concentration and purity of total RNA samples using a NanoDrop ND-1000 spectrophotometer.
3. Use one microgram of total RNA for reverse transcription reactions with PrimeScript RT reagent kit. Polymerase chain reaction (PCR) is performed with ExTaq.
4. Primer sequences were described previously (online only, available at www.exphem.org, *Experimental Hematology*).

3.3. Maintenance of ES Cell-Like Colonies or iPS Cell Colonies Cultured on Feeder Cells

3.3.1. Verification of "Stemness" of Reprogrammed Cells

1. Culture ES cell-like colonies in 6-well plates until colonies reach 70–80% confluency (Fig. 2d). Aspirate the medium, and wash the cells (iPS cell colony and SNL feeder cells) with 2 mL of PBS per well.
2. Remove PBS completely, add 0.5 mL per well of CTK solution, and incubate at 37°C for 3–5 min (see Note 10).
3. Aspirate CTK solution, and wash the well twice with 2 mL of PBS.
4. Remove PBS completely and add 3 mL of human ES medium.
5. Detach ES cell-like colonies with a cell scraper and break up the colonies to small clumps by pipetting up and down several times (use Pipetman with a volume set to 1,000 µL) (see Note 11).
6. Add 10 mL of human ES medium and resuspend the cell clumps. Centrifuge at $170 \times g$ (1,000 rpm) for 5 min, and discard the supernatant.
7. Resuspend the cell clump in 10 mL of human ES medium and transfer the cell clump suspension onto a 100-mm dish pre-seeded with MMC-SNL cells (Sect. 3.2.1).
8. Incubate in a 37°C, 5% CO_2 incubator until cells reach 70–80% confluency (Fig. 2d) (see Note 12).
9. ES cell-like colonies are classified by morphology. Those in good condition (Fig. 3a) contain small homogenous cells, while those in bad condition are overgrown with signs of differentiation (large cells gathered in the center; Fig. 3b). Pick up only those colonies in good condition and discard the remainder. Results of ALP staining (Fig. 3c) and immunohistochemical staining of ES cell-like colonies (Fig. 3d–i) are shown.

3.3.2. Preparation of Frozen Stocks

1. When iPS cells reach 70–80% confluency in 100-mm dishes, frozen stocks can be prepared (Fig. 2d for cell colony density).
2. Add 10 µL of 10 mM Y-27632, a specific inhibitor of p160-Rho-associated coiled-coil kinase (Rock inhibitor), to 10 mL of human ES medium (final concentration, 10 µM), and incubate at 37°C for 1 h.
3. Follow steps 2–6 in Sect. 3.3.1.
4. Suspend the cells in 200 µL of DAP213 solution by pipetting gently several times (see Note 13).
5. Transfer 200 µL of the cell suspension to a freezing stock vial (see Note 14).
6. Place the vials immediately into liquid nitrogen (see Note 15).

3.4. Embryoid Body (EB)-Mediated Differentiation of Reprogrammed Cells

Reprogrammed cells from an ES cell-like colony should demonstrate the capacity to differentiate to all three germ layers by responding to various differentiation conditions. This ability is particularly important to confirm that the reprogrammed cells are

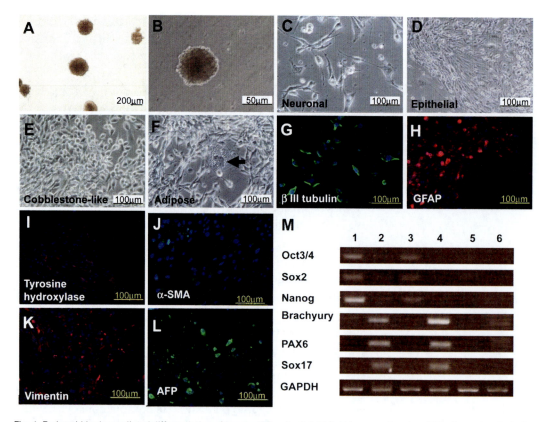

Fig. 4. Embryoid body-mediated differentiation of human iPS cells. Bright field images of embryoid bodies generated after 8 days of culture (**a, b**). Embryoid bodies were transferred to gelatin dishes or to PA-6 co-culture conditions and differentiated for a further 8 days to induce un-directed or guided differentiation. Phase-contrast images of neuron-like (**c**), epithelial (**d**), cobblestone-like (**e**), and adipose-like cells (**f**) after differentiation on gelatin. Differentiated cells were stained with antibodies against βIII-tubulin (**g**), GFAP (**h**), tyrosine hydroxylase (**i**), α-SMA (**j**), vimentin (**k**), and AFP (**l**) to identify specific cell lineages. (**m**) Expression of genes associated with pluripotency and differentiated progeny determined by RT-PCR. Lane1: un-differentiated iPS clone #22, Lane2: differentiated #22, Lane3: un-differentiated iPS clone #23, Lane4: differentiated #23, Lane5: mononuclear cells, Lane6: CD34+ cord blood cells.

"induced pluripotent stem (iPS) cells". The following assay is performed to determine the pluripotency of reprogrammed cells in vitro:

1. Harvest ES cell-like colonies using CTK solution and transfer them (1.0×10^6 cells) to 6-well, ultra-low attachment plates.
2. Grow the cell clumps in human ES medium in the absence of bFGF for 6–8 days to form ball-like cell clumps (i.e., EB) with a size of 200–250 µm.
3. Transfer EBs to gelatin-coated plates (6-well plates) and culture in the same medium for another 8 days (Fig. 4c–f).

4. Confirm the differentiation of reprogrammed cells into ectodermal, mesodermal, or endodermal tissues by immunochemically detecting βIII-tubulin and GFAP, or α-SMA and vimentin, or AFP, respectively (Fig. 4g, h, j–l). Suppression of pluripotency-related genes and detection of gene expression for each of the three germ layers in the course of differentiation can be determined by RT-PCR (Fig. 4m). For dopaminergic (ectodermal) differentiation of reprogrammed cells, small clumps of EBs can be plated on feeder cells in PA6 medium as reported previously (Fig. 4i) (17).

3.5. Teratoma Formation Assay with SCID Mice

Reprogrammed cells should demonstrate differentiation potential reflecting all three germ layers in vivo. To this end, one million reprogrammed cells are injected into the testis capsule of SCID mice to determine their ability to form teratomas containing tissues of all three germ layers.

3.5.1. Preparation of Reprogrammed Cells for Grafting into Testis of SCID Mice

1. Follow steps 2–6 in Sect. 3.3.1.
2. Resuspend approximately 2.5×10^6 cells in 50 μL of HBSS buffer, transfer it to a 1.5-mL Eppendorf tube, and place it on ice (see Note 16).

3.5.2. SCID-Mouse Operation

1. Mix 100 μL of 10× Nembutal solution with 900 μL of distilled PBS and expel any air bubbles from the syringe. Inject 200 μL of Nembutal solution using a 26-G needle into the abdominal cavity of an SCID mouse (Fig. 5a). Turn the mouse on its back, and fix the four limbs with tape and cover with gauze (Fig. 5b).
2. Pick up the skin of the abdomen using tweezers and use scissors (Fig. 5c) to cut approximately 1 cm of the peritoneum.
3. Expose both testes using tweezers (Fig. 5d).
4. Inject cell suspension into the testes using a 1-mL syringe (20 μL/testis) with a 28-G needle (Fig. 5e).
5. Return the testis and sew the mouse abdomen back together (Fig. 5f–h).
6. Observe mice on a weekly basis. Teratoma may form as soon as 6 weeks after injection, but typically, after 8–10 weeks (Fig. 5i).

3.5.3. Observations and Tumor Removal

1. Remove the teratoma using sterile surgical scissors and forceps (Fig. 5j).
2. Fix 75% of the harvested tissues by placing it in 4% PFA. These sections will be used for paraffin embedding and haematoxylin/eosin staining.
3. Determine whether reprogrammed cells are able to form all three germ layers (Fig. 5k–o) (see Note 17).

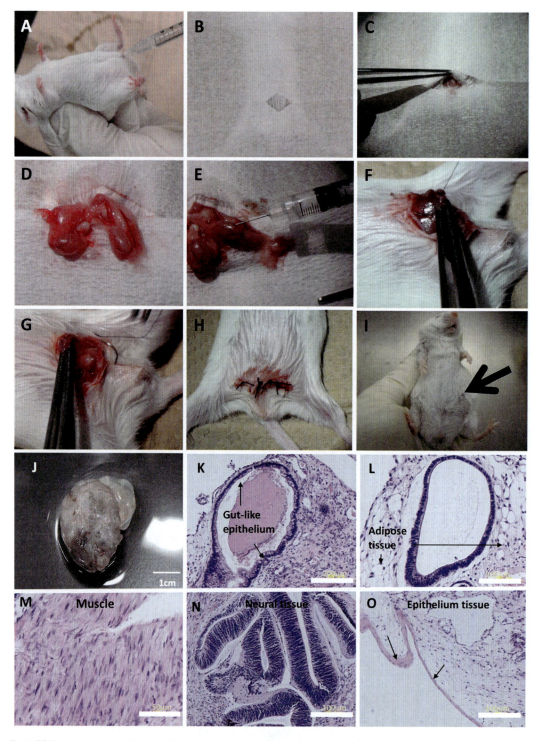

Fig. 5. SCID-mouse operation (teratoma formation assay) and histology of teratoma. Inject Nembutal (**a**), cut skin with scissors (**b**), pull the skin with tweezers (**c**), expose both testes using tweezers (**d**), inject cell suspension into the testes (**e**), return the testis and sew the mouse abdomen back together (**f,g,h**). Teratoma formation in 8–10 weeks (**i**). Teratoma removed (**j**). Hematoxylin and eosin staining of a teratoma tissue that includes a gut-like epithelium ((**k**) endoderm), adipose-like tissue ((**l**) mesoderm), muscle-like tissue ((**m**) mesoderm), immature neural tissue ((**n**) ectoderm), as well as skin-like epithelial cells ((**o**) ectoderm).

4. Notes

1. Mice must be male, SCID mice, white, and 4–6 weeks of age upon arrival. Acclimate the mice for at least 1 week prior to the procedure. Perform all cell preparations using aseptic techniques. Perform all animal work in a laminar flow hood.
2. Infect six viral constructs in six different 100-mm dishes (one viral construct for each one of the 100-mm dishes). Seed 293T cells 15 h prior to infection to ensure 293T cells are in exponential growth phase.
3. GFP-positive cell percentage should be >70%.
4. Change the medium every other day.
5. If the cell number is less than 5×10^5, seed the cells on 60-mm dishes.
6. Transfer the cell suspension to a 15-mL tube, and then add 10 mL of SNL medium and centrifuge at $170 \times g$ for 5 min at 4°C. Discard the supernatant and resuspend the cell pellet in 10 mL of SNL medium followed by a cell count. Seed cells at approximately 7.5×10^5 cells per 100-mm dish (usually seed in three or four dishes from one frozen vial) and add SNL medium up to 10 mL per 100-mm dish to continue culture. Use the MMC-SNL cells (start co-culturing with reprogrammed cell or iPS cells) within 1–4 days after seeding in dishes or plates.
7. Usually, two 60-mm dishes are used to culture transfected cord blood cells. Change the medium every other day (Fig. 2a).
8. Prepare a 96-well plate with 180 µL of human ES medium per well without MMC-SNL prior to picking the colonies. Place a selected colony in a 96-well plate (one cell clump from one colony per well) and verify that the clump was transferred to the well. Then, pipet the cell clump gently to a size of 50–100 µm in diameter, but not to a single cell suspension. Then, transfer the cell clumps to a 24-well plate pre-seeded with MMC-SNL and make up the volume with human ES medium to 500 µL per well.
9. Change the medium every day.
10. Make sure that feeder cells are dissociated, while ES cell-like colonies remain attached on the plate. Observe the detachment frequently and remove CTK solution as soon as possible to maintain good viability of the harvested cells.
11. Pipet colonies gently until clumps reach 50–100 µm in diameter. Do not form a single cell suspension.
12. Change the medium every day until the next passage step.
13. Pipet the colonies gently to achieve 50–100 mm cell clumps. Do not form a single cell suspension.

14. Transfer approximately $1.0–5.0 \times 10^4$ cells to one stock freezing vial.
15. Place the vials into liquid nitrogen within 30–60 s. Steps 4–6 have to be performed within 30 s for good viability after thawing. For long-term storage, it is recommended to store the frozen vials in the gas phase of liquid nitrogen or in a −150°C deep freezer.
16. For good cell transplantation, always keep the cells on ice and follow the aforementioned steps as quickly as possible.
17. Figure 5k–o illustrates hematoxylin and eosin staining of a teratoma derived from reprogrammed cells implanted in the testis of an SCID mouse. The teratoma includes a gut-like epithelium ((K) endoderm), adipose-like tissue ((L) mesoderm), muscle-like tissue ((M) mesoderm), immature neural tissue ((N) ectoderm), as well as skin-like epithelial cells ((O) ectoderm).

References

1. Takahashi K, Tanabe K, Ohnuki M, Narita M, Ichisaka T, Tomoda K, Yamanaka S, Induction of Pluripotent Stem cells form Adult human fibroblasts by defined factors, Cell, 2007, 131, 861–872.
2. Stadtfeld M, Nagaya M, Utikal J, Weir G, Hochedlinger K. Induced pluripotent stem cells generated without viral integration. Science. 2008, 322, 945–49.
3. Fusaki N, Ban H, Nishiyama A, Saeki K, Hasegawa M, Efficient induction of transgene-free human pluripotent stem cells using a vector based on Sendai virus, an RNA virus that does not integrate into the host genome., Proc Jpn Acad Ser B Phys Biol Sci. 2009, 85, 348–62.
4. Okita K, Nakagawa M, Hyenjong H, Ichisaka T, Yamanaka S. Generation of mouse induced pluripotent stem cells without viral vectors. Science. 2008, 322, 949–953.
5. Woltjen K, Michael IP, Mohseni P, et al. piggyBac transposition reprograms fibroblasts to induced pluripotent stem cells. Nature. 2009, 458, 766–770.
6. Kaji K, Norrby K, Paca A, Mileikovsky M, Mohseni P, Woltjen K. Virus-free induction of pluripotency and subsequent excision of reprogramming factors. Nature. 2009, 458, 771–775.
7. Soldner F, Hockemeyer D, Beard C, et al. Parkinson's disease patient-derived induced pluripotent stem cells free of viral reprogramming factors. Cell. 2009, 136, 964–977.
8. Yu J, Hu K, Smuga-Otto K, et al. Human induced pluripotent stem cells free of vector and transgene sequences. Science. 2009, 324, 797–801.
9. Zhou H, Wu S, Joo JY, et al. Generation of induced pluripotent stem cells using recombinant proteins. Cell Stem Cell. 2009, 4, 381–84.
10. Kim D, Kim CH, Moon JI, et al. Generation of human induced pluripotent stem cells by direct delivery of reprogramming proteins. Cell Stem Cell. 2009, 4, 472–76.
11. Li W, Wei W, Zhu S, et al. Generation of rat and human induced pluripotent stem cells by combining genetic reprogramming and chemical inhibitors. Cell Stem Cell. 2009, 4, 16–9.
12. Huangfu D, Maehr R, Guo W, et al. Induction of pluripotent stem cells by defined factors is greatly improved by small-molecule compounds. Nat Biotechnol. 2008, 26, 795–7.
13. Aasen T, Raya A, Barrero MJ, et al. Efficient and rapid generation of induced pluripotent stem cells from human keratinocytes. Nat Biotechnol. 2008, 26,1276–84.
14. Liu H, Ye Z, Kim Y, Sharkis S, Jang YY., Generation of endoderm-derived human induced pluripotent stem cells from primary hepatocytes. Hepatology. 2010, 5, 1810–9.
15. Loh YH, Agarwal S, Park IH, et al. Generation of induced pluripotent stem cells from human blood. Blood. 2009, 113, 5476–9.
16. Seki T, Yuasa S, Oda M, et al. Egashira T, Generation of induced pluripotent stem cells from human terminally differentiated circulating T cells. Cell Stem Cell. 2010, 2, 11–4.

17. Takenaka C, Nishishita N, Takada N, Jakt LM, Kawamata S., Effective generation of iPS cells from CD34+ cord blood cells by inhibition of P53. Exp Hematol. 2010, 38, 154–62.
18. Alessandra Giorgetti, Nuria Montserrat, Trond Aasen *et al.*, Generation of Induced Pluripotent Stem Cells from Human Cord Blood Using OCT4 and SOX2., Cell Stem Cell, 2009, 5, 353–357.
19. Nakatsuji N, Nakajima F, Tokunaga K. HLA-haplotype banking and PS cells. Nat Biotechnol. 2008, 26, 739–740.
20. Ting A, Morris P.J, A technique for lymphocyte preparation from stored heparinized blood. Vox Sang. 1971, 20, 561.
21. Bertolini F, Battaglia M, Zibera C et al, A new method for placental/cord blood processing in the collection bag. I. Analysis of factors involved in red blood cell removal, Bone Marrow Transplant. 1996, 18, 783–6.
22. Kawamata S, Du C, Li K, Lavau C. Overexpression of the Notch target genes Hes *in vivo* induces lymphoid and myeloid alterations. Oncogene. 2002, 21, 3855–63.

Chapter 11

Generation of iPS Cells from Human Umbilical Vein Endothelial Cells by Lentiviral Transduction and Their Differentiation to Neuronal Lineage

Maria V. Shutova, Ilya V. Chestkov, Alexandra N. Bogomazova, Maria A. Lagarkova, and Sergey L. Kiselev

Abstract

Substantial progress has been made in somatic cell reprogramming through ectopic expression of four transcription factors to yield induced pluripotent stem (iPS) cells. We have used the robust viral-based modification procedure to generate iPS cells from human umbilical vein endothelial cells (HUVEC), an attractive source of the cells for reprogramming. Our method uses a multistep protocol in which reprogramming cells are selected by culturing in defined conditions on Matrigel, which may facilitate potential clinical applications. HUVEC-derived iPS cells show pluripotency in vivo and can differentiate into many cell types in vitro, including neuronal lineages. Here we describe an efficient protocol for generating iPS cells from HUVEC and differentiating these iPS cells into neurons.

Key words: Induced pluripotent stem cells, Endothelial cells, Differentiation, Neural cells

1. Introduction

Four key factors Oct4 (*Pou5f1*), Sox2, c-Myc, and Klf4 have been shown to reestablish the pluripotent state in adult fibroblasts (1). Later this technique has been modified and improved (2–4). Several reports have demonstrated that some of these four factors can be replaced or omitted without affecting the reprogramming efficiency (5–8). In the original report, the induced pluripotent stem (iPS) cells were generated using viral methods of gene delivery with the subsequent integration of the transgenes into the cell genome. Subsequently, non-integrating approaches exploiting adenoviruses, plasmids, or protein transduction protocols have been developed (9, 10). However, the viral-based reprogramming procedure seems to be the most effective and reliable.

Since the first report of induced pluripotency in mouse embryonic fibroblasts was published, a number of cell types have been shown to be amenable to reprogramming (11, 12). The pluripotency state has also been induced in human fibroblasts, keratinocytes, neural cells, blood cells, and other cell types (11–16). All these studies have shown a strong influence of cell type selected on the reprogramming efficiency and timing (2, 13). Additionally, the choice of cell type is important for any possible therapeutic application of reprogrammed cells. Cell accessibility and availability as well as cell age might be the critical factors in determining the optimal cell type for a therapeutic use. Although adult skin cells are readily accessible and easy to manipulate, they are exposed to UV and other environmental factors and may therefore accumulate mutations limiting their therapeutic application. Additional passaging in vitro or growth factor application required to expand the starting amount of cells prior to iPS generation could further contribute to epigenetic and genetic alterations in parental cells (17). Therefore, the choice of cell type for reprogramming is depending on the ability of the cell type to undergo complete reprogramming without accumulating multiple DNA abnormalities. Tissues of placenta complex contain significant amount of cells that have never been exposed to the environmental factors; therefore, they can be used as an attractive source of cells for reprogramming. The growing popularity of cord blood and cord tissue banking offers a unique possibility to use these tissues for cell reprogramming.

In this chapter, we describe a procedure for establishing induced pluripotent stem cells from HUVECs. HUVECs are very well-characterized cells that are easy to handle and cultivate in vitro (18). A significant number of cells (up to 10^6) can be isolated from human umbilical vein and passaged if needed. HUVECs can be preserved by freezing until required without loosing their proliferation capacity. We found that iPS cells could be established from HUVECs and subjected to reprogramming relatively fast, limiting acquisition of possible DNA rearrangements during in vitro passaging to a minimum (19). Successful cultivation of HUVEC-generated iPS cells on MEF feeders and especially in defined conditions (mTeSR1) makes them suitable for experimental studies and therapeutic applications. Moreover, we found that the success of passaging cells in defined conditions depends on how completely reprogramming has occurred, thus offering additional selection pressure for obtaining genuine iPS cells. The iPS cells generated from endothelial cells can be efficiently differentiated into numerous specialized cell types including neurons. Here we offer protocols for the generation of iPS cells from HUVECs and for the differentiation of produced cells into neural lineage.

2. Materials

2.1. Cell Culture

1. Dulbecco's Modified Eagle's Medium (DMEM), high glucose (Hyclone)
2. Dulbecco's Modified Eagle's Medium: nutrient mixture F-12 (DMEM/F12) (Hyclone)
3. Knockout™ DMEM (Invitrogen)
4. Fetal Bovine Serum (FBS, Hyclone)
5. Knockout Serum Replacement (KO SR, Invitrogen)
6. L-Glutamine (200 mM) (Invitrogen)
7. MEM nonessential amino acids (NEAA) (10 mM) (Invitrogen)
8. β-Mercaptoethanol (Sigma)
9. Penicillin–Streptomycin (100×) (Invitrogen)
10. 0.25% Trypsin–EDTA (Hyclone)
11. Mitomycin C (Sigma)
12. Collagenase Type IV (Sigma): used at 1 mg/mL in DMEM
13. Dispase (Invitrogen): used at 1 mg/mL dissolved in DMEM
14. Gelatin (Sigma)
15. Matrigel™ matrix (BD)
16. mTeSR™1 (StemCell Technologies)
17. 1× Phosphate-buffered saline (PBS) (Hyclone)
18. 35-, 60-, and 100-mm tissue culture dish (Corning)
19. 35-mm ultralow adhesion Petri dish (Corning)
20. 15- and 50-mL polystyrene tubes (Greiner Bio-One)
21. Polybrene (Sigma)
22. B27 supplement 50× (Invitrogen)
23. N2 supplement 100× (Invitrogen)
24. Recombinant human basic fibroblast growth factor (bFGF) (PeproTech)
25. Recombinant human vascular endothelial growth factor (VEGF165) (PeproTech)
26. Recombinant human brain-derived neurotrophic factor (BDNF) (PeproTech)
27. Recombinant human glia-derived neurotrophic factor (GDNF) (PeproTech)
28. Recombinant human epidermal growth factor (EGF) (PeproTech)
29. Ascorbic acid (Sigma)
30. Recombinant human Noggin (PeproTech)
31. SB431542 (Stemgent)

	32. Heparin (Sigma)
	33. Recombinant human sonic hedgehog (SHH) (PeproTech)
2.2. Lentivirus Production (see Note 1)	1. Phoenix cells (Nolan Lab) 2. TurboFect™ transfection reagent (Fermentas) 3. Reprogramming retroviral vectors: pMX-Oct4, pMX-Sox2, pMX-Klf4, and pMX-c-Myc (from Addgene) 4. Packaging plasmids: gag/pol and pCMV-VSV-G (all from Addgene) 5. 0.45 μm PES syringe filters (Corning)
2.3. Metaphase Chromosome Preparation	1. Demecolcine solution 10 μg/mL (Sigma) 2. KCl 0.075 M 3. Fixative 1: methanol:glacial acetic acid (6:1) 4. Fixative 2: methanol:glacial acetic acid (3:1) 5. Slides (Thermo Scientific)

3. Methods

3.1. Culture Media and Cell Passaging

3.1.1. HUVEC Medium and Culture Conditions

1. Maintain HUVECs in DMEM/F12 with 20% fetal bovine serum, 5 ng/mL bFGF, 20 ng/mL VEGF165, 10 ng/mL EGF, 1% nonessential amino acids, 2 mM L-glutamine, and 50 units/mL penicillin, 50 g/mL streptomycin ("HUVEC medium") in 0.1% gelatin-coated Petri dishes. Media containing bFGF, VEGF165, and EGF can be stored no more than 3 days at 4°C (see Note 2).

2. Pass cells with 0.25% trypsin using tissue culture plates pre-coated with 0.1% gelatin. Exchange medium every other day. One day before virus transduction, VEGF and EGF should be omitted from the medium.

3.1.2. Medium for Phoenix Cells

For virus production, use "Phoenix medium" consisting of DMEM (high glucose) supplemented with 5% heat-inactivated (56°C) fetal bovine serum, 2 mM L-glutamine, 50 units/mL penicillin, 50 g/mL streptomycin. Exchange medium every other day. Aliquot 50 mL and store at 4°C.

3.1.3. Medium for iPS Cells (iPSC Medium)

Propagate iPS cell lines on the mitomycin C–treated (10 μg/mL) mouse embryonic fibroblasts (MEF) in tissue culture plates pre-coated with 0.1% gelatin in the standard hES cells media, consisting of 80% KO DMEM, 20% KO SR, 1 mM glutamine, 1% nonessential amino acids, 50 units/mL penicillin, 50 g/mL streptomycin, 0.1 mM beta-mercaptoethanol, and 10 ng/mL of bFGF (see Note 3). Aliquot to 50 mL tubes, add bFGF, store at 4°C, and use within 1 week.

Exchange medium every day. Mechanical propagation of iPS cells can be used in early passaging, whereas collagenase IV (1 mg/mL) (for maintenance on MEFs) or dispase (1 mg/mL) treatment (for maintenance on Matrigel) can be employed in later passaging.

3.1.4. Neural Differentiation Medium

DMEM/F12 supplemented with N2 and B27 supplements, 100 ng/mL recombinant human Noggin, and 5 nM SB431542. Store no more than 2 days at 4°C.

3.1.5. Neurospheres Expansion Medium

DMEM/F12 supplemented with N2 supplement, 20 ng/mL bFGF, 10 ng/mL EGF, 50 ng/mL SHH, and 2 µg/mL heparin. Store no more than 2 days at 4°C.

3.1.6. Neurospheres Differentiation Medium

DMEM/F12 supplemented with N2 supplement, 20 ng/mL BDNF, 20 ng/mL GDNF, and 150 µM ascorbic acid. Store no more than 2 days at 4°C.

3.1.7. Medium for hES Cells

Cultivate hESC lines in mTeSR1 medium in Petri dishes coated with Matrigel. Pass hESCs every 5–7 days by exposure to 1 mg/mL dispase for 5–10 min at 37°C. Store according to the manufacturer recommendations.

3.2. Lentiviral Production

Production of lentiviruses at a high titer is a key step for successful reprogramming of human somatic cells. pMXs-based retroviral vectors encoding the cDNA for four Yamanaka's factors: Oct4, Sox2, c-Myc, and Klf4 can be used. The standard protocol includes transfection of packaging cells with viral vectors, virus concentration, storage, and use at appropriate concentrations. It is strongly recommended to use freshly made viruses for transduction, but in some cases, ultracentrifugation may be used for lentiviruses concentration. Detailed protocol for lentivirus production with the use of Phoenix packaging cells is written below.

3.2.1. Production of Lentiviruses

1. Day 0: Plate 4×10^6 Phoenix cells per 100-mm 0.1% gelatin-coated Petri dish in Phoenix medium, so cells would be 70–80% confluent at the time of transfection. It is important to use heat-inactivated fetal bovine serum, because viral particles would be rapidly destroyed in the medium containing non-inactivated FBS. Incubate cells at 37°C with 5% CO_2.

2. Day 1: On the next day, pMX-based retroviral vectors can be introduced into Phoenix cells using TurboFect™ transfection reagent. The transfection cocktail should be prepared for each lentiviral vector (individual 100-mm dish) as following: in a sterile polypropylene tube mix 8.9 µg of gag/pol, 1.5 µg VSV-G, and 5.9 µg of lentiviral expression plasmid into 500 µL DMEM. In a separate tube, dilute 22 µL of TurboFect into 500 µL DMEM. Add diluted TurboFect reagent dropwise to the DNA solution and gently mix by pipetting. Incubate the mixture for 20 min at room temperature to allow the DNA-TurboFect

complexes to form. In the meantime, exchange fresh media in the Phoenix culture in 100-mm dish to a total volume of 10 mL. Finally, add the DNA-Turbofect complexes directly to the dish and gently mix by swirling the dish. Incubate cells in a CO_2 incubator at 37°C overnight (8–10 h).

3. Day 2: Replace the culture medium with fresh Phoenix medium.

4. Day 3 (48 h post-transfection): Collect virus-containing culture medium into sterile 50-mL tubes. Replace the culture medium in all dishes. Centrifuge virus-containing tubes at $500 \times g$ for 10 min to get rid of cell debris. Filter the obtained supernatant through 0.45 μm low protein-binding filters. The supernatant containing lentiviral particles can be used immediately for titer determination and further cell transduction.

5. Days 4–7: Continue to collect culture medium daily, then concentrate viruses as described above. Harvest medium into the same 50-mL tube you used on day 3 and store at 4°C. By the end of day 7, you should get about 30 mL of combined supernatant per vector. To improve cell transduction, we recommend performing the infection of the cells four times at 12-h intervals.

3.2.2. Usage of Recombinant Lentiviruses

1. For transduction, we recommend to use freshly made viruses as stated above. However, if it is not possible, the viruses from the supernatant could be concentrated by ultracentrifugation, and small aliquots of the virus suspension can be prepared from the pellet and stored at −70°C. For the concentrating step, you should use SW-28 rotor (Beckman) or similar one. Add 36 mL of filtered supernatant to the sterile SW-28 centrifuge tubes. If necessary, bring the volume to 36 mL with DMEM media. Spin at $122,000 \times g$ for 120 min at 4°C. Carefully aspirate the supernatant from the tubes. Add 700 μL of DMEM (no serum, no antibiotic) to the pellet. Cover the entire tube holder rack with saran wrap and store overnight at 4–8°C. Resuspend the virus pellet carefully by pipetting it up and down, make aliquots, and store them at −70°C.

2. The use of the appropriate amount of viruses is critical for the successful reprogramming. The viral titer has to be determined for each batch. This can be done by transduction, the known number of cells by serial dilutions of a viral vector (see Note 4). The functional viral titer can be detected in these cells by immunocytochemistry. We recommend starting from 1/10 of obtained 30 mL lentivirus solution per $4–5 \times 10^6$ HUVEC cells plated into 100-mm culture dish within first 48 h of the virus preparation.

3.3. Basic Reprogramming Procedure

The timeline of the reprogramming procedure is shown in Fig. 1a.

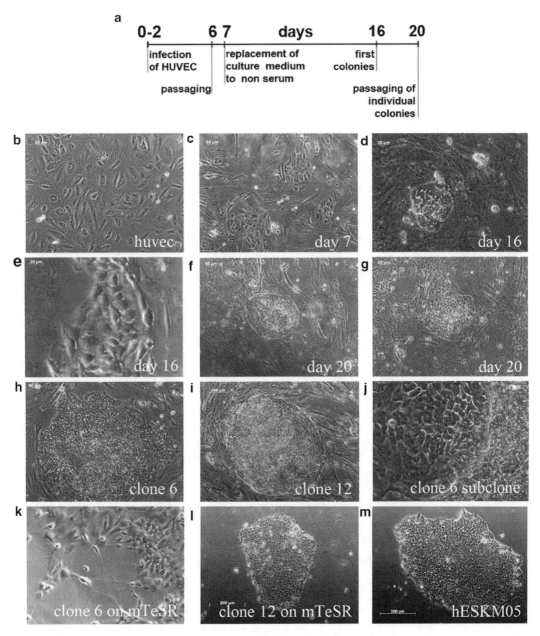

Fig. 1. Generation of human endo-iPS cells. (a) Scheme of the iPSC generation. iPSC colonies were picked based on hESC-like morphology (B–L). Morphological changes during direct reprogramming. Representative colonies are shown. (b) HUVEC after viral infection. (c) After transfer on feeder cells, some of the HUVECs can form iPSC look-alike colonies, and others die or become indistinguishable from feeder cells. (d and e) First colonies of reprogrammed cells appear on day 16 and have different morphology. (f and g) Colonies before passaging on day 20. (h and i) The clones 6 and 12 growing on MEF feeder form ES-like colonies, (j) after 5 passages on MEF, clone 6 changes cell morphology toward ES-like one, however culturing on Matrigel starts to differentiate (k), while clone 12 retains ES-like morphology (l and m).

Here we describe the protocol and specific details.

1. Day −1: For virus transduction, seed HUVEC at passage 2–3 at a density of 3×10^6 cells per 100-mm culture dish in HUVEC medium. The cells should be evenly distributed across the plate (Fig. 1b).

2. Day 0: Add 8 μg/mL of polybrene to the cells and incubate them for an hour. Aspirate the medium and incubate the cells with virus-containing supernatants (MOI 5 for each vector) supplemented with 8 μg/mL polybrene for the next 12 h. Replace the virus-containing solution with 8 μg/mL polybrene every 12 h. You should maintain cells in the medium containing heat-inactivated FBS.

3. Day 2: Replace the virus-containing medium with 10 mL of fresh HUVEC medium.

4. Day 4: Aspirate the old media off the dish and add fresh aliquot of 10 mL HUVEC medium. The cells should be approximately 80% confluent at this point.

5. Day 5: Plate MEF cells that were previously treated with 10 μg/mL mitomycin C into tissue culture plates pre-coated with 0.1% gelatin. Routinely, in iPSC generation process, we use 100-mm dishes for passing the cells and 12-well plates for monitoring the state of the cells and making assays during reprogramming process. MEF cells should be seeded at a density of 80% monolayer (this is a lower density compared to that routinely used for feeding hES cells) to simplify colony formation process.

6. Day 6: Passage virus-treated HUVECs using the 0.25% trypsin in the MEF dishes according to the following procedure: Aspirate the medium off the HUVECs, wash twice with PBS, then add 3 mL of the 0.25% trypsin, and keep at 37°C (e.g., in CO_2 incubator) for 5 min. Gently mix detached cells by pipetting them up and down and destroying any possible clumps, then add 10 mL of HUVEC media without growth factors to inactivate trypsin, and plate them into a 100-mm MEF dish for the further experiment and 12-well plate for subsequent assays. Optional but recommended: put aside a small aliquot of the cell suspension for RT-PCR to determine the expression of a few second order pluripotency genes, for example, FoxD3 or Nanog. Distribute the cells evenly and place into the CO_2 incubator for overnight attachment of the cells. We use a small cell split ratio (almost 1:1) benefiting from the fact that the iPS medium is not optimal for HUVEC cells, and mainly reprogramming cells would selectively survive.

7. Days 7–8: Replace the HUVEC medium with the iPS medium daily. Use 1 and 10 mL of medium per a well of 12-well plate and per a 100-mm dish correspondingly.

On day 7 you can observe small "colonies," resulting from non-destroyed clumps of HUVEC after replating (Fig. 1c).

8. Days 9–14: Repeat the feeding procedure every day. Feed cells by replacing half the culture medium volume with fresh cell culture medium. If cell medium becomes intensively yellow (acidic) within 24 h of cell incubation, you should completely exchange the medium. Monitor the plates for colony formation. First colonies should normally appear at days 15–20. Most of the under-reprogrammed HUVECs have undergone apoptosis by this point of the experiment, and therefore, the other cells are proliferating faster. Use the cells seeded in the 12-well tissue culture plates for the number of control procedures (ICC, FACS, RT-PCR) to monitor the reprogramming process.

9. Days 15–21: Repeat the feeding procedure every day. By day 15, the feeder becomes 7-day old and less productive. If cell proliferation rate is decreasing, add the solution of bFGF to the final concentration of 10 ng/mL to the culture medium. Check the plates every day to detect the appearance of new iPS cell colonies. Colony morphology at this stage could vary from the colonies with the irregular shape consisted of small dark cells to the colonies with the ES-like morphology (see Fig. 1d–g).

10. Day 22 onward: Repeat the feeding procedure every day. Monitor the cell morphology with the inverted microscope and check for new colony formation. Harvest the colonies when they have 100–200 cells.

3.4. Human iPS Colony Picking and Expansion

1. Twenty days after virus transduction, pick up colonies mechanically. Replace the medium with fresh iPS medium and mark the chosen cell colony position on the culture dish surface with permanent marker (see Note 5).

2. Gently cut the colony to the clumps of 20–30 cells each using Pasteur pipette needle or a 21-G needle. The experienced person can do it simply with a yellow tip.

3. Aspirate the small amount of the medium containing these clumps using micropipette under the light microscope, and transfer a half of the clump suspension into a 24-well plate covered by MEF cell layer and the other half to a Matrigel-coated well of a 24-well plate (Fig. 1h, i, k, l).

4. Add 0.5 mL of iPS or mTeSR1 medium per a well of MEF cell covered or Matrigel-coated plates. This step is required for the selection of completely reprogrammed clones as the mTeRS1 medium is a good selective medium for complete cell reprogramming (Fig. 1k, l). In case of a smaller colony size, you should pick up the whole colony and seed it onto MEFs layer in iPS medium. We advise to harvest 20–50 ES-like colonies,

and a few colonies with other morphology for further passaging and characterization. The fully reprogrammed iPS cell line should morphologically be indistinguishable from hESCs (Fig. 1l, m). In the same culture dish, you can find quite a number of colonies with varying morphology. The dynamics of endothelium-derived iPS cells reprogramming is similar to that of other published iPS cell lines. Thus, the ES-like colonies can be detected approximately at the same time point as those of the iPS cells generated from $CD34^+$ cells, and complete cell reprogramming can be achieved within 3–4 weeks (14). Expansion of a single colony to a confluent cell layer in 35-mm Petri dishes usually takes about 3 weeks. We recommend 1:3–1:5 dilution ratios during the period of iPS propagation. Starting from passage 2, propagate fully reprogrammed iPS colonies using mTesR1 medium in Matrigel-covered cell culture dishes following the manufacturer instructions.

3.5. Characterization of iPS Cells

Several criteria have been set forth to ascertain whether a fully reprogrammed cell state has been achieved comprising an array of unique features associated with pluripotency, encompassing molecular and functional attributes. First of all, at the molecular level, iPSCs must display gene expression profiles that are characteristic for hESCs, including protein expression of key pluripotency factors (e.g., Oct4, Nanog, FoxD3) and hESC-specific surface antigens (e.g., SSEA-3/-4, Tra-1-60/-81) (Fig. 2); on the other hand, they should not have transgene expression and maintain genetic integrity; and, finally, functionally pluripotent cells should be able to differentiate in vitro into cells representing three germ layer lineages.

3.5.1. Antigenes Immunostaining

Stain surface antigen using a standard immunohistochemistry procedure:

1. Aspirate the medium, wash twice with PBS, and fix cells in 4% paraformaldehyde (PFA) solution in PBS for 20 min at room temperature.

2. Incubate the fixed cells in blocking buffer (PBS, 0.1% Tween-20, 2% Goat Serum, 5% FBS) for an hour and apply primary antibody diluted in the blocking buffer.

3. Incubate cells overnight at 4°C. Wash cells in three exchanges of PBS mixed with 0.1% Tween-20.

4. Incubate with an appropriate secondary fluorescent antibody labeled with "The Alexa Fluor Dye Series" (Invitrogen). Nuclei could be counterstained with DAPI. For intracellular staining, add Triton-X-100 to the blocking buffer to the final concentration of 0.25% (refer Fig. 2a).

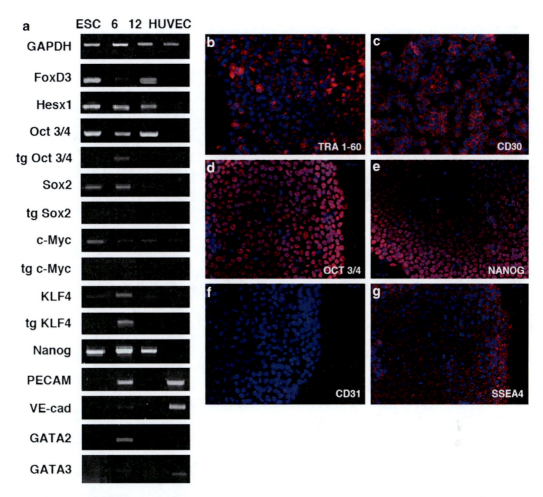

Fig. 2. RT-PCR and immunohistochemical characterization of human endo-iPSCs: partially (clone 6) and fully (clone 12) reprogrammed endo-iPSCs. (**a**) RT-PCR analysis for the expression of the pluripotency genes (FoxD3, Nanog, Oct3/4, Hesx1), the endothelial-specific markers (PECAM, VE-cadh, GATA2, GATA3), and the exogeneous expression (tg Oct3/4, tg Sox2, tg c-Myc, tg Klf4) in iPSC clones, hESCs (positive control for the pluripotency genes), as well as the parental HUVECs (negative control). GAPDH serves as a loading control. (**b–g**) Undifferentiated iPSC express markers common to hESCs. 4,6-Diamidino-2-phenylindole (DAPI) staining (*blue color*) shows the total cell content per field. Immunostaining with specific antibodies for Tra-1-81 (**b**), CD30 (**c**) Oct3/4 (**d**), Nanog (**e**), SSEA4 (**g**). HUVEC marker CD31 is absent in these cells. (**f**) Endo-iPSCs are different from HUVECs. Representative colonies of endo-iPS-12 cell line are shown.

3.5.2. Transgene Expression

1. Fully reprogrammed cells must be independent of transgene expression (Fig. 2b). This can be verified by RT-PCR with specific primers: the upstream primer should be complementary to the sequence of cDNA of the gene under study (e.g., Oct3/4, Sox2, Klf4, c-Myc), and the downstream primer should be a sequence between the viral vector LTR and the 3'end of the cDNA.

Table 1
Primer sets for RT-PCR reactions

Genes		Sequence (5′–3′)	Product size, bp
Oct3/4 (total)	Forward	CGACCATCTGCCGCTTTGAG	570
	Reverse	CCCCCTGTCCCCCATTCCTA	
Oct3/4 (exogene)	Forward	CCCCAGGGCCCCATTTTGGTACC	410
	Reverse	TTATCGTCGACCACTGTGCTGCTG	
Sox2 (total)	Forward	TCCTGATTCCAGTTTGCCTC	480
	Reverse	GCTTAGCCTCGTCGATGAAC	
Sox2 (exogene)	Forward	GGCACCCCTGGCATGGCTCTTGGCTC	365
	Reverse	TTATCGTCGACCACTGTGCTGCTG	
c-Myc (total)	Forward	AGTAATTCCAGCGAGAGGCA	389
	Reverse	AGGCTGCTGGTTTTCCACTA	
c-Myc (exogene)	Forward	CAACCGAAAATGCACCAGCCCCAG	315
	Reverse	TTATCGTCGACCACTGTGCTGCTG	
Klf4 (total)	Forward	ACCCTGGGTCTTGAGGAAGT	324
	Reverse	AGGTTTCTCACCTGTGTGGG	
Klf4 (exogene)	Forward	ACGATCGTGGCCCCGGAAAAGGACC	390
	Reverse	TTATCGTCGACCACTGTGCTGCTG	
Nanog	Forward	AGCATCCGACTGTAAAGAATCTTCAC	430
	Reverse	CGGCCAGTTGTTTTTCTGCCACCT	
FoxD3	Forward	CAAGCCCAAGAACAGCCTAGTGAA	200
	Reverse	TGACGAAGCAGTCGTTGAGTGAGA	
Hesx1	Forward	ACCTGCAGCTCATCAGGGAAAGAT	200
	Reverse	AAAGCAGTTCTTGGTCTTCGGCCT	

2. For RT-PCR analysis, we purify total RNA using Rneasy Mini Kit (QIAGEN) and make cDNA using the M-MuLV Reverse Transcriptase (Fermentas), according to the manufacturer recommendations. The sets of primers for pMX vectors are listed in Table 1.

3.5.3. Karyotyping

Multiple passaging of primary cells in vitro eventually results in chromosomal and genetic variations. Therefore, testing an iPSC line at least for the karyotypic stability at the start of the experiment and at different time points of propagation is important for the proper maintenance of the line. Prepare metaphase chromosomes from iPSCs cultured on 35-mm tissue culture dish (see Note 6).

1. Exchange mTeSR1 medium (use 2 mL per well) 8–12 h before harvesting the iPSCs. The best results can be obtained if cells are 50–80% confluent.

2. Add 20–40 µL of demecolcine to a final concentration of 0.1–0.2 µg/mL 2 h before fixation.

3. Rinse cells twice with PBS and incubate in 0.5 mL of 0.05% trypsin solution for 1–2 min at room temperature.

4. Collect trypsin solution into a 15 mL centrifuge tube, add 1 mL of mTeSR1 medium to the culture dish, scrape off iPSC colonies by the wide end of the P-1000 tip and carefully resuspend cells in the medium by pipetting them up and down and collect them to the same centrifuge tube.

5. Add 12 mL of 0.075 M KCl (brought to room temperature). Incubate in the 42°C water bath for 16–20 min.

6. Add several drops (about 200 µL) of the fixative 1. Mix the cells by inverting the tube several times. Spin down at 200–400×g for 6 min at 4°C.

7. Remove the supernatant, leaving about 400–500 µL of liquid in the tube. Resuspend the cell pellet by pipetting it up and down with P-1000 pipette. Add 2 mL of ice-cold fixative 1. Mix by inverting the tube 2–3 times and spin down at 600×g for 4 min at 4°C.

8. Remove the supernatant, leaving about 400–500 µL of the liquid in the tube. Resuspend the cell pellet by pipetting as described above. Add 2 mL of ice-cold fixative 2. Mix by inverting the tube 2–3 times and spin down at 600×g for 4 min at 4°C. Remove supernatant and add 1 mL of fixative 2.

9. For GTG or DAPI staining, centrifuge the cell suspension at 600×g for 4 min at 4°C. Remove the supernatant and add 100–500 µL of fresh fixative 2. Resuspend the cell pellet by pipetting. Place 10–30 µL of cell suspension on a cold wet microscope slide. Let the cell suspension run all over the slide. Air-dry the slide or put it on a thermo-plate at 40–50°C. Perform GTG or DAPI banding.

10. Carefully transfer the cell suspension into a 1.5-mL tube. Store metaphase preparations at −20°C.

3.6. Differentiation of iPSC to Neuronal Lineage

At a functional level, iPSCs must demonstrate the ability to differentiate to the desired cell lineage in vitro, forming embryoid bodies in vitro and teratomas in vivo (20). Here we present a detailed protocol for differentiation of endothelium-derived iPS cells into neural lineages. Our protocol is based on the previously published protocol of Chambers et al. (21) with modifications.

3.6.1. Differentiation of iPS Cells into Neuroepithelial Cells

1. Culture iPSC clone in the Matrigel-coated 35-mm or 60-mm tissue culture dishes in the mTeSR1 medium. Exchange the medium daily for 4–5 days until it reaches 40–50% of the monolayer. The colonies should be in good shape without signs of spontaneous differentiation (Fig. 3a).

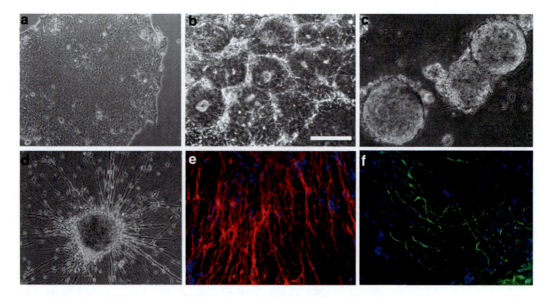

Fig. 3. Differentiation of iPS cells into neurons. (**a–d**) iPSC colony on Matrigel (*bright field images*). (**b**) Neural-tube-like rosettes within differentiating iPSC colonies at day 15 of differentiation. (**c**) Neurospheres cultured in suspension. (**d**) Neurite outgrowth from attached neurosphere at day 7 after plating. (**e–f**) Immunostaining with specific antibodies for tubulin beta III (**e**, *red*), and GABA (**f**, *green*). Nuclei are counterstained with DAPI (*blue*).

2. Remove the medium and rinse twice with 1–2 mL DMEM/F12.

3. Add 2.5 and 5 mL of neural differentiation medium to 35- and 60-mm dish, respectively.

4. Replace half the amount of medium daily; if the medium becomes too acidic (it would turn yellow), completely exchange the medium.

5. Daily examine the morphology of the cells. In about 6–8 days of iPSC differentiation, radially organized columnar neuroepithelial cells appear in the colony center first and then subsequently closer to the edges. These cells represent early steps of neuroectodermal differentiation and are Pax 6 positive.

6. The multilayer rosette-like structures resembling neural tubes are usually formed by day 12 (Fig. 3b). At this step, omit SB431542 from neural differentiation medium. Change the medium every other day for another 4–5 days.

3.6.2. Neurospheres Forming and Expansion

1. Place 2 and 4 mL of neurosphere expansion medium for 35- and 60-mm dishes, respectively. Gently scrap the clusters of rosette-like structures with a P-200 pipette tip or a glass needle.

2. Collect the rosettes into a 15-mL tube, resuspend the clumps using 2 mL pipette twice. This will break rosette structures into pieces.

3. Transfer the suspension to a 35-mm ultralow adhesion tissue culture dish. Culture the cells in CO_2 incubator. Within 24 h,

the cell clumps would form floating spherical structures (neurospheres) (Fig. 3c). Feed the cells by replacing 2/3 of the medium every other day (see Note 7). To exchange the medium completely, transfer neurospheres to a 15-mL tube, let them sit for 3–5 min, then replace the medium and transfer them back to the tissue culture dish.

4. If neurospheres become bigger than 400 μm, break them into pieces by pipetting with a Pasteur pipette or a P-200 micropipette. Continue to cultivate neurospheres in the neurosphere expansion medium.

3.6.3. Differentiation of Neurospheres into Neurons

1. On day 22, exchange the cell culture medium to the neurosphere differentiation medium. Plate the neurospheres into a Matrigel-coated tissue culture dish (see Note 8). The spheres would attach to the surface within 12–24 h. Feed the cells every other day for long-term differentiation.

2. From day 22, cells would start to migrate out of the spheres. Long neurites can be observed in 4–5 days after plating (Fig. 3d).

This protocol is quite reliable for neuron production. None or a few glial cells could be obtained. To determine cell phenotypes by immunocytochemistry, we recommend the use of the following antibodies from Abcam:

- Pax 6 and Sox1 for determination of early neuroepithelial cells
- Beta III tubulin, synapsin, neurofilament protein, MAP2 for observation of early neuronal markers
- GABA, GAD65, TH, and others for determination of neuronal subtypes

4. Notes

1. We strongly recommend that you work with lentiviral stocks as you do with Biosafety Level 2 organisms. You should follow Biosafety Level 2 guidelines and decontaminate waste properly.

2. HUVECs should be used at passage 2–3 for generation of iPSCs.

3. Alternatively, iPS cells could be cultivated on Matrigel in mTeSR1 medium according to manufacturer instructions.

4. To determine the titer of virus, we recommend starting from 1/1,000 of the viral stock.

5. Not all colonies having flat morphology and small cells with high nucleus to cytoplasm ratio are ES-like colonies. Many non-hES-looking colonies that form similar but not identical compact clusters of cells are capable of sustained self-renewal and successive passaging under hESC culture conditions on MEF. At the same time, such partially reprogrammed cells are not capable of

maintaining pluripotent state on Matrigel in mTeSR1 medium. Under latter conditions, partially reprogrammed cells start to differentiate into neuronal-like cells, fibroblast-like cells, etc. Therefore, we recommend the mTeSR1 medium for the selection of reprogrammed iPS clones. For some clones, a process of reprogramming from partial stage to more advanced takes few additional passages on MEF in iPSC medium (Fig. 1j).

6. The metaphase preparation protocol for the pluripotent cells differs from the standard procedure for adherent cells by three main points. First, very gentle treatment with trypsin should be used to avoid the formation of cell aggregates. Second, centrifugation should not be applied after cell detachment from tissue culture dish step, trying to treat vulnerably after colony demolition cells with extra care. And finally, after hypotonic treatment step, the cells are fixed using a mix of methanol and glacial acetic acid solution of 6:1 ratio instead of the standard 3:1 ratio. Using a higher concentration of acetic acid in the first fixative could cause cell membrane breakage and loss of chromosomes from cells. Additionally, the increase of temperature during hypotonic treatment to 42°C could help improve metaphase chromosome spreading.

7. Do not aspirate the medium with the vacuum pump, because you can aspirate off the neurospheres.

8. We have also tried to use poly-ornitine and laminin matrices for iPSC neural differentiation, but the efficiency of neural differentiation was approximately the same as on Matrigel matrix.

References

1. Takahashi K., Yamanaka S. (2006) Induction of pluripotent stem cells from mouse embryonic and adult fibroblast cultures by defined factors. Cell 126, 663–76.
2. Maherali N., Ahfeldt T., Rigamonti A. et al. (2008) A high-efficiency system for the generation and study of human induced pluripotent stem cells. Cell Stem Cell 3, 340–5.
3. Okita K., Nakagawa M., Hyenjong H. et al. (2008) Generation of mouse induced pluripotent stem cells without viral vectors. Science 322, 949–53.
4. Meissner A., Wernig M., Jaenisch R. (2007) Direct reprogramming of genetically unmodified fibroblasts into pluripotent stem cells. Nat Biotechnol 25, 1177–81.
5. Blelloch R., Venere M., Yen J. et al. (2007) Generation of induced pluripotent stem cells in the absence of drug selection. Cell Stem Cell 1, 245–7.
6. Nakagawa M., Koyanagi M., Tanabe K. et al. (2008) Generation of induced pluripotent stem cells without Myc from mouse and human fibroblasts. Nat Biotechnol 26, 101–6.
7. Kim J.B., Zaehres H., Wu G. et al. (2008) Pluripotent stem cells induced from adult neural stem cells by reprogramming with two factors. Nature 454, 646–50.
8. Kim J.B., Sebastiano V., Wu G., Araúzo-Bravo M.J. et al. (2009) Oct4-induced pluripotency in adult neural stem cells. Cell 136, 411–9.
9. Zhou H., Wu S., Joo J. et al. (2009) Generation of Induced Pluripotent Stem Cells Using Recombinant Proteins. Cell Stem Cell 4, 381–4.
10. Patel M., Yang S. (2010) Advances in reprogramming somatic cells to induced pluripotent stem cells. Stem Cell Rev Rep 6, 367–80.
11. Maherali N., Hochedlinger K. (2008) Guidelines and techniques for the generation of induced pluripotent stem cells. Cell Stem Cell 3, 595–605.
12. Yu J., Vodyanik M.A., Smuga-Otto K. et al. (2007) Induced pluripotent stem cell lines derived from human somatic cells. Science 318, 1917–20.

13. Aasen T., Raya A., Barrero M.J. et al. (2008) Efficient and rapid generation of induced pluripotent stem cells from human keratinocytes. Nat Biotechnol 26, 1276–84.
14. Loh Y.H., Agarwal S., Park I.H. et al. (2009) Generation of induced pluripotent stem cells from human blood. Blood 113, 5476–9.
15. Kim J.B., Sebastiano V., Wu G. et al. (2009) Oct4-Induced Pluripotency in Adult Neural Stem Cells. Cell 136, 411–19.
16. Utikal J., Maherali N., Kulalert W. et al. (2009) Sox2 is dispensable for the reprogramming of melanocytes and melanoma cells into induced pluripotent stem cells. Journal of Cell Science, 122, 3502–10.
17. Nagler A., Korenstein-Ilan A., Amiel A. et al. (2004) Granulocyte colony-stimulating factor generates epigenetic and genetic alterations in lymphocytes of normal volunteer donors of stem cells. Exp Hematol 32, 122–30.
18. Baudin B., Bruneel A., Bosselut N. et al. (2007) A protocol for isolation and culture of human umbilical vein endothelial cells. Nat Protoc 2, 481–5.
19. Lagarkova M.A., Shutova M.V., Bogomazova A.N. et al. (2010) Induction of pluripotency in human endothelial cells resets epigenetic profile on genome scale. Cell Cycle 9, 937–46.
20. Lin G., Martins-Taylor K., Xu R.H. (2010) Human embryonic stem cell derivation, maintenance, and differentiation to trophoblast. Methods Mol Biol 636, 1–24.
21. Chambers S.M., Fasano C.A., Papapetrou E.P. et al. (2009) Highly efficient neural conversion of human ES and iPS cells by dual inhibition of SMAD signaling. Nat Biotechnol 27, 275–80.

Chapter 12

Generation of Human Induced Pluripotent Stem Cells from Endoderm Origin Cells

Hua Liu, Su Mi Choi, and Yoon-Young Jang

Abstract

Human induced pluripotent stem (iPS) cells have been derived mostly from cells originating from mesoderm and in a few cases from ectoderm. This has prevented comprehensive comparative investigations of the quality of human iPS cells of different origins. We have recently reported for the first time the reprogramming of human endoderm-derived cells (i.e., primary hepatocytes) to iPS cells. In this chapter, we describe the methods for generating and characterizing iPS cells from human primary hepatocytes.

Key words: Human induced pluripotent stem (iPS) cells, Endoderm, Primary hepatocytes, Reprogramming, The origin of iPS cells

1. Introduction

Recent advances in induced pluripotent stem (iPS) cell research have shown great potential for these somatic cell–derived stem cells as sources for cell replacement therapy and for establishing disease models (1–15). However, several key issues have to be addressed in order for iPS cells to be used for clinical purposes. One of the most important issues is the generation of safe and functional cell types for therapy. A comprehensive study using various mouse iPS cells has demonstrated that the origin of the iPS cells has a profound influence on the tumor-forming propensities in a cell transplantation therapy model (3). Mouse tail-tip fibroblast derived iPS cells (mesoderm origin) have shown the highest tumorigenic propensity, whereas gastric epithelial cell derived and hepatocyte derived iPS cells (both are endoderm) have shown lower propensities (3). Recent studies have suggested that mouse iPS cells of different origins possess different capacities to differentiate into blood cells (16, 17). Although it has been demonstrated that human iPS cells retain certain gene expressions of the parent cells (2), it remains largely unclear if the cell origin could affect the safety and function of human iPS cells. It is therefore extremely important to establish human iPS cell lines of multiple developmental origins and thoroughly examine the source impact on both the safety aspects and their differentiation potentials. In the mouse, iPS cells have been generated from derivatives of all three embryonic germ layers, including mesodermal fibroblasts (6), epithelial cells of endodermal origin (7), and ectodermal keratinocytes (8), whereas human iPS cells have been produced mostly from mesoderm (fibroblasts and blood cells) or ectoderm (keratinocytes and neural stem cells) (9–13, 18, 19). The reprogramming of human primary hepatocytes (endoderm) to pluripotency has recently been demonstrated (20). Human hepatocyte-derived iPS cells appear indistinguishable from human embryonic stem (ES) cells with respect to colony morphology, growth properties, expression of pluripotency-associated transcription factors and surface markers, and differentiation potential in embryoid body (EB) formation as well as teratoma assays. The ability to reprogram human hepatocytes is crucial for developing liver disease models with iPS cells, especially for certain liver diseases involving acquired somatic mutations that occur only in hepatocytes of patients and not in other cell types (21–25). More importantly, this work lays the groundwork necessary for studying the safety and efficacy of differentially originated human iPS cells in cell therapy. Here we describe the detailed methods for generating and characterizing iPS cells from human primary hepatocytes.

2. Materials

2.1. Hepatocyte Culture

1. Primary hepatocytes (Lonza, Walkersville, MD); store in liquid nitrogen.
2. Collagen Type I solution, Rat-tail (GIBCO®, Invitrogen, Carlsbad, CA); store at 4°C.
3. Acetic acid (Sigma-Aldrich, Saint Louis, MO); store at room temperature.
4. Williams' Medium E (WEM) (Invitrogen, Carlsbad, CA); store at 4°C.
5. Heat Inactivated Fetal Bovine Serum (Sigma-Aldrich, Saint Louis, MO); store at –20°C.
6. GlutaMax™ 200 mM (GIBCO®, Invitrogen, Carlsbad, CA); store at –20°C.
7. Gentamicin (GIBCO®, Invitrogen, Carlsbad, CA); store at 4°C.

8. Human Insulin (Sigma-Aldrich, Saint Louis, MO); store at −20°C.
9. Dexamethasone (Sigma-Aldrich, Saint Louis, MO); store at 4°C.
10. Human Hepatocyte Growth Factor (hHGF) (PeproTech, Rocky Hill, NJ) (optional); store at −20°C.
11. (optional) Human Epithelial Growth Factor (hEGF) (PeproTech, Rocky Hill, NJ); store at −20°C.

2.2. Retrovirus Production

1. Dulbecco's Modified Eagle's Medium (DMEM), High Glucose 4.5 g/L (GIBCO®, Invitrogen, Carlsbad, CA); store at 4°C.
2. Fetal Bovine Serum (Hyclone, Thermo Fisher Scientific, Logan, UT); store at −20°C.
3. 293T cells (ATCC, Manassas, VA); store in liquid nitrogen.
4. 15-cm tissue culture plates (BD Falcon, Franklin Lakes, NJ).
5. Poly-D-lysine (Millipore, Billerica, MA); store at −20°C.
6. Opti-MEM-I medium (GIBCO®, Invitrogen, Carlsbad, CA); store at 4°C.
7. Lipofectamine 2000 (Invitrogen, Carlsbad, CA); store at 4°C.
8. Reprogramming retroviral vectors: pMX-Oct4, pMX-Sox2, pMX-Klf4, pMX-c-Myc (Addgene, Cambridge, MA); store at −80°C.
9. Helper plasmids: one expressing vesicular stomatitis virus G protein, such as MD.G, and one expressing MLV (retroviral) gag/pol (retro-gag/pol) (Addgene, Cambridge, MA); store at −20°C.
10. Amicon® Ultra-15 Centrifugal Filter Device with a cutoff of 100,000 MWCO (Millipore, Billerica, MA).
11. Millex®-HV 0.45 μm, Durapore PVDF filters (Millipore, Billerica, MA).
12. Polybrene (Sigma-Aldrich, Saint Louis, MO); store at −20°C.

2.3. Human iPS Cell Culture

1. Knockout™ DMEM (GIBCO®, Invitrogen, Carlsbad, CA); store at 4°C.
2. FBSd (Hyclone, Thermo Fisher Scientific, Logan, UT); store at −20°C.
3. Knockout Serum Replacement (KOSR) (GIBCO®, Invitrogen, Carlsbad, CA); store at −20°C.
4. GlutaMax™ 200 mM (GIBCO®, Invitrogen, Carlsbad, CA); store at −20°C.
5. MEM nonessential amino acids (NEAA), 10 mM (100×) (GIBCO®, Invitrogen, Carlsbad, CA); store at 4°C.

6. β-Mercaptoethanol (Sigma-Aldrich, Saint Louis, MO); store at room temperature.

7. Penicillin/Streptomycin (100×) (Invitrogen, Carlsbad, CA); store at −20°C.

8. Basic Fibroblast Growth Factor (bFGF) (PeproTech, Rocky Hill, NJ); store at −20°C.

9. Collagenase Type IV (Sigma-Aldrich, Saint Louis, MO); store at −20°C. Dissolve collagenase (1 mg/mL) in DMEM/F12. Filter using a 0.22-μm sterile filter, and then store at 4°C.

10. Gelatin (Sigma-Aldrich, Saint Louis, MO); store at room temperature.

11. Matrigel™ matrix (Becton Dickinson, Franklin Lakes, NJ); store at −20°C.

12. mTeSR1 (Stemcell, Vancouver, Canada); store at 4°C after reconstitution.

13. Mouse embryonic fibroblasts (MEFs) mitomycin treated or irradiated (Millipore, Billerica, MA, or Globalstem, Rockville, MD); store in liquid nitrogen.

2.4. Human iPS Cell Characterization

1. 16% Paraformaldehyde (Alfa Aestar, Ward Hill, MA); store at room temperature.

2. Mouse (IgM) antihuman TRA-1-60 (Millipore, Billerica, MA); store at 4°C.

3. Alexa Fluor 555 anti-mouse IgM (Invitrogen, Carlsbad, CA); store at 4°C.

4. Mouse (IgG3) antihuman stage-specific embryonic antigen 4 (SSEA4) (Cell Signaling Technology, Danvers, MA): store at −20°C.

5. FITC anti-mouse IgG (Invitrogen, Carlsbad, CA); store at 4°C.

6. Mouse (IgG1) antihuman Oct4 (Millipore, Billerica, MA); store at 4°C.

7. Mouse (IgG1κ) antihuman NANOG (BD Pharmingen™, Franklin Lakes, NJ); store at 4°C.

8. Alexa Fluor 555 anti-mouse IgG1 (Invitrogen, Carlsbad, CA); store at 4°C.

9. 0.05% Trypsin-EDTA (Invitrogen); store at −20°C.

10. 1× Phosphate-buffered saline (PBS), Ca^{2+} and Mg^{2+} free (Invitrogen); store at room temperature.

11. 10% Buffered formalin (Thermo Fisher Scientific, Waltham, MA); store at room temperature.

12. Syringe 1 mL 25 G 5/8 (Becton Dickinson, Franklin Lakes, NJ).

13. Triton® X-100 (Sigma-Aldrich, Saint Louis, MO); store at room temperature.
14. Albumin, bovine (Sigma-Aldrich); store at 4°C.
15. Saponin (Sigma-Aldrich); store at room temperature.
16. Eosin Y solution (Sigma-Aldrich); store at room temperature.
17. Hematoxylin solution (Sigma-Aldrich); store at room temperature.
18. Xylene (Mallinckrodt Baker, Phillipsburg, NJ); store at room temperature.
19. Ethanol (Phamco-AAPER, Shelbyville, KY); store at room temperature.
20. Costar® ultralow attachment 6-well plates (Corning, Lowell, NY).

2.5. Others

1. Stericup™ (Millipore, Billerica, MA).
2. 6-well and 12-well tissue culture plates (Falcon, Becton Dickinson, Franklin Lakes, NJ).
3. Trypan blue 0.4% solution (Invitrogen, Carlsbad, CA).
4. Hemocytometer (Sigma-Aldrich, Saint Louis, MO).
5. 25-cm cell scrapers (Sarstedt, Nümbrecht, Germany).
6. 15- and 50-mL polystyrene tubes (Sarstedt, Nümbrecht, Germany).

3. Methods

3.1. Preparation of Cell Culture Media and Coating Solutions

3.1.1. Human Hepatocyte Culture Media (see Note 1)

WEM, 5% FBS, 2 mM GlutaMax, 15 mM Hepes, 2 µg/mL Insulin, 50 µg/mL Gentamicin, and 0.1 µM Dexamethasone.

3.1.2. Collagen I, Matrigel, and Gelatin Coating Solutions

Collagen I coating solution: Add 10–12 µL of Collagen I solution (5 mg/mL) in 1 mL 0.02N acetic acid.

Matrigel coating solution: Thaw frozen Matrigel in 4°C overnight, and dilute the Matrigel to 1:30 with cold DMEM/F12.

Gelatin coating solution: Add 0.5 g of gelatin to 500 mL (0.1% gelatin) of endotoxin-free water, autoclave it, and store it at room temperature for further use.

3.1.3. Media for 293T Cell Culture and Virus Collection

293T cell culture medium: DMEM with 4.5 g/mL high glucose, 10% FBSd, and 1× penicillin/streptomycin.

Virus collection medium: DMEM high glucose with 1% FBSd.

3.1.4. Media and Feeder Cells for Human iPS Cells

Human iPS/ES cell culture medium: Knockout DMEM supplemented with 20% KOSR, 0.1 mM NEAA, 2 mM GlutaMax, 1× penicillin/streptomycin, 0.1 mM β-mercaptoethanol, and 4–10 ng/mL of bFGF.

Feeder cells: coat culture plates with gelatin solution (see Sect. 3.1.2) for 1 h at room temperature. Aspirate the gelatin and plate MEFs onto the gelatin-coated plates 1 day before iPS cell seeding in MEF medium (high glucose DMEM supplemented with 10% FBSd and 0.1 mM NEAA).

3.1.5. Media for Embryonic Body Formation

Knockout DMEM supplemented with 20% FBSd, 0.1 mM NEAA, 2 mM GlutaMax, and 0.1 mM β-mercaptoethanol.

3.2. Retroviral Production and Concentration

1. Day 0: Dilute poly-D-lysine in PBS to 50 μg/mL, and coat 15-cm culture dishes (15 mL/dish) for 1 h at room temperature. Wash the plates twice with PBS. Plate 7–8 million 293T cells using 20 mL of 293T medium (see Sect. 3.1.3).

2. Day 1: Transfect 293T cells with three plasmids (two helper plasmids pMD.G and retro-gag/pol, one of pMX retroviral vectors expressing Oct4, Sox2, Klf4, and c-Myc) using Lipofectamine 2000. First, add 36 μL Lipofectamine 2000 to 1.2 mL Opti-MEM I medium and incubate for 5 min at room temperature. In the meantime, mix 24 μg total DNA (3 μg pMD.G, 7 μg retro-gag/pol, and 14 μg retroviral vector) into 1.2 mL Opti-MEM-I medium. After 5 min, mix the diluted DNA with diluted Lipofectamine 2000, and incubate for another 20 min at room temperature. Change 293T medium to 10 mL virus collecting medium, then add DNA-lipid complexes into the 293T cultures. Mix uniformly and place the dishes in the incubator. After 6 h, replace the medium with ~13 mL fresh virus collecting medium and culture it overnight.

3. Day 2–4: Harvest the supernatant every 24 h, and add 17 mL (at day 2) and 20 mL (at day 3) fresh virus collecting medium per dish. Store the supernatant at 4°C. At day 4, centrifuge the collected supernatant at 3,000 rpm for 10 min at 4°C to remove 293T cells and debris.

4. Add 15 mL of the retroviral supernatant to the Amicon Ultra-15 Centrifugal Filter Units, and centrifuge at 3,000 rpm for 20–30 min at 4°C.

5. Aspirate the flow through from the centrifuge tube, add more supernatant to the filter unit, and centrifuge at the same condition. Repeat the process until ~100-fold of concentration is reached.

6. Filter the concentrated viral supernatants containing Oct4, Sox2, Klf4, and c-Myc through Millex®-HV 0.45 μm filter and wash the filter with 1 mL of hepatocyte media (total volume should be <2 mL).

7. Use the virus immediately, or store it at −80°C.

3.3. Collagen I Coating and Human Hepatocyte Culture

1. Place all culture dishes in a sterile flow cabinet.
2. Add sufficient Collagen I (and/or Matrigel, see Note 2) coating solution (see Sect. 3.1.2) to completely cover the bottom of a culture dish (i.e., 1–2 mL/well of 6-well plates).
3. Place the culture dishes for 1 h in room temperature.
4. Remove the coating solution and wash the wells 3 times with sterile PBS.
5. Place primary human hepatocytes on Collagen I-coated tissue culture plates (2–5×10^5 cells/well of 6-well plates) in hepatocyte culture media (see Sect. 3.1.1).
6. Change the media at 4 h after seeding and every other day (see Note 3).

3.4. Human Hepatocyte Reprogramming and iPS Colony Identification

1. Day 0: Premix the viruses with hepatocyte media and 4–8 μg/mL polybrene to a total volume of 1–2 mL per well. Aspirate the media from the 2–4 day pre-cultured/activated hepatocyte culture dishes and dispense the viral mixture onto human hepatocytes (see Note 4 and Fig. 1a).
2. Day 1–2: Add 1–2 mL of fresh hepatocyte media to each well (see Note 5).
3. Day 2–3: Remove the retrovirus-containing media and change it to human ES cell media.
4. Day 4–6: Replace ES cell media every day (see Note 6).
5. Day 6–14: Many transformed cells can be observed (see Note 7 and Fig. 1b). Some of the colonies start to display typical hES cell-like morphology and stain positively for TRA-1-60 (20). Stain these live colonies with the TRA-1-60 antibody (1:200 dilution) in the reprogramming culture plates by incubating the

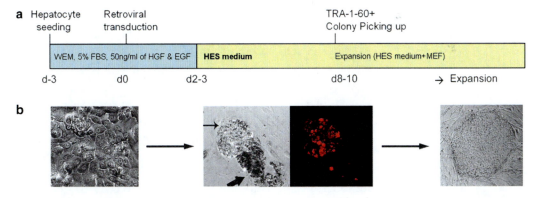

Fig. 1. (a) Diagram of the hHiPS (human hepatocyte-derived iPS cells) generation protocol. (b) Phase contrast image of primary human hepatocytes before reprogramming (*left panel*), typical example of a small ES cell-like TRA-1-60-positive (*red*) colony (*thin arrow*) adjacent to a TRA-1-60–negative non-iPS colony (*thick arrow, middle panel*), representative morphologies of hHiPS growing on MEF (*right panel*).

cells in tissue culture incubator for about 1 h. Then wash the cells with sterile PBS, and stain again with Alexa 555-conjugated anti-mouse IgM antibody (1:500 dilution). After 1 h, wash the plates with PBS, add fresh human ES cell media, and incubate for another 5 min. Identify the TRA-1-60 positive iPS cell colonies under a fluorescence microscope.

6. Day 8–20: Gently cut the colony to smaller clumps by using a colony picking tool (a J-shaped fine curve made by a glass Pasteur pipette (26)). Scrape the clumps off the surface until they are completely detached and floating. Suck up the clumps carefully with a micropipette, and put them into 96-24-well MEF plates.

3.5. Expansion and Characterization of Human iPS Cells

1. *Expansion*: Keep the picked iPS cell colonies on MEF feeder cells with human ES cell medium (or on Matrigel with MEF-conditioned medium) for at least 5 weeks after reprogramming or until stable colonies are established. Once iPS cell expansion reaches a relatively stable level without a significant level of spontaneous differentiation, a chemically defined medium such as mTeSR 1 with Matrigel can also be used for further expansion of the human iPS cells (see Note 8).

2. *Pluripotency markers*: Fix the iPS cell colonies with 4% paraformaldehyde for 30 min, and stain these cells with pluripotency related markers such as TRA-1-60, SSEA4, Oct4, and NANOG for 1 h at room temperature (or overnight in 4°C). For intracellular proteins, use an appropriate permeabilizing solution (such as 0.1–0.5% Triton or Saponin) for antibody incubation and washing steps. After staining the cells with secondary antibodies (see Sect. 2.4), counterstain with a diluted DAPI solution, and scan images under an immunofluorescence microscope.

3. *DNA fingerprinting*: Isolate genomic DNA from 293T cells, human ES cells, iPS cells and their parent cells, and perform PCR analysis to eliminate the possibility of contamination from existing ES cells and iPS cells (20).

4. *Embryoid body (EB) formation*: Collect human iPS cells by collagenase IV digestion and plate them on ultralow attachment 6-well plates with EB formation media (see Sect. 3.1.5). Replace 50% of media every day. Cystic structures can be observed within 5–9 days. After 7 days, transfer the EBs to gelatin-coated tissue culture plates and culture for another 3 days before fixation and staining. Antibodies for endoderm-, mesoderm-, and ectoderm-marker can be used to detect the spontaneously differentiated cells (20).

5. *Teratoma formation*: Collect human iPS cells from one 6-well plate (80–90% confluent) using collagenase IV and resuspend the

cell pellets in 100–200 μL of 50% Matrigel (in Knockout DMEM) on ice. Inject the cells into the hind limb of 10-week-old immunocompromised NSG (NOD.Cg-*Prkdcscid Il2rgtm1Wjl*/SzJ) mice subcutaneously. After 8–12 weeks, tumors can be detected and analyzed by hematoxylin (see Note 9) and eosin staining. Find diverse structures resembling all three germ layer tissues, such as glandular epithelia (endoderm), cartilages (mesoderm), and neural rosettes (ectoderm), for determining the in vivo pluripotency of iPS cells.

4. Notes

1. Primary human hepatocytes can survive only for 7–10 days in culture and cannot proliferate even with human HGF and EGF.
2. Either Collagen I- or Matrigel-coated culture dishes can be used for primary hepatocyte culture.
3. Human HGF (and/or EGF) can be supplemented to the hepatocyte culture medium to activate primary hepatocytes before starting the reprogramming procedure. This stimulation seems to increase the efficiency of the viral transduction (20).
4. Centrifugation of the culture plates at high speed (i.e., 3,000 rpm) may increase the transduction efficiency.
5. If the hepatocytes look very healthy 1–2 days after the first viral addition, the second round of viral mixture can be added to facilitate the reprogramming process.
6. MEF-conditioned media can be used for maintaining the cells on Collagen I once cell transformation starts to occur.
7. At this stage, many non-iPS colonies can be observed. These are TRA-1-60–negative, usually darker, and proliferate faster than TRA-1-60–positive colonies during the initial days. However, the majority of these non-iPS colonies are not able to survive for a long time, and in most cases, these non-iPS colonies are easily distinguishable by morphology from hES cell-like iPS colonies (20).
8. We observed that some of the commercially available chemically defined media promote both cell death and spontaneous differentiation of early stage human iPS cells (<10 passages) compared with the standard condition (i.e., human ES cell medium with MEF coculture).
9. Hematoxylin solution should be filtered before each use to remove oxidized particles.

References

1. Chin M.H., Mason M.J., Xie W., et al. (2009) Induced pluripotent stem cells and embryonic stem cells are distinguished by gene expression signatures. Cell Stem Cell 5, 111–123.
2. Marchetto M.C.N., Yeo G.W., Kainohana O., et al. (2009) Transcriptional signature and memory retention of human induced pluripotent stem cells. PLoS One 4, e7076.
3. Miura K., Okada Y., Aoi T., et al. (2009) Variation in the safety of induced pluripotent stem cell lines. Nat Biotechnol 27, 743–745.
4. Deng J., Shoemaker R., Xie B., et al. (2009) Targeted bisulfite sequencing reveals changes in DNA methylation associated with nuclear reprogramming. Nat Biotechnol 27, 353–360.
5. Doi A., Park I.H., Wen B., et al. (2009) Differential methylation of tissue- and cancer-specific CpG island shores distinguishes human iPSCs, embryonic stem cells and fibroblasts. Nat Genet 41, 1350–1353.
6. Takahashi K. and Yamanaka S. (2006) Induction of pluripotent stem cells from mouse embryonic and adult fibroblast cultures by defined factors. Cell 126, 663–676.
7. Aoi T., Yae K., Nakagawa M., et al. (2008) Generation of pluripotent stem cells from adult mouse liver and stomach cells. Science 321, 699–702.
8. Wernig M., Lengner C.J., Hanna J., et al. (2008) A drug-inducible transgenic system for direct reprogramming of multiple somatic cell types. Nat Biotechnol 26, 916–924.
9. Takahashi K., Tanabe K., Ohnuki M., et al. (2007) Induction of pluripotent stem cells from adult human fibroblasts by defined factors. Cell 131, 861–872.
10. Yu J., Vodyanik M.A., Smuga-Otto K., et al. (2007) Induced pluripotent stem cell lines derived from human somatic cells. Science 318, 1917–1920.
11. Lowry W.E., Richter L., Yachechko R., et al. (2008) Generation of human induced pluripotent stem cells from dermal fibroblasts. Proc Natl Acad Sci USA, 105, 2883–2888.
12. Park I.H., Zhao R., West J.A., et al. (2008) Reprogramming of human somatic cells to pluripotency with defined factors. Nature 451, 141–146.
13. Ye Z., Zhan H., Mali P., et al. (2009) Human induced pluripotent stem cells from blood cells of healthy donors and patients with acquired blood disorders. Blood 114, 5473–5480.
14. Sullivan G.J., Hay D.C., Park I.H., et al. (2010) Generation of functional human hepatic endoderm from human induced pluripotent stem cells. Hepatology 51, 329–335.
15. Saha K. and Jaenisch R. (2009) Technical challenges in using human induced pluripotent stem cells to model disease. Cell Stem Cell 5, 584–595.
16. Kim, K., Doi A., Wen B., et al. (2010) Epigenetic memory in induced pluripotent stem cells. Nature 467, 285–290.
17. Polo, J.M. Liu S., Figueroa M.E., et al. (2010) Cell type of origin influences the molecular and functional properties of mouse induced pluripotent stem cells. Nat Biotechnol 28, 848–855.
18. Kim J.B., Greber B., Arau´zo-Bravo M.J., et al. (2009) Direct reprogramming of human neural stem cells by OCT4. Nature 461, 649–643.
19. Aasen T., Raya A., Barrero M.J., et al. (2008) Efficient and rapid generation of induced pluripotent stem cells from human keratinocytes. Nat Biotechnol 26, 1276–1284.
20. Liu H., Ye Z., Kim Y., et al. (2010) Generation of endoderm-derived human induced pluripotent stem cells from primary hepatocytes. Hepatology 51, 1810–1819.
21. Rebouissou S., Amessou M., Couchy G., et al. (2009) Frequent in-frame somatic deletions activate gp130 in inflammatory hepatocellular tumours. Nature 457, 200–204.
22. Laurent-Puig P., and Zucman-Rossi J. Genetics of hepatocellular tumors. (2006) Oncogene 25, 3778–3786.
23. Bluteau O., Jeannot E., Bioulac-Sage P., et al. (2002) Bi-allelic inactivation of TCF1 in hepatic adenomas. Nat Genet 32, 312–315.
24. Wong C.M., Fan S.T., Ng I.O. (2001) Beta-catenin mutation and overexpression in hepatocellular carcinoma: clinicopathologic and prognostic significance. Cancer 92, 136–145.
25. Tanaka S., Toh Y., Adachi E., et al. (1993) Tumor progression in hepatocellular carcinoma may be mediated by p53 mutation. Cancer Res 53, 2884–2887.
26. Prashant M., Ye Z., Chou B.K., et al. (2010) An improved method for generating and identifying human induced pluripotent stem cells. Methods Mol Biol 636, 191–205.

Chapter 13

Derivation of Human Induced Pluripotent Stem Cells on Autologous Feeders

Kazutoshi Takahashi

Abstract

One of the possible applications of human induced pluripotent stem (iPS) cells is its usage for transplantation (Takahashi K et al. Cell 131:861–872, 2007; Yu J et al. Science 318:1917–1920, 2007; Park IH et al. Nature 451:141–146, 2008; Lowry WE et al. Proc Natl Acad Sci USA 105:2883–2888, 2008; Morita S et al. Gene Ther 7:1063–1066, 2000). For this purpose, animal components and undefined factors should be precluded. In this chapter, we describe a practical and effective protocol for generating and maintaining human iPS cells on autologous feeders.

Key words: Human pluripotent stem cells, Autologous feeders, Tissue regeneration, Maintenance of human pluripotent stem cells, Generation of human pluripotent stem cells

1. Introduction

Human pluripotent stem cells, both embryonic stem (ES) cells and induced pluripotent stem (iPS) cells, are generally maintained on mouse embryonic fibroblasts (MEFs) which are mitotically inactivated by treatment with mitomycin C or γ-ray irradiation. However, use of mouse cells as feeders may cause contamination problems, and they can be potential carriers of unknown virus or zoonotic pathogen. Actually, previous study demonstrated that sialic acid, which human cells could not produce, was detected on the surface of human ES cells maintained on MEF feeder. Human fibroblasts from neonatal foreskin or ES cell-derived fibroblast-like cells can support self-renewal of human ES cells. The issue of the contamination risk of xenogenic components can be overcome by using human feeder cells concerning the safety in the clinical application of stem cells.

2. Materials

2.1. Reagents

1. *pMXs retroviral vectors and Plat-E packaging cells* (5): available from Dr. Toshio Kitamura at the University of Tokyo (kitamura@ims.u-tokyo.ac.jp). Plat-E cells can be purchased from Cell Biolabs, Inc. (http://www.cellbiolabs.com/).

2. *pMXs containing cDNAs of OCT3/4, SOX2, KLF4, or c-MYC and pLenti6/UbC containing mouse Slc7a1 gene*: available from Addgene (http://www.addgene.org/Shinya_Yamanaka).

3. *Human cells*: Various types of human cells can be obtained from the following companies and organizations: Cell Applications Inc. (http://www.cellapplications.com/), Lonza (http://www.lonza.com/group/en.html), American Type Culture Collection; ATCC (http://www.atcc.org/), European Collection of Cell Cultures; ECACC (http://www.ecacc.org.uk/), RIKEN BioResource Center (http://www.brc.riken.jp/), Japanese Collection of Research Bioresources (http://cellbank.nibio.go.jp/).

4. *293FT cells*: This cell line can be purchased from Invitrogen.

5. *Dulbecco's modified Eagle's medium* contains 4.5 g/L glucose (DMEM; 14247-15, Nacalai Tesque, Japan).

6. *Phosphate-buffered saline* without calcium and magnesium (PBS; 14249-95, Nacalai tesque, Japan).

7. *DMEM/F12* (10565-018, Invitrogen).

8. *Knockout serum replacement* (KSR, 10828028, Invitrogen)

9. *L-Glutamine* (25030-081, Invitrogen).

10. *Nonessential amino acids solution* (11140-050, Invitrogen).

11. *2-Mercaptoethanol* (21985-023, Invitrogen).

12. *Sodium pyruvate* (S8636, Sigma).

13. *Penicillin/streptomycin* (15140-122, Invitrogen).

14. *Collagenase IV* (17104-019, Invitrogen), see Sect. 2.3.

15. *Recombinant basic fibroblast growth factor*, human (bFGF, 064-04541, Wako, Japan), see Sect. 2.3.

16. *Bovine serum albumin* (810-661, ICN).

17. *2.5% Trypsin* (15090-046, Invitrogen).

18. *0.25% Trypsin/1 mM EDTA solution* (25200-056, Invitrogen).

19. *0.5% Trypsin/5.3 mM EDTA solution* (25300-054, Invitrogen), see Sect. 2.3.

20. *Gelatin* (G1890, Sigma), see Sect. 2.3.

21. *Puromycin* (P7255, Sigma), see Sect. 2.3.

22. *Blasticidin S hydrochloride* (KK-400, Funakoshi, Japan), see Sect. 2.3.
23. *Virapower Lentiviral expression system* (K4990-00, Invitrogen).
24. *FuGENE 6 transfection reagent* (1 814 443, Roche).
25. *Lipofectamine 2000* (11668-019, Invitrogen).
26. *OPTI-MEM I* (31985-062, Invitrogen).
27. *Hexadimethrine Bromide* (Polybrene; 17736-44, Nacalai Tesque, Japan), see Sect. 2.3.
28. *Dimethyl sulfoxide* (D2650, Sigma).
29. *Human ES medium*, see Sect. 2.3.
30. *293FT medium*, see Sect. 2.3.
31. *FP medium* (for fibroblasts and Plat-E cells), see Sect. 2.3.
32. *Povidone iodine* (Isodine solution, Meiji Seika, Japan).

2.2. Equipment

1. 100-mm tissue culture dish (353003, Falcon).
2. 6-well tissue culture plate (353046, Falcon).
3. 24-well tissue culture plate (353047, Falcon).
4. 96-well tissue culture plate (351172, Falcon).
5. 15-mL conical tube (352196, Falcon).
6. 50-mL conical tube (352070, Falcon).
7. 1-mL plastic disposable pipette (357520, Falcon).
8. 5-mL plastic disposable pipette (357543, Falcon).
9. 10-mL plastic disposable pipette (357551, Falcon).
10. 25-mL plastic disposable pipette (357525, Falcon).
11. 0.22-μm pore size filter (Millex GP, SLGP033RS, Millipore).
12. 0.45-μm pore size cellulose acetate filter (FP30/0.45 CA-S, Schleicher & Schuell)
13. 10-mL disposable syringe (SS-10ESZ, Terumo, Japan).
14. Coulter counter (Z2, Beckman Coulter).
15. CO_2 incubator.
16. Cell scraper (9000-220, Iwaki).
17. Pipet-Aid (Falcon).
18. Pippetman and tips (GILSON).

2.3. Reagent Setup

2.3.1. Gelatin Coating

1. Prepare gelatin stock solution at 10× concentration (1% w/v) by dissolving 1 g of gelatin powder in 100 mL of distilled water, autoclave, and store at 4°C.
2. To prepare 0.1% (1×) gelatin working solution, thaw the 10× gelatin stock with a microwave and add 50 mL of 10× stock solution to 450 mL of distilled water. Filter the solution with a bottle-top filter (0.22 μm) and store at 4°C.

3. To coat a culture dish with gelatin, add enough volume of 0.1% gelatin working solution to cover the bottom of the dish. For example, 1, 3, or 5 mL of gelatin working solution is used for a 35-, 60-, or 100-mm dish, respectively. Incubate the dish for at least half an hour at 37°C. Before using, aspirate excess gelatin solution.

2.3.2. 0.05% Trypsin/0.53 mM EDTA Solution

1. To prepare 0.05% Trypsin/0.53 mM EDTA, mix 10 mL of 0.5% Trypsin/5.3 mM EDTA solution and 90 mL of PBS.
2. To prepare 0.1% Trypsin/1 mM EDTA, add 20 mL of 0.5% Trypsin/5.3 mM EDTA to 80 mL of PBS. Aliquot and store at −20°C.

2.3.3. 2× Freezing Medium

Mix 2 mL of DMSO, 2 mL of FBS, and 6 mL of DMEM, and filter through a 0.22-μm filter.

2.3.4. Puromycin

Dissolve in distilled water at 10 mg/mL and sterilize through a 0.22-μm filter. Aliquot and store at −20°C.

2.3.5. Blasticidin S Hydrochloride

Dissolve in distilled water at 10 mg/mL and sterilize through a 0.22-μm filter. Aliquot and store at −20°C.

2.3.6. Hexadimethrine Bromide (Polybrene)

Dissolve 0.8 g of polybrene in 10 mL of distilled water for a 10× stock (80 mg/mL). Dilute 1 mL of 10× stock solution with 9 mL of distilled water and filter with a 0.22-μm filter. Store at 4°C.

2.3.7. Recombinant Basic Fibroblast Growth Factor, Human

To prepare PBS containing 0.1% bovine serum albumin (BSA), add 50 μL of 10% BSA into 5 mL of PBS. Dissolve 50 μg of bFGF in 5 mL of PBS containing 0.1% BSA (10 μg/mL). Aliquot and store at −20°C.

2.3.8. Y-27632

Dissolve 5 mg of Y-27632 in 1.48 mL of distilled water (10 mM). Aliquot and store at −20°C.

2.3.9. Human ES Medium

DMEM/F12 containing 20% KSR, 2 mM L-glutamine, 1×10^{-4} M nonessential amino acids, 1×10^{-4} M 2-mercaptoethanol, and 50 units and 50 mg/mL penicillin and streptomycin. To prepare 500 mL of the medium, mix 100 mL of KSR, 5 mL of L-glutamine, 5 mL of nonessential amino acids, 1 mL of 2-mercaptoethanol, and 2.5 mL of penicillin/streptomycin, and then fill to 500 mL with DMEM/F12. Add 0.2 mL of 10 μg/mL basic fibroblast growth factor (bFGF) into 500 mL of the medium before use. Store at 4°C for a week.

2.3.10 293FT Medium

DMEM containing 10% FBS, 2 mM L-glutamine, 1×10^{-4} M nonessential amino acids, 1 mM sodium pyruvate, and 50 units and 50 mg/mL penicillin and streptomycin. To prepare 500 mL of the medium, mix 50 mL of FBS, 5 mL of L-glutamine, 5 mL of nonessential amino acids, 5 mL of sodium pyruvate, and 2.5 mL of penicillin/streptomycin, and then fill to 500 mL with DMEM. Store at 4°C for a week. Add 0.1 mL of 50 mg/mL G418 into 10 mL 293FT medium.

2.3.11. FP Medium (for Fibroblasts and Plat-E Cells)

DMEM containing 10% FBS, and 50 units and 50 mg/mL penicillin and streptomycin. To prepare 500 mL of FP medium, mix 50 mL of FBS and 2.5 mL of penicillin/streptomycin, and then fill up to 500 mL with DMEM. Store at 4°C for a week. For Plat-E cells, add 1 µL of 10 mg/mL puromycin stock and 10 µL of 10 mg/mL blastcidin S into 10 mL of FP medium.

3. Protocol

3.1. Preparation of Fibroblasts

Prepare human fibroblasts from patient's skin or as commercial products. If you obtain fibroblasts from biopsy materials, you must explain the procedure and get a patient's informed consent.

3.1.1. Thawing Fibroblasts

1. Prepare 9 mL of FP medium in a 15-mL tube.
2. Take a vial of frozen fibroblasts from the liquid nitrogen tank and put the vial into a 37°C water bath until most (but not all) cells are thawed.
3. Wipe the vial with 70% ethanol, open the cap, and transfer the cell suspension to the tube prepared in step 1.
4. Centrifuge at $160 \times g$ for 5 min, and then discard the supernatant.
5. Resuspend the cells with 10 mL of FP medium and transfer to a 100-mm dish (at least 5×10^5 cells/100-mm dish). Incubate the cells in a 37°C, 5% CO_2 incubator, until the cells become 80–90% confluent.

3.1.2. Passage of Fibroblasts

1. Discard the medium and wash the cells once with PBS.
2. Aspirate PBS, add 1 mL per dish of 0.05% trypsin/0.53 mM EDTA, and incubate for 10 min at 37°C.
3. Add 9 mL of FP medium, and break up the cells into a single cell suspension by pipetting up and down several times.
4. Adjust the cell suspension to 40 mL by addition of FP medium, and transfer to dishes (10 mL per 100-mm dish). This splits the cells 1:4. Incubate the cells at 37°C, 5% CO_2 until the cells become 80–90% confluent (approximately 2×10^6 cells per 100-mm dish). This should happen 4–5 days after passage.

3.2. Lentivirus Production

The experiment using lentivirus should be performed in a Biological Safety Level II cabinet. Wear the lab coat and gloves all the time. And wastes of the experiments must be treated with ethanol and hypochlorous acid, and then autoclaved.

3.2.1. Passaging 293FT Cells

1. Aspirate the medium and wash the cells with PBS.
2. Add 1 mL of 0.25% trypsin/1 mM EDTA and incubate the dish for 2 min at room temperature.
3. Add 10 mL of the medium and dissociate the cells by pipetting up and down (about ten times).
4. Collect the cell suspension into conical tube, and count the cell number.
5. Adjust the concentration to 4×10^5 cells per milliliter with 293FT medium without G418. Seed cells at 4×10^6 cells (10 mL) per 100-mm dish and incubate overnight at 37°C, 5% CO_2.

3.2.2. Transfection to 293FT Cells

1. Dilute 9 µg of Virapower packaging mix (pLP1, pLP2, and pLP/VSVG mixture) and 3 µg of pLenti6/UbC encoding mouse *Slc7a1* gene in 1.5 mL of OPTI-MEM I, and mix gently.
2. In a separate tube, dilute 36 µL of Lipofectamine 2000 in 1.5 mL OPTI-MEM I. Mix gently and incubate for 5 min at room temperature.
3. After incubation, combine the diluted DNA with the diluted Lipofectamine 2000. Mix gently and incubate for 20 min at room temperature.
4. During incubation, remove the medium from 293FT dish and add 9 mL of fresh medium.
5. Then add 3 mL of DNA–Lipofectamine 2000 complexes to the dish. Mix gently by rocking the dish back and forth. Incubate the dish overnight at 37°C, 5% CO_2.
6. Twenty four hours after transfection, aspirate the medium containing the transfection cocktail and add 10 mL of fresh one. Incubate the dish overnight at 37°C, 5% CO_2.

3.2.3. Collection of Virus-Containing Supernatant

Forty-eight hours after transfection, collect the supernatant of 293FT culture by using a 10-mL disposable syringe, and then filtrate it with a 0.45-µm pore size cellulose acetate filter.

3.3. Lentiviral Infection

3.3.1. Seeding Fibroblasts

1. Aspirate the medium and wash the cells with PBS.
2. Add 1 mL of 0.05% trypsin/0.53 mM EDTA and incubate the dish for 10 min at 37°C.
3. Add 9 mL of the medium and dissociate the cells by pipetting up and down.
4. Collect the cell suspension into conical tube, and count the cell number. Adjust the concentration to 8×10^4 cells/mL.
5. Seed the cells at 8×10^5 cells (10 mL of cell suspension) per 100-mm dish and incubate overnight at 37°C, 5% CO_2.

3.3.2. Transduction to Fibroblasts

1. Replace the medium with 10 mL of virus-containing supernatant supplemented with 4 µg/mL polybrene. Incubate the dish for 5 h to overnight at 37°C, 5% CO_2.
2. Twenty-four hours after transduction, aspirate off the virus-containing medium, wash the cells with 10 mL of PBS (optional), and then add 10 mL of fresh medium.

3.4. Preparation of Feeder Cells

3.4.1. Mitomycin C Inactivation of Fibroblasts

1. Add 0.3 mL of 0.4 mg/mL mitomycin C solution directly to the culture medium of subconfluent fibroblast dish, swirl it briefly, and incubate 2.25 h at 37°C, 5% CO_2. The final concentration of mitomycin C will be 12 µg/mL.
2. After incubation, remove the mitomycin C-containing medium, and wash the cells twice with 10 mL of PBS.
3. Aspirate off PBS, add 0.5 mL of 0.25% trypsin/1 mM EDTA, swirl to cover the entire surface, and let sit for 10 min at room temperature.
4. Neutralize the trypsin by adding 5 mL of FP medium, and break up the cells to a single cell suspension by pipetting up and down. Pool the cell suspension into a 50-mL tube and count the cell number. Seed the cells at 1.5×10^6 cells per 100-mm tissue culture dish, or at 2.5×10^5 cells per well of 6-well plate.
5. Cells should be nicely spread with little gaps in between. They should become ready for use by the next day. The mitomycin C-treated fibroblast dishes can be left for up to 3 days before use. And if you desire, you can make freeze stocks of mitomycin C-treated fibroblast cells with a standard technique at –80°C or in vapor phase of liquid nitrogen tank. These stocks should be waked up to gelatin-coated dish or plate within the 3 days before use. Old S feeder (over 3 days after mitomycin C treatment) may come off from the dish during over 3 weeks of iPS cell generation.

3.5. Generation of iPS Cells

3.5.1. Retrovirus Production

Day 1 *Plat-E preparation*

1. Wash the cells with PBS, add 4 mL of 0.05% trypsin/0.53 mM EDTA, and incubate for 1 min at room temperature.
2. After incubation, add 10 mL FP medium into the Plat-E dish, suspend the cells by gently pipetting, and transfer the cell suspension to a 50-mL tube.
3. Centrifuge the cells at $180 \times g$ for 5 min.
4. Discard the supernatant, break the pellet by finger tapping, and resuspend the cells in an appropriate amount of FP medium.
5. Count cell number and adjust the concentration to 3.6×10^5 cells/mL with FP medium.

6. Seed cells at 3.6×10^6 cells (10 mL) per 100-mm culture dish, and incubate overnight at 37°C, 5% CO_2.

Day 2 *Retrovirus production; Transfection into Plat-E cells*

7. Transfer 0.3 mL of OPTI-MEM I into a 1.5-mL tube.
8. Deliver 27 µL of Fugene 6 transfection reagent into the prepared tube in step 80, mix gently by finger tapping, and incubate for 5 min at room temperature.
9. Add 9 µg of pMXs plasmid DNA (encoding OCT3/4, SOX2, KLF4, and c-MYC) drop by drop into the Fugene 6/DMEM containing tube, mix gently by finger tapping, and incubate for 15 min.
10. Add the DNA/Fugene 6 complex dropwise into the Plat-E dish, and incubate overnight at 37°C, 5% CO_2.

Day 3 *Retrovirus production (continued)*

11. Aspirate the transfection reagent-containing medium, add 10 mL of fresh FP medium, and return the cells to the incubator.

Preparation of fibroblasts

12. Culture fibroblasts expressing mouse *Slc7a1* gene to 80–90% confluency in 100-mm dishes (~2×10^6 cells per dish).
13. Aspirate the culture medium and wash with 10 mL of PBS.
14. Discard PBS, add 1 mL per dish of 0.05% trypsin/0.53 mM EDTA, and incubate at 37°C for 10 min.
15. Add 9 mL of the culture medium, suspend the cells to a single cell, and transfer to a 50-mL tube.
16. Count cell numbers, and adjust the concentration to 8×10^4 cells/mL. Transfer 10 mL of cell suspension (8×10^5 cells) to a 100-mm dish. Incubate the dish overnight at 37°C, 5% CO_2.

Day 4 *Retroviral infection*

17. Forty-eight hours post-transfection, collect the medium from the Plat-E dish by using a 10-mL sterile disposable syringe, filtering it through a 0.45-µm pore size cellulose acetate filter, and transferring into a 15-mL tube.
18. Add 5 µL of 8 mg/mL polybrene solution into the filtrated virus-containing medium, and mix gently by pipetting up and down. The final concentration of polybrene will be 4 µg/mL.
19. Make a mixture of equal parts of the medium containing OCT3/4-, SOX2-, KLF4-, and c-MYC-retroviruses.
20. Aspirate the medium from a fibroblast dish, and add 10 mL of the polybrene/virus-containing medium. Incubate the cells from 4 h to overnight at 37°C, 5% CO_2.

Day 5 *TIMING 5 min*

21. After 24 h, aspirate the medium from a fibroblast dish, and add 10 mL of fresh FP medium.

Day 6–10 *TIMING 5 min each day*

22. Discard the medium, and add 10 mL of FP medium.

Day 11 *Replating fibroblasts onto mitomycin C-treated feeder. TIMING 1 h*

23. Aspirate the culture medium and wash with 10 mL of PBS.

24. Discard PBS, add 1 mL per dish of 0.05% trypsin/0.53 mM EDTA, and incubate at 37°C for 10 min.

25. Add 9 mL of the culture medium, suspend the cells to a single cell, and transfer to a 50-mL tube.

26. Count cell numbers, and adjust the concentration to 5×10^3 or 5×10^4 cells/mL. Transfer 10 mL of cell suspension (5×10^4 or 5×10^5 cells) to a 100-mm dish with mitomycin C-treated feeder cells. Incubate the dish overnight at 37°C, 5% CO_2 (see TROUBLE SHOOTING).

Day 12 ~ *TIMING 5 min. each day*

27. Replace the medium with 10 mL of human ES medium. Exchange the medium every other day until the colonies become large enough to be picked up. Colonies should first become visible approximately in 2 or 3 weeks after the retroviral infection. They should become large enough to be picked up around day 30.

Picking up the iPS colonies; TIMING 1 h

28. Aliquot 20 µL of human ES medium per well of 96-well plate.

29. Remove the medium from the dish, and add 10 mL of PBS.

30. Aspirate PBS, and add 5 mL of PBS.

31. Pick up colonies from the dish under the stereomicroscope using a P2 or P10 Pipetman set at 2 µL, and transfer it into the 96-well plate prepared in Step 61.

32. Add 180 µL of human ES medium to each well, and pipette up and down to break up the colony to small clumps carefully under the stereomicroscope.

33. Transfer cell suspension into the well of 24-well plates with mitomycin C-treated feeder cells, add 300 µL of human ES medium, and incubate in a 37°C, 5% CO_2 incubator until the cells reach 80–90% confluency. At this point, they should be passaged into 6-well plates. In general, the timing of passage has come approximately a week after picking up the colonies.

Passaging of iPS cells; TIMING 0.5 h

34. Aspirate the medium, and wash the cells with 0.5 mL of PBS.
35. Completely remove PBS, add 0.1 mL of CTK solution, and incubate at 37°C for 2–5 min. When the feeder cells are removed from the plate and iPS colonies still attach, this is time of interruption of the digestion.
36. Aspirate off CTK solution, and add 0.5 mL of PBS.
37. Aspirate off PBS, and add 0.5 mL of PBS. Most of feeder cells should be removed with discarding PBS.
38. Remove completely PBS, add 0.5 mL of human ES medium, and suspend the cells by pipetting up and down to small clumps.
39. Transfer the cell suspension to a well of 6-well plate with mitomycin C-treated feeder cells, add 1.5 mL of human ES medium, and incubate in a 37°C, 5% CO_2 incubator until cells reach 80–90% confluency in 6-well plates. Exchange the medium every day.

4. Notes

1. Fibroblasts die or undergo growth arrest after lentiviral transduction. In some cases, lentivirus is toxic for fibroblasts. This problem can be overcome by doubling dilution of the supernatant with the fresh medium or shorting exposure time from overnight to 5 h.
2. No ES-like colonies appear from fibroblast cultures after induction by the four factors. The titer of retrovirus is important as described above. The retrovirus must be prepared freshly. DO NOT FREEZE RETROVIRUS. Age (passage number) of fibroblast is also critical for iPS generation. Efficiency of retroviral transduction markedly decreases in older fibroblasts. We recommend using fibroblasts within passage 8 for iPS production. The seeding cell number after retroviral infection onto feeder cells is very important. Overgrowth of fibroblasts may block the generation of iPS cell colonies. On the other hand, if the cell number is too low, you may not be able to generate any colony. The optimal conditions could be different for individual cell types. We recommend seeding at least two different densities to test which one works better for your experiment.
3. Cells peel off the substrate during reprogramming. There are two conceivable possibilities to cause the problem. One is that feeder cells are too old. Old feeder cells that are prepared more than 3 days in advance cannot suffer by stimulation of bFGF in

human ES medium. We recommend using feeder cells within 3 days after mitomycin C treatment. Another possibility is that the seeding number of fibroblasts after transduction is too high. Overgrowth of fibroblasts can induce peeling off from the edge of the dish like a sheet. This can be overcome by reducing the cell number. In any case, optimization is required for better results.

4. iPS cells spontaneously differentiate during daily culture. In some cases, iPS cells are unstable during early passage period. When passaging, you can remove the differentiated colonies by aspirating and only undifferentiated colonies are transferred to a new dish. After repeating this procedure, majority of the clone would consist of undifferentiated colonies. In addition, the qualities of feeder cells such as density and freshness are quite important. In addition, some fibroblasts are not suitable as feeder cells. Three of 14 fibroblast cell lines we tested could not support the self-renewal of human iPS cells(6).

References

1. Takahashi, K., et al. Induction of pluripotent stem cells from adult human fibroblasts by defined factors. Cell 131, 861–72 (2007)
2. Yu, J., et al. Induced pluripotent stem cell lines derived from human somatic cells. Science 318, 1917–20 (2007)
3. Park, IH., et al. Reprogramming of human somatic cells to pluripotency with defined factors. Nature 451, 141–6 (2008)
4. Lowry, WE., et al. Generation of human induced pluripotent stem cells from dermal fibroblasts. Proc. Natl. Acad. Sci. USA. 105, 2883–8 (2008)
5. Morita, S., Kojima, T. & Kitamura, T. Plat-E: an efficient and stable system for transient packaging of retroviruses. Gene Ther. 7, 1063–6 (2000)
6. Takahashi, K., et al. Human induced pluripotent stem cells on autologous feeders. PLoS ONE 4:e8067 (2009).

Chapter 14

Human Mesenchymal Stem Cells and iPS Cells (Preparation Methods)

Hiroe Ohnishi, Yasuaki Oda, and Hajime Ohgushi

Abstract

Mesenchymal stem cells (MSCs) are adult stem cells which show differentiation capabilities toward various cell lineages, and the MSCs can be cryopreserved for a long term without noticeable loss of these capabilities. We have used bone marrow–derived MSCs for treatments of patients with osteoarthritis, bone necrosis, and bone tumor. However, their proliferation/differentiation capabilities are limited and show possible disadvantages for clinical use. On the other hand, induced pluripotent stem cells (iPSCs) have nearly unlimited proliferation and differentiation capabilities into a wide variety of cells throughout the body, and thus the iPSCs are expected to be used in regenerative medicine and drug discovery research. We have reported that the MSCs can be utilized for the generation of iPSCs. In this chapter, we describe the methods of MSC preparation from bone marrow and iPSCs generation from MSCs. We also focus on the generation from commercially available MSCs from adipose tissue.

Key words: Mesenchymal stem cell, Adipose tissue, Bone marrow, Induced pluripotent stem cell, Somatic stem cells, Regenerative medicine, Tissue engineering

1. Introduction

There are two major types of human stem cells, i.e., embryonic stem cells (ESCs) and somatic stem cells. One type of the somatic stem cells is mesenchymal stem cells (MSCs). MSCs are found in various tissues, such as bone marrow, adipose tissue, tooth germ, and synovial membranes. Human MSCs are positive for CD13, CD29, CD44, CD71, CD90, CD105, STRO-1, ALP, HLA-Class 1, AH-2, SH3, SH4, and negative for CD14, CD31, CD34, CD45, HLA-DR. The MSCs are known to differentiate into various cell lineage, including hepatocytes, neurocytes, cardiomyocytes, osteocytes, chondrocytes, and adipocytes (1–3). Usually, MSCs can be isolated by density gradient centrifugation using Ficoll and can be expanded in culture dish as adherent fibroblastic cells. We have treated many patients by transplantation of bone marrow MSCs derived from the same patients (4, 5). We did not use the gradient centrifugation for the MSC isolation but simply expanded the number of MSCs by direct culture of buffy coat together with red blood cell fraction from small amount of aspirated bone marrow. During the culture, the red blood cells can be easily eliminated after several exchanges of the culture medium. After about 10 days, the adherent cells (MSCs) become almost confluent. The primary cultured or subcultured cells can be used in the treatments of patients (4, 5).

Importantly, the MSCs can be preserved by freezing, and we reported that the viability of the cryopreserved MSCs was approximately 90%. Excellent viability was achieved regardless of the storage term (0.3–37 months) (6). The FACS analysis demonstrated that the cells were negative for hematopoietic but positive for mesenchymal characteristics. No difference of the osteogenic potential was found between cells with and without cryopreservation treatments. Furthermore, about 3 years cryopreserved cells maintained the high osteogenic potential. Therefore, even a long term cryopreserved MSCs can be expected to be used in regeneration of various tissues (6). However, the MSCs lose the high proliferation as well as differentiation capabilities after several passages of the culture; thus, we need a novel technology to recover these capabilities, aiming for the MSCs as promising cells in regenerative medicine (7).

In 2006, Takahashi and Yamanaka demonstrated that mouse somatic cells could be reprogrammed to a pluripotent state by transduction of four transcription factors (*OCT3/4, SOX2, KLF4,* and *c-MYC*) (8). Furthermore, the researchers reported that the same four factors are also effective in reprogramming human somatic cells (9). Therefore, they succeeded to generate another type of stem cells, i.e., induced pluripotent stem cells (iPSCs) (8, 9). If we can generate the iPSCs from cryopreserved patients' MSCs, the iPSCs derived from the MSCs might be available for further treatments of the same patients. We already reported the successful iPSC generation using cryopreseved MSCs from human fat tissue, bone marrow, and third molars. In this chapter, we describe the methods of MSC preparation and the generation of iPSCs from bone marrow and adipose tissue, which is available after liposuction as a waste tissue.

2. Materials

2.1. Preparation of MSCs from Human Bone Marrow and Commercially Available MSCs from Human Adipose Tissue

2.1.1. Culture Expansion of MSCs from Freshly Obtained Bone Marrow

1. Marrow collection tube; Sterilized 13 mL polypropylene tube (Assist Co. Tokyo, Japan).
2. Storage bottle; Polystyrene Storage Bottles with 45 mm Caps (Product #8390; Corning, MA).
3. Vacuum Filter (150 mL Bottle Top Vacuum Filter, 0.22 μm Pore, Product #431161; Corning, MA).
4. T75 flask (BD Falcon, NJ, USA).
5. Bone marrow aspiration needle; Illinois bone marrow aspiration intraosseous infusion needle, 15 G (Allegiance Healthcare Corporation, IL, USA).
6. Heparin; Heparin sodium injection 10,000 U/10 mL (Ajinomoto Pharma Co. Tokyo, Japan).
7. Phosphate-Buffered Saline (PBS) (Invitrogen, Carlsbad, CA, Cat. No. 10010).
8. PBS/heparin solution: Add 150 mL of the PBS and 1.5 mL of the heparin into the storage bottle (10 U/mL), then pass through the 0.22 μm filter. Take 3 mL of the PBS/heparin into the marrow collection tube. Store at –20°C.
9. Fetal Bovine Serum (FBS) (SAFC, Lenexa, KS, Cat. No. 12007) (see Note 1).
10. Gentamicin sulfate (MSD K.K., Tokyo, Japan).
11. TrypLE™ Select (Invitrogen, Carlsbad, CA).
12. TC protector (DS Pharma Biomedical Co., Ltd., Osaka, Japan, Cat. No. TCP-001).
13. Minimum Essential Medium Alpha (α-MEM; Invitrogen, Carlsbad, CA, Cat. No. 41061-029).

2.1.2. Maintenance of MSCs from Bone Marrow and Adipose Tissue

1. Cell source of MSC from bone marrow (see Sect. 3.1.2, step 13)
2. Cell source of MSC from adipose tissue:
   hADSCs (Invitrogen, Carlsbad, CA, Lot No. 1212)
   hADSCs (Lonza Biosciences, Gaithersburg, MD, Lot No. 7 F3890)
3. 100 mm dish; 10 mm cell culture dish (BD Falcon, NJ, USA, Cat. No. 353003)
4. 15 mL tube; 15 mL polypropylene centrifuge tube (BD Falcon, NJ, USA, Cat. No. 352096)
5. MSC medium:
   Minimum Essential Medium Alpha (α-MEM; Invitrogen, Carlsbad, CA, Cat. No. 12571) containing 15% FBS

(JRH Biosciences, Lenexa, KS) and 100 U penicillin/100 μg/mL streptomycin (Invitrogen, Carlsbad, CA)

6. TrypLE Express (Invitrogen, Carlsbad, CA)
7. PBS (Invitrogen, Carlsbad, CA, Cat. No. 10010)

2.2. Generation of iPSCs Derived from MSCs

2.2.1. Maintenance of Platinum-A Cells and SNL76/7 Feeder Cells

1. 100 mm dish; 10 mm cell culture dish (BD Falcon, NJ, USA, Cat. No. 353003)
2. 15 mL tube; 15 mL polypropylene centrifuge tube (BD Falcon, NJ, USA, Cat. No. 352096)
3. 50 mL tube; 50 mL polypropylene centrifuge tube (BD Falcon, NJ, USA, Cat. No. 352070)
4. Plat-A cell; Plat-A packaging cells (Cell Biolabs, San Diego, CA)
5. Plat-A medium:
    Dulbecco's Modified Eagle's Medium (DMEM; Invitrogen, Carlsbad, CA) containing 10% FBS (FBS; JRH Biosciences, Lenexa, KS), 100 U Penicillin/100 μg/mL streptomycin (Invitrogen, Carlsbad, CA), 1 μg/mL Puromycin (Sigma, St. Louis, MO), and 10 μg/mL Blasticidin S hydrochloride (Funakoshi, Tokyo, Japan)
6. SNL cell; SNL76/7 feeder cell (European Collection of Cell Cultures, Salisbury, UK)
7. SNL medium:
    Dulbecco's Modified Eagle's Medium (DMEM; Invitrogen, Carlsbad, CA) containing 10% FBS (JRH Biosciences, Lenexa, KS) and 100 U Penicillin/100 μg/mL streptomycin (Invitrogen, Carlsbad, CA)
8. Mitomycin C (MMC; Roche Diagnostics, Basel, Switzerland)
9. TrypLE Express (Invitrogen, Carlsbad, CA)
10. PBS (Invitrogen, Carlsbad, CA, Cat. No. 10010)
11. Gelatin solution:
    Ultrapure water containing 0.1% gelatin (Sigma, St. Louis, MO)
12. Cell banker; BLC-1 (Mitsubishi Chemical Medience Corporation, Tokyo, Japan)

2.2.2. Generation and Maintenance of iPSCs

1. 100 mm dish; 100 mm cell culture dish (BD Falcon, NJ, USA, Cat. No. 353003)
2. 24-well plate; 24-well cell culture plate (BD Falcon, NJ, USA, Cat. No. 353047)
3. 6-well plate; 6-well cell culture plate (BD Falcon, NJ, USA, Cat. No. 353046)
4. 15 mL tube; 15 mL polypropylene centrifuge tube (BD Falcon, NJ, USA, Cat. No. 352096)

5. Syringe; 20 mL syringe (Terumo, Tokyo, Japan, Cat. No. SS-20ESZ)
6. Syringe filter; 0.45 μm pore size filter (Sartorius Stedim Biotech, Goettingen, Germany, Cat. No. 16555)
7. Cell scraper (Iwaki, Staffordshire, UK, Cat. No. 9000-220)
8. Pipette tip; 1–200 μL pipette tip (Labcon, CA, USA, Cat. No. 1093-965-008)
9. pMXs retroviral vectors (kindly donated by Dr. Kitamura)
10. pENTR/D-TOPO Cloning Kit; directional cloning vector for gateway technology (Invitrogen, Carlsbad, CA, Cat. No. K240020)
11. Gateway Vector Conversion System, Reading Frame Cassette A; reading frame cassette for gateway technology (Invitrogen, Carlsbad, CA, Cat. No. 11828-019)
12. Gateway LR Clonase II enzyme mix (Invitrogen, Carlsbad, CA, Cat. No. 11791020)
13. FuGENE HD Transfection Reagent (Roche Diagnostics, Basel, Switzerland)
14. Opti-MEM I medium (Invitrogen, Carlsbad, CA)
15. Hexadimethrine bromide (Polybrene; Sigma, St. Louis, MO)
16. PBS (Invitrogen, Carlsbad, CA, Cat. No. 10010)
17. iPSC medium:
    DMEM/F-12 with GlutaMAX-1 (Invitrogen, Carlsbad, CA) containing 20% Knockout Serum Replacement (KSR; Invitrogen, Carlsbad, CA), 1×10^{-4} M nonessential amino acid solution (Invitrogen, Carlsbad, CA), 1×10^{-4} M 2-mercaptoethanol (Invitrogen, Carlsbad, CA), and 50 U Penicillin and 50 μg/mL streptomycin (Invitrogen, Carlsbad, CA)
18. Recombinant human basic fibroblast growth factor (bFGF)
    iPSC medium containing 10 μg/mL bFGF (Wako, Osaka, Japan)
19. Valproic acid (VPA: Wako, Osaka, Japan, Cat. No. 227-01071)
20. CTK solution (cell dissociation solution)
21. Distilled water containing 0.25% trypsin (Invitrogen, Carlsbad, CA), 0.1% collagenase type IV (Invitrogen, Carlsbad, CA), 20% KSR, and 1 mM $CaCl_2$ (Wako, Osaka, Japan)

2.3. Characterization of iPSCs

2.3.1. In Vitro Differentiation of iPSCs

1. Nonadherent culture dish (Sumilon® Celltight X: Sumitomo Bakelite, Tokyo, Japan)
2. 12-well plate; 12-well cell culture plate (Falcon, NJ, USA, Cat. No. 353043)

3. 15 mL tube; 15 mL polypropylene centrifuge tube (BD Falcon, NJ, USA, Cat. No. 352096)
4. Gelatin solution; the same solution as described in Sect. 2.2.1

2.3.2. Immunofluorescence Microscopy

1. 4% Paraformaldehyde phosphate buffer solution (Wako, Osaka, Japan)
2. 0.1% TX-100/PBS

   PBS containing 0.1% (v/v) polyoxyethylene (10) octylphenyl ether (Wako, Osaka, Japan)
3. Blocking buffer:

   PBS containing 1% (w/v) bovine serum albumin (Sigma, St. Louis, MO)
4. Hoechst 33342 (Molecular Probes, Eugene, OR)
5. First antibodies:

   Anti-SSEA-3 antibody (dilution ratio 1:200; Cat. No. MAB4303, Millipore, Billerica, MA)

   Anti-SSEA-4 antibody (1:200, MAB4304, Millipore, Billerica, MA)

   Anti-TRA-1-60 antibody (1:200, ab16288, Abcam, Cambridge, UK)

   Anti-TRA-1-81 antibody (1:200, ab16289, Abcam, Cambridge, UK)

   Anti-OCT4 antibody (1:200, ab19857, Abcam, Cambridge, UK)

   Anti-NANOG antibody (1:50, ab21624, Abcam, Cambridge, UK)

   Anti-βIII-Tubulin antibody (1:200, CBL412, Millipore, Billerica, MA)

   Anti-α-SMA antibody (1:2, N1584, Dako, Glostrup, Denmark)

   Anti-Sox17 antibody (1:200, AF1924, R&D Systems, Minneapolis, MN)
6. Second antibodies:

   Alexa488-conjugated goat anti-rat IgM (1:200, A21212, Molecular Probes, Eugene, OR)

   Alexa488-conjugated donky anti-mouse IgG (1:200, A21202, Molecular Probes, Eugene, OR)

   Alexa488-conjugated goat anti-mouse IgM (1:200, A21042, Molecular Probes, Eugene, OR)

   Alexa568-conjugated goat anti-mouse IgM (1:200, A21043, Molecular Probes, Eugene, OR)

Alexa568-conjugated goat anti-rabbit IgG (1:200, A11011, Molecular Probes, Eugene, OR)

Alexa568-conjugated donky anti-goat IgG (1:200, A11055, Molecular Probes, Eugene, OR)

2.3.3. Teratoma Formation

1. 15 mL tube; 15 mL polypropylene centrifuge tube (BD Falcon, NJ, USA, Cat. No. 352096)
2. Hamilton syringes (Osaka Chemical Co., Ltd., Osaka, Japan)
3. Y-27632 (Wako, Osaka, Japan)
4. Nembutal (Dainippon Sumitomo Pharma, Osaka, Japan)
5. Severe combined immunodeficient (SCID) mouse, 8 weeks old, male (Charles River Laboratories, Yokohama, Japan)
6. 10% Formaldehyde neutral buffer (Nacalai Tesque, Kyoto, Japan)

3. Methods

3.1. Preparation of MSCs from Human Bone Marrow and Commercially Available MSCs from Human Adipose Tissue

Bone marrow should be collected from donors with informed consents only by qualified medical doctors. The collection can be done by needle aspiration under sterile condition. Dilution with peripheral blood can be minimized and bone marrow-derived MSCs concentration maximized by limiting the aspiration volume per puncture. We recommend about 3 mL of the aspiration per puncture. Usually, 2–4 punctures at different sites are available.

3.1.1. Preparation of Fresh Human Bone Marrow

1. Under local anesthesia, insert the needle vertically through the muscle layers until it reaches bone surface over anterior or posterior iliac crest.
2. Drive the needle into the bone and make sure the stable insertion of the needle.
3. Remove stylet and attach heparin-coated 10 mL syringe.
4. Apply sharp suction to suck about 3 mL of dark red marrow into the syringe.
5. Immediately transfer the marrow into the marrow collection tube containing 3 mL of PBS/heparin.
6. The total volume of the 6-mL bone marrow/PBS/heparin solution can be used for the following primary culture procedure.

3.1.2. Culture Expansion of MSCs from Fresh Human Bone Marrow

1. Centrifuge the tube containing 6 mL bone marrow/PBS/heparin solution at $140 \times g$ for 10 min.
2. Identify that the tube shows three layers by centrifugation (From the bottom, the layers are erythrocyte-rich fraction, buffy coat, and plasma with fat fraction.) (see Note 2) (Fig. 1a).

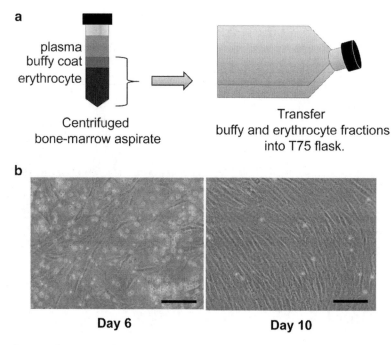

Fig. 1. (a) Schematic representation of expansion of MSCs from a human bone marrow aspirate. (b) Morphology of MSCs culture after 6 and 10 days post seeding, respectively. Scale bar = 200 μm.

3. Aspirate and discard the top layer (plasma with fat fraction) carefully.
4. Collect both erythrocytes and buffy coat layers and equally divide/transfer into two T75 flasks (Fig. 1a).
5. Add 13 mL of α-MEM containing 15% FBS and 20 μg/mL gentamicin sulfate into the flask.
6. Culture in a humidified atmosphere of 95% air with 5% CO_2 at 37°C.
7. After 2 days of culture, culture supernatants are removed, and 13 mL of the culture medium is added.
8. Change the medium (three times per week) (see Note 3).
9. The adherent cells grow and nearly reach confluence after 10–11 days.
10. Detach the adherent cells with 2 mL of TrypLE Select and collect.
11. Transfer 5×10^5 cells of the first passage (P1) to a T75 flask and further culture for several days to expand the number of adherent MSCs.
12. After 4–8 days, the morphology of most cells will be fibroblastic (see Note 4) (Fig. 1b).
13. Store 0.5–1 mL of MSCs at the concentration of 5×10^5 cells per mL of TC protector in cryovials at −80°C.

3.1.3. Maintenance of MSCs Prepared from Bone Marrow or Commercially Available Human Adipose Tissues

1. Thaw quickly the 5×10^5 cells of cryopreserved MSCs from either bone marrow or adipose tissues in water bath at 37°C and suspend with 9 mL of MSC medium in a 15-mL tube.
2. Centrifuge the 15-mL tube containing MSC suspension at $180 \times g$ for 5 min and aspirate supernatant.
3. Suspend the MSC pellet with 10 mL of MSC medium and transfer into 100 mm dish.
4. Culture in a humidified atmosphere of 95% air with 5% CO_2 at 37°C.
5. Change the medium (three times per week).
6. When the MSCs reach nearly confluence (Fig. 3a), detach the cells with 1 mL of TrypLE Express per 100 mm dish and add 9 mL of MSC medium, then transfer into a 15-mL tube.
7. Centrifuge the tube at $180 \times g$ for 5 min and aspirate supernatant.
8. Suspend the MSC pellet with MSC medium and transfer 5×10^5 MSCs into a 100-mm dish containing 10 mL of MSC medium.
9. Culture overnight in a humidified atmosphere of 95% air with 5% CO_2 at 37°C.
10. Use these MSCs in the 100-mm dish for iPSCs generation.

3.2. Generation of iPSCs Derived from MSCs

The protocol is largely based on the method described by Takahashi and coworkers (9, 10). We have used Plat-A cell line for rapid, transient production of high-titer, amphotropic retrovirus, and SNL cells as feeder cells which support the derivation and maintenance of human iPSCs (Fig. 2).

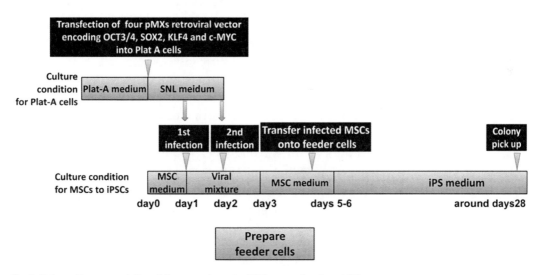

Fig. 2. Schematic representation of the procedures for iPSC generation from MSCs.

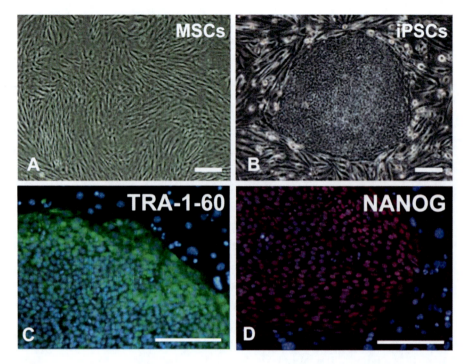

Fig. 3. (**a**) Morphology of the parental MSCs from human adipose tissue. (**b**) Morphology of ES-like colonies (iPSCs) derived from human adipose tissue. (**c**) and (**d**) Immunofluorescence microscopy of iPSCs detected using anti-TRA-1-60 antibodies and anti-NANOG antibodies, respectively. Scale bar = 200 µm. These figures reproduced from Aoki et al. (13), with permission from Mary Ann Liebert, Inc.

3.2.1. Maintenance of Plat-A Cells and SNL Cells

Maintenance of Plat-A Cells

1. Thaw quickly the 1×10^7 cells of cryopreserved Plat-A cells in a water bath at 37°C and suspend with 9 mL of Plat-A medium in a 15-mL tube.
2. Centrifuge the 15-mL tube containing Plat-A cells suspension at $180 \times g$ for 5 min and aspirate supernatant.
3. Suspend Plat-A cell pellet with 10 mL of Plat-A medium and transfer into four 100 mm dishes.
4. Culture in a humidified atmosphere of 95% air with 5% CO_2 at 37°C.
5. Change the medium every other day.
6. When the Plat-A cells reach confluence, detach the cells with 1 mL of TrypLE Express per 100 mm dish and add 9 mL of Plat-A medium, then transfer into a 15-mL tube.
7. Centrifuge at $180 \times g$ for 5 min and aspirate supernatant.
8. Suspend Plat-A cells pellet at the concentration of 8×10^6 cells per mL with Plat-A medium and transfer 1 mL of suspension into four 100 mm dishes containing 10 mL of Plat-A medium.

9. Culture overnight in a humidified atmosphere of 95% air with 5% CO_2 at 37°C.
10. Use these Plat-A cells for vector transfection required in iPSCs generation.

Maintenance and Mitomycin C (MMC) Treatment of SNL Cells

(i) MMC treatment of SNL cells

1. Add 20 mL of gelatin solution into a T225 flask and incubate for over 30 min in a humidified atmosphere of 95% air with 5% CO_2 at 37°C.
2. Aspirate geratin solution and use this dish as gelatin-coated flask.
3. Thaw quickly the 1×10^7 cells of cryopreserved SNL cells in a water bath at 37°C and suspend with 9 mL of SNL medium in a 15-mL tube.
4. Centrifuge the 15-mL tube at $180 \times g$ for 5 min and aspirate supernatant.
5. Suspend SNL cells pellet with 10 mL of SNL medium and transfer into a gelatin-coated T225 flask containing 20 mL of SNL medium.
6. Culture in a humidified atmosphere of 95% air with 5% CO_2 at 37°C.
7. Change the medium every 3 days.
8. When the SNL cells reach confluence, detach the cells with 7 mL of TrypLE Express per T225 flask and add 20 mL of SNL medium, then transfer into a 50-mL tube.
9. Centrifuge at $180 \times g$ for 5 min and aspirate supernatant.
10. Suspend SNL cell pellet with 10 mL of SNL medium and transfer into five gelatin-coated T225 flasks containing 25 mL of SNL medium.
11. When the SNL cells reach 90% confluence, add 10 μg/mL MMC and incubate for 2 h 45 min in a humidified atmosphere of 95% air with 5% CO_2 at 37°C.
12. Wash the MMC-treated SNL cells twice with 20 mL of PBS per T225 flask.
13. Detach the cells with 7 mL of TrypLE Express per T225 flask and add 20 mL of SNL medium, then transfer into a 50-mL tube.
14. Centrifuge at $180 \times g$ for 5 min and aspirate supernatant.
15. Suspend the cell pellet with 30 mL of SNL medium and centrifuge again ($180 \times g$ for 5 min).
16. Aspirate supernatant and suspend with cell banker at the cell density of 1×10^7 cells/mL.
17. Transfer 1 mL of the cell suspension into a cryovial and store at −80°C.

(ii) Prepare feeder cells

1. Add 5 mL of gelatin solution to a 100-mm dish and incubate for over 30 min in a humidified atmosphere of 95% air with 5% CO_2 at 37°C.
2. Aspirate gelatin solution and use this dish as gelatin-coated dish.
3. Thaw quickly the 1×10^7 cells of cryopreserved MMC-treated SNL cells in a water bath at 37°C and suspend with 9 mL of SNL medium in a 15-mL tube.
4. Centrifuge the tube at $180 \times g$ for 5 min and aspirate supernatant.
5. Suspend SNL cells pellet with 4 mL of SNL medium and transfer each of 1 mL cell suspension into four gelatin-coated 100 mm dishes containing 9 mL of SNL medium.
6. Culture in a humidified atmosphere of 95% air with 5% CO_2 at 37°C for 1–4 days.
7. Use for feeder cells for iPSCs generation and maintenance.

3.2.2. Generation and Maintenance of iPSCs

Plasmid Construction

Prepare plasmid constructs using Gateway technology.

1. Insert reading frame cassette A into the EcoRI site of the pMXs retroviral vector (11).
2. Amplify the open reading frames of human *OCT3/4* (*POU5f1 isoform-1*), *SOX2*, *KLF4*, and *c-MYC* by RT-PCR and clone into pENTR/D-TOPO.
3. Transfer these four open reading frames in pENTER/D-TOPO to the pMXs retroviral vector with LR clonase II, according to the manufacturer's instructions of Gateway technology.
4. Use these four vectors for retroviral production.

Retroviral Production

1. Change the medium of Plat-A cells (see Sect. 3.2.1.1, step 10) to 10 mL of fresh Plat-A medium.
2. Mix 0.3 mL of Opti-MEM I with 27 µL of FuGENE HD in 1.5 mL tube (make four tubes), then incubate for 5 min at room temperature.
3. Add 9 µg each of retroviral vector containing the open reading frames of human OCT3/4, SOX2, KLF4, or c-MYC (see above Sect. 3.2.2.1, step 4) to the Opti-MEM I/FuGENE HD mixture in 1.5 mL tube. The four different tubes with each reading frame are prepared (see Note 5).
4. Incubate these mixtures for 15 min at room temperature.
5. Transfer each of these four mixtures to four different Plat-A cultures in 100 mm dish, and incubate cells in a humidified atmosphere of 95% air with 5% CO_2 at 37°C for 24 h.

6. Aspirate the medium of the transfected Plat-A cells and wash gently with 10 mL SNL medium.
7. Add 10 mL of SNL medium and additional culture for 24 h.
8. Collect each of four supernatants from the Plat-A cell cultures to four different 15 mL tubes and transfer into a 20-mL syringe.
9. Add 10 mL of fresh SNL medium to the transfected Plat-A cells and culture continuously.
10. Filter each of the collected culture supernatant in the syringe through 0.45 μm pore size filter into 15 mL tube.
11. Add 4 μg/mL polybrene into the tube containing filtered supernatant.
12. Mix equal amount of four viral supernatants (*viral mixture #1*).
    After 24 h, repeat steps 8–12 (obtained viral supernatant is referred to as *viral mixture #2*).

iPSC Generation Through Virus Infection and Their Expansion by Passage

1. Replace the culture supernatant of MSCs (see Sect. 3.1.3, step 10) with 10 mL of *viral mixture #1* (see Sect. 3.2.2.2, step 12) (this step is called as *first infection*) and culture in a humidified atmosphere of 95% air with 5% CO_2 at 37°C for 24 h.
2. Replace *the viral mixture #1* with *viral mixture #2* (see Sect. 3.2.2.2, step 13) (this step is referred to as the *second infection*) and culture in a humidified atmosphere of 95% air with 5% CO_2 at 37°C for 24 h.
3. Replace the *viral mixture #2* with fresh MSC medium.
4. Exchange the medium to a fresh MSC medium every day until the viral infected MSCs become about 90% confluent.
5. Wash the cells with 10 mL PBS per 100 mm dish and treated with 2 mL TrypLE Express per 100 mm dish.
6. Collect detached cells and add 8 mL MSC medium, then transfer into a 15-mL tube.
7. Centrifuge at $180 \times g$ for 5 min and aspirate supernatant.
8. Suspend MSC pellets with 10 mL MSC medium and transfer into four 100 mm dish containing SNL feeder cells (see Sect. 3.2.1.2, (ii) step 7) at the cell density of 6×10^5 to 2×10^6 cells per 100 mm dish.
9. Next day, replace the medium of the MSCs on SNL feeder with iPSC medium containing 5 ng/mL bFGF and 1 mM VPA.
10. Exchange the iPSC medium supplemented with 5 ng/mL bFGF and 1 mM VPA every other day until 21 days post reseeding.
11. Exchange the iPSC medium containing 5 ng/mL bFGF every other day until 21 days to 45 days post reseeding.

12. About 18 days after reseeding, colonies of ESC-like morphology begins to appear.
13. Around 25–40 days after reseeding, scrape the colonies with 20 μL tips.
14. Transfer the scraped colonies (iPSCs) on SNL feeder cell in a well of 24-well plates. Add 0.5 mL of bFGF (5 ng/mL) containing iPSC medium to each well (see Note 6).
15. Culture the scraped iPSCs in a humidified atmosphere of 95% air with 5% CO_2 at 37°C for about 5 days.
16. Exchange the medium with the fresh iPSC medium that contains 5 ng/mL bFGF every day until cells become confluent.
17. Wash confluent iPSCs on SNL feeder cells in 24-well plate with 1 mL PBS per well.
18. Treat cells with 200 μL of CTK solution until SNL feeder cells are peeled away.
19. Wash remained iPSCs twice with 1 mL PBS.
20. Add 500 μL of iPSC medium to each well, scrape the iPSC colonies from dish with cell scraper.
21. Transfer the scraped iPSCs colonies in 500 μL iPSC medium into one well of 6-well plate containing SNL feeder cells in 2 mL iPSC medium containing 5 ng/mL bFGF.
22. Culture the cells for about 5 days and exchange the medium to fresh iPSC medium containing 5 ng/mL bFGF every day.
23. Wash confluent iPSCs on SNL feeder cells in 6-well plates with 2 mL PBS per well.
24. Treat cells with 500 μL of CTK solution until SNL feeder cells are peeled away.
25. Wash remained iPSCs twice with 2 mL PBS.
26. Add 1 mL iPSC medium to each well and scrape the iPSCs colonies.
27. Transfer the scraped iPSCs colonies in 1 mL iPSC medium into one 100 mm dish containing SNL feeder cells in 10 mL iPSC medium containing 5 ng/mL bFGF.
28. Culture the cells for about 5 days and exchange the medium every day.
29. Wash confluent iPSCs on SNL feeder cells in the 100 mm dish with 10 mL PBS.
30. Treat cells with 1 mL of CTK solution until SNL feeder cells are peeled away (see Note 7).
31. Wash the remained iPSCs twice with 10 mL PBS.

32. Add 4 mL of iPSC medium to each dish and scrape the iPSCs colonies (*iPSCs colonies suspension*).
33. Transfer the iPSCs colony suspension into another two to four 100 mm dishes containing 10 mL of iPSC medium supplemented with 5 ng/mL bFGF.
34. Culture the cells and exchange the medium every day.
35. Passage the iPSCs colonies (Fig. 3b) (steps 29–34) every 4–6 days.

3.3. Characterization of iPSCs Derived from MSCs

Embryoid body (EB) formation is one of the tests for determining the pluripotency potential of iPSCs.

3.3.1. In Vitro Differentiation of iPSCs

1. Prepare iPSCs colonies suspension (see Sect. 3.2.2, step 32) from confluent iPSCs in one 100 mm dish.
2. Centrifuge at $180 \times g$ for 1 min and aspirate supernatant.
3. Suspend cell pellet with 2 mL of iPSC medium gently and transfer into a nonadherent dish.
4. Culture in a humidified atmosphere of 95% air with 5% CO_2 at 37°C for 12 days. As the culture is done on the nonadherent dish, the culture can be called as suspension culture.
5. Exchange the medium with fresh iPSC medium every other day.
6. After 12 days of the suspension culture, iPSCs form ball-like structures (EB formation).
7. Suspend EBs in a nonadherent dish with 12 mL of iPSC medium and transfer into a gelatin-coated 12-well culture plate (see Note 8).
8. Culture EBs with 1 mL of iPSC medium per well in a humidified atmosphere of 95% air with 5% CO_2 at 37°C for 12 days and exchange the medium every other day.
9. After 12 days of cultivation in the gelatin-coated well plate, EBs adhere to well surface and differentiate into cells of three germ lineages (ectoderm, mesoderm, and endoderm).
10. Examine these cells by immunofluorescent microscopy.

3.3.2. Immunofluorescent Microscopy

iPSCs or iPSCs-derived differentiated cells in 12-well plates can be stained as follows:

1. Remove culture medium.
2. Fix cells with 1 mL of 4% PFA/PBS per well for 10 min at room temperature.
3. Wash three times with 1 mL of PBS per well.
4. Permeabilize the cells with 1 mL of 0.1% TX-100 in PBS per well for 10 min at room temperature.
5. Wash three times with PBS.

6. Incubate the cells with 1 mL of blocking buffer per well for 10 min at room temperature.
7. Incubate the cells with 500 μL of primary antibodies diluted with blocking buffer per well for overnight at 4°C (for pluripotent markers) or for 1 h at room temperature (for differentiated markers).
8. Wash three times with PBS.
9. Incubate the cells with 500 μL secondary antibodies and Hoechst 33342 diluted with blocking buffer per well for 30 min at room temperature.
10. Wash three times with PBS.
11. Observe under microscopy (Figs. 3c, d and 4a–c).

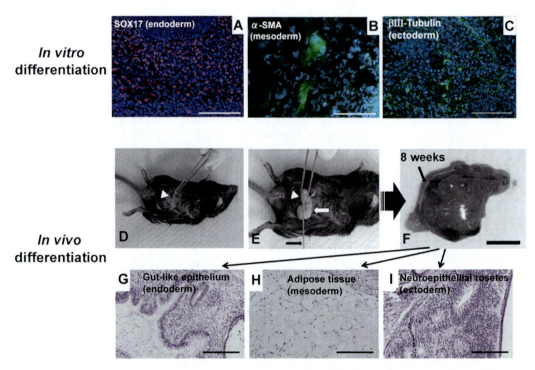

Fig. 4. In vitro and in vivo differentiation capability of iPSCs. (**a–b**) In vitro differentiation of iPSCs from human adipose tissue. Immunofluorescence microscopy showing SOX17 (**a**, endoderm marker, *red*), α-SMA (**b**, mesoderm marker, *green*), βIII-tubulin (**c**, ectoderm marker, *green*), respectively. Scale bar = 200 μm. (**d, e**) Injection of iPSCs into testis of mouse. About 1 cm skin incision over abdomen of the mouse (**d**: *arrow head*). Pulled out testis with fat tissue (**e**: *arrowhead*). Insertion of needle (**e**: *black arrow*) into the testis. For teratoma formation, the human iPSCs are injected into the testis of SCID mice. This figure shows the C57BL/6 mouse as an injection model for visual image. (**f**) Teratoma derived from iPSCs from human adipose tissue 8 weeks after injection. Scale bar = 1 cm. (**f–i**) In vivo differentiation of the iPSCs evidenced by histological section of the teratoma. (**g**) Hematoxylin and eosin staining shows gut-like epithelium (endoderm), (**h**) adipose tissue (mesoderm), (**i**) neuroepithelial rosettes (ectoderm). Scale bar = 50 μm. Panels a, b, c, g, h, i reproduced from Aoki et al. (13), with permission from Mary Ann Liebert, Inc.

3.3.3. In Vivo Differentiation of IPSCs (Teratoma Formation)

Teratoma formation is the other test of iPSCs pluripotency.

1. Prepare iPSC colonies suspension (see Sect. 3.2.2.4, step 32) from confluent iPSCs in one 100 mm dish (see Note 9).
2. Centrifuge at $180 \times g$ for 5 min and aspirate supernatant.
3. Suspend gently the cell pellet with 60 µL of iPSC medium containing 5 ng/mL bFGF and 10 µM Y-27632 and put into 1.5 mL tube.
4. Anesthetize SCID mouse by intraperitoneal injection of 0.4 mL of 5 mg/mL Nembutal/PBS per one mouse.
5. Wipe the lower abdominal area with 70% ethanol.
6. Dissect lower abdominal area about 1.5 cm width (see Fig. 4d) and draw out testis (indicated by arrowhead in Fig. 4e). Grab adipose tissue (indicated by white arrow in Fig. 4e) with tweezers.
7. Inject 20 µL of iPSC suspension per testis of SCID mouse with Hamilton syringes (see Fig. 4e).
8. Turn back the testis into original location.
9. Do the same thing to the other side of testis.
10. Suture the incision and keep the mouse warm until awake from anesthesia.
11. Tumor mass appears after 8–12 weeks after injection (Fig. 4f).
12. Harvest and fix tumors with 10% formaldehyde neutral buffer (see Note 10), followed by histological section (Fig. 4g–i).

4. Notes

1. Rot check should be done to verify the proliferation differentiation/capability supported by the selected serum using rat (12) or human MSCs culture (4).
2. If the separation is insufficient to show the three layers, additional centrifuge should be done.
3. Any nonadherent hematopoietic cells and erythrocytes can be removed, leaving only adherent cells in the dish.
4. These cells are negative for hematopoietic markers (such as CD34, 45) but positive for markers present in mesenchymal cells (such as CD44, 105).
5. Three vectors containing OCT3/4, SOX2, and KLF4 without c-MYC can be used for generation of iPSCs from selected MSCs from adipose tissue (13).
6. The cells at this stage can be referred to as Passage 1.

7. About 1–2 min.
8. 1 mL of EBs suspension per one well.
9. Add 10 μM Y-27632 to the medium of iPSCs in a 100-mm dish at least before an hour.
10. Amount of the buffer is at least tenfold volume of teratoma.

References

1. Salem, H.K., and Thiemermann, C. (2009) Mesenchymal stromal cells: current understanding and clinical status. Stem Cells **28**, 585–596.
2. Ohgushi, H., and Caplan, A.I. (1999) Stem cell technology and bioceramics: from cell to gene engineering. J Biomed Mater Res **48**, 913–927.
3. Kotobuki, N., Hirose,M., Takakura,Y., and Ohgushi, H. (2004) Cultured autologous human cells for hard tissue regeneration: preparation and characterization of mesenchymal stem cells from bone marrow. Artif Organs **28**, 33–39.
4. Ohgushi, H., Kotobuki, N., Funaoka, H., Machida, H., Hirose, M., Tanaka, Y., and Takakura, Y. (2005) Tissue engineered ceramic artificial joint--ex vivo osteogenic differentiation of patient mesenchymal cells on total ankle joints for treatment of osteoarthritis. Biomaterials **26**, 4654–4661.
5. Morishita, T., Honoki, K., Ohgushi, H., Kotobuki, N., Matsushima, A., and Takakura, Y. (2006) Tissue engineering approach to the treatment of bone tumors: three cases of cultured bone grafts derived from patients' mesenchymal stem cells. Artif Organs **30**, 115–118.
6. Kotobuki, N., Hirose, M., Machida, H., Katou, Y., Muraki, K., Takakura, Y., Ohgushi, H. (2005) Viability and osteogenic potential of cryopreserved human bone marrow-derived mesenchymal cells. Tissue Eng. **11**, 663–673.
7. Go, M.J., Takenaka, C., and Ohgushi, H. (2008) Forced expression of Sox2 or Nanog in human bone marrow derived mesenchymal stem cells maintains their expansion and differentiation capabilities. Exp Cell Res **314**, 1147–1154.
8. Takahashi, K., and Yamanaka, S. (2006) Induction of pluripotent stem cells from mouse embryonic and adult fibroblast cultures by defined factors. Cell **126**, 663–676.
9. Takahashi, K., Tanabe, K., Ohnuki, M., Narita, M., Ichisaka, T., Tomoda, K., and Yamanaka, S. (2007) Induction of pluripotent stem cells from adult human fibroblasts by defined factors. Cell **131**, 861–872.
10. Ohnuki, M., Takahashi, K., and Yamanaka, S. (2009) Generation and characterization of human induced pluripotent stem cells. Curr Protoc Stem Cell Biol *Chapter 4*, Unit 4A 2.
11. Kitamura, T., Koshino, Y., Shibata, F., Oki, T., Nakajima, H., Nosaka, T., and Kumagai, H. (2003) Retrovirus-mediated gene transfer and expression cloning: powerful tools in functional genomics. Exp Hematol **31**, 1007–1014.
12. Kihara, T., Oshima, A., Hirose, M., Ohgushi H. (2004) Three-dimensional visualization analysis of in vitro cultured bone fabricated by rat marrow mesenchymal stem cells. Biochem and Biophys Res Commun **316**, 943–948.
13. Aoki, T., Ohnishi, H., Oda, Y., Tadokoro, M., Sasao, M., Kato, H., Hattori, K., and Ohgushi, H. (2010) Generation of induced pluripotent stem cells from human adipose-derived stem cells without c-MYC. Tissue Eng Part A **16**, 2197–2206.

Chapter 15

Retroviral Vector-Based Approaches for the Generation of Human Induced Pluripotent Stem Cells from Fibroblasts and Keratinocytes

Athanasia D. Panopoulos, Sergio Ruiz, and Juan Carlos Izpisua Belmonte

Abstract

The ability of somatic cells to be induced to pluripotency by ectopic expression of defined transcription factors has altered the course of research in developmental biology and regenerative medicine. Somatic cell reprogramming has now been performed with numerous somatic sources with variable kinetics and efficiencies. In this chapter, we describe a protocol for generating induced pluripotent stem (iPS) cells from human fibroblasts or keratinocytes. We provide a step-by-step procedure detailing how to culture the initial somatic cells (i.e., fibroblasts or keratinocytes), produce retrovirus encoding each of the four reprogramming factors (i.e., Klf-4, Oct4, Sox2, c-Myc, or KOSM), and infect the somatic cells to produce iPS cells.

Key words: Induced pluripotent stem cells, Somatic cell reprogramming, Fibroblasts, Keratinocytes, Retroviral infection

1. Introduction

Somatic cells can be "reprogrammed" to pluripotency by a number of approaches including somatic cell nuclear transfer, treatment of differentiated cells with pluripotent cell extracts, or cell fusion (1). Seminal more recent studies demonstrated that fibroblasts could be reprogrammed to a pluripotent state by ectopic expression of defined factors including either the combination of Klf-4, Oct4, Sox2, and c-Myc (2, 3) or Oct4, Sox2, Nanog, and Lin28 (4). Induced pluripotent stem (iPS) cells appear to be morphologically, epigenetically, and functionally very similar to embryonic stem (ES) cells (2–4), although ongoing studies continue to more carefully

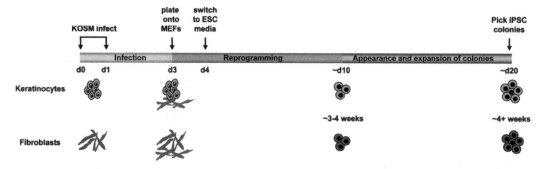

Fig. 1. Schematic representation and timeline of the described protocol to generate induced pluripotent stem cells from human keratinocytes and fibroblasts (*d* day; *KOSM* Klf-4, Oct4, Sox2, c-Myc encoding retroviruses; *iPSC* induced pluripotent stem cell; *MEFs* mouse embryonic fibroblasts; *ESC* embryonic stem cell).

examine the equivalence of these cell types. iPS cells have now been reported from numerous somatic sources, including fibroblasts (2–4), keratinocytes (5), cord blood cells (6, 7), neural stem cells (8), and adipocytes (9).

Here we describe a protocol for generating iPS cells from keratinocytes or fibroblasts by retroviral transduction of the originally described reprogramming transcription factors Klf-4, Oct4, Sox2, and c-Myc (KOSM) (Fig. 1). Fibroblasts and keratinocytes were chosen for their varied reprogramming efficiencies, ability to be manipulated in culture, and accessibility. Fibroblasts are common to most labs, are amenable to culture manipulations, but do not reprogram with great efficiency or speed. Keratinocytes, in contrast, reprogram with a reported 100-fold greater efficiency and two-fold faster kinetics than fibroblasts (5), but are not always common to every lab, and cannot be passaged as continuously or routinely as fibroblasts. Thus, both have been included to enable some flexibility in each individual researcher's experimental design. The protocol described encompasses three main steps: (1) culturing the somatic cells of origin, (2) producing retrovirus for each of the reprogramming factors, and (3) infecting the somatic cells to generate iPS cells. Importantly, the methods described utilize common or easily obtainable reagents to make iPS cell generation accessible to laboratories inexperienced in reprogramming.

2. Materials

2.1. Human Somatic Cell Culture

1. IMR90 (ATCC, CCL-186) or BJ human fibroblasts (ATCC, CRL-2522).
2. Neonatal human epidermal keratinocytes (NHEK) pooled (Lonza, C2507A) (see Note 1).

3. Dulbecco's Modified Eagle Medium (D-MEM) (1×) high glucose (Invitrogen, 11965).
4. Fetal bovine serum (FBS) (Atlanta Biologicals, S11550).
5. Nonessential amino acids (NEAA) (Gibco, 11140).
6. KGM-2 media (Lonza, CC-3107).
7. Gelatin (Sigma, G1890).
8. Tissue culture plates (60 mm × 15 mm (i.e., 6-well plate) and 100 mm × 20 mm (i.e., 10 cm dish); e.g., Nunc, 140685 and Corning 430167, respectively).
9. 15 mL Falcon tubes (BD, 352096).

2.2. Preparation of Retrovirus

2.2.1. Plasmid Preparation

1. pMX-Oct4, pMX-Sox2, pMX-Klf-4, pMX-c-Myc, pBABE-GFP, pCMV-VSV-G, and MSCV-gag/pol plasmids can be obtained from Addgene (plasmids 17217, 17218, 17219, 17220, 10668, 8454, and 14887, respectively).
2. Maxiprep Kit (Invitrogen, K2100-07).

2.2.2. Transfection to Obtain Retroviral Particles

1. 293T cells (ATCC, CRL-11268).
2. Dulbecco's Modified Eagle Medium (D-MEM) (1×) high glucose (Invitrogen, 11965).
3. Fetal bovine serum (FBS) (Atlanta Biologicals, S11550).
4. Nonessential amino acids (NEAA) (Gibco, 11140).
5. Lipofectamine 2000 (Invitrogen, 11668).
6. Syringes (BD, 309604).
7. 0.45 μm filters (Pall Corporation, PN4614).
8. TrypLE Express (Gibco, 12604).
9. OPTI-MEM (Gibco, 31985).
10. 50 mL Falcon tubes (BD, 352070).

2.2.3. Infection of Primary Cells

1. Hexadimethrine bromide (polybrene) (Sigma, 107689).
2. PBS$^{Mg\text{-}Ca\text{-}}$ (PBS−/−) (Gibco, 14190).

2.3. iPS Cell Generation

2.3.1. Preparation of Mouse Embryonic Fibroblast Feeder Layers

1. Mouse embryonic fibroblasts (MEF), Mytomycin C treated (CF-1 MEFs, Millipore/Chemicon, PMEF-CF).
2. Dulbecco's Modified Eagle Medium (D-MEM) (1×) high glucose (Invitrogen, 11965).
3. Fetal bovine serum (FBS) (Atlanta Biologicals, S11550).
4. Nonessential amino acids (NEAA) (Gibco, 11140).
5. Gelatin (Sigma, G1890).
6. Tissue culture plates (60 mm × 15 mm (i.e., 6-well plate) and 100 mm × 20 mm (i.e., 10 cm dish); e.g., Nunc, 140685 and Corning, 430167, respectively).
7. 15 mL Falcon tubes (BD, 352096).

2.3.2. Obtaining iPS Cell Colonies

Plating Infected Cells onto MEFs

1. MEF-coated plates as prepared in Sect. 3.3.1.
2. TrypLE Express (Gibco, 12604).
3. PBS$^{Mg\text{-}Ca\text{-}}$ (PBS-/-) (Gibco, 14190).
4. Keratinocyte or fibroblast media, see Sect. 2.1.
5. 15 mL Falcon tubes (BD, 352096).

Switching to hES Media for Colony Formation

1. Dulbecco's Modified Eagle Medium Nutrient Mixture F-12 (1×) (DMEM/F12) (Invitrogen, 11330).
2. Knockout serum replacement for ES and iPS cells (Invitrogen, 10828).
3. GlutaMAX-1 (Invitrogen, 35050).
4. Nonessential amino acids (NEAA) (Gibco, 11140).
5. β-mercaptoethanol 1,000×, 55 mM (Invitrogen, 21985).
6. Fibroblast growth factor-basic (FGFb), human recombinant (Stemgent, 03-0002).

3. Methods

3.1. Human Somatic Cell Culture

3.1.1. Human Fibroblasts

1. Place all cell culture plates in a sterile flow cabinet.
2. Add sufficient amount of autoclaved gelatin solution (0.1% gelatin in water) to completely cover the plates. Place the gelatin-coated plates in a 37°C incubator for 30–60 min.
3. Prepare fibroblast media: DMEM, 10% FBS, and 1× NEAA.
4. Thaw the cells by placing the vial of cells in a 37°C water bath until the last sliver of ice is melted and then place the vial at room temperature.
5. Transfer the contents of the vial into a 15 mL Falcon tube and add 5 mL of fibroblast media dropwise while flicking the tube to mix the contents.
6. Pellet the cells by centrifugation at 1,000 rpm at room temperature.
7. Aspirate the supernatant and resuspend the cells in fibroblast media at a density of ~1 × 10^4 cells/cm^2.
8. Remove the gelatin solution by aspiration from each well immediately prior to adding the fibroblast-containing media to culture plates.
9. Place the cells in a 37°C CO$_2$/O$_2$ incubator and allow them to attach overnight.

3.1.2. Human Keratinocytes

1. Place all cell culture plates in a sterile flow cabinet.
2. Thaw the cells by placing the vial of cells in a water bath at 37°C until the last sliver of ice is melted and place the vial at room temperature.

3. Transfer the contents of the vial into a 15 mL Falcon and add 5 mL of KGM-2 media dropwise while flicking the tube to mix the contents.

4. Pellet the cells by centrifugation at 1,000 rpm at room temperature.

5. Aspirate the supernatant and resuspend the cells in KGM-2 media at a density of 3,500 cells/cm^2.

6. Place the cells in a 37°C CO_2/O_2 incubator and allow them to attach overnight.

3.1.3. 293T Cells

For 293T cells, follow protocol described in Sect. 3.1.1, for human fibroblasts (see Notes 2 and 3).

3.1.4. Splitting Cells

1. To split the cells, wash once with PBS−/− and add sufficient solution of TrypLE Express to cover the plate.

2. Place the plate at 37°C and examine the layer of cells under the microscope 3–5 min after the addition of the TrypLE Express solution.

3. Incubate until most of the cells (>90%) have detached. This should take between 3 and 10 min (e.g., ~3–5 min for 293T cells, ~5–7 min for fibroblasts, and ~7–10 min for keratinocytes).

4. Gently tap the plate against the palm of your hand to detach the cells. If only a few cells detach, allow for a longer incubation of the cells in TrypLE Express solution.

5. Add either fibroblast or KGM-2 media correspondingly, transfer the cells to a 15 mL Falcon tube, and centrifuge the cells at 1,000 rpm at room temperature.

6. Aspirate the supernatant and resuspend the cells at the appropriate plating concentration as described above.

7. Place the cells in a 37°C, CO_2/O_2 incubator and allow them to attach overnight.

3.2. Preparation of Retrovirus

3.2.1. Plasmid Preparation

Prepare sterile plasmid DNA diluted at 1 µg/µL in distilled water using Maxiprep Kit. Store plasmid stocks at −20°C.

3.2.2. Transfection to Obtain Retroviral Particles

1. Culture 293T cells in 100 mm plates until they reach ~80% confluence. Each 100 mm plate will be used to generate a total of 8 mL of each individual retrovirus. Thus, for a reprogramming experiment, at least five different 100 mm plates will be necessary to generate Klf-4, Oct4, Sox2, c-Myc, and GFP-encoding retroviral particles. Ideally, the transfection is performed in the late afternoon.

2. 30–60 min before the transfection, aspirate off the media from the 293T cells and gently add 5 mL of fresh fibroblast media (DMEM, 10% FBS, 1× NEAA). Place the plates back into a 37°C CO_2/O_2 incubator while preparing the transfection reagents.

3. For each plate to be transfected, prepare two 15 mL Falcon tubes containing 2.5 mL OPTI-MEM media each.

4. To the first Falcon tube containing OPTI-MEM media, add 15 µg of the retroviral plasmid (e.g., pMX-Oct4), 10 µg of MSCV-gag/pol, and 5 µg of pCMV-VSV-G.

5. To the second Falcon tube containing OPTI-MEM media, add 50 µL of Lipofectamine 2000. Flick the tube and leave it for 5 min at room temperature.

6. After the 5 min incubation, mix the DNA/OPTI-MEM solution with the Lipofectamine 2000/OPTI-MEM solution by pipetting up and down slowly to have a final volume of 5 mL. Leave the mix for 15–20 min at room temperature to allow for the formation of the DNA/Lipofectamine 2000 complexes.

7. Add the 5 mL of DNA/Lipofectamine 2000 mix in OPTI-MEM to the 100 mm plate of 293T cells *dropwise and gently* (otherwise the 293T cells will detach from plate) and place the culture dishes back in the 37°C CO_2/O_2 incubator overnight.

8. The next day (~16 h after the transfection), aspirate the DNA/Lipofectamine solution and add 8 mL of fresh fibroblast media (DMEM, 10%FBS, 1× NEAA). Incubate the plates at 37°C for 24 h.

9. Collect the supernatant containing the active retroviral particles and pool the different retroviral particles (Klf-4, Oct4, Sox2, c-Myc, and GFP) at a 1:1:1:1:1 ratio in a 50 mL Falcon tube. Pellet the cell debris by centrifugation at 1,000 rpm at room temperature.

10. Filter the retroviral solution with a 0.45 µm filter. Retroviral solution can be stored up to 5 days at 4°C. It may also be frozen at –80°C for single aliquot use (i.e., avoiding repeated freeze/thaws), but freezing the virus will likely cause a decrease in viral activity, so use of fresh virus is preferred.

3.2.3. Infection of Primary Cells

1. Culture the cells to reach the desired confluence: for keratinocytes ~40–50%; for fibroblasts ~70%. Usually, a 6-well format is the ideal format to perform the viral transduction.

2. Allow the pool of retrovirus mix (Klf-4, Oct4, Sox2, c-Myc, and GFP) to reach room temperature. Prepare a sterile solution of polybrene (4 mg/mL in PBS–/– which can be stored in aliquots at –20°C).

3. Remove the media from the culture plates where keratinocytes or fibroblasts are growing and wash twice with PBS–/–.

Add 2 mL of the retroviral mix supplemented with 4 μg–8 μg/mL polybrene (i.e., 1:500-1:1000 of stock solution) per well of a 6-well plate.

4. Spin the tissue culture plate in a tabletop centrifuge (e.g., Beckman Coulter, Allegra X15R) at 800×g for 1 h at room temperature. This is performed using a swinging bucket centrifuge equipped with carriages for tissue culture plates (e.g., Plate carriers, SX-4750). Make sure plates are properly balanced prior to spinning.

5. Following completion of the spinfection, remove the retroviral mix from the culture plates and wash twice with PBS–/–. Add corresponding media (fibroblast or KGM-2 media) to each infected well. Allow the cells to recover in the incubator for 24 h before repeating the infection.

6. A total of two rounds of spinfection are performed (see Note 4).

7. Following the second spinfection, supplement the cells with fresh media. Allow the cells to recover in the incubator for 48 h (feeding with fresh media again after 24 h) before plating them as described in Sect. 3.3.

3.3. iPS Cell Generation

3.3.1. Preparation of Mouse Embryonic Fibroblast Feeder Layers

1. Place all cell culture plates in a sterile flow hood.

2. Add sufficient amount of gelatin solution (see Sect. 3.1.1. for preparation) to completely cover the plates. Leave the plates at 37°C in the incubator for at least 30–60 min (Gelatin-coated plates may also be prepared and placed at 37°C up to 2-3 days in advance).

3. Prepare fibroblast media: DMEM, 10% FBS, and 1× NEAA.

4. Thaw the cells by placing the vial of cells in a 37°C water bath until the last sliver of ice is melted and then place the vial at room temperature.

5. Transfer the contents of the vial into a 15 mL Falcon tube and add 5 mL of fibroblast media dropwise while flicking the tube to mix the contents.

6. Pellet the cells by centrifugation at 1,000 rpm at room temperature.

7. Aspirate the supernatant, resuspend the cells in fibroblast media (for example, ~5 mL/vial of MEFs), and using a hemacytometer, count the number of cells.

8. Add media to the cells so that 200,000 cells are plated per 6-well, or 1.2×10^6 cells per 6-well plate or 100 mm dish. For example, if the vial contains 4.8×10^6 cells, then bring up the volume to be 48 mL, and disperse 2 mL/6-well into a total of four 6-well plates.

9. Remove the gelatin-coated plates from the incubator, and aspirate solution.

10. Plate MEFs at appropriate density.

11. Place the cells in a 37°C CO_2/O_2 incubator and shake plate gently front-to-back, and left-to-right. This is to ensure that the MEFs disperse and attach throughout the well.

12. Allow the cells to attach overnight.

3.3.2. Obtaining iPS Cell Colonies

Plating Infected Cells onto MEFs

1. Remove cell culture plates containing infected somatic cells from incubator, and place in sterile hood.

2. Wash each well once with PBS-/- and add sufficient solution of TrypLE Express to cover the well.

3. Place the plate at 37°C and examine the layer of cells under the microscope 5–7 min after adding the TrypLE Express solution.

4. Incubate until most of the cells (>90%) have detached. This should take between 5 and 10 min.

5. Gently tap the plate against the palm of your hand to detach the cells. If only a few cells detach, allow for a longer incubation of the cells in TrypLE Express solution.

6. Add 1–2 mL of either fibroblast or KGM-2 media correspondingly, transfer the cells to a 15 mL Falcon tube, and centrifuge the cells at 1,000 rpm at room temperature.

7. Aspirate the supernatant, resuspend each somatic cell type in their appropriate media, and count the cells.

8. Remove MEF-coated plates prepared as described in Sect. 3.3.1 from incubator and place in sterile flow cabinet.

9. Wash MEF-coated wells once with PBS-/- and plate infected somatic cells in their corresponding medias onto wells containing MEFs. The density to plate will depend on infection efficiency and somatic cell source. For example, if >80% of the cells are infected (this infection efficiency is feasible for keratinocytes and fibroblasts), then some examples of suggested densities are ~10,000–50,000 infected keratinocytes per well of a 6-well plate, or ~one 6-well of fibroblasts into a 100 mm dish (see Note 5).

10. Place the cells in a 37°C CO_2/O_2 incubator and shake plate gently front-to-back, and left-to-right. This is to ensure that the infected cells disperse and attach evenly throughout the well of MEFs. (Be sure to gently close the incubator doors after the shaking plate. If the incubator door is slammed after the cells are freshly split, the cells will congregate to the middle of the well, which could not only negatively influence the reprogramming process but also cause complications at later stages of picking and expanding iPS cell colonies).

11. Allow the cells to attach overnight.

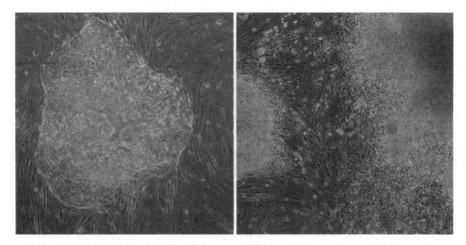

Fig. 2. KOSM-infected keratinocytes were plated onto MEFs and treated with ES cell media. Twelve days after the initial infection, colonies demonstrating an ES cell-like morphology (*left panel*) and colonies resembling a transformed morphology (*right panel*) were present.

Switching to hES Media for Colony Formation

1. Prepare ES cell media (DMEM/F12, 20% KOSR, 1 mM GlutaMAX-1, 0.1 mM nonessential amino acids, 55 μM β-mercaptoethanol, 10 ng/mL FGFb) (media can be stored 2–3 weeks at 4°C).
2. Remove plates containing infected somatic cells on MEFs from incubator and place into sterile flow cabinet.
3. Aspirate media, and wash each well once with PBS–/–.
4. Add ES cell media to each well.
5. Feed cells daily with ES cell media.
6. Keratinocyte-generated iPS cell colonies should start to appear ~10–12 days after the first infection. Fibroblast-generated iPS cell colonies should start to appear ~3–4 weeks following the initial infection (see Notes 5 and 6).
7. Monitor daily for colony formation. iPS cell colonies should have an ES cell-like morphology, with clear colony borders and a large nuclear:cytoplasmic ratio (Fig. 2). Retroviruses should be silenced in iPS cells, so GFP expression may also be used to monitor efficient transgene silencing (see Note 7).

4. Notes

1. The use of pooled cells has the advantage of generating iPS cells from varied genetic sources to account for possible differences that may exist between individuals. However, take note if the end experiment would benefit from the use of a clonal somatic source of origin.

2. It is strongly suggested that mycoplasma testing be performed (e.g., MycoAlert Mycoplasma Detection Kit, Lonza, LT07-218). Generating and complete characterization of iPS cells is a labor-intensive and time-consuming process, and in some cases, the somatic source of origin may be a valuable patient-specific source. Pluripotent cells are especially sensitive to mycoplasma, so iPS cell stocks would be at significant risk to be permanently destroyed. Since 293T cells are common throughout laboratories, 293T cell stocks have often exchanged many hands and laboratories, increasing the chance of mycoplasma. It is worth performing a mycoplasma test on the stock of 293T cells used to make virus, and again in the later stages of reprogramming to ensure that iPS cell colonies picked, expanded, and stored are free of mycoplasma.

3. The quality of virus generated can decrease as the passage number of 293T cells increases. It is generally recommended to perform the described transfections on 293T cells at passage 30 or lower.

4. More infections do not always translate to a higher reprogramming efficiency, as the balance of viral incorporation and tolerance to infection will vary depending on the somatic source and the quality/titer of virus. However, the two spinfections suggested in this chapter consistently generate an adequate number of iPS cell colonies using the somatic cells and virus preparation described.

5. The kinetics of iPS colony appearance will vary depending on virus quality and somatic source of origin. However, keratinocytes generally reprogram more rapidly and efficiently (~1% reprogramming efficiency) than fibroblasts (~0.01% reprogramming efficiency). Furthermore, there may be differences within similar types of somatic cells. For example, BJ fibroblasts generally reprogram with a greater efficiency than IMR90 fibroblasts. Also note that transformed colonies (with a non ES cell-like morphology) in many cases appear before iPS cell colonies and will often dissociate in culture over time rather than acquire a defined colony border.

6. Since the reprogramming process takes place over several weeks, a contamination at later stages can cause great delay in experimental progression. The antibiotic Normocin (InvivoGen, #ant-nr-1) is a very reliable reagent to "rescue" contaminated cultures and has proven invaluable when contaminations occur in rare or irreplaceable samples.

7. For researchers new to identifying iPS cell colonies by morphology, a live-cell pluripotent marker staining kit (e.g., Stemgent StainAlive DyLight 488 mouse anti-human TRA-1-81 antibody, 09-0069) can be employed.

References

1. Patel M, Yang S. (2010) Advances in reprogramming somatic cells to induced pluripotent stem cells. Stem Cell Rev. 2010; 6:367–380.
2. Takahashi K, Yamanaka S. Induction of pluripotent stem cells from mouse embryonic and adult fibroblast cultures by defined factors. Cell. 2006; 126:663–676.
3. Takahashi K, Tanabe K, Ohnuki M, Narita M, Ichisaka T, Tomoda K, Yamanaka S. Induction of pluripotent stem cells from adult human fibroblasts by defined factors. Cell. 2007; 131:861–872.
4. Yu J, Vodyanik MA, Smuga-Otto K, Antosiewicz-Bourget J, Frane JL, Tian S, Nie J, Jonsdottir GA, Ruotti V, Stewart R, Slukvin II, Thomson JA. Induced pluripotent stem cell lines derived from human somatic cells. Science. 2007; 318:1917–1920.
5. Aasen T, Raya A, Barrero MJ, Garreta E, Consiglio A, Gonzalez F, Vassena R, Billio J, Pekarik V, Tiscornia G, Edel M, Boue S, Izpisua Belmonte JC. Efficient and rapid generation of induced pluripotent stem cells from human keratinocytes. Nat Biotechnol. 2008; 26: 1276–1284.
6. Giorgetti A, Montserrat N, Aasen T, Gonzalez F, Rodriguez-Piza I, Vassena R, Raya A, Boue S, Barrero MJ, Corbella BA, Torrabadella M, Veiga A, Izpisua Belmonte JC (2009) Generation of induced pluripotent stem cells from human cord blood using OCT4 and SOX2. Cell Stem Cell. 5:353–357.
7. Haase A, Olmer R, Schwanke K, Wunderlich S, Merkert S, Hess C, Zweigerdt R, Gruh I, Meyer J, Wagner S, Maier LS, Han DW, Glage S, Miller K, Fischer P, Scholer HR, Martin U. Generation of induced pluripotent stem cells from human cord blood. Cell Stem Cell. 2009; 5:434–441.
8. Kim JB, Greber B, Araúzo-Bravo MJ, Meyer J, Park KI, Zaehres H, Schöler HR. Direct reprogramming of human neural stem cells by OCT4. Nature. 2009; 461:649–653.
9. Sugii S, Kida Y, Kawamura T, Suzuki J, Vassena R, Yin YQ, Lutz MK, Berggren WT, Izpisúa Belmonte JC, Evans RM (2010) Human and mouse adipose-derived cells support feeder-independent induction of pluripotent stem cells. PNAS. 107:3558–3563.

Chapter 16

Generation of Nonviral Integration-Free Induced Pluripotent Stem Cells from Plucked Human Hair Follicles

Ann Peters and Elias T. Zambidis

Abstract

Human induced pluripotent stem cells (hiPSCs) provide a unique experimental reagent for dissecting the complex transcriptional, regulatory, and epigenetic mechanisms of pluripotency, as well as for studying normal and diseased human development. However, the utility of current iPSC lines may be limited by the utilization of integrating viral vectors for transgenic ectopic expression of oncogenic reprogramming factors (e.g., *SOX2, OCT4, KLF4, MYC, NANOG, LIN28*, and *SV40 T antigen*). Leaky expression of integrated pluripotency factor transgenes may inhibit completion of the somatic cell reprogramming process, pose great potential for subsequent malignant transformation, and ultimately limit differentiation strategies and their future clinical application. hiPSCs generated with transgene and vector-free approaches may more faithfully resemble human embryonic stem cells (hESCs) and obviate some or all of these important caveats. In this chapter, we describe a simple and reproducible methodology for generating clinically safe nonviral, integration-free hiPSCs from keratinocytes noninvasively obtained and expanded from a donor's single plucked hair follicle. Nonintegrated hiPSCs free of viral and transgene sequences should provide a potent tool for studies of pluripotency, and ultimately be more clinically useful in regenerative medicine.

Key words: Induced pluripotent stem cells, Keratinocytes, Hair follicle, Integration free, Viral free

1. Introduction

Although the derivation of autologous human induced pluripotent stem cells (hiPSCs) promises to advance the field of patient-specific regenerative medicine (1, 2), the use of viral vectors for their generation poses several limiting caveats. For example, the use of genome-integrating viral vectors to generate hiPSCs results in an increased propensity for random insertional mutagenesis, which ultimately limits their clinical application (3). More importantly, despite the general silencing of viral vector promoters, low-level expression of reactivated pluripotency transgenes (e.g., *SOX2, OCT4, KLF4, MYC, NANOG, LIN28,* and *SV40 T antigen*) remains problematic in downstream differentiation applications since they likely skew faithful lineage commitment, as well as pose great risk for subsequent malignant transformation. Not surprisingly, chimeric mice prepared from murine iPSCs generated with virally expressed pluripotency factors increasingly formed malignant tumors starting at 1 month postnatally, even in the absence of ectopic Myc expression (4–6). It has become evident that iPSCs made with viral transgenic approaches have molecular signatures and differentiation potentials that although similar to human embryonic stem cells (hESCs) are significantly distinct a variable (7–9). These differences may result from an epigenetic memory inherent to the somatic cells that the iPSCs were generated from (10). Alternatively, leaky transgene expression from partially silenced integrated viral vectors likely influences their molecular signatures and differentiation potential following completion of the reprogramming process.

There is much anticipation that iPSCs generated with nonviral integration-free approaches will produce human pluripotent stem cells devoid of these problems and may potentially be more akin to hESCs. A new wave of approaches for transgene-free reprogramming of somatic cells has recently been described. For example, a proof of principle experiment with Cre recombinase excisable lentiviral vectors demonstrated that continuous transgene expression was not necessary for reprogramming and maintenance of pluripotency (11). However, since Cre recombinase does not completely excise all viral sequences, concerns for insertional mutagenesis remained in this approach. Other alternative integration-free reprogramming approaches have also been described for circumventing this obstacle. These approaches have included delivering defined factors with (1) nonintegrating minus strand RNA viruses, (2) transfection with *oriP*/EBNA1-based episomal plasmids containing reprogramming factors (e.g., *SOX2, OCT4, KLF4, C-MYC, NANOG, LIN28,* and *SV40LT*), or (3) transfection with plasmid minicircle vectors. These approaches still require verification that vector DNA sequences have indeed not been retained (e.g., by Southern blot and PCR), but have successfully been employed to generate transgene-free hiPSC from peripheral blood-derived T cells, neonatal fibroblasts, and adipose stem-progenitor cells (12–14). Additionally, another theoretically safer approach (that avoids transiently expressed DNA constructs altogether) utilizes cell-permeant peptide (CPP)-anchored reprogramming proteins (15). In general, these approaches all suffer from extremely low reprogramming efficiencies, and much effort has recently been invested to improve these technologies and to make them more practical.

One important issue in the generation of clinically useful hiPSCs has been the debate regarding the optimal choice of somatic cell used for reprogramming. While almost all somatic cell types have been shown to be amenable to reprogramming, most are obtained from donors with at least a need for minimally invasive biopsy, or requirement for cell purification. However, hair follicle-derived keratinocytes obviate these caveats since they are the most noninvasive cell type that can be obtained from a donor, and represent a highly proliferative cell type that can be acquired in abundant supplies. The derivation of hiPSCs from hair follicle-derived keratinocytes with viral vectors has already been demonstrated (16). In this chapter, we describe a reproducible methodology developed in our lab for generating nonintegrated virus-free hiPSCs from keratinocytes expanded from plucked hair follicles. The protocol relies on direct nucleofection of hair follicle-derived keratinocytes with a three plasmid, seven-factor (*SOX2, OCT4, KLF4, MYC, NANOG, LIN28,* and *SV40 T Ag*) EBNA1-based episomal system previously described for generating nonintegrating hiPSCs from neonatal fibroblasts (13). We outline a step-by-step roadmap for the detailed generation and characterization of hair follicle-derived hiPSC clones free of transgene and vector sequences using a previously described seven-factor EBNA1-based plasmid episomal system. hiPSCs are generated at frequencies comparable to fibroblasts (~0.001% of input cells), which although low in efficiency, can be scaled up to derive enough hiPSCs for downstream applications. This protocol produces high quality bona fide hiPSCs with molecular signatures and differentiation potentials that are highly similar to hESCs (17), form multilineage teratomas (18), and can be differentiated robustly to clinically relevant hematopoietic, vascular, cardiac, and neural lineages.

2. Materials

2.1. Reagents

1. Phosphate-buffered saline (PBS), pH 7.4, without $CaCl_2$ and $MgCl_2$ (Invitrogen, Carlsbad, CA, Cat. No. 10010-023); store at room temperature (RT).
2. 0.05% Trypsin-EDTA (Invitrogen, Cat. No. 25300-054), make 5 mL aliquots of 0.5% Trypsin and store at −20°C. Before use, thaw and add 45 mL of PBS. Store 0.05% Trypsin at 4°C and immediately prior to application warm to 37°C in water bath for 5 min.
3. 0.1% Gelatin (Sigma-Aldrich, St. Louis, MO, Cat. No. G1890, 100 mg/L in dH_2O). Make 1 L bottles, autoclave, and store at 4°C.
4. Dulbecco's Modified Eagle's Medium (1×), liquid (high glucose) (DMEM) (Invitrogen, Cat. No. 11995-065); store at 4°C.

5. Fetal bovine serum (FBS), (HyClone Thermo Scientific, Logan, UT, Cat. No. SH30071.03). Store at −20°C, thaw at 4°C, and once thawed store at 4°C.

6. MEM nonessential amino acids (NEAA, 100× solution) (Invitrogen, Cat. No. 11140-050); store at 4°C.

7. 2 β-Mercaptoethanol (55 mM) (Invitrogen, Cat. No. 21985-023); store at 4°C.

8. Penicillin-Streptomycin, liquid (Invitrogen, Cat. No. 15140-122). Make 5 mL aliquots and store at −20°C; once thawed store at 4°C.

9. DMEM/F-12 (1×), liquid, 1:1 (with GlutaMAX™-I) (Invitrogen, Cat. No. 10565–018); store at 4°C.

10. KnockOut™ Serum Replacement (KSR) (Invitrogen, Cat. No. 10828-028). Store at −20°C, thaw at 4°C, and once thawed store at 4°C.

11. Human FGF2 (FGF basic, R&D Systems, Minneapolis, MN, Cat. No. 233-FB-025). Dilute to 100 ng/µL in DMEM/F-12 + 0.5% BSA, make 50 µL aliquots, and store at −20 °C; once thawed store aliquots at 4°C for up to 2 weeks.

12. Collagenase IV (Invitrogen, Cat. No. 17104-019). Dilute to 1 mg/mL in DMEM/F12 and make 5 mL aliquots; store at −20°C; once thawed store aliquots at 4°C, and warm to 37°C in water bath for 5 min before use.

13. Accutase (Sigma-Aldrich, Cat. No. A6964). Store 5 mL aliquots at −20°C; once thawed store aliquots at 4°C, and warm to 37°C in water bath for 5 min before use.

14. Dimethyl sulfoxide (DMSO) (Sigma-Aldrich, Cat. No. D2650).

15. 2-Propanol (C_3H_8O) (Fisher Scientific, Cat. No. 67-63-0).

16. Keratinocyte AOF Growth Kit (Invitrogen, Cat. No. A1051501): 500 mL EpiLife Basal Medium, Supplement S7, and Coating Matrix Kit (Cascade Biologics™, Cat. No. M-EPI-500-CA, S-017-5, R-011-K, R-011-05).

17. Growth Factor Reduced (GFR) Basement Membrane BD Matrigel™ (BD Bioscience, Cat. No. 354230).

18. Primary mouse embryonic fibroblasts (PMEF), derived from E13.5 matings of ♀ DR4 X ♂ CF-1 mice, used at passage 1–2. The DR4 strain is obtained from Jackson Labs and is homozygous for neomycin, puromycin, hygromycin, and HPRT resistance transgenes, which provides a useful coculture stroma for hiPSC if subsequent genetic modification experiments are planned.

2.2. Equipment

1. 6-Well plates (Greiner Bio-one, Monroe, NC, Cat. No. 657160, through ISC Bioexpress, Cat. No. T-3026-3)

2. T-175 flasks (BD Biosciences, San Jose, CA, Cat. No. 353112)

3. 15 and 50 mL Conical tubes (BD Biosciences, Cat. No. 352097 and 352098)

4. 5, 10, 25, and 50 mL Plastic pipettes (BD Biosciences, Cat. No. 357543, 357551, 357525, and 357550)

5. 250, 500 mL and 1 L PES media filters (Millipore, Billerica, MA, Cat. No. SCGPU02RE, SCGPU05RE, SCGPU10RE)

6. 0.22 μm Millex GP PES filters (Millipore, Cat. No. SLGP033RS)

7. 100 μm Sterile Cell Strainer (Fisherbrand/Fisher Scientific, Cat. No. 22363549)

8. Hemocytometer (Hausser Scientific, Horsham, PA, Cat. No. 3200)

9. Mr. Frosty™ Cryo 1°C Freezing Container (Nalgene Labware, San Diego, CA, Cat. No. 5100-0001)

10. Cesium source irradiator

11. Nanodrop Spectrophotometer (NanoDrop Technologies, Inc., Wilmington, DE, Cat. No. ND1000)

12. E-Gel® Power Base™ v.4 (Invitrogen, Cat. No. G6200-04) and 0.8% E-Gel® precast gel (G5018-08) (or use regular Electrophoresis Apparatus and 0.8% Agarose Gel)

13. Universal surgical forceps and tweezers (Roboz, Gaithersburg, MD)

2.3. Tissue Culture

2.3.1. Media

hiPSC Culture Reagents

1. hiPSC Medium: DMEM/F-12 (1×) with GlutaMAX™-I, 20% KSR, 1% MEM-NEAA, 100 μM 2β-mercaptoethanol, 4–40 ng/mL FGF2. Filter-sterilize and store at 4°C for no longer than 2 weeks.

2. hiPSC Freezing Medium: 50% KSR, 40% hiPSC Medium, 10% DMSO. Filter-sterilize the solution (see Note 1).

Primary Mouse Embryonic Fibroblast Culture Reagents

1. Primary Mouse Embryonic Fibroblast (PMEF) medium (FDMEM): DMEM, 10% FBS, 1% NEAA, 50 units Penicillin/50 μg Streptomycin/mL, 55 μM 2β-mercaptoethanol. Filter sterilize the solution and store at 4°C for up to 2 weeks.

2. PMEF freezing medium: 50% FBS, 40% PMEF medium, 10% DMSO. Filter sterilize the solution (see Note 1).

3. PMEF Conditioned Medium (CM): DMEM/F-12 (1×) with GlutaMAX™-I, 20% KSR, 1% MEM-NEAA, 100 μM 2-mercaptoethanol, and 4 ng/mL FGF2 conditioned on 5,000 cGy irradiated PMEF for 24 h. Filter sterilize and spike with fresh 40 ng/mL FGF2 for immediate use or store at 4°C up to 24 h. For long-term storage, aliquot and store at −20°C. Filter sterilize and add fresh FGF2 after thawing at 4°C.

Keratinocyte Culture Reagents	1. Keratinocyte Derivation Medium (KDM): PMEF-conditioned medium, 10 ng/mL FGF2, ±50 units Penicillin/50 μg Streptomycin/mL. Filter sterilize, store at 4°C up to 24 h.
2. Keratinocyte Growth Medium (KGM): EpiLife Basal Medium, Supplement S7. Filter sterilize, wrap in aluminum foil to protect from light, and store at 4°C up to 2 weeks.
3. Keratinocyte Plate Coating Matrix (KCM): 0.67 μL/cm² coating matrix in 66.67 μL/cm² Dilution Medium (1:100 dilution).
4. Keratinocyte Freezing Medium: 50% KSR, 40% KGM, 10% DMSO. Filter sterilize the solution (see Note 1).

2.4. Nucleofections

Human Keratinocyte Nucleofector Kit (Lonza, Walkersville, MD, Cat. No. VPD-1002), and AMAXA II Nucleofector® device (Lonza, Cat. No. AAD-1001).

2.5. Reprogramming Plasmids

1. QIAquick Gel Extraction Kit (Qiagen, Cat. No. 28704).
2. Three oriP/EBNA1-based episomal plasmids for expression of seven reprogramming factors as described in (13): (a) pEP4 E02S EM2K, (b) pEP4 E02S ET2K, and (c) pEP4 E02S CK2M EN2L (Addgene, Cambridge, MA, Cat. No. 20923, 20927, and 20924). Plasmids are prepared with standard maxiprep kits (e.g., Qiagen), and repurified and concentrated (see Sect. 3.2.3).

2.6. hiPSC Characterization

1. Primers were ordered from Integrated DNA Technologies, Inc. (IDT, Coralville, IA).
2. RNeasy Mini Kit (50) (Qiagen, Cat. No. 74104).
3. RNase-Free DNase Set (50) (Qiagen, Cat. No. 79254).
4. SuperScript® First-Strand Synthesis System for RT-PCR (50) (Invitrogen, Cat. No. 11904-018).
5. DNeasy Blood & Tissue Kit (50) (Qiagen, Cat. No. 69504).
6. Platinum® Pfx DNA polymerase (250 reactions) (Invitrogen, Cat. No. 11708021).
7. Restriction Enzymes: BamHI-HF, SpeI, NotI, NruI, and AluI (New England Biolabs, Ipswich, MA, Cat. No. R3136L, R0133M, R0189M, R0192M, and R0137L, respectively).
8. pCEP4 (Invitrogen, Cat. No. V044-50).
9. DIG-High Prime DNA Labeling and Detection Starter Kit II (Roche Applied Science, Indianapolis, IN, Cat. No. 11585614910).
10. Nylon Membranes, positively charged (Roche Applied Science, Cat. No. 11417240001).
11. Lumi-Film Chemiluminescent Detection Film 8×10 in., 20.3×25.4 cm (Roche Applied Science, Cat. No. 11666657001).

3. Methods

3.1. Primary Culture of Hair Follicle-Derived Keratinocytes

The following is a modified protocol from our laboratory for the derivation of keratinocytes expanded from plucked hair follicles. Other excellent protocols without our modifications have also been described (19) (see Fig. 1a).

1. One day prior to collecting hair follicle samples, cover one 35-mm plate per hair follicle with undiluted Matrigel and incubate overnight at 4°C. On average 10 plates should be sufficient to derive keratinocytes from at least one to two hair follicles.

2. Using tweezers pluck single hair follicles from the occipital region of the scalp and cut 5 mm distal from the hair follicle bulb.

3. Aseptically transfer the hair follicle immediately to a PBS Penicillin/Streptomycin solution for 1 min.

4. Using sterile tweezers, gently place the hair follicle directly onto Matrigel-coated plate (see Note 2).

5. Dropwise add enough KDM supplement with antibiotics to cover the bottom of the 35 mm dish (see Note 3).

6. The next day, add 1 additional mL KDM supplement with antibiotics. Repeat every other day thereafter, but discontinue use of antibiotics after 2 days. Outgrowth from the hair follicle bulk should become visible after 1 week (Fig. 2a).

7. Once the keratinocyte colony surrounding the bulb has reached 5–10 mm in diameter, trypsinize the sample until most of the cells around the hair follicle have detached. Stop the reaction with two volumes of FDMEM.

8. Gently dissociate the cells into single cells using a P1000 pipette.

9. Pass the cells through a 100 μm cell strainer to remove the hair. Wash the cell strainer once with two volumes of PBS.

10. Centrifuge the cells at $90 \times g$ for 10 min.

11. Resuspend cells in KGM and plate cells derived from one hair follicle in two wells of a KCM-coated 6-well plate (Passage 1) (see Note 4).

12. Once 80–90% confluent, passage two wells of healthy, proliferating keratinocytes onto one KCM-coated T-175 flask, and culture in KGM (Fig. 2b). Keratinocyte identity can be confirmed by flow cytometry and straining for $\alpha 6$-integrin^{high} and CD71^{low} (Fig. 2g).

13. One T-175 flask usually is sufficient for one hiPSC experimental condition. However, keratinocytes should be passaged 1:4 once more (Passage 2) to prepare frozen cell stocks.

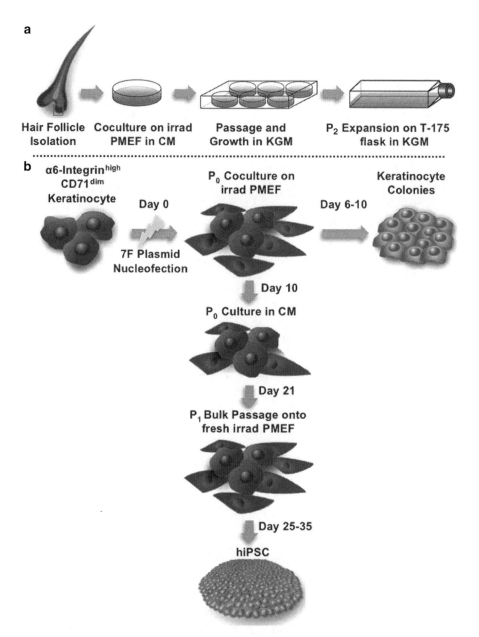

Fig. 1. Summary of methodology for the generation of transgene and vector-free human induced pluripotent stem cells from hair follicle-derived keratinocytes. (a) Overview of the major steps involved in the derivation of human keratinocytes from plucked hair follicles. (b) Chronological outline of the protocol utilized to produce hiPSC from keratinocytes.

3.2. hiPSC Derivation

An overview of the major steps involved in this protocol is summarized in Fig. 1b. A step-by-step description of the expansion, culture, and characterization of hiPSC clones generated with this method is given in Fig. 3. Additionally, basic hiPSC cell culture techniques that must be mastered for successful execution of this protocol are described in Sect. 3.4 and (20).

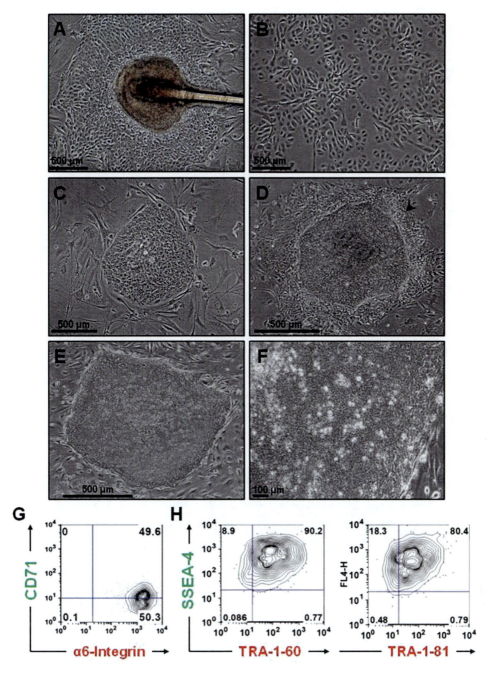

Fig. 2. Comparative morphological differences between hair follicle-derived keratinocytes and bona fide transgene and vector-free hiPSC. (**a**) Outgrowth of keratinocytes from the bulb of a plucked hair follicle 1 week after culture on Matrigel-coated 35-mm dishes in KDM. (**b**) Typical morphologic of keratinocytes on passage 1 in KGM. (**c**) Characteristic appearance of keratinocyte colonies 6–10 days post-nucleofection culture in hiPSC medium. (**d**) Passage 1 keratinocyte-derived hiPSC colony. Fibroblastic differentiation is visible around the colony edges (*arrow*). (**e**) Low power view of a fully reprogrammed (Type III) hiPSC colony on passage 5. (**f**) High power view of a Type III hiPSC colony shows classic hESC morphology with tightly packed cells that have high nuclear:cytoplasmic ratio and prominent nucleoli. (**g**) Flow cytometric analysis of keratinocyte markers ($\alpha 6$ *integrin* + *CD71*^{lo}) on hair follicle-derived cells which expanded on passage 2. (**h**) Pluripotency surface marker (SSEA4, Tra-1-81) analysis of fully reprogrammed (Type III) nonintegrating hiPSC on passage 5 by flow cytometry.

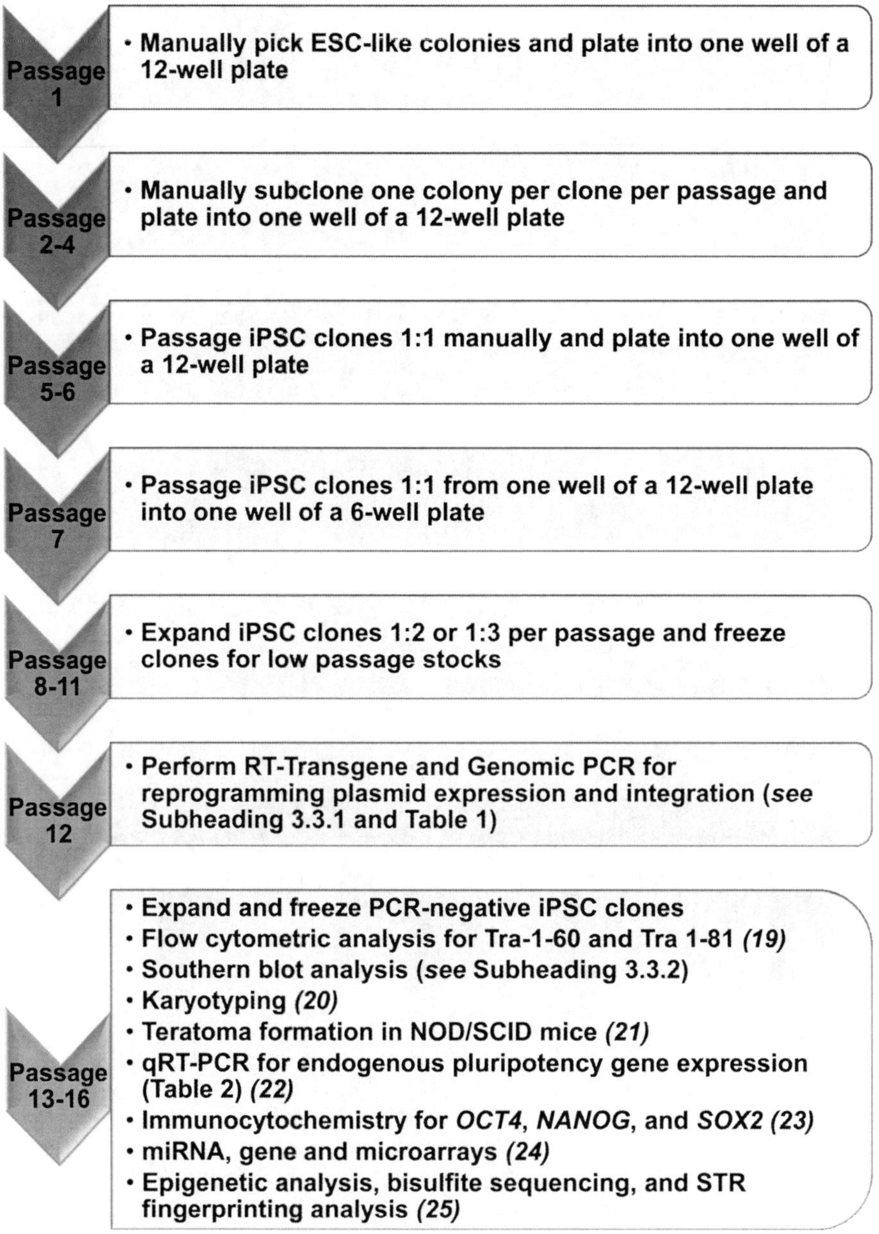

Passage 1
- Manually pick ESC-like colonies and plate into one well of a 12-well plate

Passage 2-4
- Manually subclone one colony per clone per passage and plate into one well of a 12-well plate

Passage 5-6
- Passage iPSC clones 1:1 manually and plate into one well of a 12-well plate

Passage 7
- Passage iPSC clones 1:1 from one well of a 12-well plate into one well of a 6-well plate

Passage 8-11
- Expand iPSC clones 1:2 or 1:3 per passage and freeze clones for low passage stocks

Passage 12
- Perform RT-Transgene and Genomic PCR for reprogramming plasmid expression and integration (see Subheading 3.3.1 and Table 1)

Passage 13-16
- Expand and freeze PCR-negative iPSC clones
- Flow cytometric analysis for Tra-1-60 and Tra 1-81 *(19)*
- Southern blot analysis (see Subheading 3.3.2)
- Karyotyping *(20)*
- Teratoma formation in NOD/SCID mice *(21)*
- qRT-PCR for endogenous pluripotency gene expression (Table 2) *(22)*
- Immunocytochemistry for *OCT4*, *NANOG*, and *SOX2* *(23)*
- miRNA, gene and microarrays *(24)*
- Epigenetic analysis, bisulfite sequencing, and STR fingerprinting analysis *(25)*

Fig. 3. Experimental summary for the complete characterization of integration-free hiPSC.

3.2.1. Day −4: Pre-nucleofection Culture of Hair Follicle-Derived Keratinocytes

1. Split and seed passage two to three keratinocyte cultures at a density of 2.5×10^3 cells/cm² onto KCM-coated flasks for 2–4 days before nucleofection (see Notes 5 and 6).
2. Replace medium the next day and every other day thereafter.

3.2.2. Day −3: Culture PMEF

1. Thaw one vial of 5×10^6 PMEF at 37°C in a water bath.
2. When only a small piece of ice remains to be thawed, remove the vial from the water bath and spray with EtOH.

3. Add the vial's contents to a 50 mL conical tube and add 9 mL 37°C prewarmed FDMEM dropwise to the PMEF.

4. Mix the two media layers by gently inverting the tube once.

5. Centrifuge at $200 \times g$ for 5 min.

6. Remove the supernatant, resuspend in 25 mL FDMEM and add cell suspension to a 0.1% gelatin-coated T-175 flask (see Note 7).

7. Change the medium the next day, and every other day thereafter.

3.2.3. Day −1: Plasmid Preparation and Irradiation of PMEF

1. Mix 20 μg of plasmids pEP4 E02S EM2K, pEP4 E02S ET2K, and pEP4 E02S CK2M EN2L each from a maxi prep in an eppendorf tube.

2. Repurify and concentrate the three-plasmid combination with Qiagen DNA gel extract kit (see Note 8).

3. Quantitate the repurified DNA concentration using a Nanodrop or standard spectrophotometer and confirm DNA quality by running 1 μL on a 0.8% Agarose Quick E-Gel.

4. Store plasmid mix at 4°C for short-term use, or store long-term at −20°C.

5. Coat one 6-well plate per nucleofection with KCM (see Notes 4 and 5).

6. Prepare one 6-well plate of irradiated PMEF (see Note 4 and Sect. 3.4.1) for post-nucleofection culture of keratinocyte, and plate on KCM-coated 6-well plates.

3.2.4. Day 0: Nucleofection

1. On the day of nucleofection, add the entire supplement to the keratinocyte nucleofector solution. Equilibrate solution and plasmid DNA to RT.

2. Aspirate off media from PMEF plates, wash with two volumes of PBS, and add 2 mL KGM. Equilibrate at 37°C for 15 min (see Note 9).

3. Aliquot 500 μL of KGM in an eppendorf tube per nucleofection of 7×10^5 keratinocytes. Equilibrate in 37°C for 5 min (see Note 9).

4. Aspirate off medium from keratinocyte flask, wash once with one volume of PBS, and add 40 μL/cm² prewarmed 0.05% Trypsin-EDTA.

5. Incubate at 37°C for 3–5 min, and tap the side of the flask occasionally to loosen cells. When cells have visibly detached from the flask, neutralize the Trypsin-EDTA by adding 75 μL/cm² FDMEM directly to the flask.

6. Using a 10 mL pipette, gently pipette the cell suspension one to two times to generate a single cell suspension, and add it into a 50 mL conical tube.

7. Wash the flask with 10 mL PBS and add the solution to the same conical tube.

8. Centrifuge at 90×g for 7 min.
9. Aspirate the supernatant and resuspend cell pellet in 10 mL of prewarmed KGM.
10. Count cells with hemocytometer. 7×10^5 keratinocytes/nucleofection should each be transferred to a new 15 mL conical tube.
11. Centrifuge at 90×g for 7 min.
12. Aspirate off the entire supernatant.
13. Resuspend cell pellet in 100 μL nucleofection solution.
14. Add 3 μg of plasmid DNA directly to cells resuspended in nucleofection solution (see Note 10).
15. Transfer the cell/DNA mixture to a nucleofection cuvette quickly, but without air bubbles. Tap the cuvette to ensure complete mixing of DNA and cells.
16. Insert the cuvette into the Amaxa II Nucleofection device and select program T-024.
17. While nucleofecting, remove the 500 μL pre-equilibrated KGM from the water bath.
18. Using a supplied nucleofection pipette, add 500 μL KGM directly into the cuvette, remove cell suspension gently, and add dropwise in equal amount to the 6-well plate of PMEF.
19. Repeat steps 10–18 and add keratinocytes to the same 6-well plate as the first nucleofection.
20. After 4 hrs, harvest any floating cells in a 15 mL conical tube. Add 2.5 mL of fresh prewarmed KGM per well.
21. Centrifuge cell suspension at 90×g for 7 min.
22. Aspirate off supernatant (see Note 11).
23. Resuspend cell pellet in 3 mL of prewarmed KGM and add 500 μL per well.

3.2.5. Day 2: Change Medium

Aspirate off KGM and add 3 mL of fresh medium per well.

3.2.6. Day 4: Change Medium to hiPSC Medium

Aspirate off medium, wash once with one volume PBS per well, and add 3 mL hiPSC medium supplemented with 40 ng/mL FGF2. Replace medium with fresh hiPSC medium every other day.

3.2.7. Day 6–10: Keratinocyte Colonies

1. Once cultured in hiPSC medium, keratinocytes will cease to proliferate as single cells and form clonal keratinocyte colonies (Fig. 2c). These colonies are often hard to distinguish from a typical hiPSC colony as they have sharp demarcated borders, a high nuclear:cytoplasmic ratio, and prominent nucleoli.

2. On average, keratinocyte colonies will subside growth within 4–5 days and undergo apoptosis within 2–3 days thereafter. In contrast, bona fide keratinocyte hiPSC colonies are rare, appear several weeks after plating, and usually after replating on fresh PMEF on day 21 (see Sect. 3.2.12).

3.2.8. Day 10: Change Medium to CM

Replace medium with 3 mL CM supplemented with 40 ng/mL FGF2 every other day (see Note 12).

3.2.9. Day 15–21: Cell Changes

During this time period, cells sometimes start changing morphology and decrease in size. Smaller cells may also cluster together.

3.2.10. Day 19: Preparation of New PMEF Plates

0.1% gelatin-coated two to three 6-well plates depending on the expansion of cells.

3.2.11. Day 20: PMEF Plates

Plate 3×10^5 irradiated PMEF per well of 6-well plate (see Sect. 3.4.1). Prepare two 6-well plates.

3.2.12. Day 21: Passaging Cells

1. Aspirate off medium, wash once with one volume PBS, and add 1 mL Accutase per well.
2. Incubate at 37°C for 1–2 min. Observe the cells periodically during this incubation period under the microscope to confirm that single cells have rounded up and are detaching from the plate (see Note 13).
3. When most cells are rounded up, stop treatment by adding 2 mL FDMEM per well.
4. Using a 5 mL pipette as a cell scraper, gently scrape the bottom of the dish by making vertical cuts followed by horizontal cuts. Swirl the plate to lift off cells into the media and repeat the procedure once more to detach any remaining adherent cells or pre-iPSC colonies.
5. Slowly, using a 10 mL pipette, remove the cell suspension and add it dropwise into a 15 mL conical tube. Using a P1000 and 1 mL PBS gently scrape off any remaining cells and add them dropwise to the same conical tube.
6. Centrifuge at $200 \times g$ for 5 min. While centrifuging, aspirate off medium from PMEF plate, wash once with one volume PBS, and add 1 mL hiPSC medium supplemented with 40 ng/mL FGF2.
7. Carefully, remove the supernatant, and using a P1000 resuspend the pellet in 1 mL of hiPSC medium supplemented with 40 ng/mL FGF2 (see Note 14).
8. Dropwise add 11 mL hiPSC medium to the cell suspension.
9. Using a 10 mL pipette add 1 mL of the cell suspension per well of previously prepared PMEF plates (1:2 split). Allow 24 hrs for cell clumps to attach.

10. The next day, remove the medium, wash once with one volume PBS to remove any debris, and replace medium with 2 mL per well of fresh hiPSC medium.

3.2.13. Days 22–35: Picking ESC-Like hiPSC Colonies

1. hiPSC colonies will arise in two waves:

    (a) First, disorganized partially reprogrammed SSEA4$^+$ Tra-1-81$^-$ Type I hiPSC colonies with loosely packed cells become visible as described in (6). These cells have a higher nuclear: cytoplasmic ratio than the starting cell population, but lack demarcated, sharp edges as observed in human embryonic stem cells (hESCs). The majority of colonies will take on this morphology and do not acquire a true ESC-like phenotype.

    (b) The second, rarer population of colonies represents bona fide hiPSC and start emerging approximately 4–10 days after bulk passaging onto fresh P_1 PMEF. These SSEA4$^+$ Tra-181$^{\text{low-high}}$ Type II hiPSC can easily be distinguished from Type I hiPSC by their compact morphology, sharply demarcated edges and high nuclear: cytoplasmic ratio. Fibroblastic differentiation may be apparent around the colonies (Fig. 3d).

2. As hiPSC colonies grow in size, prepare 1×10^5 irradiated PMEF per well on 0.1% gelatin-coated 12-well plates. A single experiment will yield on average 0–20 colonies (see Note 15).

3. The next day, remove FDMEM, wash once with one volume PBS, and add 750 µL of hiPSC medium supplemented with 40 ng/mL FGF2.

4. Under the microscope in a sterile laminar flow hood, manually pick only Type II hiPSC colonies using a P200 by dissociating colonies into clumps of 50–100 cells (see Notes 16 and 17). Gently pipette one colony into one well of a 12-well plate. Allow colonies to attach for at least 24–36 h.

5. Once cells have attached to plates, wash wells once with one volume PBS and add fresh hiPSC medium supplement with 40 ng/mL FGF2.

3.2.14. Day 36 and Onward: Subcloning and Expanding hiPSC

1. Passages 2–4 are usually reserved for subcloning while passages 5–11 serve to expand frozen stocks of subcloned hiPSC. 3–5 clones are subsequently characterized for absence of episomal DNA and integration, karyotype, and pluripotency potential by in vivo teratoma formation in NOD/SCID mice (Fig. 3).

2. During passage 2–4, the best morphological hiPSC colony should be manually picked as described above and transferred in clumps to a new well of 12-well plates with irradiated PMEF feeder cells (see Notes 18 and 19).

3. For passages 5–6, hiPSC clones should be manually passaged 1:1 into new wells of 12-well plates with PMEF feeder cells (see Sect. 3.4.3.1 and Note 19).

4. At the end of passage 6, hiPSC clones should be manually passaged 1:1 into one well of 6-well plates of PMEF feeder cells (see Sect. 3.4.3.1, Notes 20 and 21).

5. On passage 7, when colonies are confluent hiPSC should be manually passaged 1:1 or 1:2, if permissible based on density, onto new wells of PMEF feeder 6-well plates (see Sect. 3.4.3.2).

6. Passages 8–11, continue to expanding hiPSC clones 1:2 or 1:3.

7. Freeze 1.5 confluent hiPSC wells of a 6-well plate per vial (see Sect. 3.4.4).

8. On passage 11, freeze all but 3–5 clones for characterization. By the end of these passages, hiPSC have acquired a Type III SEEA4$^+$ Tra-181$^+$ phenotype as previously described (6) (Fig. 2e, f, h).

3.3. hiPSC Characterization

The detailed description of the characterization steps required for validation that bona fide hiPSCs have been generated exceeds the scope of this chapter. However, we have provided an outline of each step and detailed references required for the full characterization of nonviral integration-free hiPSC in Fig. 3. The user who utilizes this protocol should adhere to the recommended passage number and order in which each characterization should be performed as some steps will lead to false positive results if performed too early. For example, the important verifications that hiPSC are devoid of transgene and episomal vector sequences (see Sect. 3.3.1) should be executed no earlier than passages 11–12 to assure episomal DNA has been sufficiently diluted. Furthermore, no resources should be unnecessarily expended on other validations such as karyotyping and teratoma formation before hiPSC have acquired a true Type III phenotype as shown by Tra-1-60 and Tra-1-81 staining analyzed by flow cytometry (21).

3.3.1. Genomic PCR and RT-PCR of Vector Sequences and Transgene Expression

The primers required for validating the loss of episomal vectors and their transgene expressions are listed in Table 1.

1. Briefly, harvest one well of a 6-well plate of confluent hiPSC using a cell scraper and centrifuge at 200 ×g for 5 min.

2. Remove supernatant and place cell pellet on ice.

3. For RT-Transgene PCR analysis, use RNeasy Kit (Qiagen) to isolate RNA and elute from spin columns with 30 µL DEPC-treated water. Be sure to perform optional on column DNase digestion when isolating RNA. Quantitate RNA concentration using a Nanodrop instrument, and use 1 µg of RNA per reaction for the production of cDNA using First-Strand cDNA Synthesis. Dilute final cDNA 1:3 in DEPC-treated water.

4. For Genomic PCR, use DNeasy Blood and Tissue Kit to isolate genomic DNA and elute DNA with 100 µL DEPC-treated water.

Table 1
Primer sequences for RT-PCR and Genomic PCR detection of expression of *oriP*/EBNA1 episomal transgene and vector sequences

PCR	Genes	5′ → 3′ primer sequence	Amplicon size, bp	Annealing temperature, °C
RT-transgene PCR	SOX2	FOR: TGGCTCTCCTCAAGCGTATT REV: GCTTAGCCTCGTCGATGAAC	498	55
	KLF4	FOR: TGGCTCTCCTCAAGCGTATT REV: GTGGAGAAAGATGGGAGCAG	253	55
	LIN28	FOR: TGGCTCTCCTCAAGCGTATT REV: GCAAACTGCTGGTTGGACAC	245	55
RT-transgene and genomic PCR	OCT4	FOR: AGTGAGAGGCAACCTGGAGA REV: AGGAACTGCTTCCTTCACGA	657	55
	C-MYC	FOR: TCAAGAGGCGAACACACAAC REV: AGGAACTGCTTCCTTCACGA	546	55
	NANOG	FOR: CAGAAGGCCTCAGCACCTAC REV: AGGAACTGCTTCCTTCACGA	732	55
	SV40LT	FOR: TGGGGAGAAGAACATGGAAG REV: AGGAACTGCTTCCTTCACGA	491	55
	EBNA1	FOR: ATCGTCAAAGCTGCACACAG REV: CCCAGGAGTCCCAGTAGTCA	666	55
	GAPDH	FOR: GTGGACCTGACCTGCCGTCT REV: GGAGGAGTGGGTGTCGCTGT	152	55
Genomic PCR	SOX2	FOR: ACCAGCTCGCAGACCTACAT REV: CCCCCTGAACCTGAAACATA	534	55
	KLF4	FOR: CCCACACAGGTGAGAAACCT REV: CCCCCTGAACCTGAAACATA	401	55
	LIN28	FOR: AAGCGCAGATCAAAAGGAGA REV: CCCCCTGAACCTGAAACATA	447	55
	oriP	FOR: TTCCACGAGGGTAGTGAACC REV: TCGGGGGTGTTAGAGACAAC	544	55

5. For Genomic PCR, quantitate concentration of DNA samples with Nanodrop instrument.
6. For both RT-Transgene and Genomic PCR use normal hESC samples as a negative control and an early plasmid post-nucleofection (i.e., Day 4) cell sample as positive control.
7. Set up the following PCR reactions in final concentration in 20 μL PCR reaction for each sample and primer:

   3× Enhancer solution

   2× Amplification solution

   0.3 mM dNTP mixture

   1 mM $MgSO_4$

   0.3 μM of forward and reverse primers

0.4 units of Platinum Pfx DNA polymerase (0.16 μL)

100 ng Genomic DNA or 2 μL of diluted cDNA

DEPC to 20 μL volume

8. Run PCR reactions using the following cycling conditions:

Initialization hot start:		94°C for 5 min
35 cycles:	Denaturing:	94°C for 15 s
	Annealing:	55°C for 30 s
	Elongation:	68°C for 1 min
Final elongation:		68°C for 10 min
Final hold:		4°C

9. Run 3 μL of PCR reaction on a 2% Agarose Quick E-Gel to visualize PCR products.

3.3.2. Southern Blotting

1. For Southern blotting, 5 μg of genomic DNA is required. In general, two 6-well plates of confluent hiPSCs are sufficient to isolate this amount of DNA. Alternatively, two confluent wells of a 6-well plate of hiPSC can be split onto one 0.1% gelatinized T-75 flask and allowed to differentiate by culturing in FDMEM.

2. Once confluent, harvest cells by cell scraping, isolate genomic DNA as described above and elute in 50 μL of DEPC-treated water. Quantitate DNA to ensure DNA concentrations are at least 1 μg/μL. If necessary, ethanol-precipitate DNA to purify and concentrate it in case a lower yield was obtained.

3. Digest 5 μg of genomic DNA from hiPSCs and negative control hESCs with high fidelity restriction enzymes BamHI and SpeI (50 units/reaction each) at 37°C overnight.

4. Run digested samples and 10 and 100 pg of nucleofection plasmids (see Sect. 3.2.3; 10 and 100 pg correspond to 0.4× and 4× integrations per genome, respectively) as positive controls on a 0.8% agarose gel at 30 V overnight.

5. Transfer onto positively charged nylon membrane with 20× SSC buffer, and auto-UV-crosslink.

6. As probe, digest pCEP4 vector with NotI and NruI to release the 7.3 kb episomal vector backbone contained in *oriP*/EBNA-1-based reprogramming vectors. Gel purify digested vector using a 0.8% agarose gel, and gel extract using the QIAquick Gel Extraction Kit.

7. Label backbone as described in DIG-High Prime DNA Labeling and Detection Starter Kit II.

8. Digest Dig-labeled pCEP4 vector backbone probe with AluI and purify again.

9. Determine the amount of DIG-label probe as described by manufacturer's protocol.

10. Perform all prehybridization, hybridization, and wash steps as described by manufacturer's protocol. Use 42.3°C as hybridization temperature and wash with high stringency buffers as indicated.

11. Visualize DIG staining membrane as described by manufacturer's protocol and develop on Lumi-Film Chemiluminescent Detection Film for 1–25 min as needed.

3.3.3. qRT-PCR for Endogenous Gene Expression

The fully reprogrammed (Type III) hiPSC phenotype can be quickly confirmed by performing quantitative real-time qRT-PCR analysis for high expression of endogenously expressed pluripotency genes. Particular attention should be paid to high expression of *ABCG2*, *DNMT3B*, and *REX1* (*ZFP42*) as they have been shown to successfully distinguish Type II and Type III hiPSC (6). These studies should always be run with comparative expression of a control hESC line. PCR primers in Table 2 were optimized for this purpose using standard SYBR protocols. Results should be interpreted by the $2^{-\Delta\Delta CT}$ method (22).

In our experience, hiPSC derived from various sources expand more robustly and maintain an undifferentiated morphology in 20–40 ng/mL FGF2. Following completion of all hiPSC characterization steps as outlined in Fig. 3, these higher doses of FGF2 may be reduced as tolerated over several passages (see Note 22).

3.4. hiPSC Cell Culture

3.4.1. Preparation of PMEF Feeder Layers for Cultivation and Derivation of hiPSC

1. Thaw 3×10^6 ♂ CF-1 × ♀ DR4-derived PMEF in PMEF medium onto a 0.1% gelatin-coated T-175 flask (see Note 7). Change medium the day after thawing and every other day thereafter.

2. After approximately 2–3 days, PMEF will reach 95% confluency (see Note 23).

3. Aspirate off the medium, and wash the flask with one volume of PBS, and add 7 mL prewarmed 0.05% Trypsin-EDTA.

4. Incubate at 37°C for 2–5 min, and tap the side of the flask occasionally to loosen cells.

5. When cells have visibly detached from the flask, neutralize the Trypsin-EDTA by adding 13 mL PMEF medium directly to the flask.

6. Using a 10 mL pipette, gently pipette the cell suspension 1–2 times to generate a single cell suspension and add into a 50 mL conical tube.

7. Wash the flask with 10 mL PBS and add the solution to the same conical tube.

8. Centrifuge at $200 \times g$ for 5 min.

Table 2
Primer sequences for quantitative real-time (qRT-PCR) for analysis of endogenously expressed pluripotency genes

PCR	Genes	5′ → 3′ primer sequence	Amplicon size, bp	Annealing temperature, °C
qRT endogenous PCR	SOX2	FOR: CCCAGCAGACTTCACATGT REV: CCTCCCATTTCCCTCGTTTT	151	60
	OCT4	FOR: CCTCACTTCACTGCACTCTA REV: CAGGTTTTCTTTCCCTAGCT	164	60
	KLF4	FOR: GACCACCTCGCCTTACACAT REV: TGGGAACTTGACCATGATTG	161	60
	C-MYC	FOR: AAGAGGACTTGTTGCGGAAA REV: CTCAGCCAAGGTTGTGAGGT	179	60
	NANOG	FOR: CTCCATGAACATGCAACCTG REV: GGCATCATGGAAACCAGAAC	157	60
	LIN28	FOR: CACAGGGAAAGCCAACCTAC REV: TGCACCCTATTCCCACTTTC	162	60
	ABCG2	FOR: AGCTGCAAGGAAAGATCCAA REV: TCCAGACACACCACGGATAA	286	60
	TP53	FOR: GGCCCACTTCACCGTACTAA REV: GTGGTTTCAAGGCCAGATGT	156	60
	DNMT3B	FOR: GCGTTTGGTGGATGATTTCT REV: CAGGGCCTCGTCTTCTACAG	151	60
	hTERT	FOR: AAGTTCCTGCACTGGCTGAT REV: TTGCAACTTGCTCCAGACAC	129	60
	REX1	FOR: GGCGGAAATAGAACCTGTCA REV: CTCACCCCTTATGACGCATT	130	60
	UTF1	FOR: AGCTGCTGACCTTGAACCAG REV: GTGGGAAGGCAGCAGGAG	204	60
	ACTIN	FOR: GGCATCCTCACCCTGAAGTA REV: GGGGTGTTGAAGGTCTCAAA	203	60

9. Aspirate the supernatant and resuspend the cell pellet in 20 mL PMEF medium.
10. Count the cell using a hemocytometer. A density of 2×10^5 and 3×10^5 cells per well of a 6-well plate are needed for routine hiPSC culture or hiPSC derivation and culture until passage 2, respectively.
11. Irradiate PMEF at 5,000 cGy with a Cesium source.
12. Wash with 30 mL PBS to the 50 mL conical, and centrifuge at $200 \times g$ at RT for 5 min. Centrifuge at $200 \times g$ for 5 min.
13. Resuspend PMEF in FDMEM and plate onto 0.1% gelatin-coated 6-well plates (see Note 7).

3.4.2. Preparation of Conditioned Media (CM) from PMEF

1. Thaw passage 1–2 PMEF into T-175 flask in 25 mL PMEF medium and grow to 90–100% confluency (see Note 23).
2. Gelatinize T-75 or T-175 overnight at 37°C the day before PMEF reach confluency.
3. Perform steps 3–9 as described in Sect. 3.4.1.
4. Count the cell using a hemocytometer. PMEF are seeded at a density of 6×10^4 cells/cm^2 (4.5 million cells per T-75 flask; 10.5 million cells per T-175 flask).
5. Irradiate PMEF at 5,000 cGy.
6. Add 30 mL PBS to the 50 mL conical and centrifuge at $200 \times g$ at RT for 5 min.
7. Aspirate off the supernatant, resuspend cells in PMEF medium, and plate on 0.1% gelatinized flasks (see Note 7).
8. After 24 h, aspirate off PMEF medium and wash with one volume of PBS.
9. Add 0.5 mL/cm^2 hiPSC medium supplemented with 4 ng/mL FGF2.
10. Harvest PMEF-CM every 22–26 h for a maximum of 7 days, aliquot and freeze at −20 °C.
11. Upon need, thaw CM overnight at 4°C and filter sterilize the next day.
12. Add fresh FGF2 at desired concentrations of 4–40 ng/mL.

3.4.3. Passaging hiPSC

Manual Passaging of Colonies

1. Prepare new irradiated PMEF plate as described in Sect. 3.4.1, wash with PBS the next day, and add enough hiPSC medium to cover the bottom.
2. Aspirate off medium from hiPSC plate and wash with 2–3 mL PBS, and add 1 mL of fresh hiPSC medium per well.
3. With 1 mL of fresh hiPSC medium in the tip of a P1000 pipette, while slowly releasing the medium from the pipette, gently scrape the bottom of the well in a grit-like fashion, first vertically then horizontally, to obtain medium-sized 50–100 cell clumps of hiPSC.
4. Gently transfer cells dropwise to a new well of irradiated PMEF.
5. Repeat step 3 with 500 µL hiPSC to transfer any remaining colonies.
6. Allow colonies to attach for 12–24 h.
7. The following day, wash once with PBS and replace with fresh hiPSC medium.
8. Replace medium every day.

Enzymatic Passage of hiPSC Colonies

1. Remove medium from 80% to 90% confluent hiPSC, wash with one volume PBS, and add 1 mL of 1 mg/mL Collagenase IV or Accutase per well of a 6-well plate.
2. Incubate at 37°C for 3–5 min for Collagenase IV and 1–2 min for Accutase treatment. Periodically during the incubation period, under the microscope observe whether the hiPSC colonies' edges are detaching from the plate while the colonies' centers are still intact.
3. When edges are detaching from the plate, stop the treatment by adding two volumes of FDMEM per well.
4. Using a 5 or 10 mL pipette as cell scraper, gently scrape the bottom of the dish by making vertical cuts followed by horizontal cuts. Swirl the plate to lift off any detached colonies into the media and repeat the procedure once more to detach any remaining adherent colonies (see Note 24).
5. Slowly, using a 10 mL pipette, remove the cell suspension and add it dropwise into a 15 mL conical tube.
6. Centrifuge at $200 \times g$ for 5 min.
7. Carefully, remove the supernatant and resuspend the pellet in 1 mL per well of hiPSC medium.
8. Add cell suspension 1:2 or higher depending on density and viability of hiPSC cell clumps to previously prepared PMEF plate. Allow 24 h for cell clumps to attach.
9. The next day, remove the medium, wash the plate vigorously with PBS to remove any differentiating areas and debris.
10. Aspirate the PBS and gently add 2 mL of fresh hiPSC medium to each well.

3.4.4. Freezing hiPSC

1. Harvest hiPSC manually or by enzymatic methods as described in Sect. 3.4.3 (see Note 25).
2. Resuspend cell pellet in 1 mL of hiPSC freezing medium per vial.
3. Label vial with name, starting cell population, passage number, FGF2 concentration, number of wells frozen per vial, and date.
4. Freeze hiPSC using 1°C/min freezing containers.
5. The next day, transfer hiPSC vials to liquid nitrogen for storage.

3.4.5. Thawing hiPSC

1. Prepare one well of a 0.1% gelatin-coated 6-well plate with 2×10^5 irradiated PMEF for each vial of hiPSC to be thawed, as described in Sect. 3.4.1.
2. The following day, prewarm 11 mL of hiPSC supplemented with 4–40 ng/mL at 37°C for 5 min (see Note 26).
3. Thaw one vial of hiPSCs at 37°C until only a small ice cube remains. Remove vial and prewarmed medium from water bath, spray with ethanol (EtOH).

4. Using a P1000, dropwise transfer contents to a 15 mL conical tube (see Note 27).

5. Dropwise add 9 mL of hiPSC medium to the cell suspension. Mix by gently inverting the conical tube once.

6. Centrifuge at $200 \times g$ for 5 min.

7. During the centrifugation step, wash PMEF plate once with one volume PBS and add 1 mL of hiPSC. Briefly, equilibrate plate at 37°C.

8. Carefully aspirate supernatant (see Note 28). Without dissociating the cell pellet, dropwise add 10 mL of PBS.

9. Centrifuge at $200 \times g$ for 3 min.

Using a P1000, resuspend cell pellet in 1 mL hiPSC and add dropwise onto PMEF feeders using a 10 mL pipette (see Note 27).

4. Notes

1. Freezing solution should be made up fresh for each use. Freezing solution should not be aliquoted or stored for later use.

2. Following overnight incubation at 4°C, warm dishes to 37°C in the incubator for 5 min. Do not remove or wash off Matrigel from 35 mm dishes.

3. It is essential to allow the hair follicle to attach to the plate within the first 24–48 h. Any disturbances should be minimized.

4. To allow proper coating, KCM should be applied to plates at RT for at least 30 min.

5. The low reprogramming efficiency of nonviral, transgene-free approaches can often be overcome by nucleofecting more starting cells. For this protocol 7×10^5 cells are used per nucleofection. Thus, it is advised to perform at least two nucleofections per condition with a total of 1.4×10^6 keratinocytes.

6. On the day of nucleofection, keratinocytes should have reached 70–80% confluency. Cells should not be allowed to overgrow as this may reduce cell viability and plasmid transfer.

7. To allow proper coating, plates should be allowed to gelatinize at 37°C for at least two or preferably overnight.

8. The protocol should be followed directly starting with step 2, skipping steps 3–4, and performing all optional steps. 30 μL of DEPC-treated water should be used to elute DNA. The last step should be performed in a tissue culture hood using tissue culture grade DEPC-treated water.

9. The best nucleofection results are obtained when medium is aliquoted for each nucleofection, equilibrate to 37°C, and removed from water bath or incubator immediately before use.

10. DNA volume should not exceed 5 μL. If DNA concentrations derived as described in Sect. 3.2.3 exceed this volume, we recommend ethanol precipitation to concentrate plasmids further.

11. Continued exposure to nucleofection solution, even when diluted in medium, can be toxic. Ensure that supernatant is removed completely.

12. If cells, other than irradiated PMEF, become too crowded, feed cells every day with 3 mL of CM.

13. If pre-iPSC colonies or other cell changes have occurred that formed cell clusters incubate with Accutase until most of the edges have detached, but the cells in the center are only rounded up or still attached. Whenever possible, pre-iPSC should be passaged in 30–50 cell clumps.

14. Do not resuspend vigorously. It is important to preserve these pre-iPSC colonies as clumps whenever possible.

15. A beginner should pick as many ESC-like colonies that arise, while 12 colonies are usually sufficient for screening for more experienced individuals.

16. Type I hiPSCs do not complete reprogramming fully and thus never take on a true hESC phenotype. These colonies should not be picked.

17. Depending on the number of hiPSC colonies that arises following bulk passage on day 21, Type I hiPSCs usually overgrow the plate and often reduce the space available for Type II hiPSC to expand. In this situation, Type II colonies should be picked earlier to ensure that no Type I hiPSC are transferred to subsequent cultures.

18. Subcloning one hiPSC colony not only ensures clonality, but selecting the highest quality colony also reduces changes of expanding a clone that spontaneously differentiates at later passages when exogenous pluripotency factor expression from episomal vectors is lost.

19. Early passage (e.g., less than passage 10–12) clones are generally unstable and have a tendency to spontaneously differentiate (Fig. 2d). In our hands, these low passage clones require great individual care and manual passaging techniques using a sterile hood and an inverted microscope. We recommend plating early passage nonviral hiPSC in 12-well plates at higher densities compared to culture in 6-well plate to increases hiPSC stability and maintenance of pluripotency.

20. hiPSCs should be passaged at 50–75 cell clump compared to hESCs which are often viable as smaller 15–30 cell clumps.

21. When hiPSCs are not plated at high enough densities in the transition from 12 to 6-well plates, spontaneous differentiation is often observed. In this case, it is advisable to manually subclone all undifferentiated colonies again back into a 12-well process and to repeat the process with adequate cell numbers.

22. In some cases, hiPSC also spontaneously differentiate as FGF2 concentrations are reduced (Fig. 2d). Subcloning is advisable under those circumstances until cultures are stable. For some hiPSC clones, it may not be possible to reduce FGF2 concentrations to 4–10 ng/mL without spontaneous differentiation despite subcloning. These unstable hiPSC lines should not be used for directed differentiation experiments.

23. PMEF should not be overgrown at any point as this may compromise viability and introduce inconsistency into the hiPSC cultures.

24. Pipetting hiPSC should be minimized to ensure large cell clumps. Following centrifugation, when hiPSC are resuspended in fresh medium, clumps should be reduced in size to 50–75 cells per cell clump. When hiPSCs are dissociated too vigorously prior to centrifugation, the final resuspension step often dramatically reduced cell viability.

25. Since hiPSC clumps will be manipulated further upon thawing, hiPSCs should be frozen down in larger cell clumps of 75–150 cells.

26. Use the appropriate FGF2 concentration that hiPSCs were fed with prior to freezing. If using low FGF concentrations, temporary higher FGF concentrations for 1–2 days may increase viability and reduce differentiation upon thawing.

27. Minimize pipetting while resuspending in order to preserve cell clumps.

28. It is important to remove all supernatant to reduce trace amounts in DMSO in culture to reduce toxicity.

References

1. Takahashi K., Tanabe K., Ohnuki M. et al. (2007) Induction of pluripotent stem cells from adult human fibroblasts by defined factors. Cell 131, 861–872.
2. Hanna J., Wernig M., Markoulaki S. et al. (2007) Treatment of sickle cell anemia mouse model with iPS cells generated from autologous skin. Science 318, 1920–1923.
3. McCormack M.P. and Rabbitts T.H. (2004) Activation of the T-cell oncogene LMO2 after gene therapy for X-linked severe combined immunodeficiency. N.Engl.J.Med. 350, 913–922.
4. Nakagawa M., Koyanagi M., Tanabe K. et al. (2008) Generation of induced pluripotent stem cells without Myc from mouse and human fibroblasts. Nat.Biotechnol. 26, 101–106.
5. Miura K., Okada Y., Aoi T. et al. (2009) Variation in the safety of induced pluripotent stem cell lines. Nat.Biotechnol. 27, 743–745.

6. Chan E.M., Ratanasirintrawoot S., Park I.H. et al. (2009) Live cell imaging distinguishes bona fide human iPS cells from partially reprogrammed cells. Nat.Biotechnol. 27, 1033–1037.
7. Chin M.H., Mason M.J., Xie W. et al. (2009) Induced pluripotent stem cells and embryonic stem cells are distinguished by gene expression signatures. Cell.Stem Cell. 5, 111–123.
8. Feng Q., Lu S.J., Klimanskaya I. et al. (2010) Hemangioblastic derivatives from human induced pluripotent stem cells exhibit limited expansion and early senescence. Stem Cells 28, 704–712.
9. Hu B.Y., Weick J.P., Yu J. et al. (2010) Neural differentiation of human induced pluripotent stem cells follows developmental principles but with variable potency. Proc.Natl.Acad.Sci.U.S.A. 107, 4335–4340.
10. Kim K., Doi A., Wen B. et al. (2010) Epigenetic memory in induced pluripotent stem cells. Nature. doi:10.1038/nature09342.
11. Soldner F., Hockemeyer D., Beard C. et al. (2009) Parkinson's disease patient-derived induced pluripotent stem cells free of viral reprogramming factors. Cell 136, 964–977.
12. Seki T., Yuasa S., Oda M. et al. (2010) Generation of induced pluripotent stem cells from human terminally differentiated circulating T cells. Cell.Stem Cell. 7, 11–14.
13. Yu J., Hu K., Smuga-Otto K. et al. (2009) Human induced pluripotent stem cells free of vector and transgene sequences. Science 324, 797–801.
14. Jia F., Wilson K.D., Sun N. et al. (2010) A nonviral minicircle vector for deriving human iPS cells. Nat.Methods 7, 197–199.
15. Kim D., Kim C.H., Moon J.I. et al. (2009) Generation of human induced pluripotent stem cells by direct delivery of reprogramming proteins. Cell.Stem Cell. 4, 472–476.
16. Aasen T., Raya A., Barrero M.J. et al. (2008) Efficient and rapid generation of induced pluripotent stem cells from human keratinocytes. Nat.Biotechnol. 26, 1276–1284.
17. Wesselschmidt R.L., Loring J.F. (2007) Classical Cytogenetics: karyotyping. In: Loring J.F., Wesseschmidt R.L., Schwartz P.H (ed) Human stem cell manual, 1st edn. Elsevier, New York, pp. 59–70.
18. Gertow K., Przyborski S., Loring J.F. et al. (2007) Isolation of human embryonic stem cell-derived teratomas for the assessment of pluripotency. Curr.Protoc.Stem Cell.Biol. Chapter 1, Unit1B.4.
19. Aasen T. and Belmonte J.C. (2010) Isolation and cultivation of human keratinocytes from skin or plucked hair for the generation of induced pluripotent stem cells. Nat.Protoc. 5, 371–382.
20. McWhir J., Wojtacha D. and Thomson A. (2006) Routine culture and differentiation of human embryonic stem cells Methods. Mol. Biol. 331, 77–90.
21. Laslett A.L., Fryga A., Pera M.F. (2007) Flow Cytometric Analysis of Human Embryonic Stem Cells. In: Loring J.F., Wesseschmidt R.L., Schwartz P.H (ed) Human stem cell manual, 1st edn. Elsevier, New York, pp. 96–107.
22. Zambidis E.T., Peault B., Park T.S. et al. (2005) Hematopoietic differentiation of human embryonic stem cells progresses through sequential hematoendothelial, primitive, and definitive stages resembling human yolk sac development. Blood 106, 860–870.

Part III

Generation of Patient-Specific iPS Cells for Clinical Application

Chapter 17

Generation of iPS Cells from Human Skin Biopsy

Katie Avery and Stuart Avery

Abstract

The reprogramming of human somatic cells to induced pluripotent stem (iPS) cells has offered the opportunity to derive patient-specific cells with embryonic stem cell (ESC) properties. Human iPS cells demonstrate the ability to self-renew and to differentiate into any cell type of the body. They provide vast opportunities for in vitro modeling of development and disease and have the potential to be a source for cell replacement therapies. This protocol describes the culture and reprogramming of primary human fibroblasts from skin biopsies using retroviral transduction of OCT4, SOX2, KLF4, and c-MYC with the addition of the histone deacetylase (HDAC) inhibitor, valproic acid (VPA). Characterized patient-specific iPS cell lines can be obtained within a period of 2–3 months.

Key words: Induced pluripotent stem cells, Patient-specific, Retroviral transduction, Human primary fibroblasts

1. Introduction

Human embryonic stem cells (hESCs) are derived from the inner cell mass of the embryo and have the ability to self-renew while retaining their potential to differentiate into all somatic cell types in the human body (1). ESCs have an array of applications, including uses in drug screening, disease modeling, and tissue engineering for degenerative diseases. However, hESC lines are not patient or disease specific and may lead to immunological rejection if used in cell therapies. In addition, there are ethical considerations and governmental restrictions surrounding the use of these cells.

To eliminate the possibility of immune rejection in clinical applications, a variety of methods have been investigated to produce patient-specific pluripotent stem cells. These techniques include somatic cell nuclear transfer (SCNT) and fusion of somatic cells with hESCs (2), although these approaches have resulted in limited success. The generation of induced pluripotent stem cells (iPSCs) demonstrates that defined factors can reprogram human somatic cells to pluripotency, and therefore achieves the aim of establishing patient-specific cells in culture.

A crucial breakthrough was achieved by the reprogramming of mouse embryonic fibroblasts (MEF) cells to an ESC-like state after retroviral transduction of four transcription factors: Oct4, Sox2, c-myc, and Klf4 (3). The generation of iPSCs was subsequently achieved by using human fibroblasts and human homologues of these four factors (4) and by using different combinations of genes, such as OCT4, SOX2, NANOG, and LIN28 (5). The iPSCs generated resemble ESCs in their morphology, growth properties, gene expression, and their ability to form teratomas in immunodeficient mice. Reprogramming viruses were found to be silenced in human iPSCs, demonstrating that continuous transgene expression is not necessary for the maintenance of pluripotency (4–6).

More recently, iPSC lines have been obtained using two or three defined factors. OCT4, SOX2, and KLF4 were found to be sufficient to reprogram dermal fibroblasts; however, the exclusion of c-MYC greatly reduced the efficiency of reprogramming (7, 8). Subsequent studies demonstrated the generation of iPSCs using lentiviral vectors, doxycycline-inducible vectors, adenoviral and plasmid vectors, and excisable genetic inserts to deliver the exogenous genes (9–13). In addition to these delivery systems, recombinant proteins and small molecules have been used to replace specific transcription factors or enhance the efficiency of reprogramming (14–19). Exposure to a hypoxic environment or to vitamin C has been shown to increase the frequency of reprogramming (20, 21), and the disruption of signaling pathways mediated by P53 or the cell cycle regulator, INK4a, was found to increase the speed of iPSC generation (22–26). Improved efficiency of reprogramming has also been documented with the use of microRNAs (27).

The ability to obtain patient-specific stem cells offers many new opportunities for the regenerative medicine field. However, there remain concerns regarding the use of oncogenic factors and the integration of viral vectors into the genome during reprogramming. These issues will need to be addressed before this technology could be moved toward clinical applications. Despite these limitations, iPSCs provide an invaluable tool for in vitro models of human development and disease. Cell lines have successfully been derived from fibroblasts harboring monogenic defects (such as Huntington's disease, Becker's muscular dystrophy, and adenosine deaminase-associated severe combined immunodeficiency) and complex defects (including Parkinson's disease and juvenile diabetes mellitus) (28). In addition, combining the iPS cell technology with gene therapy to reprogram genetically corrected somatic cells can generate disease-free progenitor cells for use in autologous cell therapy applications (29).

To date, the most efficient method to generate iPSCs is to use retroviral transduction of a combination of the four transcription factors: OCT4, SOX2, KLF4, and c-MYC. This protocol describes the culturing and reprogramming of human primary cells using a refined version of the original method and is conducted in the presence of a histone deacetylase (HDAC) inhibitor, valproic acid (VPA), to enhance efficiency. This small molecule was chosen as the addition of VPA had been shown to increase the number of ES-like colonies by up to 100-fold (15), indicating that chromatin modification is a key step in the reprogramming process.

This protocol uses primary fibroblast lines derived from a skin punch biopsy (which should be performed by a trained clinician and with informed consent from the donor) and demonstrates the feasibility of establishing iPSCs from patients with specific diseases to be used for research purposes. In this chapter, instruction is provided on how to establish a primary culture from patient biopsies, the preparation of 293T and MEF cell cultures, production of retroviral supernatants, and their use in the reprogramming of somatic cells. This is followed by directions for the picking and expansion of iPS cell colonies and subsequently provides information on the characterization and verification of iPS cell lines. Figure 1 depicts the time line of events in iPSC generation from patient biopsies.

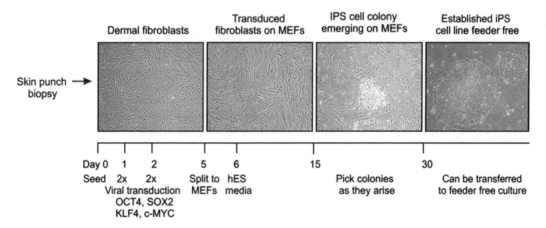

Fig. 1. Time line for reprogramming of human dermal fibroblasts into human iPS cells.

2. Materials

2.1. Establishing Primary Cell Cultures from Patient Biopsies

1. A 6-mm skin punch biopsy. This should be obtained with informed consent from the patient under a protocol approved by the relevant institutional review board, and by a trained physician. The generation of human iPS cell lines may require permission by an oversight committee in some countries.

2. hFib medium. To prepare hFib media, the following reagents are required: DMEM (Invitrogen, Cat. No. 21969–035) containing 10% heat-inactivated FBS (Invitrogen, Cat. No. 10270–106), 2 mM L-glutamine (Invitrogen, Cat. No. 25030–081), 50 U/mL penicillin and 50 mg/mL streptomycin (Invitrogen, Cat. No. 15140–163). To prepare 500 mL of medium, add 50 mL heat-inactivated FBS (55°C for 30 min), 5 mL L-glutamine and 5 mL penicillin/streptomycin together, and then add DMEM for a total volume of 500 mL. The medium should then be filtered with a 0.22-µm bottle top filter. Medium can be kept for 1 month at 4°C.

3. Sterile dissection forceps.

4. Sterile scalpel blades.

5. 10-mm Sterile coverslip.

6. Petri dish.

7. 24-well Tissue culture-treated plastic dish.

8. 6-well Tissue culture-treated plastic dish.

9. 100-mm Tissue culture-treated plastic dish.

10. 70-µm Cell strainer (BD Biosciences, Cat. No. 352350).

11. 0.05% Trypsin/EDTA (Invitrogen, Cat. No. 25300–054).

12. Freezing medium (Invitrogen, Cat. No. 12648-010).

13. PBS without Ca^{++}/Mg^{++} (Invitrogen, Cat. No. 14190).

14. DMEM/F12 (Invitrogen, Cat. No. 11330–32).

15. Dispase II neutral protease grade II (Roche, Cat. No. 04942078001) dissolved in sterile PBS to a concentration of 3 mg/mL.

2.2. Preparing 293T Cells

1. 293 T cells (American Type Culture Collection, Cat. No. CRL-11268).

2. 293 T medium (Invitrogen, Cat. No. 12338–018).

3. 100-mm tissue culture-treated plastic dish.

2.3. Production of Reprogramming Retroviruses

1. Retroviral vectors. Moloney-based retroviral vectors (pMXs) containing the human cDNAs for OCT4, SOX2, c-MYC, and KLF4 (4) can be obtained from Addgene. Packaging

and envelope plasmids, pUMVC and pVSVG (30), can also be obtained from Addgene (www.addgene.org, pMX plasmid numbers 17217–17220, pUMVC plasmid number 8449, and pVSVG plasmid number 8454).

2. DMEM (Invitrogen, Cat. No. 11965-092).
3. SuperFect transfection reagent (Qiagen, Cat. No. 301305).
4. 0.45-µm Filter (Sartorius Stedim Biotech, Cat. No. 16555).

2.4. Preparing Mouse Embryonic Fibroblasts

1. Day 13. Five pregnant female mice (strain MF-1 or CD-1 is suitable).
2. Mouse embryonic fibroblast (MEF) medium. To prepare MEF media, the following reagents are required: DMEM (Invitrogen, Cat. No. 21969–035) containing 10% heat-inactivated FBS (Invitrogen, Cat. No. 10270–106), 2 mM L-glutamine (Invitrogen, Cat. No. 25030–081), 50 U/mL penicillin and 50 mg/mL streptomycin (Invitrogen, Cat. No. 15140–163). To prepare 500 mL of medium, add 50 mL heat-inactivated FBS, 5 mL L-glutamine and 5 mL penicillin/streptomycin to 440 mL of DMEM. The medium should then be filtered with a 0.22-µm bottle top filter. Medium can be kept for 1 month at 4°C.
3. Mytomycin-C (Sigma-Aldrich, Cat. No. M4287) is very hazardous (see Note 1). Mytomycin-C comes in an ampoule (2 mg) and should be dissolved by adding 1 mL MEF media using an 18G needle. Transfer the dissolved material into 200 mL MEF media using an 18G needle and filter sterilize it using a 0.22-µm bottle top filter.
4. Gelatin (Sigma-Aldrich, Cat. No. G1890). Gelatin is dissolved in PBS by autoclaving before use. To gelatin-coat dishes, add 0.1% gelatin solution to cover the bottom of the dish. Incubate the dish for at least 2 h at 37°C. To use plates, remove gelatin solution.
5. Sterile scalpel blades.
6. Freezing medium (Invitrogen, Cat. No. 12648–010).
7. PBS without Ca^{++}/Mg^{++} (Invitrogen, Cat. No. 14190).
8. 100-mm tissue culture-treated plastic dish.
9. 75-cm^2 tissue culture-treated plastic flask.
10. Petri dish.

2.5. Reprogramming Somatic Cells

1. PBS without Ca^{++}/Mg^{++} (Invitrogen, Cat. No. 14190).
2. 0.05% Trypsin/EDTA (Invitrogen, Cat. No. 25300–054).
3. hFib media (as previously described).
4. Retroviral infection medium. To prepare retroviral infection medium, the following reagents are required: minimum essential media α (MEMα, Sigma-Aldrich, Cat. No. M4526) containing

10% heat-inactivated FBS (Invitrogen, Cat. No. 10270–106), 50 U/mL penicillin and 50 mg/mL streptomycin (Invitrogen, Cat. No. 15140–163). To prepare 500 mL of medium, mix 50 mL of heat-inactivated FBS with 5 mL of penicillin/streptomycin and 440 mL MEMα. The medium should then be filtered with a 0.22-μm bottle top filter. Medium can be kept for 1 month at 4°C.

5. 0.25% Trypsin/EDTA (Invitrogen, Cat. No. 25200–056).
6. Polybrene (Millipore, Cat. No. Tr-1003-G).
7. Valproic acid (EMD Biosciences, Cat. No. 676380).

2.6. Picking and Expanding of iPS Colonies

1. Gelatin (Sigma-Aldrich, Cat. No. G1890) 0.1% solution (as previously described).
2. MEF medium (as previously described).
3. hES cell medium. To prepare hES medium, the following reagents are required: Knockout (KO) DMEM (Invitrogen, Cat. No. 10829–018) containing 20% KO serum replacement (KOSR; Invitrogen, Cat. No. 10828–028), 8 ng/mL bFGF (Invitrogen, Cat. No. 13256–029), 1 mM l-glutamine, 100 mM nonessential amino acids (Invitrogen, Cat. No. 11140–050), and 0.1 mM 2-mercaptoethanol (Sigma-Aldrich, Cat. No. M3148). To prepare bFGF stock, reconstitute bFGF in sterile PBS (Invitrogen, Cat. No. 14190) containing 0.1% BSA (Invitrogen, Cat. No. 15260–037) to a concentration of 8 μg/mL; store 500μL aliquots at −80°C. To prepare 500 mL of the medium, mix 100 mL KOSR, 2.5 mL l-Glutamine, 5 mL nonessential amino acids, 3.5 μL Sigma Cat M3148, 5 mL penicillin/streptomycin, and 500 μL bFGF and then add Knockout DMEM to a total volume of 500 mL. The medium should then be filtered with a 0.22-μm bottle top filter. Medium can be kept for 1 week at 4°C.
4. Collagenase type IV (Invitrogen, Cat. No. 17104–019). Dissolve to 1 mg/mL in DMEM/F12 (Invitrogen, Cat. No. 11330) and filter sterilize using a 0.2-μm filter.
5. iPS cell freezing medium. To prepare freezing medium, the following reagents are required: 10% DMSO (Sigma-Aldrich, Cat. No. D2650), 30% KOSR (Invitrogen, Cat. No. 10828–028), and 60% hES medium. The medium should be made fresh directly before use.
6. 12-Well tissue culture-treated plastic dish.
7. 6-Well tissue culture-treated plastic dish.

3. Protocol

3.1. Establishing Primary Cell Cultures from Patient Biopsies (3 Weeks)

1. Obtain a 6-mm skin punch biopsy from the desired patient (acquired by a trained physician, with informed consent from the patient).
2. Place specimen in hFib media for transport on ice to the laboratory.
3. In a tissue culture hood, disinfect specimen by dipping into 70% ethanol for 5 s and wash in sterile PBS.
4. Incubate tissue in 3 mg/mL Dispase for at least 2 h at 37°C in a 1.5-mL tube.
5. Wash the biopsy by dipping in PBS, place it on petri dish, and peel away the epidermis from the dermis using sterile forceps.
6. Mince the dermis tissue into small pieces using sterile scalpel blades.
7. Score the bottom of each well of a 24-well dish with a cross in the center. Add a single piece of minced material to each scored well and press to the bottom with a 10-mm coverslip.
8. Gently (so as not to lift the coverslip) add 500 µL hiFib to each well and incubate at 37°C, 5% CO_2.
9. Culture the attached tissue for 1–2 weeks; change media once a week.
10. When cell outgrowth from the tissue becomes ~80%, confluent cells are ready to be split.
11. Aspirate the media and gently remove the coverslip (which should remove the explant, but leave the outgrowth of dermal fibroblasts). Gently wash cells in PBS; aspirate and cover cells in 0.05% trypsin/EDTA. Incubate for 5 min at 37°C.
12. Add 500 µL of hFib media to inactivate trypsin and pipette to remove cells from the bottom of wells. Pool cells from all wells that provide fibroblast outgrowths into a 15-mL polypropylene conical tube.
13. If tissue aggregates remain, the cell sample can be passed through a 70-µm strainer.
14. Centrifuge the cell suspension $250 \times g$ for 5 min.
15. Resuspend the cells in appropriate volume of hFib media (2 mL per well of 6-well dish) and pipette into wells of a 6-well plate (cells from one well of the 24-well dish can be split onto one well of a 6-well dish).
16. Cells can be split at a ratio of 1:3 every 6–7 days (it is crucial to maintain cells at high density, otherwise they stop proliferating) and expanded into larger culture vessels. Splitting is

performed by incubation with 0.05% trypsin for 5 min at 37°C, followed by trypsin inactivation by addition of hFib media. Cells are pelleted by centrifugation at 250×g for 5 min and resuspended in an appropriate volume of fresh hFib media to seed culture vessels.

17. Human primary fibroblasts can be stored in liquid nitrogen. This is performed by trypsinization of the cells (as described above), followed by centrifugation at 250×g for 5 min. The supernatant is discarded, and the pellet is resuspended in freezing medium. Cells are distributed into cryovials in 1 mL aliquots and frozen overnight at –80°C in a cryo-freezing container before transfer to liquid nitrogen for storage. Typically, one confluent 100-mm dish provides three aliquots, which can be resurrected into one 100-mm dish.

3.2. Preparing 293T Cells (2 days)

1. Defrost a cryovial of 293T cells by swirling in a water bath until the majority of the contents are thawed (see Note 2).
2. Spray and wipe vial with 70% ethanol before placing in the hood.
3. Prepare a 15-mL polypropylene conical tube by adding 4 mL of 293T media.
4. Add 1 mL of 293T media to the contents of the cryovial and mix before transferring to the prepared conical tube.
5. Centrifuge cells at 250×g for 3 min at room temperature and discard supernatant.
6. Resuspend cells in 5 mL of 293T media and transfer to a 15-mL polypropylene conical tube.
7. Count cells using a hemocytometer and adjust the concentration of cells to 2×10^5 cells per mL.
8. Transfer 10 mL of solution to a 100-mm dish (three to five dishes should be prepared per transfection with viral constructs).
9. Incubate cells at 37°C, 5% CO_2 overnight (see Note 3).

3.3. Production of Reprogramming Retroviruses (3–4 days)

1. Prepare DNA transfection mixture in a 15-mL polypropylene conical tube. For each 100-mm plate to be transfected, add 300 μL of DMEM (without FBS or antibiotics), 5 μg of retroviral vector, 4.5 μg pUMVC, and 0.5 μg pVSVG into a 15-mL conical tube and mix (see Note 4).
2. To this mixture, add 60 μL of SuperFect transfection reagent per plate to be transfected. Mix and incubate at room temperature for 10 min.
3. Add transfection mixture to each plate in a dropwise manner and swirl dish to ensure even distribution.

4. Incubate cells at 37°C, 5% CO_2 for 48–72 h (see Note 5).

5. Collect supernatant and filter through a 0.45-μm filter.

6. Transfer to a centrifuge tube and concentrate virus by ultracentrifugation at $100,000 \times g$ at 8°C for 90 min.

7. Remove supernatant and cover viral pellet with 500 μL of DMEM and store overnight at 4°C.

8. To ensure the pellet is completely dissolved, pipette the DMEM up and down using a 100-μL tip.

9. Virus can be stored in 100 μL aliquots at –80°C for several months.

3.4. Preparing Mouse Embyronic Fibroblasts (MEFs) (5–7 days)

The preparation of MEFs is provided as follows. However, frozen vials of MEFs are commercially available.

1. Perform cervical dislocation on a 13.5-day pregnant mouse (strain MF-1 or CD-1).

2. In a laminar flow hood, cover the mouse in 70% alcohol, then open the abdomen and remove the uterine horns, which should be placed in a petri dish containing PBS (without Ca^{2+} and Mg^{2+}).

3. Remove the embryos from the embryonic sac. Discard the placenta and then decapitate and eviscerate the embryo.

4. Wash the carcass three times in PBS and transfer to a clean petri dish and mince with a scalpel blade.

5. Add 2 mL of 0.25% trypsin/EDTA solution to the tissue and incubate at 37°C for 15 min.

6. Stop digestion with 5 mL of DMEM/10% FBS and transfer to a universal tube before allowing the large aggregates to settle by gravity.

7. Transfer the supernatant to a 75-cm^2 flask and add a further 15 mL of DMEM/10% FBS and incubate at 37°C, 5% CO_2 overnight.

8. Remove cellular debris by aspirating the media. Replace with fresh MEF media and culture cells until 90% confluent. Cells can be expanded over several passages; however, they should not be used as feeders beyond passage 6 (see Note 6).

9. Inactivate MEFs (typically P6 or earlier) by replacing media with 2 mg/mL Mitomycin-C-DMEM/10% FBS to cover cells and incubate for 2–3 h at 37°C, 10% CO_2 (see Note 1).

10. Remove Mitomycin-C and wash the cells three times with PBS.

11. Add 2 mL 0.25% trypsin/EDTA and incubate at 37°C until cells start to dissociate. Knock cells off by tapping the flask and inactivate trypsin by adding 10 mL DMEM/FBS. Count the

cells using a hemocytometer and centrifuge at 250×*g* for 3 min. Either freeze cells in pre-counted aliquots (as described previously) or resuspend them in appropriate volume to seed gelatin-coated 100-mm dishes (previously described) at a density of 1×10^6 cells/dish. Incubate at 37°C, 10% CO_2 overnight (see Note 7).

3.5. Reprogramming Somatic Cells (7–8 days)

1. When hFib cells have reached 80–90% confluence, aspirate medium and wash in PBS. Cover cells with 0.05% trypsin for 3–5 min at 37°C.
2. Inactivate trypsin with hFib medium and transfer cells to a 15-mL polypropylene conical tube.
3. Centrifuge cells at 250 ×*g* at room temperature for 5 min and remove supernatant.
4. Resuspend cells in 1 mL of hFib medium and count cells using a hemocytometer. Dilute cells to 5×10^4 cells per mL.
5. Seed 1×10^5 patient fibroblasts (2 mL of suspension) into each well of a 6-well plate and incubate overnight at 37°C, 5% CO_2.
6. Remove medium from plate; replace it with 2 mL of retroviral infection medium supplemented with 4 µg/mL polybrene.
7. Infect fibroblasts with 50 µL of each of the concentrated viruses (OCT4, SOX2, KLF4, and c-MYC). Infect one well with empty vectors as a control (see Note 8).
8. Over a period of 2–3 days, perform a total of four infections by changing media and adding concentrated virus to each well (see Note 9).
9. On the fifth day after the first infection, passage infected cells using 0.25% trypsin and plate each well of cells onto a 100-mm dish of MEFs (1×10^6 cells/dish) using hFib media. Incubate at 37°C, 5% CO_2 overnight.
10. Change media to human ES media supplemented with valproic acid (0.5 mM). Replace ES media daily for 30 days, adding valproic acid for the first 10 days. ESC-like colonies tend to appear from around 15 days post infection (see Note 10).

3.6. Picking and Expanding of iPS Colonies (3–4 weeks)

1. One day before picking colonies, prepare 12-well plates of MEFs. Gelatinize plates (as described earlier). Either prepare fresh mitotically inactivated MEFs (as described) and proceed to step 5 or defrost pre-mitotically inactivated MEFs from liquid nitrogen by swirling in a 37°C water bath until the majority of the contents are thawed.
2. Spray and wipe vial with 70% ethanol before placing in the hood.

3. Add 1 mL of warm MEF medium and mix with contents of cryovial before transferring to a 15-mL polypropylene conical tube containing 4 mL of pre-warmed (37°C) MEF media.

4. Centrifuge cells at 250 ×g at room temperature for 3 min and aspirate supernatant.

5. If cells have not been pre-counted, resuspend cells in 1 mL of MEF medium and count cells using a hemocytometer. Dilute cells to a concentration of ~84,000 cells per mL.

6. Add 1 mL of cell suspension to each well of a 12-well plate (~1 × 10^6 cells per plate) and incubate overnight at 37°C, 5% CO_2.

7. Remove MEF media in 12-well plate and replace with 1.5 mL hES media per well.

8. To the 100-mm dish containing reprogrammed iPS colonies, remove media and add 1.5 mL of 1 mg/mL collagenase. Incubate at 37°C for about 5 min (or until colony edges appear to be dissociating from the feeders). Aspirate collagenase and replace with 10 mL of hES media. Use a needle or fine tip plastic Pastette (Alpha Laboratories, Cat. No. LW4636) to pick iPS colonies. Lift a colony from the edge using the needle or Pastette and transfer the entire colony into a well of the prepared 12-well plates containing MEFs. Place one iPS colony in each well (see Note 11).

9. Incubate cells overnight at 37°C, 5% CO_2. Do not disturb cells for 24 h.

10. Feed cells daily using 1.5 mL of hES media.

11. After 5–7 days, the colonies should be large enough to be passaged mechanically onto fresh 12-well plates of MEFs (see Note 12). After another 5–7 days, all colonies present in each individual well of the 12-well plates should be passaged into a well of a 6-well plate. Passage should be performed for hES cells. Cells are treated with collagenase as per step 8; however, after removal of collagenase and replacement with media, colonies can be cut into smaller pieces by use of the plastic Pastette or pipette tip.

12. Once the iPS cell line is established in 6-well plates, it is, at this stage, possible to transfer to a feeder-free culture system (see Note 13).

3.7. Characterization of iPS Cell Lines

Once iPS cell lines are generated, they should be characterized to ensure their similarity to hESCs. These new lines should be morphologically identical to ES cell lines, share similar gene expression profiles, and functionally be able to undergo lineage differentiation in a similar manner to ES cells. The following list exemplifies typical assays with which to characterize the reprogrammed cell lines:

1. Pluripotency should be confirmed by immunostaining for markers, such as OCT4, NANOG, SSEA-3, SSEA-4, TRA-1-60, and TRA-1-81. Alkaline phosphatase staining should also be performed.

2. Gene expression levels can be compared to hESCs using quantitative PCR for a panel of pluripotency genes, including endogenous *OCT4*, *NANOG*, and *SOX2*.

3. DNA methylation can be confirmed by bisulphate sequencing. Reprogramming is accompanied by demethylation of the promoters of critical pluripotency genes. Typically, the promoter regions of *OCT4* and *NANOG* are found to be hypomethylated in both iPS and ES cells. This is in contrast to their hypermethylated state in primary fibroblasts.

4. Teratoma formation by injection of stem cells into immunodeficient mice can be used to demonstrate pluripotency. Fully reprogrammed iPSCs are expected to form tumors that contain differentiated elements of each of the three primary embryonic germ layers.

5. Other potential assays to characterize iPS cell lines include karyotyping, ChIP studies to examine ES cell-like histone modifications, and X chromosome reactivation.

Figure 2 highlights some of the typical characteristics of iPSCs. It is beyond the scope of this protocol to provide detailed accounts of each of the methods.

In brief, all staining was performed upon cells fixed in PBS containing 4% paraformaldehyde (Sigma-Aldrich, Cat. No. 158127). Primary antibodies for SSEA3, SSEA4, TRA-1-60, and TRA-1-81 were a kind gift from PW Andrews (CSCB University of Sheffield, UK) but can be purchased from AbCam (ab-16286, ab-16287, ab-16288, and ab-16289). Signal was detected using goat anti-mouse IgG+IgM (H+L) FITC-conjugate secondary antibody (Invitrogen, Cat. No. M30801). Alkaline phosphatase was detected with Alkaline Phosphatase Substrate Kit 1 (Vector Laboratories, Cat. No. SK-5100). For the detection of nuclear proteins, cells were first permeabilized in PBS containing 0.1% Triton X-100 (BDH, Cat. No. 306324N). Primary antibodies NANOG (Abcam, Cat. No. ab-21624) and Oct-3/4 (Santa Cruz Biotechnology, Inc., Cat. No. SC-5272) were used in combination with secondary antibodies Alexa Flour 488 chicken anti-rabbit IgG (H+L) and chicken anti-mouse IgG (H+L) (Invitrogen, Cat. No. A21441 and A21200), respectively. Quantitative PCR was performed using the following primers: endogenous *OCT4* sense gtgcctgcccttctaggaat, antisense ggcacaaactccaggttttct; endogenous *SOX2* sense ttgctgcctctttaagactagga, antisense ctggggctcaaacttctctc; *NANOG* sense atgcctcacacggagactgt, antisense agggctgtcctgaataagca; and *GAPDH* (reference

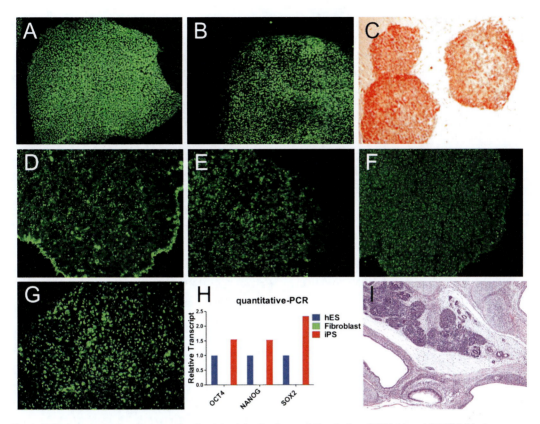

Fig. 2. Cell characterization. Images of cells stained for the transcription factors OCT4 (**a**) and NANOG (**b**), the enzyme alkaline phosphatase (**c**), and surface markers indicative of the pluripotent state SSEA3 (**d**), SSEA4 (**e**), TRA-1-60 (**f**), and TRA-1-61 (**g**). Quantitative PCR demonstrating strong endogenous expression of OCT4, NANOG, and SOX2; expression is similar to hESCs but absent in dermal fibroblasts (**h**). Formation of well-differentiated teratoma in a SCID mouse, exhibiting tissues derived from the three germ lineages (**i**).

gene) sense agccacatcgctcagacac, antisense gcccaatacgaccaaatcc. These primer pairs were used in combination with probes #52, #35, #69, and #60, respectively, from the Universal ProbeLibrary (Roche) (but should also be applicable to use with SYBR green).

4. Notes

1. Mitomycin-C is carcinogenic and cytotoxic. As such, it should be handled carefully by wearing both gloves and safety glasses. Mitomycin-containing media and subsequent PBS washes should be inactivated in bleach before disposal as hazardous waste.

2. It is essential that the 293T cells are below passage 20 to enable good transfection efficiency and production of virus. The use of higher passage 293T cells will greatly reduce the titer of the virus.

3. The confluency of the 293T cells at transfection can play a critical role. In our hands, the optimal density was found to be 50–60% at transfection. This allowed the cells space to grow during the viral production stage and led to less cells becoming unattached due to overconfluence. Cells should not reach 100% confluency during this period. Optimal density was achieved with 8–12 h growth following plating. Cells can be left overnight or alternatively plated in the morning and transfected in the evening. This timing may need to be altered between different laboratories and will be affected by the passage number of the 293T cells used.

4. The production, concentration and use of retroviruses must be performed in BL2 biosafety containment facilities. Retroviral work may also require approval from institutional and/or governing bodies.

5. If viral yield is low, it may enhance production to incubate the 293T cells at 32°C since virus is more stable at this temperature.

6. MEFs can be expanded and frozen (using same protocol as human primary fibroblasts) in batches prior to mitotic inactivation (at passage 6). Typically, we freeze stocks at passages P0 and P3. P3 stocks are typically expanded to P6 before mitotic inactivation, after which cells can also be frozen in pre-counted aliquots and plated directly after thawing for convenience.

7. When plating MEFs, the plates should not be swirled. This causes cells to concentrate in the center of the wells. Tipping plates from side to side leads to better MEF distribution within wells. Plated MEFs can be stored at 37°C, 10% CO_2 for up to a week before use.

8. This protocol does not establish the titer of the virus produced. If, after viral infection, a large amount of cell death is witnessed or cells appear unable to divide, it may be due to the viral concentration being too high. In this case, it is advisable to reduce the amount of each retrovirus used for infection. If the problem persists, it is advisable to titer the virus to establish a suitable concentration for your specific cell type. It is also possible to follow this protocol with different combinations of retrovirus, to use novel genes or to eliminate the use of c-MYC and/or KLF-4. Removal of these factors will however greatly reduce the number of iPS colonies derived and may not be successful in generating iPS colonies with all cell types. It is therefore advised to infect a control well with all four factors when determining the efficiency of iPS generation in your cell type.

9. It is possible to vary the interval between infections. The cells are able to begin infection 24 h after plating and should be infected with retrovirus on four separate occasions on days 2–4.

On the day after plating (day 2), this laboratory performs two infections (morning and evening) and a further two infections on day 3. It has been found to be favorable to remove the media containing the retrovirus on day 4 and change to hFib media for the final day of growth before passage. This method promotes expansion of the fibroblasts and leads to lower levels of cell death in infected cultures. Cells infected with low-titer viruses may benefit from the media containing the last infection to remain on the cells until passage. With cells exhibiting slow growth, it may be preferential to change media on day 4 and leave cells to grow for a further 2–3 days before passage to ensure cells are of a good confluency to be passaged (70–90%).

10. iPSCs should initially appear as aggregates that form colonies with defined borders and exhibit the same morphology as hESCs with a high nuclear-to-cytoplasmic ratio. When these colonies are 2–4 mm in diameter, they are ready to be picked. It is important to note that the infected cells will need to be split to avoid over-confluence. Slower growing cell lines do not need to be split until the appearance of iPS colonies. No colonies should appear in the wells transfected with empty vectors. If no colonies appear in wells infected with the four retroviruses, the most common problem is that the titer of the retroviral supernatant is not sufficient. It is critical that the viral supernatant is aliquoted since freeze–thaw cycles are detrimental to the viral titer. If no colonies appear, it may be necessary to prepare fresh plasmids using an endotoxin-free kit for purification of plasmid DNA. If colonies are witnessed in wells infected with four factors but not in wells infected with fewer or different factors, it is possible that the cell type is not easily reprogrammed; this is commonly due to the age of the patient, the skin biopsy that was taken, or the passage number of the fibroblasts being too high.

11. iPS colonies do not appear at the same time period, and therefore after picking one batch of colonies, maintain the incubation of the plate and take further colonies as they appear.

12. If no colonies are witnessed at this stage, it is possible that the colonies picked were not iPSCs. To avoid this confusion, it is essential to identify colonies with a high nuclear-to-cytoplasmic ratio and form colonies with defined borders. If colonies appear but differentiate rapidly, this may be due to the use of poor quality MEFs.

13. After iPS colonies have undergone the initial expansion step, it is possible to passage the colonies onto a feeder-free system. In this laboratory, this has been achieved using the combination of Matrigel (BD Biosciences, Cat. No. 354234)-coated plates (with Matrigel diluted 1:60 in DMEM/F12) and mTeSR™1

media (STEMCELL Technologies, Cat. No. 05850). It was however found to be critical to the efficiency of derivation of iPS colonies, for the derivation to occur on MEFs rather than on Matrigel and to the passage onto the feeder-free system after the picking of iPS colonies.

References

1. Thomson JA, Itskovitz-Eldor J, Shapiro SS, et al. Embryonic stem cell lines derived from human blastocysts. Science 1998;282:1145–7.
2. Cowan CA, Atienza J, Melton DA, Eggan K. Nuclear reprogramming of somatic cells after fusion with human embryonic stem cells. Science 2005;309:1369–73.
3. Takahashi K, Yamanaka S. Induction of pluripotent stem cells from mouse embryonic and adult fibroblast cultures by defined factors. Cell 2006;126:663–76.
4. Takahashi K, Tanabe K, Ohnuki M, et al. Induction of pluripotent stem cells from adult human fibroblasts by defined factors. Cell 2007;131:861–72.
5. Yu J, Vodyanik MA, Smuga-Otto K, et al. Induced pluripotent stem cell lines derived from human somatic cells. Science 2007;318:1917–20.
6. Park IH, Zhao R, West JA, et al. Reprogramming of human somatic cells to pluripotency with defined factors. Nature 2008;451:141–6.
7. Nakagawa M, Koyanagi M, Tanabe K, et al. Generation of induced pluripotent stem cells without Myc from mouse and human fibroblasts. Nat Biotechnol 2008;26:101–6.
8. Wernig M, Meissner A, Cassady JP, Jaenisch R. c-Myc is dispensable for direct reprogramming of mouse fibroblasts. Cell Stem Cell 2008;2:10–2.
9. Hanna J, Markoulaki S, Schorderet P, et al. Direct reprogramming of terminally differentiated mature B lymphocytes to pluripotency. Cell 2008;133:250–64.
10. Maherali N, Hochedlinger K. Guidelines and techniques for the generation of induced pluripotent stem cells. Cell Stem Cell 2008;3:595–605.
11. Soldner F, Hockemeyer D, Beard C, et al. Parkinson's disease patient-derived induced pluripotent stem cells free of viral reprogramming factors. Cell 2009;136:964–77.
12. Woltjen K, Michael IP, Mohseni P, et al. piggyBac transposition reprograms fibroblasts to induced pluripotent stem cells. Nature 2009;458:766–70.
13. Yusa K, Rad R, Takeda J, Bradley A. Generation of transgene-free induced pluripotent mouse stem cells by the piggyBac transposon. Nat Methods 2009;6:363–9.
14. Feng B, Ng JH, Heng JC, Ng HH. Molecules that promote or enhance reprogramming of somatic cells to induced pluripotent stem cells. Cell Stem Cell 2009;4:301–12.
15. Huangfu D, Maehr R, Guo W, et al. Induction of pluripotent stem cells by defined factors is greatly improved by small-molecule compounds. Nat Biotechnol 2008;26:795–7.
16. Huangfu D, Osafune K, Maehr R, et al. Induction of pluripotent stem cells from primary human fibroblasts with only Oct4 and Sox2. Nat Biotechnol 2008;26:1269–75.
17. Shi Y, Desponts C, Do JT, Hahm HS, Scholer HR, Ding S. Induction of pluripotent stem cells from mouse embryonic fibroblasts by Oct4 and Klf4 with small-molecule compounds. Cell Stem Cell 2008;3:568–74.
18. Shi Y, Do JT, Desponts C, Hahm HS, Scholer HR, Ding S. A combined chemical and genetic approach for the generation of induced pluripotent stem cells. Cell Stem Cell 2008;2:525–8.
19. Zhou H, Wu S, Joo JY, et al. Generation of induced pluripotent stem cells using recombinant proteins. Cell Stem Cell 2009;4:381–4.
20. Esteban MA, Wang T, Qin B, et al. Vitamin C enhances the generation of mouse and human induced pluripotent stem cells. Cell Stem Cell 2010;6:71–9.
21. Yoshida Y, Takahashi K, Okita K, Ichisaka T, Yamanaka S. Hypoxia enhances the generation of induced pluripotent stem cells. Cell Stem Cell 2009;5:237–41.
22. Hong H, Takahashi K, Ichisaka T, et al. Suppression of induced pluripotent stem cell generation by the p53-p21 pathway. Nature 2009;460:1132–5.
23. Kawamura T, Suzuki J, Wang YV, et al. Linking the p53 tumour suppressor pathway to somatic cell reprogramming. Nature 2009;460:1140–4.

24. Li H, Collado M, Villasante A, et al. The Ink4/Arf locus is a barrier for iPS cell reprogramming. Nature 2009;460:1136–9.
25. Marion RM, Strati K, Li H, et al. A p53-mediated DNA damage response limits reprogramming to ensure iPS cell genomic integrity. Nature 2009;460:1149–53.
26. Utikal J, Polo JM, Stadtfeld M, et al. Immortalization eliminates a roadblock during cellular reprogramming into iPS cells. Nature 2009;460:1145–8.
27. Judson RL, Babiarz JE, Venere M, Blelloch R. Embryonic stem cell-specific microRNAs promote induced pluripotency. Nat Biotechnol 2009;27:459–61.
28. Park IH, Arora N, Huo H, et al. Disease-specific induced pluripotent stem cells. Cell 2008;134:877–86.
29. Raya A, Rodriguez-Piza I, Guenechea G, et al. Disease-corrected haematopoietic progenitors from Fanconi anaemia induced pluripotent stem cells. Nature 2009;460:53–9.
30. Stewart SA, Dykxhoorn DM, Palliser D, et al. Lentivirus-delivered stable gene silencing by RNAi in primary cells. RNA 2003;9:493–501.

Chapter 18

Generation of Induced Pluripotent Stem Cells from Human Amnion Cells

Masashi Toyoda, Shogo Nagata, Hatsune Makino, Hidenori Akutsu, Takashi Tada, and Akihiro Umezawa

Abstract

Induced pluripotent stem (iPS) cells have been generated through nuclear reprogramming of somatic cells via retrovirus- or lentivirus-mediated transduction of exogenous reprogramming factors *OCT3/4*, *SOX2*, *KLF4*, and *c-MYC*. The extraembryonic amnion is considered to be a promising candidate cell source for cellular therapeutics and the generation of iPS cells because it contains a large number of cells that do not require any genetic or epigenetic modifications. In this chapter, we describe how to generate human amniotic membrane (hAM) primary cells derived from placenta and how to establish the iPS cells from these hAM cells using four reprogramming factors. The hAM-derived iPS cells could be useful for personal regenerative medicine for future infants and useful as genetic disease models and for disease-specific drug discovery.

Key words: Human placenta, Amniotic membrane cells, Congenital disease, Induced pluripotent stem cells, Cell bank, Disease model

1. Introduction

The human placenta is a large discoid organ with a diameter of around 20 cm and a weight of approximately 500 g. It contains a large number of cells possessing a wide range of phenotypes and potentials (1, 2). Because the human placenta is usually considered as medical waste, it can be collected at birth with no risk to the individual and is an easily accessible cellular source without ethical problems. The placenta consists of two components: (a) a fetal portion, derived from the amniotic membrane, the chorionic plate, the smooth chorion (chorion laeve), and the villous chorion (chorion frondosum); and (b) a maternal portion, derived from the decidua basalis. Each part of the placenta has recently been a candidate for cell source for cell-based therapies because of the variety of cell types that become available (3–5).

It is evident that generation of human iPS cells from adult somatic cells is much harder than that from fetal cells. In fact, analysis with a secondary doxycycline-inducible transgene system shows that the efficiency varies between different somatic cell types (6). In addition, iPS cells are generated through epigenetic reprogramming of somatic cells. Recent reports show that early passage iPS cells retain a transient epigenetic memory of the tissue of origin. This may influence efforts to use iPS cells for applications in disease modeling and in cell therapies aimed to enhance differentiation in desired cell lineages (7, 8). Information on the base sequence of genome is unchanged through the reprogramming and passaging. In general, mutations of genome accumulate through aging and cell culture with cellular divisions and DNA misrepair. Young somatic cells are therefore more suitable for iPS cell generation than aged somatic cells. This suggests that the fetal portion of placenta-derived cells, which have accumulated less genetic mutations, is safer than that of the adult somatic cells as a cell source for iPS cell generation. Thus, identification of reprogramming-sensitive cell types is a key issue.

Because human amniotic membrane (hAM) is one of the fetal portions of the placenta, these hAM-derived cells do not express the major histocompatibility complex (MHC) class I molecules and may be expected to demonstrate immunologic tolerance. In fact, hAM-derived cells were reported to have potential for transdifferentiation into cells of various organs; specifically, cardiomyogenic and myogenic differentiation potential have been shown in vitro and in vivo (9–11).

We have demonstrated that newborn mouse and human extraembryonic amniotic membrane and yolk-sac cells can be efficiently reprogrammed to pluripotency (12). Our findings illustrate that hAM cells are a strong candidate cell source for collection and banking that could be retrieved on demand and used for generating personalized genetic modification-free iPS cells applicable for clinical treatment and drug screening. Furthermore, iPS cells derived from human disease-model mouse- or patient-specific AM have enhanced the promise for finding the causes of and potential cures for many genetic diseases.

2. Materials

2.1. Cells and Vectors

1. 293FT cells (Invitrogen)
2. pMXs retrovirus vectors containing cDNA of *OCT3/4*, *SOX2*, *KLF4*, or *c-MYC*

3. VSV-G protein expression vector (for example, pCMV-VSV-G; http://www.brc.riken.go.jp/lab/cfm/Subteam_for_Manipulation_of_Cell_Fate/Lentiviral_Vectors.htmL)

4. Gag–pol protein expression vector (for example, pCL-Gag/Pol provided by T. Kiyono, National Cancer Center Research Institute, Tokyo, Japan)

2.2. Reagents

1. Dulbecco's modified phosphate-buffered saline without calcium and magnesium, DPBS (14190-144, Invitrogen)
2. Dulbecco's modified Eagle's medium (DMEM), high glucose (D6429, Sigma-Aldrich Co.,)
3. Antibiotic–Antimycotic (100×), liquid (10,000 U/mL penicillin, 10,000 μg/mL streptomycin, and 25 μg/mL of amphotericin B) (15240-062, Invitrogen)
4. Fetal bovine serum (FBS)
5. Penicillin/streptomycin (15140-050, Invitrogen)
6. 0.25% Trypsin/1 mM EDTA (23315, IBL)
7. Knockout DMEM (10829-018, Invitrogen)
8. Knockout serum replacement (KSR; 10828028, Invitrogen)
9. Collagenase II (LS004176, Worthington Co.)
10. Dispase II (4942078, Roche Applied Science) for generation of AM-derived cells
11. Dispase II (GD 81070, EIDIA Co., Ltd., Japan) for passage of iPS cells
12. Collagenase IV (17104-019, Invitrogen)
13. $CaCl_2$ (C7902, Sigma-Aldrich Co.,)
14. Y-27632 (253-00513, Wako Pure Chemical Industries, Ltd., Japan)
15. Gelatin, type A (G1890, Sigma-Aldrich Co.,)
16. 293IT transfection reagent (MIR 2700, Mirus Bio LLC)
17. Hexadimethrine bromide (polybrene, 52495, Sigma-Aldrich Co.,)
18. TC-protector (freezing medium, TCP-001, DS Pharma Biomedical Co., Ltd., Japan)
19. Freezing medium (for iPS cell, RCHEFM001, ReproCELL, Inc.)
20. iPSeed (MEF medium, 007101, Cardio Inc., Japan)
21. iPSellon (human iPS medium, 007001, Cardio Inc.)
22. Recombinant basic fibroblast growth factor, human (bFGF; 064-04541, Wako Pure Chemical Industries, Ltd., Japan)
23. HEPES buffer solution (1 M) (15630-106, Invitrogen)

2.3. Reagent Setup

2.3.1. Culture Medium for Primary AM-Derived Cells

To prepare 500 mL of the culture medium for AM cell derivation, mix 5 mL of Antibiotic–Antimycotic and 50 mL of FBS, and then fill to 500 mL with DMEM. For the preparation of 500 mL of culture medium for AM-derived cell maintenance, mix 5 mL of penicillin–streptomycin and 50 mL of FBS, and then fill to 500 mL with DMEM.

2.3.2. Dispase II Solution for Generation of AM-Derived Cells

Add Dispase II to DMEM at a concentration of 2.4 U/mL. Aliquot and store at −20°C.

2.3.3. Collagenase II Solution

Add collagenase II to DMEM at a concentration of 270 U/mL. Aliquot and store at −20°C.

2.3.4. Gelatin-Coated Dishes

- 0.1% Gelatin solution: Dissolve 0.5 g gelatin powder in 500 mL of distilled water, autoclave, and store at room temperature.
- To coat a culture dish, add enough volume of 0.1% gelatin solution to cover the entire bottom of the dish. Incubate the dish for at least 30 min at 37°C. Before using, aspirate excess gelatin solution.

2.3.5. iPS Cell Medium

- Dissolve bFGF with iPSellon or DMEM at a concentration of 10 μg/mL. Store at −20°C.
- Add bFGF stock solution to iPSellon at a final concentration of 10 ng/mL. iPS medium should be kept at 4°C for no more than 1 week.

2.3.6. Collagenase IV Solution

- 0.1 M $CaCl_2$: Dissolve 0.11 g of $CaCl_2$ in 10 mL of distilled water, and pass through a 0.45-μm pore filter. Store at 4°C.
- Dissolve 100 mg collagenase IV in 59 mL DPBS, and add 1 mL of 0.1 M $CaCl_2$ and 20 mL KSR. Aliquot 8 mL per 15-mL tube, and store at −20°C. Before using, add 2 mL of 0.25% Trypsin/1 mM EDTA solution, and store at 4°C. Avoid freezing and thawing.

2.3.7. Dispase II Solution for Passage of iPS Cells

Add 1 g of Dispase II in 100 mL of DPBS, and filter the solution with a 0.45-μm pore filter. Aliquot 1 mL per tube, and store at −20°C. Before using, dilute the dispase solution (1 mL) in a tube containing 9 mL of Knockout DMEM.

2.4. Equipment

1. 150-mm Cell culture dishes (353025, Falcon)
2. 100-mm Cell culture dishes (353003, Falcon)
3. 60-mm Cell culture dishes (3010–060, IWAKI)
4. 35-mm Cell culture dishes (3000–035, IWAKI)
5. 6-Well cell culture plates (353046, Falcon)
6. 15-mL Tubes (352196, Falcon)
7. 50-mL Tubes (352070, Falcon)
8. 5-mL Plastic disposable pipettes (357543, Falcon)
9. 10-mL Plastic disposable pipettes (357551, Falcon)

10. 25-mL Plastic disposable pipettes (357525, Falcon)
11. 0.45-μm Pore size filters (Millex HV, SLHV033RS, Millipore)
12. 40-μm Cell strainers (352340, Falcon)
13. 100-μm Cell strainers (352360, Falcon)
14. 2-mL Cryogenic vials (430488, Corning)
15. Polystyrene tubes (352235, Falcon)
16. Scissors
17. Tweezers
18. Forceps
19. Needles
20. Disposable syringes
21. Cell counter (Vi-CELL, Beckman Coulter)
22. CO_2 incubator
23. X-ray source (MBR-1520A-3, Hitachi Medical, Japan)
24. Liquid nitrogen storage tank
25. Pipette aid
26. Pasteur pipettes
27. Pipetman pipettors (2, 10, 20, 200, and 1,000 μL)
28. Dissection microscope
29. Inverted microscope

3. Methods

3.1. Isolation and Culture of AM-Derived Primary Cells

3.1.1. Isolation of AM-Derived Primary Cells

1. Peel the amniotic membrane from placenta (Fig. 1).
2. Wash the amniotic membrane with DPBS containing Antibiotic–Antimycotic (1×) (Fig. 2a).
3. Cut amniotic membrane into small pieces (approximately 5 mm^3 in size) with dissection scissors (Fig. 2b–d).
4. Collect a piece of amniotic membrane in a 50-mL tube and wash it with DPBS containing Antibiotic–Antimycotic (1×). Centrifuge amniotic membrane at 300×g for 5 min and remove and discard the supernatant (see Note 1).
5. Add 20 mL of Dispase II solution to the amniotic membrane and incubate at 37°C for 60 min by gently shaking (160 min^{-1}).
6. Centrifuge amniotic membrane pieces at 300×g for 5 min; remove and discard the supernatant.
7. Incubate the amniotic membrane with 20 mL of Collagen type II solution at 37°C for 60 min by gently shaking (160 min^{-1}).

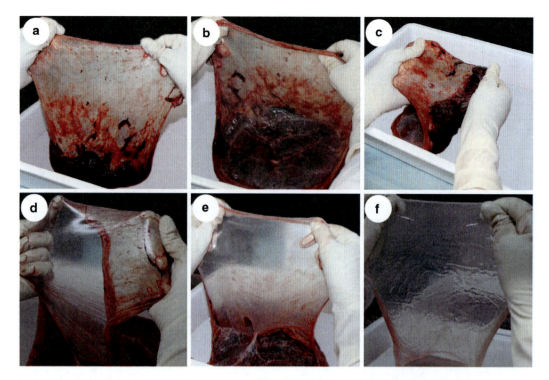

Fig. 1. Separation of amniotic membrane from human placenta. The fetal membrane viewed from the maternal side (**a**) and fetal side (**b**) is shown. To obtain amniotic membrane, peel the fetal membrane from the placenta (**c**) and then peel the amniotic membrane from fetal chorion and maternal decidua (**d**, **e**, **f**).

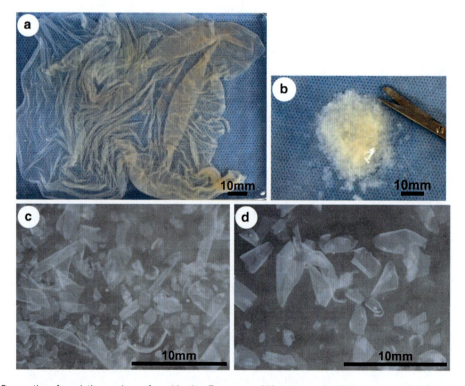

Fig. 2. Preparation of amniotic membrane for cultivation. To generate AM-derived cells, AM is washed in DPBS supplemented with penicillin–streptomycin and amphotericin B (**a**). The AM is cut into pieces approximately 5 mm^3 in size (**b**, **c**, **d**).

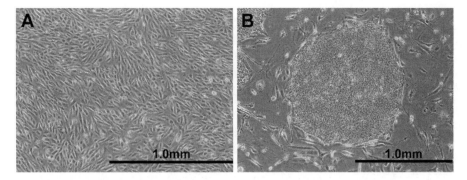

Fig. 3. Photomicrograph of AM-derived cells (**a**) and AM-derived iPS cells on MEF feeder layers (**b**).

8. Transfer the cell suspension to a new 50-mL tube after passing through a 100-μm and then a 40-μm cell strainer.
9. Centrifuge the cell suspension at 300×g for 5 min; remove and discard the supernatant.
10. Resuspend the cell pellet in culture medium, and mix and transfer to new culture dishes.
11. Incubate at 37°C, 5% CO_2.

3.1.2. Culture of AM-Derived Primary Cells

1. Grow the adherent AM-derived primary cells to approximately 80% confluence in 6-cm dishes (Fig. 3a).
2. Aspirate the culture medium and wash once with DPBS containing Antibiotic–Antimycotic (1×).
3. Add 1 mL of 0.25% Trypsin/1 mM EDTA, and incubate the dish at room temperature for 5 min or until cells are detached.
4. Add 4 mL of the culture medium and dissociate the cells by pipetting up and down.
5. Collect the cell suspension into a 50-mL tube and determine the total cell number.
6. Centrifuge the cells at 200×g at room temperature for 5 min.
7. Aspirate and discard the medium, and resuspend the cell pellet in culture medium. Mix and transfer to new culture dishes.

3.1.3. Freezing AM-Derived Primary Cells

1. When AM-derived primary cells reach 80–90% confluence in 10-cm dishes, frozen stocks should be prepared.
2. Aspirate and discard the medium, and wash the cells with 4 mL of DPBS.
3. Completely remove DPBS, add 2 mL of 0.25% Trypsin/1 mM EDTA solution and incubate at 37°C for 2–5 min.
4. Add 8 mL of culture medium and dissociate the cells by pipetting up and down.

5. Collect the cell suspension into a 50-mL tube and count the cells.
6. Centrifuge the cells at 200×*g* at room temperature for 5 min.
7. Aspirate and discard the medium, resuspend the cell pellet in 1 mL of freezing medium (TC-protector) per 1×10^6 cells, and transfer 1 mL aliquots into cryogenic vials.
8. Store at −80°C overnight, then place in liquid nitrogen for long-term storage.

3.2. Mouse Embryonic Fibroblast Culture Methods

3.2.1. Derivation of MEFs from Pregnant Mice

1. Mate ICR mice overnight at designated day 0.5 of pregnancy (E0.5) following plug discovery. Sacrifice one pregnant ICR mouse at E13.5 by cervical vertebral dislocation.
2. Wash abdomen with 70% ethanol and dissect the abdominal cavity to expose the uterine horns.
3. Remove the uterine horns into a 10-cm dish and wash three times with DPBS.
4. Using two pairs of tweezers, open each uterine wall and release all embryos carefully.
5. Wash retrieved embryos with DPBS and place each embryo into a 3.5-cm dish.
6. Use the same tweezers to dissect each embryo from the placenta and membranes, and remove and discard as much soft tissue (head, liver, heart, intestines, and all viscera) as possible.
7. Transfer clean embryos to new dishes and mince thoroughly using shape scissors.
8. Add 6 mL of Mouse Embryonic Fibroblast (MEF) medium and transfer the two to three minced embryo equivalents into each 15-cm culture dish. Add MEF medium to a final volume of 30 mL.
9. Grow the MEFs up to 2 days or until 80–90% confluence.
10. Aspirate the medium and wash the cells once with DPBS.
11. Add 3 mL of 0.25% Trypsin/1 mM EDTA and 2 mL of DPBS solution into the dish and incubate for 3 min at room temperature.
12. Add 10 mL of MEF medium into the dish and collect the cell suspension into a 50-mL tube through a 100-μm filter. Use 10 mL of DPBS to wash the plate, and add that material to the tube.
13. Centrifuge the cells for 5 min at 200×*g* at room temperature.
14. Aspirate and discard the medium; resuspe nd the cell pellet in MEF medium.
15. Divide the cell suspension evenly between three 15-cm culture dishes. Add MEF medium to a final volume of 30 mL.

16. Grow the MEFs up to 3 days or until 80–90% confluence.
17. Aspirate and discard the medium, and wash the cells once with DPBS.
18. Add 3 mL of 0.25% Trypsin/1 mM EDTA and 2 mL of DPBS solution into the dish and incubate for 3 min at room temperature.
19. Add 10 mL of MEF medium into the dish and collect the cell suspension into a 50-mL tube. Use 10 mL of DPBS to wash the plate and add the wash solution to the tube. Count the number of viable cells.
20. Centrifuge the cells at $200 \times g$ for 5 min at room temperature.
21. Aspirate and discard the medium; resuspend the cells in TC-protector (freezing medium) at a concentration of 4×10^6 cell/mL.
22. Dispense 1 mL of cell mixture into cryogenic vials and freeze at $-80°C$ overnight.
23. After overnight incubation, rapidly remove the vials and place in liquid nitrogen for long-term storage.

3.2.2. Thawing MEFs and Inactivation

1. Remove a vial from liquid nitrogen and thaw briefly in a 37°C water bath.
2. Pipette up and down the cell solution gently in the vial, and then transfer the cells into a 15-mL tube.
3. Add 9 mL of MEF medium dropwise.
4. Centrifuge at $200 \times g$ for 5 min. Discard the supernatant.
5. Resuspend the cell pellet in 10 mL of MEF medium.
6. Plate the cell suspension on a 15-cm culture dish. Add MEF medium to a final volume of 30 mL.
7. Incubate the dish overnight at 37°C, 5% CO_2.
8. Aspirate and discard the medium, add 20 mL of fresh MEF medium, and incubate overnight at 37°C, 5% CO_2.
9. Grow MEFs to approximately 80% confluence in 15-cm culture dishes.
10. Aspirate and discard the medium and wash the cells once with DPBS.
11. Add 3 mL of 0.25% Trypsin/1 mM EDTA and 2 mL of DPBS solution to the dish and incubate for 3 min at room temperature.
12. Add 10 mL of MEF medium into the dish and collect the cell suspension into a 50-mL tube. Use 10 mL of DPBS to wash the dish and add the wash solution to the tube.

13. Centrifuge the cells at 200×g for 5 min at room temperature.
14. Aspirate the medium and resuspend the cell pellet in MEF medium.
15. Divide the cell suspension evenly between three 15-cm culture dishes. Add MEF medium to a final volume of 30 mL.
16. Incubate the dish overnight at 37°C, 5% CO_2.
17. Aspirate and discard the medium, add 30 mL of fresh MEF medium, and incubate overnight at 37°C, 5% CO_2.
18. Grow MEFs to approximately 80% confluence in 15-cm culture dishes.
19. Aspirate and discard the medium and wash the cells once with DPBS.
20. Add 3 mL of 0.25% Trypsin/1 mM EDTA and 2 mL of DPBS solution to each dish and incubate for 3 min at room temperature.
21. Add 10 mL of MEF medium into each dish and collect the cell suspension into a 50-mL tube. Use 10 mL of DPBS to wash each and add the wash solution to the tube.
22. Centrifuge the cells at 300 ×g at room temperature.
23. Aspirate the medium and resuspend the cell pellet in 8 mL of MEF medium.
24. Transfer the cell suspension evenly into two polystyrene tubes and add 40 µL of 1 M HEPES buffer solution per polystyrene tube. Cover the polystyrene tube with aluminum foil.
25. Irradiate the cells with 30 Gy at a dose-rate of 1 Gy/min for 30 min with an X-ray source.
26. Transfer the cell suspension into a 50-mL tube, use 5 mL of MEF medium to wash the polystyrene tubes, and add the wash solution to the tube.
27. Add 38 mL of MEF medium to a 50-mL tube containing the cell suspension, mix gently, and count the cells.
28. Centrifuge the cells at 200×g for 5 min at room temperature.
29. Aspirate and discard the medium and resuspend the cell pellet in TC-protector (freezing medium) at a concentration of 4×10^6/mL.
30. Dispense 1 mL of cell mixture per cryogenic vial and freeze at −80°C overnight.
31. After overnight incubation, rapidly remove the vials and place in liquid nitrogen for long-term storage.

3.3. Generation of AM-Derived iPS Cells

3.3.1. Virus Production

1. Grow 293FT cells to approximately 80% confluence in 10-cm dishes.
2. Aspirate and discard the medium and wash once with DPBS.

3. Add 2 mL of 0.25% Trypsin/1 mM EDTA, and incubate the dish at room temperature for 5 min.

4. Add 9 mL of the culture medium and dissociate the cells by pipetting up and down several times.

5. Collect the cell suspension into a 50-mL tube and count the cells.

6. Centrifuge the cells at $200 \times g$ for 5 min at room temperature.

7. Aspirate and discard the medium and resuspend the cell pellet in culture medium.

8. Seed the cells at 2×10^6 cells per 10-cm dish and incubate overnight at 37°C, 5% CO_2.

9. Transfer 500 µL of DMEM into a 1.5-mL tube.

10. Deliver 30 µL of TransIT-293 transfection reagent into the prepared tube from step 9, mix gently, and incubate for 5 min at room temperature.

11. Mix 5 µg of pMXs plasmid (respectively encoding *OCT3/4, SOX2, KLF4, or c-MYC*), 3.3 µg of pCL-Gag/Pol, and 1.7 µg of pHCMV-VSV-G vectors in a fresh 1.5-mL tube.

12. Add the DNA mixture (step 11) into the transfection reagent/DMEM (step 10), mix gently, and incubate at room temperature for 5 min.

13. Add the DNA–reagent complex into the 293FT cell dish. Incubate overnight at 37°C, 5% CO_2.

14. After 24 hours of transfection, aspirate the transfection reagent-containing medium, add 10 mL of fresh medium to the plate, and incubate at 37°C, 5% CO_2.

15. After 48–72 h of step 14, collect the supernatant from the 293FT cell culture with a 10-mL disposable syringe, filter it with a 0.45-µm pore size filter, and transfer it to a 50-mL tube.

16. Centrifuge the collected supernatant at $8,000 \times g$ at 4°C overnight.

17. Aspirate and discard the supernatant, add 1 mL of DMEM to the 50-mL tube, and resuspend gently the virus pellet overnight at 4°C.

18. Use the virus solution for infection experiment at next step (3.3.2). Otherwise remove the virus solution into cryogenic vials and freeze at −80°C until use.

3.3.2. Virus Infection

1. Grow AM-derived cells to approximately 80% confluence in 10-cm dishes.

2. Aspirate and discard the medium and wash once with DPBS.

3. Add 2 mL of 0.25% Trypsin/1 mM EDTA, and incubate the dish at room temperature for 5 min.

4. Add 9 mL of the culture medium and dissociate the cell by pipetting up and down.

5. Collect the cell suspension into a 50-mL tube and count the cells.

6. Centrifuge the cells at $200 \times g$ at room temperature.

7. Aspirate and discard the medium and resuspend the cell pellet in fresh culture medium.

8. Seed the cells at 1×10^5 cells per a well of 6-well plate and incubate overnight at 37°C, 5% CO_2.

9. Aspirate and discard the medium and wash once with PBS.

10. Make a mixture of equal parts (1 mL) of the retrovirus supernatant containing OCT3/4, SOX2, KLF4, and c-MYC (final volume, 4 mL) (see Note 2).

11. Add the 2 mL of virus supernatant and 4 µL of 2 mg/mL polybrene per well. The final concentration of polybrene is 4 µg/mL.

12. Incubate the virus and polybrene for 8 h at 37°C, 5% CO_2.

13. After 8 h of infection, aspirate and discard the medium and add 2 mL of fresh medium to each well.

14. After 48 h of step 12, seed the irradiated MEFs in 10-cm dishes (see Note 3).

15. After 72 h of step 12, aspirate and discard the medium, and wash once with PBS.

16. Add 0.5 mL of 0.25% Trypsin/1 mM EDTA, and incubate the well at room temperature for 5 min.

17. Add 5 mL of the MEF medium and dissociate the cells by pipetting up and down several times.

18. Collect the cell suspension into a 50-mL tube and count the cells.

19. Centrifuge the cells at $200 \times g$ at room temperature for 5 min.

20. Aspirate and discard the medium and resuspend the cell pellet in 10 mL culture medium.

21. Transfer 4–6 mL of cell suspension per 10-cm dish with the irradiated MEFs, and add fresh MEF medium to a final volume of 8 mL per dish. Incubate the dishes overnight at 37°C, 5% CO_2.

22. Aspirate and discard the medium, and replace with 10 mL of iPS medium. Change the medium every other day until the colonies are large enough to be picked.

3.3.3. Picking the AM-iPS Colonies

1. Seed the irradiated MEFs in 3.5-cm gelatin-coated dishes and incubate overnight at 37°C, 5% CO_2 (see Note 3).

2. Aspirate and discard the medium, and replace with 2 mL of iPS cell medium for at least 1 h.

3. Carefully cut hES-like colonies from the dish under the stereomicroscope using a syringe with a 27-G needle. Avoid any but areas of differentiated cells and leave them attached to the dish.

4. After the colony has been cut, transfer by a P20 pipetman set at 20 µL to a preequilibrated dish in step 2.

5. Place the dish in 37°C, 5% CO_2. Once on the shelf, move the dish up and down and sideways twice to evenly distribute colony pieces.

6. After 2 days of transfer, remove and replace 50% of the medium daily for about a week.

3.3.4. Passage of AM-iPS Cells (see Note 4)

1. Seed the irradiated MEFs in gelatin coated-dishes and incubate overnight at 37°C, 5% CO_2 (see Note 2).

2. Aspirate and discard the medium, and replace with iPS cell medium for at least 1 h.

3. Aspirate and discard the medium from a 60–80% confluent 6-cm dish of AM-derived iPS cells, and wash the cells with DPBS.

4. Completely remove DPBS, add 0.5 mL of collagenase IV solution or Dispase II solution, and incubate at 37°C for 2–5 min.

5. Aspirate and discard the collagenase or dispase solution, and add 2 mL of iPS cell medium.

6. Using a P20 or P200 pipetman with a 20–200-µL tip, gently pipette the cell colonies up and down several times until most of the colonies become detached from the MEFs, leaving the iPS cells in suspension. Aim to have up to ten colonies per 6-cm dish (see Note 5).

7. Remove the supernatant containing the iPS cells to a 1.5-mL tube, and gently pipette up and down several times until cell colonies are too small to be apparent.

8. Transfer the cell supernatant to a preequilibrated dish.

9. Place the dish in 37°C, 5% CO_2. Move dishes crosswise once they are placed on the shelf to ensure even distribution.

10. Check daily for colony formation. Colonies may take as long as 2–3 days to appear (Fig. .4).

11. After 2 days of transfer, remove and replace 50% of the media daily until they are ready to passage again, in 7–10 days.

3.3.5. Freezing AM-iPS Cells

1. When iPS cells reach 80–90% confluence in the 6-cm dishes, frozen stocks should be prepared.

2. Before making frozen stocks, add 4 µL of 10 mM Y-27632 to each dish (final concentration, 10 µM) and incubate at 37°C, 5% CO_2 for 1 h.

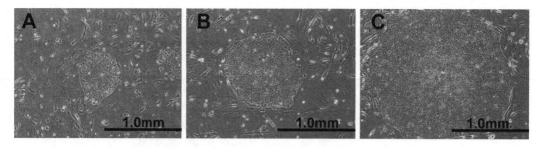

Fig. 4. AM-derived iPS cell colony at 1 (**a**), 5 (**b**), and 7 (**c**) days of culture after passage.

3. Aspirate and discard the medium, and wash the cells with 2 mL of DPBS.
4. Completely remove DPBS, add 0.5 mL of collagenase IV or Dispase II solution, and incubate at 37°C for 2–5 min.
5. Aspirate and discard the collagenase or dispase solution, and add 2 mL of iPS cell medium.
6. Using a P20 or P200 pipetman with a 20–200-µL tip, gently pipette the cell colonies up and down several times until most of the colonies become detached from the MEFs, leaving the iPS cells in suspension, and transfer to a 1.5-mL tube.
7. Remove the supernatant containing the iPS cells, and place in a 15-mL tube.
8. Centrifuge the cells at $170 \times g$ for 3 min at room temperature.
9. Remove the supernatant.
10. Suspend the pellet in 0.2 mL of freezing medium per tube by gently pipetting up and down several times.
11. Transfer all of the cell suspension to a cryogenic vial.
12. Put the vial quickly into liquid nitrogen.
13. Store in liquid nitrogen tank.

3.3.6. Thawing AM-iPS Cells

1. Prepare 10 mL of pre-warmed (37°C) iPS medium in a 15-mL tube per frozen stock vial of iPS cells to be thawed.
2. Take frozen stocks of iPS cells out from liquid nitrogen tank.
3. Add 0.8 mL of iPS medium prepared in step 1 into each vial of frozen stocks, and thaw quickly by pipetting up and down gently 2–3 times.
4. Transfer the cell suspension to the tube in step 1.
5. Centrifuge the cells at $170 \times g$ for 5 min at room temperature.
6. Remove the supernatant and add 4 mL of iPS medium to each 15-mL tube.

7. Transfer the cell suspension to a 6-cm dish per 15-mL tube seeded with the irradiated MEFs. Add 4 μL of 10 mM Y-27632 to each dish (final concentration, 10 μM) and incubate overnight at 37°C, 5% CO_2.

8. Remove and replace 50% of the media daily until they are ready to passage again, in 7–10 days.

4. Notes

1. To isolate AM-derived cells, we also use the explant culture method in which the cells are outgrown from pieces of amniotic membrane attached to dishes. Briefly, the amniotic membrane from the fetal part is cut into pieces approximately 5 mm^3 in size. The pieces are washed in culture medium until the supernatant is free of erythrocytes. Some pieces are attached to the substratum in a 10-cm dish. Culture medium is added. The cells migrate out from the cut ends after approximately 20 days of incubation at 37°C in 5% CO_2. The migrated cells are harvested with DPBS with 0.1% trypsin and 0.25 mM EDTA for 5 min at 37°C and counted. The harvested cells are reseeded at a density of 3×10^5 cells in a 10-cm dish. Confluent monolayers of cells are subcultured at a 1:8 split ratio onto new 10-cm dishes. The culture medium is replaced with fresh culture medium every 3 or 4 days.

2. If iPS cells are generated by three factors, make a mixture of equal parts (1 mL) of the retrovirus supernatant containing OCT3/4, SOX2, and KLF4, and 1 mL of DMEM (final volume, 4 mL).

3. Thaw the irradiated MEF stock cells (4×10^6 cells/vial) rapidly in a 37°C water bath until only a small piece of ice remains. Clean the vial with 70% ethanol and transfer the thawed cells to a 50-mL tube. Add 9 mL of MEF medium to the cell suspension. Centrifuge for 5 min at $200 \times g$ at room temperature. Resuspend the cell pellet in MEF medium. Plate the irradiated MEF cells in a gelatin-coated culture dish at the density desired: in four 10 cm dishes/vial, twelve 6 cm dishes/vial, or thirty 3.5 cm dishes/vial.

4. If almost all of the iPS cell colonies in a dish are normal in morphology, it is possible to passage them manually with a StemPro EZ passage tool (23181-010, Invitrogen).

5. Mark the dish where iPS cell colonies have good morphology (Fig. 3b). Avoid any undesirable areas of differentiated cells and leave them attached to the dish.

References

1. Cunningham, F. G., Williams, J. W. (2005) Implantation, embryogenesis, and placenta development. Williams obstetrics. 22nd ed. New York: McGraw-Hill Professional. pp. 39–90.
2. Sadler, T. W. (2006) Third month to birth : the fetus and placenta. Langman's Medical Embryology. 10th ed. Philadelphia: Lippincott Williams & Wilkins. pp. 89–109.
3. Portmann-Lanz, C. B., Schoeberlein, A., Huber, A., Sager, R., Malek, A., et al. (2006) Placental mesenchymal stem cells as potential autologous graft for pre- and perinatal neuroregeneration. American journal of obstetrics and gynecology 194: 664–673.
4. Fukuchi, Y., Nakajima, H., Sugiyama, D., Hirose, I., Kitamura, T., et al. (2004) Human placenta-derived cells have mesenchymal stem/progenitor cell potential. Stem Cells 22: 649–658.
5. Kawamichi, Y., Cui, C. H., Toyoda, M., Makino, H., Horie, A., et al. (2010) Cells of extraembryonic mesodermal origin confer human dystrophin in the mdx model of Duchenne muscular dystrophy. J Cell Physiol 223: 695–702.
6. Wernig, K., Griesbacher, M., Andreae, F., Hajos, F., Wagner, J., et al. (2008) Depot formulation of vasoactive intestinal peptide by protamine-based biodegradable nanoparticles. J Control Release 130: 192–198.
7. Polo, J. M., Liu, S., Figueroa, M. E., Kulalert, W., Eminli, S., et al. (2010) Cell type of origin influences the molecular and functional properties of mouse induced pluripotent stem cells. Nat Biotechnol 28: 848–855.
8. Kim, K., Doi, A., Wen, B., Ng, K., Zhao, R., et al. (2010) Epigenetic memory in induced pluripotent stem cells. Nature 467: 285–290.
9. Zhao, P., Ise, H., Hongo, M., Ota, M., Konishi, I., et al. (2005) Human amniotic mesenchymal cells have some characteristics of cardiomyocytes. Transplantation 79: 528–535.
10. Miki, T., Lehmann, T., Cai, H., Stolz, D. B., Strom, S. C. (2005) Stem cell characteristics of amniotic epithelial cells. Stem Cells 23: 1549–1559.
11. Tsuji, H., Miyoshi, S., Ikegami, Y., Hida, N., Asada, H., et al. (2010) Xenografted human amniotic membrane-derived mesenchymal stem cells are immunologically tolerated and transdifferentiated into cardiomyocytes. Circ Res 106: 1613–1623.
12. Nagata, S., Toyoda, M., Yamaguchi, S., Hirano, K., Makino, H., et al. (2009) Efficient reprogramming of human and mouse primary extraembryonic cells to pluripotent stem cells. Genes Cells 14: 1395–1404.

Part IV

Lineage-Specific Differentiation of hES and iPS Cells

# Chapter 19

In Vitro Two-Dimensional Endothelial Differentiation of Human Embryonic Stem Cells

Xiaolong Lin, Hua Jiang, Zack Zhengyu Wang, and Tong Chen

Abstract

Pluripotent human embryonic stem cells (hESCs) comprise of cells from all three germ layers in vivo. They have been proved to differentiate in vitro into a variety of cell lineages. Endothelial cells derived from hESCs could potentially contribute to cellular treatment of vascular diseases and regenerative development of engineered vessels. As different germ cells are intimately linked during hESC differentiation, embryonic endothelial cells can be isolated based on specific cell surface markers and further be identified by distinct properties, including expression of endothelial markers, LDL uptake, Matrigel tubular formation, and in vivo microvessel generation. Because of inefficient yield from three-dimensional (3D) embryoid body (EB) differentiation, we sought to develop a simple two-dimensional (2D) culture system for differentiating hESCs into endothelial lineages that bypasses EB formation. In this chapter, we describe a procedure for the 2D differentiation of hES cells toward endothelial cells. It is an efficient and reproducible method with potential for large-scale cultures in future applications.

Key words: Human embryonic stem cells (hESCs), Vascular cells, Stem cell differentiation, Endothelial cells

1. Introduction

Some vascular diseases result from the loss of vascular-specific cells in tissues and organs, such as ischemic diseases that are primarily caused by endothelial dysfunction. Cellular therapies to treat these diseases are hampered by the lack of donor cells. Therefore, the successful derivation of hESCs (1) and efficient endothelial differentiation (2–6) could provide an unlimited source of endothelial cells for therapeutic application toward regeneration of ischemic tissues and preparation of tissue-engineered vascular grafts.

During vertebrate embryogenesis, the endothelium, which develops from the mesoderm, is the first tissue to differentiate and form primitive vascular networks by aggregation, a process termed vasculogenesis (7). Based on the close relationship between the vascular and hematopoietic systems during organogenesis, it is hypothesized that hematopoietic and endothelial cells are derived from a common precursor, the hemangioblast (8). The sequential angioblasts are committed to the endothelial lineage differentiation (9).

Endothelial differentiation of hESCs can be induced by two methods: 2D culturing of the cells on extracellular matrix (ECM) or a feeder layer, which can induce directed differentiation toward endothelial lineages (3), or growing hESCs in a 3D system in a differentiation medium to form embryoid bodies (EBs), which induce spontaneous differentiation into the various cell types of the three germ layers (6). Endothelial cells can then be isolated for further differentiation and maturation. The hESC-derived endothelial cells were first identified and isolated in Langer's laboratory (3). Studies from adult circulating cells indicated that vascular endothelial growth factor receptor-2 (VEGFR2), CD133, CD31, and CD34 are expressed in endothelial and hematopoietic progenitor cells in relating human tissues (10–13), whereas VEGFR2 and CD133 are also expressed in undifferentiated hESCs. With respect that CD31 and CD34 are not expressed (or minimally expressed) in undifferentiated hESCs (3, 13, 14), they are chosen to be candidate markers for isolating hESC-derived endothelial progenitor cells (EPCs). The proportion of CD34$^+$ cells reaches a peak at day 10 during 2D or 3D differentiation (5). In spite of similar percentage of CD34$^+$ expression between 2D and 3D procedures, the number of CD34$^+$ cells yielded was five to eightfold higher than that in 2D culture systems (5). Therefore, a 2D cultivation is a more efficient and reproducible method than the 3D system for endothelial differentiation.

Isolated CD34$^+$ cells can be cultured in endothelial growth media containing VEGF and basic fibroblast growth factor (bFGF) for additional 7–10 days. The hESC-derived endothelial cells are characterized by specific endothelial markers, including CD31, vascular/endothelial (VE)-Cadherin, von Willebrand factor (vWF), VEGFR2 (KDR), and Tie-2. They are capable of uptaking DiI-acetylated low-density lipoprotein (DiI-Ac-LDL) and form vascular network-like structures when placed on Matrigel. Endothelial cells isolated and differentiated under the described conditions may constitute a potential cell source for further application.

2. Materials

2.1. Two-Dimensional Differentiation of hESCs

1. hESC lines: H9 and H1 (WiCell).
2. Tissue culture dishes, 100×20 mm (BD Falcon).
3. L-Glutamine solution, 100×solution (Gibco). Make aliquots of 5 mL and store at −20°C.
4. 2-Mercaptoethanol, 1,000×solution (Gibco) (see Note 1).
5. Nonessential amino acid solution, 100×solution (NEAA, Gibco).
6. Knockout serum replacement (KSR, Gibco). Make aliquots of 50 mL and store at −20°C.
7. Dulbecco's modified eagle medium: Nutrient mixture F-12 1:1 (DMEM/F12, Gibco).
8. Dulbecco's PBS, Ca^{2+} and Mg^{2+} free (Gibco).
9. Recombinant human FGF basic (bFGF, R&D Systems) is dissolved in DPBS at a concentration of 25 μg/mL. Make 50 μL aliquots and store at −80°C (see Note 2).
10. Penicillin/streptomycin, 100×solution (P/S, Gibco). Make 2.5 mL aliquots and store at −20°C.
11. Human ES cell growth medium was prepared with DMEM/F12 supplemented with 20% KSR, 1% NEAA, 2 mM L-glutamine, 0.55 mM 2-mercaptoethanol, 100 units/mL penicillin, 100 mg/mL streptomycin, 4 ng/mL bFGF. Sterilize by 0.22-μm filtration. Media can be stored at 4°C for up to 7–10 days.
12. Fetal bovine serum (FBS, Hyclone). Make aliquots of 25 mL and store at −20°C.
13. Human ES cell differentiation medium was prepared with IMDM, supplemented with 15% FBS, 1% NEAA, 2 mM L-glutamine, 0.55 mM 2-mercaptoethanol, 100 units/mL penicillin, 100 mg/mL streptomycin. Sterilize by 0.22-μm filtration.

2.2. Isolation of CD34+ Cells by Magnetic-Activated Cell Sorting

1. 0.25% Trypsin/EDTA (Gibco).
2. Magnetic-Activated Cell Sorting (MACS) columns, separators, pre-separation filters (Miltenyi Biotec).
3. CD34 MultiSort Kit (Miltenyi Biotec).
4. Antibody murine antihuman CD34-PE (Miltenyi Biotec).
5. Bovine serum albumin (BSA, Sigma-Aldrich). PBS/0.5% BSA (without EDTA) was used as the buffer for cell isolation.
6. Ethylenediaminetetraacetic acid (EDTA, Gibco).

2.3. Endothelial Induction	1. Milli-Q purified water (Millipore).
	2. Gelatin from porcine skin (Sigma-Aldrich). Dissolve 0.5 g of gelatin in 500 mL of warm (50–60°C) Milli-Q water. Cool at room temperature and sterilize by 0.22-μm filtration.
	3. Recombinant human $VEGF_{165}$ (R&D Systems) is dissolved in DPBS at a concentration of 50 ng/μL. Make 50 μL aliquots and store stocks at –80°C.
	4. Endothelial induction medium was prepared with IMDM supplemented with 15% FBS, 1% NEAA, 2 mM L-glutamine, 0.55 mM 2-mercaptoethanol, 100 units/mL penicillin, 100 mg/mL streptomycin, 50 ng/mL $hVEGF_{165}$, 5 ng/mL bFGF. Sterilize by 0.22-μm filtration.
2.4. Immunofluorescence Staining	1. Fixative solution: paraformaldehyde (PFA, Sigma-Aldrich), 4% (w/v) in PBS. Dissolve 4 g PFA in 100 mL of PBS, heat the mixture in a 60°C water bath, and stir until the solution becomes clear. Cool to room temperature before use. Store at 4°C and use within 1 week (see Note 3).
	2. Permeabilization solution: Triton X, 0.01% (v/v) (Sigma-Aldrich) in PBS.
	3. Nonspecific binding block: goat serum, 4% (v/v) (Sigma-Aldrich) in PBS.
	4. Primary antibodies: • Mouse antihuman CD31 (BD PharMingen) • Mouse antihuman CD34 (BD PharMingen) • Mouse antihuman VE-cad (BD PharMingen) • Mouse antihuman KDR (R&D Systems)
	5. Secondary antibody: goat anti-mouse IgG FITC (BD PharMingen).
2.5. LDL Uptake Assay	DiI-acetylated low-density lipoprotein (DiI-LDL, Molecular Probes).
2.6. Matrigel Assay	1. Matrigel matrix (BD Biosciences): concentration is ~10 mg/mL in 10 mL from commercial source. Prepare 200 μL aliquots and store at –20°C (see Note 4).
	2. 24-well plate (BD Falcon).
2.7. Reverse Transcriptase-Polymerase Chain Reaction (RT-PCR)	1. Total RNA extraction: TRIzol (Invitrogen). Additional reagents required: absolute ethanol (Sigma-Aldrich), trichloromethane (Sigma-Aldrich), isopropanol (Sigma-Aldrich).
	2. cDNA synthesis: PrimeScript cDNA synthesis kit (Takara).
	3. PCR reactions: Master Mix System (Takara).

Table 1
Primer sequence and conditions for endothelial-related markers used in reverse transcriptase-polymerase chain reaction (RT-PCR)

Genes	Primer sequence(5'–3')	Products (bp)	Condition
CD31	(F)CAACGAGAAAATGTCAGA (R)GGAGCCTTCCGTTCTAGAGT	260	50°C, 35 cycles
CD34	(F)GCCATTCAGCAAGACAACAC (R)AAGGGTTGGGCGTAAGAGAT	152	52°C, 35 cycles
VE-cad	(F)CAGCCCAAAGTGTGTGAGAA (R)TGTGATGTTGGCCGTGTTAT	162	50°C, 35 cycles
vWF	(F)TCGGGCTTCACTTACGTTCT (R)CCTTCACTCGGACACACTCA	178	52°C, 35 cycles
Tie-2	(F)ATCCCATTTGCAAAGCTTCTGGCTGGC (R)TGTGAAGCGTCTCACAGGTCCAGGATG	512	60°C, 35 cycles
KDR	(F)ATGCACGGCATCTGGGAATC (R)GCTACTGTCCTGCAAGTTGCTGTC	537	53°C, 35 cycles

Abbreviations: *F* forward primer; *R* reverse primer; *bp* base pair

4. Sterile RNase-free reagents, polypropylene tubes, tips, and other materials.
5. UV-2000 spectrophotometer (UNICO).
6. PCR thermocycler Applied Biosystems 9700 (Applied Biosystems).
7. Gel electrophoresis and standard apparatus (Bio-Rad).
8. Gel Imaging System (Bio-Rad).
9. Primers for genes CD31, CD34, VE-Cad, vWF, Tie-2, and KDR. For details, see Table 1.

3. Methods

The hESC lines, H1 and H9, can be acquired from the WiCell Research Institute. hESCs (passages 29–60) can be grown on either irradiation (60 Gy) or mitomycin C (1 mg/mL) inactivated mouse embryonic fibroblasts (MEFs) in hESC medium (1). To maintain hESCs in a long-term undifferentiated status, hESCs should be cultured at ~500 hESC colonies per 100-mm dish at 37°C and 5% CO_2. The plating density of MEFs has a significant effect on the differentiation of hESCs into $CD34^+$ cells. The optimal MEF plating density is $~1 \times 10^4$ cells/cm² on a gelatin-coated dish (5–6×10^5 cells/100-mm). MEFs can be used up to passage 3 for hESC cultures (5).

272 X. Lin et al.

The methods described here include two-dimensional differentiation of hESCs, isolation of CD34+ cells by MACS, endothelial induction, and analysis of endothelial specific makers. Detailed procedures are described as follows.

3.1. Two-Dimensional Differentiation of hESCs

After 6–7 days of hESC culture (see Note 5), exchange the growth medium to a differentiation medium, and allow hESCs to have spontaneous differentiation directly on MEF feeders for additional 10–12 days (Fig. 1a, b).

3.2. Isolation of CD34+ Cells by MACS

1. Aspirate medium and wash with DPBS. Add 2 mL 0.25% Trypsin/EDTA to each 100-mm dish.
2. After being incubated for 2 min at 37°C, trypsin is inactivated by addition of 10% FCS, and the cells are dissociated by gentle pipetting. Pass through 40-μm cell strainers to get single cells. To minimize the loss of cells, the strainers should be rinsed extensively with PBS/1%FBS or an additional strainer be used.
3. Count cell number.
4. Follow the manufacturer's instruction for magnetic labeling and isolation.
5. Collect the CD34+ cells in a 15-mL conical tube.

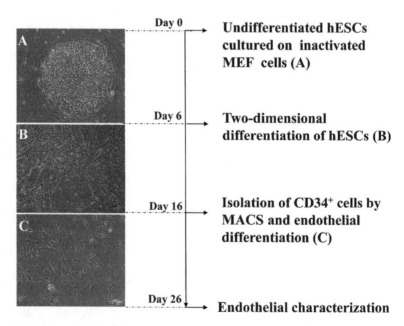

Fig. 1. Culture scheme of hESC endothelial differentiation. Undifferentiated hESCs (H1) were cultured on MEF feeder at day 0 (**a**). After 6 days of cultivation, media were changed to hESC differentiation medium for additional 10 days (**b**). At day 16, CD34+ cells were selected from differentiated hESCs and plated onto gelatin-coating plates in endothelial induction medium for 10 days, and the cells can be applied to further characterization (**c**).

3.3. Induction of Endothelial Cells

1. Coat 24-well plate by adding 500 µL 0.1% gelatin to each well (see Note 6).
2. Centrifuge the collected cells, resuspend in endothelial inducing medium, and transfer to gelatin-coated wells at a density of $1.5–2 \times 10^4$ cells/cm^2.
3. Change medium every 2 days. Add fresh VEGF and FGF every day. Culture the cells for additional 7–10 days (Fig. 1c).

3.4. Immunofluorescence Staining

1. Cells are cultured in 24-well plates. Remove the medium completely from the wells. Rinse once with 1 mL of PBS.
2. Fix the cells with 4% PFA for at least 20 min at room temperature.
3. Wash the cells three times with PBS.
4. Permeabilize the cells with 0.01% Triton X-100/PBS for 15 min at room temperature.
5. Block with 4% goat serum/PBS for 45 min.
6. Incubate with primary antibody, which is diluted in 4% goat serum/PBS for 4 h at room temperature.
7. Wash the cells three times with PBS.
8. Incubate with diluted secondary antibody in PBS for 1 h at room temperature in the dark.
9. Wash the cells three times with PBS.
10. Antibody isotypes or no antibody sera served as the negative control.
11. Cells are examined under a fluorescence microscope.

3.5. DiI-LDL Uptake Assay

1. Cells are incubated with 10 µg/mL of DiI-LDL for at least 4 h.
2. Wash the cells twice with PBS.
3. Cells are examined under a fluorescence microscope.

3.6. Matrigel Tubular Formation Assay

1. Coat 24-well plates with 200 µL/well Matrigel at room temperature for more than 30 min.
2. Aspirate medium from the endothelial cells in 24-well plates. Add 200 µL/well 0.25% Trypsin/EDTA to make single cell suspensions.
3. Plate single cell suspensions (5×10^4 cells/well) onto Matrigel-coated plates in differentiation media. Incubate for 16 h.
4. Photograph the structures under phase contrast microscope.

3.7. RT-PCR

3.7.1. Total RNA Extraction

1. Remove the medium from the cultures and wash once with PBS. Extract total RNA using the 1-mL TRIzol/10 cm² of TRIzol following the manufacturer's instructions.
2. Homogenize the cell suspension and spin down to obtain the homogenate. The homogenate can be frozen and stored at −80°C for subsequent RNA extraction.
3. Continue the RNA extraction protocol according to the manufacturer's instructions and dissolve the sample RNA with RNase-free water.
4. Determine the RNA quantity and quality by measuring the absorbance at 260 and 280 nm with the UNICO UV-2000 spectrophotometer. Calculate the RNA concentration. Store the RNA samples at −80°C if cDNA synthesis is not carried out at once.

3.7.2. RT Reaction

cDNA synthesis is performed using 1 μg of total RNA per 20 μL reaction volume. RT reaction is carried out at 42°C for 30 min, followed by enzyme inactivation at 95°C for 5 min.

3.7.3. PCR

1. PCR primers are listed in Table 1.
2. Perform PCR amplifications using 5 μL of the resulting cDNA samples per 50 μL reaction volume. First denature at 95°C for 5 min, followed by 35 cycles of denaturing at 95°C for 30 s, annealing at 50–60°C for 45 s, and extending at 72°C for 60 s, and followed by a final extension at 72°C for 5 min.
3. Analyze the PCR-amplified products on 2% (w/v) agarose gel electrophoresis. The results are developed by ultraviolet detection of the ethidium bromide-stained gel. Images are taken using the Bio-rad Light Imaging System.

4. Notes

1. 2-Mercaptoethanol is toxic. Avoid inhalation, ingestion, or contact with skin or mucous membranes. However, at this concentration, it is not necessary to use a fume hood while preparing cell culture medium.
2. Avoid repeated freeze–thaw cycles of the growth factors. Once thawed, keep the growth factor aliquot at 4°C and use within 1 week.
3. PFA is toxic. Work inside hood and use gloves when handling. Prepare fresh before use to avoid formic acid formation.
4. Cool cryovials and 200-μL tips to 4°C before use. One day before use, transfer from −20°C to 4°C for overnight thawing.

5. hESC cultures (H1 and H9; NIH stem cell registry: http://stemcells.nih.gov/research/registry); also see reference [1] and follow exactly as recommended by the WiCell protocol. Refer to website http://www.wicell.org/ for protocols describing the expansion and propagation of hESCs on murine embryonic feeder cells.

6. Coat the wells with 0.1% gelatin at 37°C for at least 30 min. We found that the attachment of CD34+ cells was greater on collagen I–coated wells (BD Labware) than on gelatin-coated wells. However, gelatin has the advantage of being considerably less expensive than collagen I–coated wells.

References

1. Thomson, J. A., J. Itskovitz-Eldor, S. S. Shapiro, M. A. Waknitz, J. J. Swiergiel, V. S. Marshall, and J. M. Jones. (1998). Embryonic stem cell lines derived from human blastocysts. *Science* 282:1145–1147.

2. Wang, L., L. Li, F. Shojaei, K. Levac, C. Cerdan, P. Menendez, T. Martin, A. Rouleau, and M. Bhatia. (2004). Endothelial and hematopoietic cell fate of human embryonic stem cells originates from primitive endothelium with hemangioblastic properties. *Immunity* 21:31–41.

3. Levenberg, S., J. S. Golub, M. Amit, J. Itskovitz-Eldor, and R. Langer. (2002). Endothelial cells derived from human embryonic stem cells. *Proc Natl Acad Sci USA* 99:4391–4396.

4. Gerecht-Nir, S., A. Ziskind, S. Cohen, and J. Itskovitz-Eldor.(2003). Human embryonic stem cells as an in vitro model for human vascular development and the induction of vascular differentiation. *Lab Invest* 83:1811–1820.

5. Wang, Z. Z., P. Au, T. Chen, Y. Shao, L. M. Daheron, H. Bai, M. Arzigian, D. Fukumura, R. K. Jain, and D. T. Scadden. (2007). Endothelial cells derived from human embryonic stem cells form durable blood vessels in vivo. *Nat Biotechnol* 25:317–318.

6. Ferreira, L. S., S. Gerecht, H. F. Shieh, N. Watson, M. A. Rupnick, S. M. Dallabrida, G. Vunjak-Novakovic, and R. Langer. (2007). Vascular progenitor cells isolated from human embryonic stem cells give rise to endothelial and smooth muscle like cells and form vascular networks in vivo. *Circ Res* 101:286–294.

7. Hirashima, M. (2009). Regulation of endothelial cell differentiation and arterial specification by VEGF and Notch signaling. *Anat Sci Int* 84:95–101.

8. Pardanaud, L., F. Yassine, and F. Dieterlen-Lievre. (1989). Relationship between vasculogenesis, angiogenesis and haemopoiesis during avian ontogeny. *Development* 105:473–485.

9. Pardanaud, L., D. Luton, M. Prigent, L. M. Bourcheix, M. Catala, and F. Dieterlen-Lievre. (1996). Two distinct endothelial lineages in ontogeny, one of them related to hemopoiesis. *Development* 122:1363–1371.

10. Peichev, M., A. J. Naiyer, D. Pereira, Z. Zhu, W. J. Lane, M. Williams, M. C. Oz, D. J. Hicklin, L. Witte, M. A. Moore, and S. Rafii. (2000). Expression of VEGFR-2 and AC133 by circulating human CD34(+) cells identifies a population of functional endothelial precursors. *Blood* 95:952–958.

11. Asahara, T., T. Murohara, A. Sullivan, M. Silver, R. van der Zee, T. Li, B. Witzenbichler, G. Schatteman, and J. M. Isner. (1997). Isolation of putative progenitor endothelial cells for angiogenesis. *Science* 275:964–967.

12. Gehling, U. M., S. Ergun, U. Schumacher, C. Wagener, K. Pantel, M. Otte, G. Schuch, P. Schafhausen, T. Mende, N. Kilic, K. Kluge, B. Schafer, D. K. Hossfeld, and W. Fiedler. (2000). In vitro differentiation of endothelial cells from AC133-positive progenitor cells. *Blood* 95:3106–3112.

13. Kennedy, M., S. L. D'Souza, M. Lynch-Kattman, S. Schwantz, and G. Keller. (2007). Development of the hemangioblast defines the onset of hematopoiesis in human ES cell differentiation cultures. *Blood* 109:2679–2687.

14. Kaufman, D. S., E. T. Hanson, R. L. Lewis, R. Auerbach, and J. A. Thomson. (2001). Hematopoietic colony-forming cells derived from human embryonic stem cells. *Proc Natl Acad Sci USA* 98:10716–10721.

Chapter 20

Feeder-Free Culture for High Efficiency Production of Subculturable Vascular Endothelial Cells from Human Embryonic Stem Cells

Kumiko Saeki

Abstract

Conventional methods for vascular endothelial differentiation of human embryonic stem (hES) cells had suffered from subculture incompetence of the final products due to dominant expansion of contaminating pericytic components. We have overcome this problem by adding a "hematopoietic cytokine cocktail" to the differentiation medium. No pericytes are produced by our method, and the vascular endothelial cells are expanded purely by repetitive subculture up to 10–20 passages, depending on the lines of hES cells. The hES-derived vascular endothelial cells undergo senescence thereafter as in the case of primary cultured human vascular endothelial cells. During the course of subculture, vascular endothelial functions are well preserved. Because our system is entirely feeder-free, including the step to maintain undifferentiated hES cells, contamination by xenogeneic cells is completely excluded. In this chapter, we describe our method in detail with troubleshooting guidelines.

Key words: Feeder-free, Human ES cells, Subculturable, Vascular endothelial cells

1. Introduction

Several methods had been reported regarding the vascular endothelial differentiation of hES cells; for example, coculture with murine OP9 feeder cells (1) and feeder-free systems with three-dimensional (3D) (2) or two-dimensional (2D) culture (3). Irrespective of the culture techniques, differentiation efficiencies were not high enough, and thus, cell-sorting processes are required for the purification of vascular endothelial cells. Moreover, there was an additional outstanding issue: The vascular endothelial cells generated from hES cells were incapable of in vitro expansion. This is probably due to the dominant expansion of pericyte-lineage cells, which were major components generated from hES cells by conventional methods (1, 4). Thus, an establishment of a new method for the production of pure vascular endothelial cells without cogeneration of pericytic components had been an essential and urgent task.

During our trial to develop a system to produce hematopoietic stem cells from cynomolgus monkey ES cells, we realized that our system provided an almost pure population of vascular endothelial cells rather than producing hematopoietic stem cells (5). Eventually, it provided a very small population of non-adherent cells that consisted of myeloid progenitor cells and macrophages; however, the majority of the products were adherent populations. Interestingly, the non-adherent hematopoietic cells seemed to be emerging from a part of adherent populations. Because hematopoietic stem cells are known to emerge from a type of vascular endothelial cell (6–9), and also because the adherent cells showed a uniform cobblestone-like appearance, a morphological characteristic of vascular endothelial cells, we hypothesized that the adherent populations might be vascular endothelial cells. This was confirmed by gene and protein expression analyses along with in vitro and in vivo functional assays (5). Interestingly, no pericytes were cogenerated, and the vascular endothelial cells were easily expanded by an ordinary subculture technique.

By adding slight modifications to the original protocol for the vascular endothelial differentiation of monkey ES cells, we successfully established a method for the vascular endothelial differentiation of hES cells (10). Our protocol is a two-tiered system: The former step is a 3D culture to form spheres from hES cells, and the latter step is an ordinary 2D culture on gelatin-coated dishes. In our original protocol, we applied a hanging drop culture technique to the 3D culture because it guaranteed the formation of even-sized spheres. By contrast, a simple floating culture using ultralow-attachment dishes resulted in a production of spheres of uneven sizes. Although the hanging drop culture is superior to the floating culture in the case of monkey ES cells, the former technique is not applicable to hES cells due to the apoptosis-prone nature of hES cells after disaggregation. In any events, our modified system has enabled the production of subculturable vascular endothelial cells without co-producing pericytic components from hES cells. Our system can be reproduced even under serum-free conditions, such as in serum replacer KSR™-supplemented situations, although the maximal passage number before undergoing senescence might be reduced by about half.

2. Materials

2.1. Culture Medium for Mouse Embryonic Fibroblasts

1. Low glucose Dulbecco's Modified Eagle's Medium (DMEM) with L-glutamine (Cat 041-29775, WAKO Pure Chemical Industries, Osaka, Japan); store at 4°C.
2. 100× Penicillin/streptomycin solution (Cat P0781, Sigma Chemical Co., St. Louis, MO, USA); store at −20°C.
3. Fetal Bovine Serum (FBS) (PAA Laboratories GmbH, Pasching, Austria); store at −20°C.

2.2. Preparation of Mouse Embryonic Fibroblasts

1. A 12.5-week pregnant female mouse (ICR strain, for example).
2. Absorbent cottons.
3. Sevofrane (WAKO Pure Chemical Industries); store at room temperature.
4. A stereo microscope (Model LG-PS2, Olympus Optical Co. Ltd., Tokyo, Japan).
5. Sterilized phosphate-buffered saline (PBS); store at 4°C.
6. Sterilized ordinary tweezers and scissors.
7. Sterilized ophthalmic tweezers and scissors.
8. Four 60-mm culture dishes.
9. Five 100-mm culture dishes.
10. Five 50-mL conical tube.
11. Five 2.5-mL syringe.
12. Five 18-G needle.
13. 0.25% Trypsin-EDTA (Sigma Chemical Co.); store at −20°C.
14. Bambanker™ (Lymphotec Inc., Tokyo, Japan); store at −20°C.
15. 1.8-mL Inner-capped cryotubes (Nunc A/S, Roskilde, Denmark).
16. An X-ray irradiation device (Model MBR-1520R-3, Hitachi Medical Corporation, Tokyo, Japan) (see Note 1).

2.3. hES Cells

khES-1 and khES-3 lines were established by Dr. Suemori at Kyoto University in Japan (11).

2.4. Culture Medium for hES Cells

1. Dulbecco's Modified Eagle's Medium: nutrient mixture F-12 Ham 1:1 (WAKO Pure Chemical Industries, Cat 048-29785); store at 4°C.
2. Knockout™ Serum Replacement (KSR™) (Invitrogen Corp., Carlsbad, CA, USA, Cat 10828-028); store at −20°C.
3. Nonessential Amino Acid (NEAA) Solution (100×) (Sigma Chemical Co., M7145); store at 4°C.

4. Gelatin type A (Sigma Chemical Co., G2625); store at room temperature.
5. 2-Mercaptoethanol (Sigma Chemical Co., M7522); store at room temperature.
6. Fibroblast growth factor (FGF)-2 (PeproTech Inc., Rocky Hill, NJ, USA); store at –20°C.

2.5. Dissociation Liquid for hES Cells

1. KSR™ (Invitrogen Corp.); store at –20°C.
2. 100 mM $CaCl_2$; store at 4°C.
3. Sterilized PBS; store at 4°C.
4. 2.5% trypsin (Invitrogen Corp., Cat 15090-046); store at –20°C.
5. Type IV collagenase (Invitrogen Corp., Cat 17104-019); store at –20°C.

2.6. Freezing Medium for hES Cells

Freezing Medium for Human ES/iPS cells (ReproCELL Inc., Tokyo, Japan, Cat RCHEFM001); store at –80°C.

2.7. Layers for the Feeder-Free Culture of hES Cells

Matrigel™ Basement Membrane Matrix, Phenol Red-free (Cat 356237, BD Biosciences, San Jose, CA, USA); store at 4°C.

2.8. Differentiation Medium

1. 60 mm HydroCell® (CellSeed Inc. Tokyo, Japan, Cat CS2014) or 60 mm 2-methacryloyloxyethyl phosphorylcholine (MPC)-coated dish (Nalge Nunc International, Tokyo, Japan, Cat 145389).
2. Iscove's modified Dulbecco's medium (IMDM) (Sigma Chemical Co.); store at 4°C.
3. FBS (PAA Laboratories GmbH, Linz, Austria); store at –20°C.
4. 2-Mercaptoethanol (Sigma Chemical Co., M7522); store at 4°C.
5. 200 mM L-glutamine (Sigma Chemical Co., G7513); store at –20°C.
6. 100× Penicillin/streptomycin solution (Sigma Chemical Co., P0781); store at –20°C.
7. Vascular Endothelial Growth Factor A (VEGFA) (PeproTech Inc.); store at –20°C.
8. Bone Morphogenetic Protein 4 (BMP4) (R&D Systems Inc., Minneapolis, MN, USA); store at –20°C.
9. Stem Cell Factor (SCF) (PeproTech Inc.); store at –20°C.
10. FMS-related tyrosine kinase-3 ligand (Flt3-L) (PeproTech Inc.); store at –20°C.
11. Interleukin 3 (IL3) (PeproTech Inc.); store at –20°C.
12. IL6 (PeproTech Inc.); store at –20°C.
13. Gelatin, type A (Sigma Chemical Co.,G2625); store at room temperature.

14. 2.5% Trypsin (Invitrogen Corp. Cat 15090–046); store at −20°C.

15. Ethylenediaminetetraacetic acid (EDTA) (WAKO Pure Chemical Industries); store at room temperature.

2.9. Dissociation Liquid for hES-Derived Vascular Endothelial Cells

Trypsin-EDTA solution (Invitrogen Corp. Cat 25300–062); store at −20°C.

2.10. Freezing Solution for hES-Derived Vascular Endothelial Cells

Bambanker™ (Lymphotec Inc., Tokyo, Japan); store at −20°C.

2.11. Cord Formation Assays

1. Matrigel™ Basement Membrane Matrix, Phenol Red-free (Cat 356237, BD Biosciences, San Jose, CA, USA); store at 4°C.
2. EGM®-2 BulletKit (Lonza Group Ltd.); store at 4°C.
3. A light microscope (Olympus Optical Co. Ltd).

2.12. Acetylated Low-Density Lipoprotein (Ac-LDL)-Uptaking Assays

1. A 4-well chamber slide system (Nalge Nunc International Corp., Naperville, IL).
2. Sterile Hank's balanced salt solution (HBSS).
3. Low-density lipoprotein from human plasma, acetylated, DiI complex (DiI Ac-LDL) (Invitrogen Corp.).
4. EGM®-2 BulletKit (Lonza Group Ltd.); store at 4°C.
5. Hoechst 33342 (Sigma Chemical Co.).
6. Fluorescence microscope (Olympus Optical Co. Ltd).

2.13. Matrigel Plug Assay

1. FGF-2 (PeproTech Inc.); store at −20°C.
2. Matrigel™ Basement Membrane Matrix, Phenol Red-free (Cat 356237, BD Biosciences); store at 4°C.
3. Severe combined immunodeficient (SCID) mice.
4. Fluorescein isothiocyanate (FITC)-dextran (500,000 average molecular size, Sigma Chemical Co., St. Louis, MO, USA).
5. 10% Formalin Neutral Buffer Solution (WAKO Pure Chemical Industries, Cat 062–01661).
6. Paraffin for embedding of tissues, for example, Pathoprep®568 (WAKO Pure Chemical Industries, Cat 162–18961).
7. A machine for tissue processing for paraffin embedding, for example, Semi-enclosed Benchtop Tissue Processor (Leica Microsystems GmbH, Wetzlar, Germany, Product code TP1020).
8. A microtome, for example, Semi-automated Rotary Microtome (Leica Microsystems GmbH, Product code RM2245).

3. Methods

The aforementioned two-tiered differentiation method has enabled the production of subculturable vascular endothelial cells from hES cells without a cell-sorting technique. The differentiation medium contains a hematopoietic cytokine cocktail, which is not expected to be of use to vascular endothelial differentiation from a conventional viewpoint. Although removal of one of the cytokines, other than VEGF, from the differentiation medium may not significantly influence differentiation efficiencies, a combined usage of all hematopoietic cytokines guarantees the high reproducibility of the experiments (5).

Total differentiation procedure requires about a month if hES cells are prepared on Mouse Embryonic Fibroblasts (MEF) layers. This includes 1 week for maintaining hES cells under feeder-free conditions after the removal of MEFs, 3 days to form spheres by floating culture, and 3 weeks to promote maturation of vascular endothelial cells by adherent culture. A global gene expression analysis by microarray assay indicates that the expressions of vascular endothelial cell-specific markers, such as VE-cadherin, are sufficiently induced as early as at the sphere-forming stage. Thus, vascular endothelial commitment has occurred within a few days after the start of differentiation.

The hES-derived vascular endothelial cells are easily expanded by an ordinary subculture technique. The cells are freeze-and-thaw tolerable. They enter senescence after 10–20 passages, depending on the lines, as in the cases of primary culture of normal human vascular endothelial cells. We have not yet observed any tumorigenicity of hES-derived vascular endothelial cells in in vivo studies.

3.1. Culture Medium for Mouse Embryonic Fibroblasts (MEFs)

Mix the following materials: 445 mL of low glucose Dulbecco's Modified Eagle's Medium (DMEM) with L-glutamine, 5 mL of 100× penicillin/streptomycin solution, and 50 mL of heat-inactivated (56°C, 30 min) FBS.

3.2. Preparation of MEFs

1. Administer an anesthetic to a 12.5 days pregnant female mouse by placing in a 50-mL conical tube containing sevofrane-moistened absorbent cottons to the mouth of the mouse.

2. Euthanize the mouse by cervical dislocation with compliance to an animal protection guideline or law.

3. Place the mouse body in the sterilized tray on the clean bench and wash it with 70% ethanol.

4. Take out the uteri through abdominal incision and transfer them into 5 mL sterile PBS in a 60-mm culture dish.

5. Cut the walls of uteri and transfer them into 5 mL sterile PBS in a 60-mm culture dish.

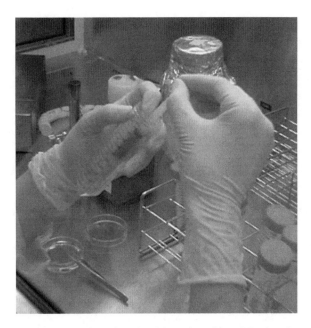

Fig. 1. A homogenizing procedure of murine fetuses by using a 2.5-mL syringe and an 18-G needle.

6. Remove the fetuses by cutting amniotic membranes and transfer them into 5 mL sterile PBS in a 60-mm culture dish.
7. Remove the heads and entire internal organs including lungs, heart, liver, spleen, guts, and kidneys using sterilized ophthalmic tweezers and scissors under a stereo microscope.
8. Transfer the ablated fetuses into 5 mL sterile PBS in a 60-mm culture dish.
9. Wash the fetuses with sterile PBS three times. Remove the plunger from 2.5-mL syringes fixed with 18-G needles.
10. Transfer each fetus into the syringe, reinsert the plunger, push out the fetuses into a 50-mL conical tube containing 2 mL of medium (i.e., DMEM supplemented with 10% FBS) (Fig. 1). This fragmented fetus suspension was further homogenized by expelling and reloading the medium two or three times through a 18-G needle (see Note 2).
11. Then transfer the fetal tissue slurry into a 100-mm culture dish containing 8 mL medium and incubate it in a 100% humidified 5% CO_2 incubator at 37°C.
12. Collect the cells using 0.25% trypsin-EDTA treatment and seed them on new culture dishes at the density of $7.5 \times 10^5/100$ mm dish after reaching confluence.
13. At confluence, collect the cells using trypsin-EDTA treatment; suspend the cells in Bambanker solution at the density of

1×10^6 cells/mL and transfer 1 mL aliquots into cryotubes and store them at −80°C (Passage 1 stock).

14. Thaw the stocks when an expansion of MEFs is required.
15. Subculture the cells at the density of 7.5×10^5/100 mm and make frozen stocks at Passage 3 ($1–3 \times 10^6$ cells/cryotube).

3.3. Culture Medium for hES Cells

Mix the following materials: 500 mL of DMEM-F12, 5 mL of NEAA solution, 6.25 mL of 200 mM L-glutamine, 125 mL of KSR, and 5 μL of 2-mercaptoethanol. This medium can be stored at 4°C up to 2 weeks.

3.4. Dissociation Liquid for hES Cells

1. Mix the following materials: 10 mL of 2.5% Trypsin, 10 mL of 10 mg/mL type IV collagenase, 20 mL of KSR, 1 mL of 100 mM $CaCl_2$, and 59 mL of $1 \times PBS$.
2. Prepare aliquots and store at −20°C. After thawing, the solution should be stored at 4°C and used within 1 week.

3.5. Preparation of Frozen Stocks of hES Cells

1. Prepare a confluent culture of hES cells on a 60-mm dish.
2. After washing with PBS, incubate the cells with 1 mL of dissociation liquid at room temperature.
3. Observe the cells under microscope; add 5 mL of hES culture medium at the time point when more than half of hES colonies begin to detach from their margins.
4. Transfer the cells en bloc into a 15-mL conical tube. Avoid cell-dissociating procedures such as pipetting and shocking of dishes (see Note 3).
5. Add 5 mL of hES culture medium to the original dish; transfer the residual cells into the same conical tube.
6. After centrifuge at $170 \times g$ for 5 min at 4°C, suspend the cell pellets with 200 μL of freezing medium and transfer it into a cryotube.
7. Soak the cryotube in liquid nitrogen as soon as possible, at least within 15 s (see Note 4). After soaking in liquid nitrogen for 1 min (see Note 5), set the cryotube into a liquid nitrogen tank.

3.6. Maintenance of hES Cells on MEF Layers

1. *Preparation of gelatin-coated dishes*: Incubate a 60-mm culture dish with 1.5–2 mL of 0.1% gelatin solution for 10 min at room temperature; aspirate the gelatin solution and wash with sterile PBS.
2. *Preparation of MEF layers*: Thaw the frozen stock of MEFs (passage 3) as immediately as possible with a mild shaking in a 37°C water bath. Transfer the thawed MEF suspension into 10 mL of MEF culture medium contained in a 15-mL conical tube.

After centrifuge, suspend the MEFs in 5 mL of MEF culture medium and seed them onto a 60-mm gelatin-coated dish above prepared at the density of $1.8–2.4 \times 10^4$ cell/60-mm dish. After overnight culture in a 100% humidified 5% CO_2 incubator at 37°C, irradiate the cells at 46 Gy. These MEF-layered dishes should be used within 4 days.

3. *Thawing of hES cells*: Add 1 mL of 37°C pre-warmed hES culture medium to a frozen hES stock tube. With pipetting, thaw the hES cells as immediately as possible. Transfer the thawed cells into a 15-mL conical tube containing 10 mL of 37°C pre-warmed hES culture medium. After centrifuging at $170 \times g$ for 5 min, aspirate the supernatant, resuspend the cells with 5 mL of hES culture medium, and seed them on the MEF layer. Culture the hES cells in a 100% humidified 3% CO_2 incubator at 37°C (see Note 6). Change the culture medium every day.

4. *Passage of hES cells*: Aspirate the culture medium; add 1 mL of dissociation liquid that is pre-warmed at room temperature; incubate the cells at 37°C in a CO_2 incubator for 5 min; add 2 mL of hES cell culture medium; suspend the cells by using Gilson P-1000 Pipetman® several times to fragment the hES cell clots into 100 cell clumps; centrifuge at $170 \times g$ for 5 min, aspirate the supernatant as much as possible (see Note 7); mildly resuspend the cells in 2 or 3 mL of hES culture medium; pour 1 mL aliquot into 4 mL of hES cells culture medium contained in a 60-mm MEF-layered dish; add 5 ng/mL of FGF2 into each dish; culture the diluted hES cell culture in a 100% humidified 3% CO_2 incubator at 37°C. Exchange the culture medium every day. The hES cells should be maintained by regular passages twice a week (Figs. 2a and 3a).

3.7. Maintenance of Feeder-Free hES Cells

1. *Preparation of Matrigel™-coated dishes*: Dilute the Matrigel™ Basement Membrane Matrix with DMEM/F12 at a dilution ratio of 1:30; add 2 mL of the diluted Matrigel™ solution into 100-mm culture dishes; incubate the dishes for 30 min at room temperature. The Matrigel™-coated dishes should be used on the same day.

2. *Transferring the hES cells onto Matrigel™-coated dishes*: Collect the hES cells from a MEF layer as described above (Sect. 3.6, (4)); resuspend the cells in hES culture medium without FGF2; seed the fragmented hES cell clumps on Matrigel™-coated dishes at a splitting ratio of 1:2–1:3. See to it that the hES cells should be maintained as a crop of colonies of similar sizes with diameters between 200 and 1,000 mm at the densities of less than 8 colonies/cm² (see Note 8). Incubate the hES cells in a 100% humidified 3% CO_2 incubator at 37°C. Change the culture medium every day.

Maintenance of undifferentiated hES cells

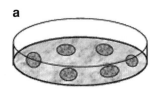

1° Culture the hES cells on MEF layers in a 100 mm dish.

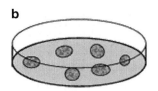

2° Culture the hES cells on a Magtrigel™ Matrix-coated 100 mm dish. For elimination of MEFs, it is recommended to perform at least two passages, preferably four passages, on Magtrigel™ Matrix-coated dishes before starting differentiation.

Differentiation procedure

Day 0

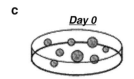

3° Collect the feeder-free hES cells by a dissociation liquid treatment. To form spheres, culture the cells in a 60 mm low attachment dish for three days using the differentiation medium.

Day 3

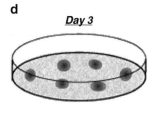

4° Collect the spheres, re-suspend them in a fresh differentiation medium and culture the spheres on a 0.1% gelatin-coated 100 mm dish.

Day 6

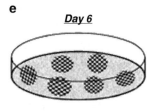

5° Refresh the culture medium. By this time, spheres are flattened and extended, showing a cobblestone appearance. Continue the culture. Change the medium twice a week.

Day 12-13

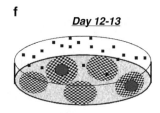

6° Around this time, sac-like structures containing hematopoietic cells may emerge in the centers of some flattened spheres. In that case, cut the sac walls by using microknives and release the inner hematopoietic cells. Floating hematopoietic cells will be removed by a medium-changing procedure.

Fig. 2. Schematic diagram of the procedures used for maintaining and differentiating hES cells into vascular endothelial cells.

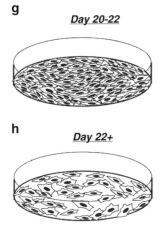

g *Day 20-22*

7° The cells, which are reaching confluence, show cobblestone appearances in some areas and cord-like forms in other areas.

h *Day 22+*

8° Subculture the cells at splitting ratios of 1:2 to 1: 3 by a trypsin-EDTA treatment.

Fig. 2. (continued)

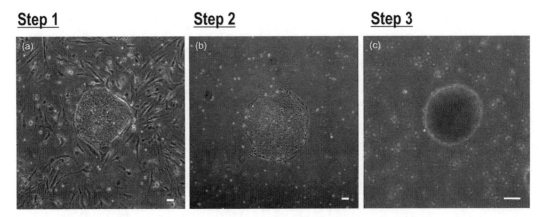

Fig. 3. Phase contrast microscopy of hES cells on MEF layers (*left*), feeder-free hES cells (*middle*), and hES-derived spheres (*right*). Scale bar = 100 μm. (**a**) hES cells on MEF layers, (**b**) feeder-free hES cells, and (**c**) hES cells-derived spheres.

3. *Maintenance of hES cells on Matrigel™-coated dishes without FGF2*: Aspirate the culture medium; add 1 mL of dissociation liquid that is pre-warmed at room temperature (see Note 9); incubate the cells in a CO_2 incubator for 5–15 min; add 2 mL of hES cell culture medium; transfer the cells into a 15-mL conical tube; centrifuge it at $170 \times g$ for 5 min; aspirate the supernatant as much as possible (see Note 7); gently resuspend the cells in 2–4 mL of hES culture medium without FGF2; transfer 1 mL of the cell suspension onto a new 100 mm Matrigel™-coated dish containing 9 mL of hES cell culture medium without FGF2. Incubate the hES cells in a 100% humidified 3% CO_2 incubator at 37°C. Again, see to it that the hES cells should be maintained as a crop of colonies of similar sizes with diameters between 200 and 1,000 mm at the

densities of less than 8 colonies per 1 cm². Change the culture medium every day. hES cells should be maintained by regular passages twice a week (Figs. 2b and 3b). If ES colonies with undesirable morphologies, such as flattened, extended, or dissociated, are observed, scrape them out from the dishes by using microknives under microscopy (see Note 10).

3.8. Vascular Endothelial Differentiation of hES Cells

1. *Sphere formation*: Prepare a 100-mm dish of feeder-free hES cell culture (see Note 11); aspirate the culture medium; add 1 mL of dissociation liquid that is pre-warmed at room temperature; incubate the cells in a CO_2 incubator for 15 min; add 2 mL of a differentiation medium; transfer the cells into a 15-mL conical tube; centrifuge it at $170 \times g$ for 5 min; aspirate the supernatant as much as possible (see Note 7); gently resuspend the cells in 5 mL of a differentiation medium; transfer the cell suspension into a 60-mm low-attachment sterile dish such as HydroCell® (CellSeed Inc) and a MPC-coated dish (Nalge Nunc International) (see Note 12); culture the cells in a 100% humidified 5% CO_2 incubator at 37°C. After 3 days, spheres of uneven sizes will be generated (see Note 13) (Figs. 2c and 3c).

2. *Maturation of vascular endothelial cell by attachment culture*: Transfer the suspension of hES-derived spheres into a 15-mL conical tube; centrifuge it at $170 \times g$ for 3 min; aspirate the supernatant; resuspend the cells in 10 mL of differentiation medium; seed the cells on a 100-mm 0.1% gelatin-coated dish; incubate the cells in a 100% humidified 5% CO_2 incubator at 37°C (Fig. 2d). Change the culture medium twice a week. Within a couple of days, the spheres will be flattened and show disk-like morphologies (Fig. 2e). After 2 weeks from the start of differentiation, the cells will show a cobblestone-like appearance, which is a characteristic of vascular endothelial cells. Around this time, spheroid cells may be floating in the culture supernatant. These non-adherent cells are hematopoietic cells, the majority of which are myelomonocytic lineages. Also, sac-like structures may emerge in the middle of the flattened spheres (Fig. 2f). In such cases, cut the sac walls by using microknives under microscope (see Note 14). After 3 weeks from the start of differentiation, confluent culture of cobblestone-like vascular endothelial cells will be obtained (Fig. 2g and 4a). Cord-like forms, which are also characteristics of vascular endothelial cells, may be observed in some regions (Fig. 4b). At this time, almost all the cells express VE-cadherin at intercellular junctions (Fig. 4c). Harvest the cells by trypsin-EDTA treatment for 5 min at 37°C. Seed the cells on new gelatin-coated 100-mm dishes at a split ratio of 1:2 (Fig. 2h). Subcultured cells show fairly homogeneous morphologies (Fig. 5a) that are similar to primary human vascular endothelial cells. Almost all the cells express von Willebrand factor (Fig. 5b)

Step 7

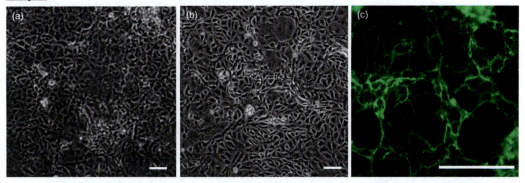

Fig. 4. Phase contrast microscopy of the differentiated hES cells (**a–b**) and an immunostaining study on VE-cadherin expression (**c**) were shown. Scale bar = 100 μm. (**a**) Areas with cobblestone appearances, (**b**) areas with cord-like forms, and (**c**) VE-cadherin immunostaining.

Step 8

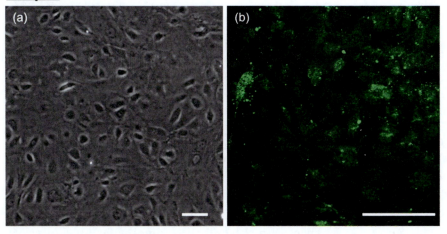

Fig. 5. Phase contrast microscopy of the subcultured hES-derived vascular endothelial cells at passage number 1 (**a**) and an immunostaining study on the expression of von Willebrand factor (**b**) were shown. Scale bar = 100 μm. (**a**) Morphology of the subcultured cells and (**b**) VWF immunostaining.

and NOS3 (data not shown). After 3–4 days, subculture the cells at split ratios of 1:2–1:3. Continue subculture at a split ratio of 1:2–1:3 twice a week.

3. *Preparation of frozen stock*: Harvest the hES cell-derived vascular endothelial cells by trypsin-EDTA treatment; add 10 mL of differentiation medium; transfer the cell suspension into a 15-mL conical tube; centrifuge it at $170 \times g$ for 5 min; suspend the cells in Bambanker at the density of 1×10^6 cells/mL; prepare 1 mL aliquots into cryotubes and store them at –80°C.

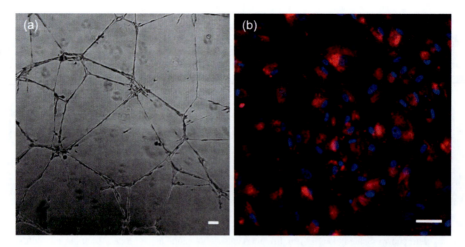

Fig. 6. The in vitro function assays. Cord-forming activities (**a**) and Ac-LDL-uptaking capacities (**b**) were shown: *Red*, DiI-Ac-LDL, *Blue*, nuclei. Scale bar = 100 μm. (**a**) Cord formation assays and (**b**) Ac-LCL-uptaking assays.

3.9. Cord Formation Assays

1. Load 95 μL of Matrigel™ Basement Membrane Matrix into the required number of wells of a 24 multi-well dish; incubate the dish for 30 min at 37°C; suspend hES-derived vascular endothelial cells in EGM®-2 BulletKit at the density of 1×10^4 cells/mL (see Note 15); transfer the cell suspension into the Matrigel™ Matrix-coated wells (1 mL/well). After overnight culture, observe cell morphologies under an inverted light microscope (Fig. 6a).

3.10. Acetylated Low-Density Lipoprotein (Ac-LDL)- Uptaking Assays

1. Transfer hES-derived vascular endothelial cells into 4-well chamber slide system.
2. After overnight culture, wash the cells twice using HBSS and incubate them with EGM®-2 BulletKit containing 10 μg/mL of DiI-Ac-LDL for 4 h at 37°C.
3. Then wash the cells three times with HBSS, and deliver the cells into 0.5 mL HBSS containing 10 nM Hoechst 33342 for 5 min at room temperature.
4. Examine the cells under a fluorescence microscope (Fig. 6b) after washing the cells once by HBSS.

3.11. Matrigel Plug Assays

1. Suspend the 3×10^6 of hES-derived vascular endothelial cells in 10 μL of differentiation medium supplemented with 150 ng/mL of FGF-2.
2. Mix the cell suspension with 500 μL of Matrigel™ Basement Membrane Matrix.
3. Transplant the mixture subcutaneously into SCID mice. For control, mix 10 μL of FGF-2-containing differentiation medium with 500 μL of Matrigel™ Matrix and transplant in the same way.

Matrigel™ plug assay

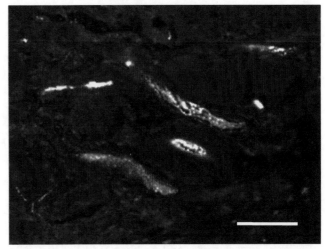

Fig. 7. The in vivo function assay. Matrigel plug assay was performed and a sliced specimen was observed under fluorescence microscope. FITC-dextran-stained lumens of the neovascularities were detected. The majority of the cells lining on the lumens are hES-derived vascular endothelial cells as we confirmed previously (10). Scale bar = 100 μm.

4. After 3 weeks, inject a 0.2 mL of FITC-dextran solution (100 mg of FITC-dextran is suspended in 5 mL of PBS) via the tail vein to detect the neovascularities connected with the systemic circulation (see Note 16).
5. A few minutes later, sacrifice the mice and remove the transplanted material.
6. Soak it in 10% formalin neutral buffer for fixation.
7. Embed it in paraffin by using, for example, Pathoprep®568.
8. Slice the paraffin-embedded sample by 4 μm.
9. Observe the sliced specimen under fluorescent microscope to detect FITC-dextran-packed lumens in the plugs (Fig. 7). The specimens can be further subjected to immunostaining studies, for example, using a rabbit polyclonal antihuman PECAM1 antibody (H-300) (Santa Cruz Biotechnology Inc.), which does not recognize murine vascular endothelial cells, or mouse monoclonal anti-HLA-A, B, C antibody (BD Biosciences) to determine human-oriented cells.

4. Notes

1. Mitomycin C (MMC) treatment is an alternative measure to block proliferation of MEFs if an X-ray irradiation device is not available. Although there may be a risk that a trace amount of

MMC, a genotoxic agent, is released to hES cells, and thus, X-ray irradiation seems to be superior to MMC treatment, we have not yet detected chromosomal abnormalities in the hES cells maintained on MMC-treated MEFs. MMC treatment of MEFs is as follows: Prepare a 2-mg/mL MMC stock solution; add 10 μg/mL of MMC on the confluent culture of MEFs in a 100-mm dish; mix gently; incubate the cells in 100% humidified 5% CO_2 incubator at 37°C for 90–120 min. Wash the cells twice with PBS; add 10 mL of MEF culture medium; incubate the cells in a 100% humidified 5% CO_2 incubator at 37°C for 3 h to overnight.

2. Too much homogenation reduces the viability of MEFs. Thus, top homogenation when small fragments of tissues are still visible by the naked eye.

3. Procedure to dissociate or fragment the hES cell clots is unnecessary at this step because hES cell clots will be automatically fragmented during the cell-thawing process.

4. To minimize cytotoxic effect of freezing medium of hES cells, the cell suspension should be soaked in liquid nitrogen as fast as possible, within 15 s at longest.

5. Ensure that cell suspension is fully frozen including inner portions.

6. Because of an acidification-prone nature of the hES culture medium, it is recommended that the CO_2 concentration of the incubator is set at 2–3%. Alternatively, pH of the medium can be adjusted by adding NaOH.

7. Collagenase and trypsin, which are added to dissociation liquid, will not be inactivated by the addition of culture medium. Thus, it is recommended that dissociation liquid should be fully aspirated. Alternatively, washing the cell pellets by culture medium will be advantageous to complete elimination of the enzymes.

8. Ensure that the sizes and numbers of the crops of hES cells do not exceed the range described. Too many hES cell crops per dish, an existence of oversized clots of hES cells, or fusion of two or three clots of hES cells will induce spontaneous differentiation.

9. Occasionally for an unknown reason, hES cells will not detach from the Matrigel™ layer when a dissociation liquid is used. In that case, try Dispase (BD Biosciences, San Jose, CA, USA, Cat. No. 354235). There is no significant difference between the vascular endothelial differentiation of Dispase-treated hES cells and that of dissociation buffer-treated hES cells.

10. The requirement of scraping of the hES colonies with undesirable morphologies differs depending on the lines. For example,

only one-time scraping is sufficient in the case of khES-1 line, whereas scraping three or four times sequentially is required in the case of khES-3 after thawing. Even during the maintenance of hES cells on MEF layers, performing a similar scraping procedure is recommended.

11. It is recommended to perform at least four passages on the Matrigel™ Matrix layers before starting the differentiation procedures to eliminate MEFs.

12. For best performance, HydroCell® (CellSeed Inc) is recommended; it guarantees highly efficient sphere formation without the least attachment of hES cells. The MPC-coated dish (Nalge Nunc International) is also usable, although it occasionally permits hES cells to attach at its margin. Low-attachment dishes by other manufactures are not recommended.

13. Spheres of uneven sizes, some of which may fuse with one another, will be formed by this floating culture technique. Because the sizes of spheres do not significantly affect the differentiation efficiency, efforts to form even-sized spheres are not required. On the other hand, the number of spheres subjected to an attachment culture (i.e., spheres per gelatin-coated dish) influences the quality and efficiency of differentiation: the more spheres, the better the results. Thus, if the differentiation efficiency is not as high as expected, use two dishes of feeder-free hES cell cultures per one dish of attachment culture.

14. Within the sac-like structures, there may be monocytes and macrophages. Because macrophages are highly adhesive when activated, contamination of the cobblestone cell population by macrophages may impair the quality of differentiation. So, cut off the walls of the sac-like structure and release the inner hematocytes into culture supernatant. The hematocytes will be removed after changing of the culture medium.

15. The number of cells per well is a critical factor for cord formation assays. Usually, 1×10^4 cells per well is the optimal condition; it can vary depending on the lots of Matrigel™ Matrix and the type of hES cells. Thus, perform the assay at various densities around 1×10^4 cells per well; for example, 0.5×10^4 or 2×10^4 cells per well.

16. In a previous study by Kaufman et al. utilizing rhesus monkey ES cell-derived vascular endothelial cells (12), FITC-dextran was injected after 28–32 days from the transplantation. However, the optimal time point should be determined in each case. From our experiences in hES-derived vascular endothelial cells, we recommend to inject after 21 days from the transplantation. Longer incubation in vivo may possibly induce an

immunologic rejection due to preservation of NK cell activities in SCID mice. In the case where longer-term observations are required, we recommend to use NOG mice (13), which have multifunctional defects including NK activities, macrophage functions, complement activities and functions of dendritic cells, instead of SCID mice.

References

1. Sone, M., Itoh, H., Yamahara, K., Yamashita, J.K., Yurugi-Kobayashi, T., Nonoguchi, A., Suzuki, Y., Chao, T.H., Sawada, N., Fukunaga, Y., Miyashita, K., Park, K., Oyamada, N., Sawada, N., Taura, D., Tamura, N., Kondo, Y., Nito, S., Suemori, H., Nakatsuji, N., Nishikawa, S., Nakao, K. (2007). Pathway for differentiation of human embryonic stem cells to vascular cell components and their potential for vascular regeneration. Arterioscler. Thromb. Vasc. Biol. 27, 2127–2134.

2. Levenberg, S., Golub, J.S., Amit, M., Itskovitz-Eldor, J., Langer, R. (2002). Endothelial cells derived from human embryonic stem cells. Proc. Natl. Acad. Sci. U.S.A. 99, 4391–4396.

3. Wang, Z.Z., Au, P., Chen, T., Shao, Y., Daheron, L.M., Bai, H., Arzigian, M., Fukumura, D., Jain, R.K., Scadden, D.T. (2007). Endothelial cells derived from human embryonic stem cells form durable blood vessels in vivo. Nat. Biotechnol. 25, 317–318.

4. Sone, M., Itoh, H., Yamashita, J., Yurugi-Kobayashi, T., Suzuki, Y., Kondo, Y., Nonoguchi, A., Sawada, N., Yamahara, K., Miyashita, K., Park, K., Shibuya, M., Nito, S., Nishikawa, S., Nakao, K. (2003). Different differentiation kinetics of vascular progenitor cells in primate and mouse embryonic stem cells. Circulation 107, 2085–2088.

5. Saeki, K., Yogiashi, Y., Nakahara, M., Nakamura, N., Matsuyama, S., Koyanagi, A., Yagita, H., Koyanagi, M., Kondo, Y., Yuo, A. (2008). Highly Efficient and Feeder-Free Production of Subculturable Vascular Endothelial Cells from Primate Embryonic Stem Cells. J. Cell. Physiol. 217, 261–280.

6. Lugus JJ, Park C, Choi K. (2005). Developmental relationship between hematopoietic and endothelial cells. Immunol. Res. 32, 57–74.

7. Wu, X., Lensch, M.W., Wylie-Sears, J., Daley, G.Q., Bischoff, J. (2007). Hemogenic endothelial progenitor cells isolated from human umbilical cord blood. Stem Cells 25, 2770–2776.

8. Taoudi, S., Gonneau, C., Moore, K., Sheridan, J.M., Blackburn, C.C., Taylor, E., Medvinsky, A. (2008). Extensive hematopoietic stem cell generation in the AGM region via maturation of VE-cadherin+CD45+ pre-definitive HSCs. Cell Stem Cell 3, 99–108.

9. Lancrin, C., Sroczynska, P., Stephenson, C., Allen, T., Kouskoff, V., Lacaud, G.(2009). The haemangioblast generates haematopoietic cells through a haemogenic endothelium stage. Nature 457, 892–895.

10. Nakahara M, Nakamura N, Matsuyama S, Yogiashi Y, Yasuda K, Kondo Y, Yuo A, Saeki K. (2009) High-efficiency production of subculturable vascular endothelial cells from feeder-free human embryonic stem cells without cell-sorting technique. Cloning Stem Cells 11, 509–522.

11. Suemori, H., Yasuchika, K., Hasegawa, K., Fujioka, T., Tsuneyoshi, N., Nakatsuji, N. (2006). Efficient establishment of human embryonic stem cell lines and long-term maintenance with stable karyotype by enzymatic bulk passage. Biochem. Biophys. Res. Commun. 345, 926–932.

12. Kaufman, D.S., Lewis, R.L., Hanson, E.T., Auerbach, R., Plendl, J., Thomson, J.A. (2004). Functional endothelial cells derived from rhesus monkey embryonic stem cells. Blood 103, 1325–1332.

13. Ito M, Hiramatsu H, Kobayashi K, Suzue K, Kawahata M, Hioki K, Ueyama Y, Koyanagi Y, Sugamura K, Tsuji K, Heike T, Nakahata T. NOD/SCID/gamma(c)(null) mouse: an excellent recipient mouse model for engraftment of human cells. Blood 100, 3175–3182.

Chapter 21

Feeder-Independent Maintenance of Human Embryonic Stem Cells and Directed Differentiation into Endothelial Cells Under Hypoxic Condition

Xiuli Wang

Abstract

Human embryonic stem cells (hESCs) represent a promising source of cells for modern regenerative medicine due to their highly self-renewal capability and pluripotency to differentiate into almost all cell types from three different germ layers. Recent advances in the stem-cell field have enabled generation of hESC-derived endothelial cells (or endothelial progenitor cells) that might be potentially employed to build up a vascular network in engineered tissue or reconstruct artificial vascular grafts to repair ischemic tissue. However, the differentiation efficiency is relatively low. Considering the important roles of both growth factors and "physiological hypoxia" environment in facilitating endothelial-lineage differentiation and development in the embryos, we studied combinatorial effect of VEGFs and a hypoxic culture condition on in vitro differentiation of hESCs toward endothelial cells. We show that hESC endothelial-lineage differentiation efficiency can be significantly improved under hypoxic conditions, indicated by the increased expression of marker proteins and genes. These experimental results suggest a synergy effect of VEGF and hypoxic environment on the generation of hESC-derived endothelial cells. In this chapter, we describe our protocols for the maintenance of undifferentiated hESCs with feeder-free medium (mTeSR™1 medium), and for the directed differentiation of hESCs toward endothelial cells with specific reference to growth factor supplementation under hypoxic culture condition in vitro.

Key words: Embryonic stem cells (ESCs), Hypoxia, Self-renewal, Differentiation, Endothelial cells

1. Introduction

Human embryonic stem cells (hESCs) represent a promising cell source for modern regenerative medicine due to their highly self-renewal capability and pluripotency to differentiate into all cell types from three different germ layers (1). Particularly, hESCs are capable of differentiating into hemangioblast cells, including hematopoietic and endothelial precursor cells (2, 3), and could be potentially employed to build up vascular network in engineered tissue or reconstruct artificial vascular grafts to repair ischemic tissue (4, 5). Moreover, these hESC-derived endothelial cells also provide an advantageous experimental system to recapitulate the complex molecular and cellular events during early vascular development, disease progression, and epigenetics (6).

An early study by Levenberg et al. first described an identification of endothelial lineage differentiation of hESCs when the cells were cultivated as embryoid bodies (EBs) in hESC growth medium without requiring additional growth factors (7). An assembly of developing vascular-like structures was also observed during this spontaneous differentiation, suggesting that hESC-derived endothelial cells could organize into vessel-like structures in vitro in a pattern of resembling embryonic vascularization (7). To date, some progress has been made in promoting endothelial lineage differentiation of hESCs with an improved efficiency. Co-culture of hESCs with OP9 or S17 cells, cell lines of bone marrow-derived mouse stromal cells, has been shown to induce an efficient differentiation of hESCs into endothelial progenitor cells other than the lineage of monocytes and macrophages (8). No exogenous growth factors or a complex embryoid structure was involved in this induction system, offering its advantages in low cost and ease of operation. However, it should be noted that incorporation of heterogeneous cells not only poses a risk of pathogenic contamination, but also complicates the process of isolation/purification of target cells in this co-culture system. This would hamper its extensive application in stem cell–based therapy in clinic. In contrast, the addition of growth factors, such as endothelial growth factor (VEGF), and/or bone morphogenic protein-4 (BMP-4) partially address this issue, and provides a more optimized option to promote hESC differentiation into vascular cells, as demonstrated by the expression of multiple markers specific to progenitor or fully differentiated endothelial cells and their capability of integrating with the microvessel in vivo (9).

On the other hand, it has been known that early stage mammalian embryos develop in a hypoxic environment of the reproductive tract, in which the oxygen concentration ranges from 1.5% to 5.3% (10). This resultant "physiologic hypoxia" has been shown not only to play a central role in keeping ESCs from differentiation (11), but also to stimulate the formation of a vascular system during the development of embryos (12). A detailed mechanism under this effect has not been fully elucidated so far; however, accumulated evidence suggests that the hypoxia-inducible factor (HIF) signaling pathway, a transcription factor that responds to changes in available oxygen in the cellular environment, participates in this event (13). Particularly, HIF could regulate *vegf* or *vegf*-receptor gene expression level via hypoxia-responsive elements (HREs) that bind the HIF complex, and serve as a stimulus to initiate the morphogenesis and development of the cardiovascular system (14). Mouse ESCs deficient in HIFs failed to activate genes responsive to hypoxic environment, leading to vascular malformation and developmental arrest of those ESC-derived mice (15). More recently, one study by Stojkovic and colleagues analyzed the transcriptome profile of hESCs cultured at different O_2 levels, and revealed that reduced O_2 tension could down-regulate expression of pluripotency markers of hESCs and increase significantly the expression of genes associated with angio- and vasculogenesis (16). This further supports the critical role of hypoxic environment in facilitating hESC differentiation into endothelial or relevant progenitor cells.

Considering the important role of either growth factors or hypoxia demonstrated in those previous studies, the effect of the combined utilization of growth factors (VEGFs) and hypoxia during the in vitro directed differentiation of hESCs can be important. A number of recent studies suggest that a synergy effect of VEGF and hypoxic environment exist in efficiently promoting the lineage-specific differentiation of hESCs into endothelial cells, as evidenced by the phenotypic characteristics of those differentiated hESCs and their functional characterization (Fig. 1).

In this chapter, we outline our protocol for the in vitro differentiation of hESCs toward endothelial cells with specific reference to VEGFs supplementation under hypoxic culture condition. Our procedure includes two steps: (1) feeder-free culture of pluripotential hESCs in mTeSR medium and their characterization, and (2) directed differentiation of hESCs toward endothelial cells in vitro.

2. Materials

2.1. Cell Lines and Incubators

1. H1 human ESCs line (WiCell Research Center, Madison, Wisconsin)
2. Normal cell culture incubator with 5% CO_2, 20% O_2, 95% humidity (VWR, Model No. 2310)
3. Low oxygen incubator with 5% O_2, 5% CO_2, and 90% nitrogen (Model No. MCO-18AIC (UV), SANYO, Japan)

2.2. Media and Supplements

1. mTeSR™1 basal medium (Stemcell Technologies, Vancouver, Canada, Cat. No. 05850)
2. mTeSR™1 5× supplement (Stemcell Technologies, Cat. No. 05850); store at −20°C
3. KnockOut™ DMEM/F12 medium (Invitrogen, Cat. No.126600-012)

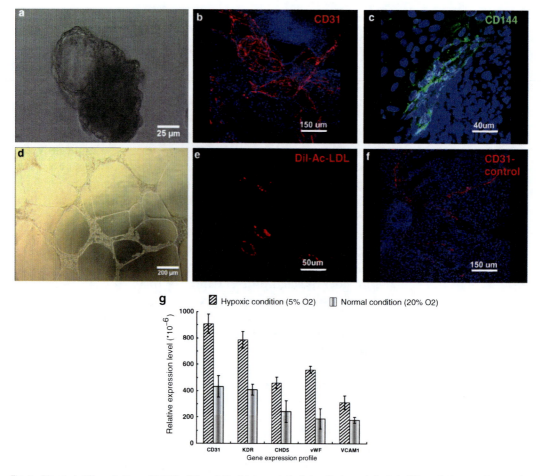

Fig. 1. Directed differentiation of hESCs (H1 cell line) into endothelial cells in endothelial differentiation medium under hypoxic condition. (**a**) EB generated from suspending culture, (**b**) and (**c**) showed the immunostaining of CD31 and CD144. The cells cultured under normal oxygen culture condition served as control (**d**). The hESC-derived endothelial cells could form the cord-like structure in Matrigel™ assay, and uptake Dil-ac-LDL in the medium (**e, f**). Moreover, compared with the control group, they also displayed an increased gene expression profile after the directed differentiation (**g**).

4. DMEM/F12 medium (Stemcell Technologies, Cat. No. 36254)
5. Knockout serum replacement (KSR, Invitrogen, Cat. No. 10828–028)
6. Cryopreservation medium, mFreSR™ (Stemcell Technologies, Cat. No. 058540); store at −20°C
7. L-Glutamine-200 mM (Invitrogen, Carlsbad, CA, Cat. No. 25030081)
8. ß-Mercaptoethanol (Invitrogen, Cat. No. 21985023)
9. Nonessential amino-acid solution (Invitrogen, Cat. No. 11140050)

10. Endothelial cell medium, EGM-2 medium (Lonza, Basel, Switzerland, Cat. No. cc-3162)

11. Adenosine 3′,5′-cyclic monophosphate (cAMP, Sigma-Aldrich, St. Louis, MO, Cat. No. A9501)

12. Human recombinant vascular endothelial growth factor (hVEGF, Invitrogen, Cat. No. PHC9391): add 1 mL of a 0.3% (w/v) BSA solution to a vial containing 100 μg of $VEGF_{165}$ to yield a 50 ng/μL solution. Aliquot the resultant VEGF165 solution and store the vials at −70°C.

13. EB culture medium: 90% KnockOut™ DMEM/F12 medium supplemented with 10% KSR, 2.5 mL L-glutamine, 1 mL β-mercaptoethanol, 5 mL nonessential amino acid solution, and filtered through a 0.2 μm filter; store at 4°C for up to 3 weeks

14. Endothelial cell differentiation medium: EGM-2 medium supplemented with 50 ng/mL $VEGF_{165}$ and 0.5 mM cAMP.

2.3. Enzymes and Other Solutions

1. Dispase solution (Stemcell Technologies, Cat. No. 07923); 1 mg/mL, store at −20°C.

2. 0.05% Trypsin-EDTA solutions (Invitrogen, Cat. No. 25300120)

3. Dulbecco's Phosphate-Buffered Saline (D-PBS) (Invitrogen, Cat. No.14190250)

4. Colcemid solution (Invitrogen, Cat. No. 15212–012); 10 μg/mL, store at 4°C

5. Matrigel™ hESC-qualified Matrix (BD Biosciences, San Jose, California, Cat. No. 354277); store at −70°C

6. BD Matrigel™ (BD Biosciences, Cat. No. 354230)

7. Gelatin solution (Sigma-Aldrich, Cat. No. G1393); diluted with sterile distilled water to yield 0.1% working solution, store at 4°C

8. Paraformaldehyde (Sigma-Aldrich, Cat. No. 185127); 4% in PBS, prepared freshly

9. Bovine serum albumin (BSA, Sigma-Aldrich, Cat. No. A2153); 3% in PBS, store at 4°C

10. Triton X-100 (Sigma-Aldrich, Cat. No. T9284); 0.3% in PBS, store at 4°C

11. Normal donkey serum (Santa Cruz Biotechnology, Santa Cruz, CA, Cat. No. sc2044)

12. SouthernBiotech fluoromount-G® slide mounting medium (Fisher Scientific, Pittsburgh, PA, Cat. No. OB100-01)

13. 4′,6-Diamino-2-phenylindole dihydrochloride (DAPI, Fisher Scientific, Cat. No. 46190)

    (a) DAPI stock solution: dissolve 10 mg DAPI in 10 mL of ultrapure water to yield a concentration of 1 mg/mL. Stock solution is stable for several months and repeated use if stored protected from light at −20°C.

    (b) DAPI working solution: dilute the DAPI stock solution 1:1,000 in ultrapure water (1 μg/mL DAPI). Filter the working solution to remove dye aggregates that can result in punctate signal.

14. FACS buffer: add 2% (v/v) fetal bovine serum (FBS) and 1% penicillin/streptomycin to PBS and filter with 0.22 μm filter; prepare the solution freshly

15. Hypotonic solution: mix 50 mL of 0.075 M potassium chloride (KCl) with 25 mL of 0.6% sodium citrate; store at 4°C

16. Cell fixation solutions for karyotype analysis:

    (a) Solution (I): 3:1 (v/v) methanol: glacial acetic acid
    (b) Solution (II): 1:1 (v/v) methanol: glacial acetic acid

2.4. Antibodies and Reagent Kits

1. Human ESCs marker antibody panel kit (R&D systems, Minneapolis, MN, Cat. No. SC008); 10 μg/mL in 3% BSA solution; store at −80°C in aliquot

2. Phycoerythrin (PE)-mouse-antihuman SSEA-4 (R&D systems, Cat. No. FAB1435P)

3. Mouse antihuman PECAM (CD31) primary antibody (Dako, Carpinteria, CA, Cat. No. M0823)

4. Mouse antihuman VE-Cadherin (CD144) primary antibody (Abcam®, Cambridge, MA, Cat. No. ab7047)

5. Alexa Fluor® 647-mouse-antihuman CD31 (BD Pharmingen, Cat. No. 558094)

6. Rabbit anti-goat IgG-TRITC (Sigma-Aldrich, Cat. No. T7028)

7. Rabbit anti-mouse IgG-FITC (Sigma-Aldrich, Cat. No. F9137)

8. Rabbit anti-goat IgG-FITC (Sigma-Aldrich, Cat. No. F7367)

9. Goat anti-mouse IgG-TRITC (Sigma-Aldrich, Cat. No. T5393)

10. Dil AcLDL solution (Invitrogen, Cat. No. L-3484); 1 mg/mL; store at 4°C in the dark

    (a) RNeasy mini kit (QIANGEN Inc. Valencia, CA, Cat. No. 74104)

11. High capacity cDNA reverse transcript kit (Applied Biosystems, Carlbad, CA, Cat. No. 4368814)

12. TagMan® gene expression assay kit (Applied Biosystems, Cat. No. 4304437)

2.5. Additional Materials

1. Falcon Conical tubes (15 and 50 mL) (Fisher Scientific, Cat. No. 02-683-173)

2. Costar® 6-well tissue culture-treated plates (Fisher Scientific, Cat. No. 07-200-80)

3. Costar® 24-well tissue culture-treated plates (Fisher Scientific, Cat. No. 07-200-80)

4. Ultralow attachment tissue culture 6-well plate (Fisher Scientific, Cat. No. 07-200-601)

5. Serological pipettes (2, 5, and 10 mL) (BD Biosciences, Cat. No. 357507, 357543, 357551)

6. Glass cover slips (VWR, West Chester, PA, Cat. No. 89015–724)

7. Glass bottom culture dishes (MatTek Corporation, Ashland, MA, Cat. No. P35G-1.5-14-C)

8. Syringe filters with pore size 0.22 μm (Fisher Scientific, Cat. No. 09719A)

9. FACS tubes (BD Sciences, Cat. No. 352235)

10. Cell strainer (40 μm, BD Sciences, Cat. No. 352340)

11. Medium filtration systems 500 and 150 mL (Corning, Lowell, MA, Cat. No. 431097, 431153)

12. Plastic disposable pipettes 1, 5, 10, and 25 mL (Fisher Scientific, Cat. No. 13-675-2C, 13–676–10 C, 13–676–10J, 13–676–10 K)

3. Methods

3.1. Feeder-Independent Maintenance of H1 hESCs

3.1.1. Preparation of Matrigel™-Coated 6-Well Plate

1. Pre-cold tips and serological pipettes (2, 5, and 10 mL) at −20°C; pre-cold DMEM/F12 medium at 4°C.

2. Thaw one aliquot of BD Matrigel™ hESC-qualified matrix (see Notes 2 and 3) on ice; add the thawed aliquot into 25 mL cold DMEM/F12 medium and keep on ice.

3. Quickly transfer the diluted Matrigel™ solution into four 6-well tissue culture-treated plates (1 mL/well) using pre-cooled serological pipettes. Swirl the plate gently to spread the solution evenly across the well surface.

4. Keep the coated plates in cell culture hood for at least 1 h at room temperature, and then carefully aspirate the coating solution by vacuum before the use (see Note 4).

3.1.2. Transitioning from Feeder Culture to Feeder-Free Culture of hESCs

1. Quickly remove one vial of hESCs (with feeder cells) from the liquid nitrogen tank and immerse it into a 37°C water bath by gently swirling. Remove the cryovial from the water bath when only a small ice crystal remains, and wipe it with 70% ethanol.

2. Gently transfer the hESC suspension into a 15-mL conical tube with a 2-mL serological pipette. Then, add 4–5 mL mTeSR™1 medium dropwise into the tube. To better mix the hESCs with the medium, keep shaking the tube gently when adding the medium.

3. Centrifuge the cells at $200 \times g$ for 5 min and gently re-suspend cell pellet into 2 mL mTeSR™1 medium.

4. Transfer the hESC suspension into one well of the pre-coated 6-well plate. Place the plate into an incubator with 5% CO_2, 95% humidity. Move the plate in quick side to side, forward to back motion to make the cell clusters distribute more evenly.

5. Refresh medium daily. Check the morphology of the cells and remove any differentiated clusters through "pick-to-remove" or remove those undifferentiated cells through "pick-to-keep."

6. When the hESC colonies are ready to passage, remove the medium by vacuum and add 1 mL dispase solution (1 mg/mL) into each well. Incubate at 37°C for 5–10 min until a slightly folded edge of the colony is observed.

7. Aspirate Dispase solution and rinse each well with 2 mL DMED/F12 medium for two times.

8. Add 2 mL mTeSR™1 medium into each well and scrape colonies off with a 10-mL serological pipette while pipeting up and down.

9. Transfer the cell suspension into 15-mL conical tube and centrifuge at $200 \times g$ for 5 min.

10. Re-suspend the cell pellet with 6–8 mL mTeSR™1 medium and distribute the cell clumps into three or four wells evenly (2 mL/well).

11. Maintain the culture with morphology check and medium change daily.

3.2. Characterization of Undifferentiated hESCs In Vitro

Although the mTeSR™1 medium developed by the Stemcell Technology has now been extensively utilized to support feeder-free culture of hESCs in vitro, few up-to-date published literatures have elucidated its long-term effect on the phenotype of hESCs, including their marker protein expression and karyotype analysis. Given the fact that in vitro culture condition (or microenvironment) of hESCs plays a central role in maintaining the cells' phenotype and pluripotency; it's imperative to characterize the phenotype of hESCs after switching to the mTeSR™1 medium from the feeder culture. This also serves as a "quality control" of the hESCs in those studies on hESC self-renewal or lineage-directed differentiation in vitro or in vivo.

3.2.1. Growth Profile of Undifferentiated hESCs in mTeSR™1 Medium

Similar to the feeder culture, undifferentiated hESCs maintained in mTeSR™1 medium grow as compact, multicellular colonies with uniform cellular morphology, as shown in Fig. 2a. Each colony displays a bright center and a well-defined border under phase-contrast microscope.

3.2.2. Immunofluorescence Staining of Nanog and Oct-3/4

Nanog and Oct-3/4 are transcription factors critically involved with the self-renewal of undifferentiated hESCs. Their expression level assayed by immunofluorescence staining has served as one of the parameters to characterize the undifferentiated phenotype of hESCs (Fig. 2b, c).

1. Prepare Matrigel™-coated glass bottom culture dishes (or glass coverslips in a 24-well cell culture plate).

2. Gently seed hESCs clumps onto the Matrigel™-coated Glass bottom culture dishes (or glass coverslips) and maintain the culture in mTeSR™1 medium for a few days to allow more cell colonies to grow up.

3. Remove the culture medium by vacuum, and gently wash the cells twice with 1 mL of PBS.

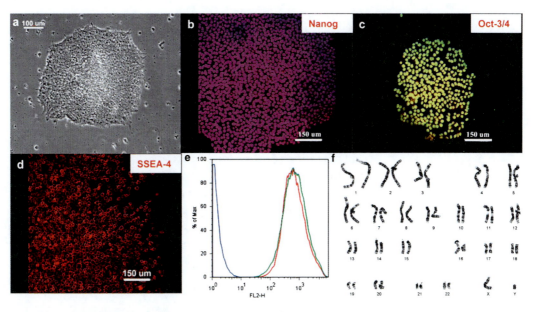

Fig. 2. Phenotypic characterization of hESCs cultured in mTeSR™1 medium. (**a**) Phase contrast image of undifferentiated hESC colony. (**b**)–(**d**) showed the marker protein expression profile (Nanog, Oct-3/4, and SSEA-4) of hESCs. Flow cytometry assay (**e**) and karyotype analysis (**f**) further support their undifferentiated phenotype and non-tumorigenesis property, respectively (Passage 40).

4. Fix the cells with 0.5–1 mL of 4% paraformaldehyde for 30 min at room temperature, and wash the cells twice with 1 mL of PBS.

5. Permeabilize the cells with 0.5–1 mL of 0.1% Triton X-100 in PBS at room temperature for 15 min.

6. Block the cells with 0.5 mL of 10% normal donkey serum in PBS at room temperature for 20 min.

7. After blocking, incubate the cells with 50 μL of diluted goat anti-Nanog antibody (1 μg/mL) or goat anti-Oct-3/4 antibody (1 μg/mL) overnight at 4°C.

8. Wash the cells three times with PBS to remove the antibody solution completely.

9. Incubate the cells with 50 μL of diluted rabbit anti-goat IgG-TRITC (1:100) at 37°C for 2 h in the dark.

10. Wash the cells with PBS, and counterstain the cell nuclei with DAPI working solution (1 μg/mL) at room temperature for 15 min.

11. Wash the cells twice with PBS, cover them with mounting medium (or load the glass coverslips onto glass slides), and visualize the cells under a confocal microscope.

3.2.3. Flow Cytometry Assay of SSEA-4

Stage specific embryonic antigen-4 (SSEA-4), an early embryonic glycolipid antigen expressed on the surface of cell membrane, is another important surface marker for pluripotent hESCs. In addition to the aforementioned immunostaining method, flow cytometry assay has been a more efficient way to quantitatively evaluate the phenotype of hESCs based on the SSEA-4 antigen expression level (Fig. 2d, e).

1. Prepare single-cell suspension by trypsinizing the cells with the pre-warmed 0.05% trypsin-EDTA solution (1 mL/well), incubate at 37°C for 5–10 min.

2. Wash the cells with 4–5 mL of FACS buffer twice through centrifugation ($300 \times g$ for 5 min for each). Resuspend the cells in 1 mL of FACS buffer.

3. Count cell number by using hemocytometer. Place aliquot cells ($2–5 \times 10^5$/sample) into 1.5 mL vials and centrifuge at $200 \times g$ for 5 min. Aspirate supernatant and resuspend the cell pellet in a small volume of FACS buffer (~10 μL/sample).

4. Add 10 μL of PE-conjugated mouse-antihuman SSEA-4 per sample, gently mix, and incubate at 4°C for 30 min in the dark.

5. Wash the cells twice with FACS buffer at room temperature, resuspend each sample in 300 μL of FACS buffer, and load it for detection immediately.

3.2.4. Karyotype Analysis

In addition to those marker proteins (or genes) expressed by the undifferentiated hESCs, a normal karyotype, as distinct from that of tumor cells, is a "gold standard" for characterizing undifferentiated hESC. This is important especially when an alteration of culture condition happens to hESCs. Before utilizing the feeder-free cultured hESCs for the study on the lineage-directed differentiation into endothelial cells, we usually maintain H1 cell line in mTeSR™1 medium up to 35–40 passages and conduct the karyotype analysis. The result showed that the feeder-free culture with mTeSR™1 medium be able to support normal karyotype of hESCs for a long-term culture in vitro (Fig. 2f).

1. Prepare the hypotonic solution and cell fixation solution (see Sect. 2).
2. Prepare dividing cells for analysis by splitting the cells 2 days before the harvesting.
3. Add 12 μL of colcemid solution (10 μg/mL) into each 6-well containing 2 mL of culture medium; incubate at 37°C for 20 min.
4. Remove the medium and wash the cells once with PBS.
5. Trypsinize the cells with 0.05% trypsin-EDTA and harvest the cell pellet after centrifugation.
6. Re-suspend the cell pellet with 20 μL of PBS in a 15-mL conical tube by gently tapping the wall of the tube, then add 3–4 mL of pre-warmed hypotonic solution into the tube; incubate at room temperature for 20 min.
7. Centrifuge at $200 \times g$ for 5 min; aspirate the supernatant and resuspend the pellet in 2–3 mL of pre-cold fixation solution (I) by inverting two to three times.
8. Remove solution I through centrifugation; resuspend the cells into 2–3 mL fixation solution (II).
9. Centrifuge to remove the supernatant.
10. Add 2 mL of cell-fixation solution II; re-suspend the cells by inverting the tube for two to three times.
11. Store the cells at 4°C for slide making; check the karyotype under microscope within 1 week.

3.3. Directed Differentiation of hESCs into Endothelial Cells Under a Hypoxic Condition

3.3.1. Embryoid Bodies Formation and Suspending Culture (Timing 0–5 Days)

1. Culture hESCs (H1 cell line) in mTeSR™1 medium as described in Sect. 3.1.
2. When hESC colonies are ready for passage, aspirate the medium and detach the colonies with 1 mL of dispase/well (6-well plate).
3. Gently rinse the plate with 2 mL of DMEM/F12 medium/well, and transfer all the suspending cell clumps into a 50-mL conical tube.

4. Centrifuge at $200 \times g$ for 4 min at room temperature, aspirate the supernatant, and resuspend the cell clumps in 12 mL of EB medium (see Sect. 2).

5. Transfer the cell suspension into a low-attachment 6-well plate (2 mL/well), and culture the cells in a low oxygen incubator with 5% O_2, 5% CO_2, and 90% nitrogen.

6. Check EB formation in 24 h (Fig. 2); refresh the culture medium daily.

3.3.2. Directed Differentiation of hESCs into Endothelial Cells (Timing 6–20 days)

1. Maintain EBs in suspending culture for 5 days with daily medium refreshment.

2. Prepare 0.1% gelatin-coated glass coverslips in 24-well plate; transfer EBs into each well (~5 EBs per well) to allow for their attachment to the surface of coverslips.

3. Carefully add a small volume (~100 μL) of differentiation medium into each well and incubate for 3–4 h to facilitate EB attachment.

4. Add 1.5 mL endothelial cell differentiation medium into each well, and culture the cells in a low oxygen incubator with 5% O_2, 5% CO_2, and 90% nitrogen. Spontaneous differentiation with H-DMEM medium containing 10% FBS serves as control.

5. Maintain the culture for 2 weeks before the cells are harvested for characterization; refresh the differentiation medium every other day during the cultivation.

3.3.3. Characterization of hESC-Derived Endothelial Cells In Vitro

1. *Immunofluorescent staining*
   (a) Collect the differentiated cells on glass cover slips, and gently rinse the slips twice with PBS.
   (b) Fix the cells with 4% paraformaldehyde for 30 min at room temperature, and wash three times with PBS.
   (c) Permeabilize and block the samples as described above (3.2.2).
   (d) Remove the excessive blocking solution, and incubate the cells with diluted antibody CD31 (1:50) and CD144 (1:50) at 4°C overnight.
   (e) Wash cells twice with PBS; incubate cells with secondary antibody TRITC or FITC-conjugated IgG for 2 h at room temperature in the dark.
   (f) Wash twice with PBS; counterstain cell nuclei with DAPI working solution (1 μg/mL) for 15 min at room temperature.
   (g) Wash the cells with PBS, and load the cover slips onto a glass slide with Fluoromount G.
   (h) Check the staining, and capture the images under confocal microscope.

Table 1
Gene probes of real-time RT-PCR assay

Genes	Cat. No.
KDR	Hs00911700 (Applied Biosystems)
CD31	Hs00169777 (Applied Biosystems)
CDH5	Hs00901463 (Applied Biosystems)
vWF	Hs00169795 (Applied Biosystems)
VCAM1	Hs 00365486 (Applied Biosystems)
GAPDH	Hs 99999905 (Applied Biosystems)
Oct-3/4	Hs03005111 (Applied Biosystems)

2. *Gene expression assay by real-time RT-PCR*

   (a) Aspirate the culture medium by vacuum and wash the differentiated cultures twice with PBS.

   (b) Add 0.5 mL of TRIZOL per well to extract total RNA using RNeasy mini kit.

   (c) Determine total RNA content by OD_{260} with a UV spectrophotometer.

   (d) Conduct reverse transcript reaction with 1 μg of RNA using high capacity cDNA reverse transcript kit based on the protocol provided by the manufacturer.

   (e) Set up real-time PCR using 2 μL of cDNA and TaqMan® gene expression assay kit to detect transcript levels of relevant genes. Probes are summarized in Table 1.

   (f) Analyze real-time RT-PCR data by ABI Prism 7000 Sequence Detection Systems version 1.0 software (17, 18). Relative expression level for each target gene is normalized by the Ct value of human GAPDH ($2\Delta^{Ct}$ formula, Perkin Elmer User Bulletin #2). Each sample should be analyzed in triplicate.

3. *Functional characterization of hESC-derived endothelial cells*

   To better characterize the functional activity of the hESC-derived endothelial cells, isolation and enrichment of the target cells (endothelial cells) by fluorescence-activated cell sorting (FACS) is needed. All the following procedures should be operated under a sterile condition.

   (a) Harvest the differentiated cells ($1–2 \times 10^6$) by incubation with 0.05% trypsin-EDTA solution for 5–10 min; prepare single-cell suspension as described before.

   (b) Resuspend the cell pellet with 50 μL of anti-CD31-Alexa Fluor®647 working solution; incubate for 30 min at 4°C in the dark.

(c) Wash the cells twice with FACS buffer; filter the cell suspension through 40 μm cell strainer to remove any cell aggregates.

(d) Transfer cells into the FACS tubes; determine optimized parameters and perform cell sorting.

(e) Collect all the sorted cells and replate them onto fibronectin-coated 6-well plate with EGM-2 medium supplementation. In 1-week cultivation, functional characterization could be performed as follows:

  i. Dil-ac-LDL uptaking

   1. Aspirate the culture medium by vacuum.
   2. Add fresh culture medium containing 10 μg/mL Dil-ac-LDL solution; incubate for 4 h at 37°C in the dark.
   3. Wash the cells twice with PBS and fix the cells with 4% paraformaldehyde for 30 min at room temperature in the dark.
   4. Evaluate the Dil-ac-LDL uptaking by the hESC-derived endothelial cells under a confocal microscope.

  ii. Matrigel assay

   1. Prepare Matrigel-coated 24-well plate and seed the cells at a density of 1×10^5 cells per well.
   2. Culture the cells for 1–2 days in a cell culture incubator.
   3. Observe the cord-like structure formation under a phase-contrast microscope or determine its lumen structure using transmission electron microscope (TEM).

4. Notes

1. It is advisable to aseptically aliquot mTeSR™1 5×supplement and dispase solution (1 mg/mL) into working volume and stored frozen at −20°C. Do not refreeze aliquots after thawing. Both complete mTeSR™ and dispase working solution are stable when stored at 4°C for up to 2 weeks.

2. For coating two 6-well plates, 1 mg of Matrigel™ is enough. Thus, a calculation is needed to determine an appropriate aliquot volume of Matrigel™. The aliquoted Matrigel™ can be stocked at −70°C.

3. The quality or concentration of Matrigel™ varies from batch to batch. It is advisable to prescreen and qualify any new batch of the Matrigel™ prior to use.

4. When preparing more Matrigel™-coated plates for future use, add 1 mL of DMEM/F12 medium into each well and seal the plate with Parafilm® to prevent dehydration. These plates can be stored at 4°C for up to 1 week.

5. No adaptation is required by the hESCs when switching into mTeSR™1 medium from mouse embryonic fibroblast (MEF) feeder culture. However, initially many feeder cells are also observed to attach to the surface of the plate. But they will be removed completely after 2–3 passages in mTeSR™1 medium due to their lost viability.

6. An increased passage number may lead to declined pluripotency of hESCs in vitro. Thus, it is advisable to make a bank of relatively low passage of cells after the cellular characterization (Sect. 3.2).

7. hESCs are sensitive to any chemical reagents. Be sure to remove any residue of dispase during cell passaging or EB preparation with an aim to maintain or improve the viability of hESCs.

8. The reason of splitting cells before colcemid solution treatment is to allow more cells in a dividing phase. This is the key point of sample preparation for karyotype analysis.

9. To improve the yield of EBs, scraping those larger cell colonies and washing the cell clumps with DMEM/F12 medium thoroughly are necessary.

10. The size and seeding density of EBs could affect the differentiation efficiency of hESCs into endothelial cells.

11. To prepare a single-cell suspension for either flow cytometry or karyotype analysis, treatment of low concentration of trypsin-EDTA and pre-resuspend the cell pellet in a small volume of appropriate solution are the key tips to prevent cell aggregation.

12. A prolonged sorting time may cause lower viability of cells. Therefore, it's advisable to finish the sorting within 2–3 h.

13. Contamination is another major concern after cell sorting. Be sure to operate all the procedure under sterile condition.

References

1. Thomson J.A., Itskovitz-Eldor J., Shapiro S.S, et al. (1998) Embryonic stem cell lines derived from human blastocysts. Science 282, 1145–1147.
2. Levenberg S., Ferreira L.S., Chen-Konak L., et al. (2010) Isolation, differentiation and characterization of vascular cells derived from human embryonic stem cells. Nat Protoc 5, 1115–1126.
3. Bai H., Wang Z.Z. (2008) Directing human embryonic stem cells to generate vascular progenitor cells. Gene Ther 15, 89–95.
4. Levenberg S. (2005) Engineering blood vessels from stem cells: recent advances and applications. Curr Opin Biotechnol 16, 516–523.
5. Wang Z.Z., Au P., Chen T., et al. (2007) Endothelial cells derived from human embryonic stem cells form durable blood

vessels in vivo. Nat Biotechnol 25, 317–318.
6. Martin-Rendon E., Snowden J.A., Watt S.M. (2009) Stem cell-related therapies for vascular diseases. Transfus Med 19, 159–171.
7. Levenberg S., Golub J.S., Amit M., et al. (2002) Endothelial cells derived from human embryonic stem cells. Proc Natl Acad Sci USA 99, 4391–4396.
8. Vodyanik M.A., Bork J.A., Thomson J.A., et al. (2005) Human embryonic stem cell-derived CD34+ cells: efficient production in the coculture with OP9 stromal cells and analysis of lymphohematopoietic potential. Blood 105, 617–626.
9. Chadwick K., Wang L., Li L., et al. (2003). Cytokines and BMP-4 promote hematopoietic differentiation of human embryonic stem cells. Blood 102, 906–915.
10. Fischer B., Bavister B.D. (1993) Oxygen tension in the oviduct and uterus of rhesus monkeys, hamsters and rabbits. J Reprod Fertil 99, 673–679.
11. Ezashi T., Das P., Roberts R.M. (2005) Low O2 tensions and the prevention of differentiation of hES cells. Proc Natl Acad Sci USA 102, 4783–4788.
12. Simon M.C., Liu L., Barnhart B.C. et al. (2008) Hypoxia-induced signaling in the cardiovascular system. Annu Rev Physiol 70, 51–71.
13. Ramírez-Bergeron D.L., Simon M.C. (2001) Hypoxia-inducible factor and the development of stem cells of the cardiovascular system. Stem Cells 19, 279–286.
14. Maltepe E., Simon M.C. (1998) Oxygen, genes, and development: an analysis of the role of hypoxic gene regulation during murine vascular development. J Mol Med 76, 391–401.
15. Adelman D.M., Simon M.C. (2001) Hypoxic Gene Regulation in Differentiating ES Cells. In: Turksen K. (ed) Methods in Molecular Biology, Embryonic Stem Cells: Methods and Protocols, Humana Press Inc., NJ, vol. 185, pp. 55–62.
16. Prado-Lopez S., Conesa A., Armiñán A., et al. (2010) Hypoxia promotes efficient differentiation of human embryonic stem cells to functional endothelium. Stem Cells 28, 407–418.
17. Wang Y., Kim H.J., Vunjak-Novakovic G., et al. (2006) Stem cell-based tissue engineering with silk biomaterials. Biomaterials 27, 6064–6082.
18. Wang X., Sun L., Maffini M.V., et al. (2010) A complex 3D human tissue culture system based on mammary stromal cells and silk scaffolds for modeling breast morphogenesis and function. Biomaterials 31, 3920–3929.

Chapter 22

Differentiation of Endothelial Cells from Human Embryonic Stem Cells and Induced Pluripotent Stem Cells

Shijun Hu, Preston Lavinghousez, Zongjin Li, and Joseph C. Wu

Abstract

Endothelial cells line the entire circulatory system and form the interface between the blood vessel intima and the circulating red blood cells. Endothelial cells are crucial to the proper function of the circulatory system and tissue viability, including their roles in coagulation, fibrinolysis, inflammation, and most specifically, vasculogenesis and angiogenesis. Given the importance of endothelial cells in vascular formation, it is essential to expand our knowledge of endothelial cell physiology and growth. The vasculogenic and angiogenic properties that enable new vascular networks to form and consequently perfuse ischemic tissues make endothelial cells an essential element of potential novel therapies. In this chapter, we describe a three-step technique to derive endothelial cells from human embryonic stem cells or induced pluripotent stem cells using a three-dimensional embryoid body formation protocol. A technique to derive a highly pure endothelial population using flow cytometry will also be discussed.

Key words: Human embryonic stem cells, Induced pluripotent stem cells, Endothelial cells, Embryoid body, Angiogenesis, Vasculogenesis

1. Introduction

As one of the main cellular components of blood vessels, endothelial cells play a vital role in vascular pathology. Human embryonic stem cells (ESCs) are pluripotent stem cells derived from the inner cell mass of the blastocyst, which are able to differentiate into all derivatives of the three primary germ layers (1). Human induced pluripotent stem cells (iPSCs) are a type of pluripotent stem cell derived from somatic cells by delivering Oct4, Sox2, Klf4, and c-Myc (2) or Oct4, Sox2, Nanog, and Lin28 (3). Human iPSCs can potentially bypass the ethical debates and immunological rejection associated with human ESCs (4).

The therapeutic applications of human ESC-derived endothelial cells (ESC-ECs) or iPSC-derived endothelial cells (iPSC-ECs) are numerous, including cell transplantation for repair of ischemic tissues and tissue-engineered vascular grafts. Most specifically, their central role in cardiovascular disease makes endothelial cells a logical therapeutic research target because of their potential vasculogenic and angiogenic properties. Adult endothelial cells have shown angiogenic properties in ischemic and infarcted myocardium by improving cardiac function (5–7). A major challenge to establishing stable vascular networks and cell replacement therapies is the limited availability of endothelial cells. Human ESCs and/or iPSCs potentially solve this issue by deriving large numbers of endothelial cells, but they are not without drawbacks. As with any human ESC or iPSC derivatives, the potential of teratoma formation and uncontrolled differentiation must be curbed (8).

Several methods have been successfully used to produce human ESC- and iPSC-derived endothelial cells, including three-dimensional (3D) embryoid body (EB) formation and two-dimensional (2D) culture system (9–14). Feraud et al. showed ESC-derived embryoid bodies developed in collagen gels recapitulate sprouting angiogenesis in vitro (15). When transplanted into SCID mice, human ESC-ECs appear to form microvessels containing mouse blood cells in vivo. In most 3D embryoid body protocols, endothelial differentiation efficiency is low, and in the 2D method by coculture with mouse cells, there is animal material contamination. We have developed an extracellular matrix culture system for increasing endothelial differentiation. This 3D embryoid body formation protocol has three stages: EB formation (stage 1), expansion of endothelial lineage by subculturing EBs in collagen (stage 2), and derivation of a highly pure endothelial population by $CD31^+/CD144^+$ double sorting using flow cytometry (stage 3) (16, 17). Figure 1a shows the diagram of protocol for the endothelial differentiation. Details of the differentiation protocol are described below.

2. Materials

2.1. Maintenance of Human ESCs and iPSCs

1. BD Matrigel™ hESC-qualified Matrix (BD Biosciences, San Jose, CA, USA)
2. Knockout™ D-MEM (Invitrogen, Carlsbad, CA, USA)
3. BD Falcon™ conical tubes (15 mL, 50 mL) (BD Biosciences, San Jose, CA, USA)
4. Tissue culture dish (BD Falcon™, San Jose, CA, USA)
5. Parafilm® M (Pechiney Plastic Packaging Company, Chicago, IL, USA)

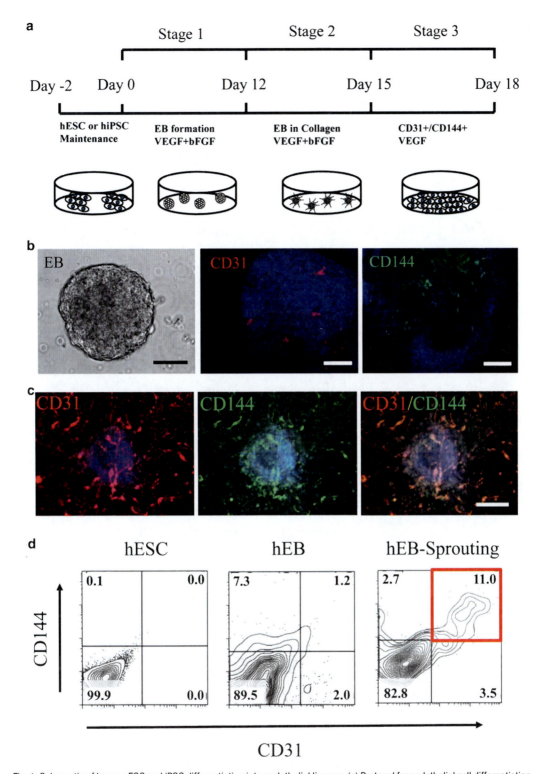

Fig. 1. Schematic of human ESC and iPSC differentiation into endothelial lineage. (**a**) Protocol for endothelial cell differentiation, which includes three stages: (1) EB formation, (2) expansion of endothelial lineage by subculturing EBs in collagen, and (3) derivation and culture of a highly pure endothelial population by CD31+/CD144+ double sorting. (**b**) Whole-mount staining shows 12-day-old EBs express endothelial specific CD31 (*red*) and CD144 (*green*) by tube structure. (**c**) Staining of endothelial differentiated sprouting EBs in collagen shows CD31+/CD144+ cells form vessel structures in the sprouting EBs. (**d**) Kinetic expression of CD31 and CD144 during endothelial cells differentiation by FACS. CD31 and CD144 expressions are not detected in undifferentiated hESC. CD31 and CD144 expression are induced after EB differentiation. The CD31+/CD144+ cells increased to ~11% after EBs are subcultured in collagen. The cells are further isolated by FACS for enrichment.

6. mTeSR®1 medium for maintenance of human ESCs and iPSCs (STEMCELL Technologies, Vancouver, BC, Canada)
7. Dulbecco's Phosphate-Buffered Saline (DPBS) (Invitrogen, Carlsbad, CA, USA)
8. Dispase (Invitrogen, Carlsbad, CA, USA)
9. Corning® cell lifter (Sigma-Aldrich, USA)

2.2. Differentiation of Human ESC-ECs and iPSC-ECs

1. Iscove's Modified Dulbecco's Medium (IMDM) (Invitrogen, Carlsbad, CA, USA)
2. Knockout™ Serum Replacement (KSR) (Invitrogen, Carlsbad, CA, USA)
3. BIT 9500 Serum Substitute (STEMCELL Technologies, Vancouver, BC, Canada)
4. Nonessential amino acids (NEAA) (Invitrogen, Carlsbad, CA, USA)
5. L-Glutamine (Invitrogen, Carlsbad, CA, USA)
6. Monothioglycerol (Sigma-Aldrich, USA)
7. bFGF (R&D Systems, Minneapolis, MN, USA)
8. VEGF (R&D Systems, Minneapolis, MN, USA)
9. Ultralow attachment dish (Corning Incorporated, Corning, NY, USA)
10. Rat tail collagen type I (BD Biosciences, San Jose, CA, USA)
11. EBM-2 basal medium (Lonza, Switzerland)
12. EGM-2 Bullet Kit (Lonza, Switzerland)
13. Collagenase I (Invitrogen, Carlsbad, CA, USA)
14. Librerase blendzyme IV (Roche, CA, USA)
15. 40-μm cell strainer (BD Biosciences, San Jose, CA, USA)
16. FITC Mouse anti-human CD31 (BD Biosciences, San Jose, CA, USA)
17. PE Mouse anti-human CD144 (BD Biosciences, San Jose, CA, USA)
18. Fibronectin (Invitrogen, Carlsbad, CA, USA)

2.3. Biological Characterization of Human ESC-ECs and Human iPSC-ECs

1. Trypsin/EDTA (Invitrogen, Carlsbad, CA, USA)
2. Trypsin Neutralizer Solution (Invitrogen, Carlsbad, CA, USA)
3. Alexa Fluor® 594 AcLDL (Invitrogen, Carlsbad, CA, USA)

3. Methods

3.1. Maintenance of Human ESCs and iPSCs in Feeder-Free Condition

3.1.1. Coating Tissue Culture Dishes with BD Matrigel™ hESC-Qualified Matrix

1. Take out an aliquot of frozen BD Matrigel™ hESC-qualified Matrix from −80°C refrigerator. Thaw it on ice until liquid.
2. Dispense 25 mL cold Knockout™ D-MEM into a 50-mL conical tube and keep on ice.
3. Add the thawed BD Matrigel™ Matrix into the cold Knockout™ D-MEM medium and mix thoroughly.
4. Coat tissue culture dishes with the diluted BD Matrigel™ solution. For 100-mm dish, use 6 mL of diluted BD Matrigel™ solution. Swirl the dish to spread the BD Matrigel™ solution evenly.
5. Coated dish should be left at room temperature for at least 1 h. If not used immediately, the coated dishes must be sealed by Parafilm® M and can be stored at 4°C for at most 1 week.
6. Remove the diluted BD Matrigel™ solution by aspiration before you use the dishes.

3.1.2. Thawing and Seeding of Human ESCs and iPSCs on BD Matrigel™-Coated Dishes in mTeSR®1 Media

1. Wear eye protection and ultralow temperature cryo gloves. Take a vial of human ESCs or iPSCs from liquid nitrogen tank.
2. Immerse the vial in a 37°C water bath without submerging the cap. Swirl the vial gently.
3. Take out the vial from the water bath when only an ice crystal remains.
4. Transfer the contents of the cryovial into a 15-mL conical tube, then add 9 mL mTeSR®1 to the tube, mixing gently as the medium is added.
5. Centrifuge the cells at 200 ×g for 5 min at room temperature.
6. Suck out the supernatant carefully by aspiration.
7. Resuspend the cell pellet with 2 mL mTeSR®1 medium gently. Take care not to break the cell colonies.
8. Remove the Matrigel solution from the 100-mm tissue culture dish and immediately add 8 mL mTeSR®1.
9. Transfer the medium containing the cells to the dishes.
10. Place the dish into the 37°C incubator and move the dish side-to-side, forward-to-backward quickly, in order to distribute the clumps evenly in the dish.
11. Culture the cells at 37°C, with 5% CO_2. Change medium every day. Examine the undifferentiated cell colonies. The cells typically have to split in 3–5 days.

3.1.3. Passaging Human ESCs or iPSCs Growing in mTeSR®1. The Split Ratio is Typically 1:3 or Great than 1:3

1. Aspirate the medium from the dish and rinse with DPBS.
2. Add 6 mL per 100-mm dish of dispase at the concentration of 1 mg/mL. Place at 37°C for 7 min.
3. Remove the dispase and gently wash the dish two times with DPBS.
4. Add 5 mL mTeSR®1 per 100-mm dish and scrape cells with a cell lifter. Transfer the detached cell colonies to a 15-mL conical tube and wash the dish with additional 5 mL of mTeSR®1.
5. Centrifuge the tube at $200 \times g$ for 5 min at room temperature.
6. Aspirate the supernatant carefully and resuspend the pellet with mTeSR®1.
7. Seed the cells into BD Matrigel™-coated dishes. Then culture the cells in an incubator at 37°C with 5% CO_2.

3.2. Differentiation of Human ESC-ECs and iPSC-ECs

3.2.1. EBs Formation (Stage 1)

1. Deattach the human ESCs or iPSCs as discussed in Sect. 3.1.3.
2. Resuspend the colonies in a differentiation medium containing Iscove's Modified Dulbecco's Medium (IMDM) and 15% Knockout™ Serum Replacement (KSR), 1×BIT, 0.1 mM NEAA, 2 mM L-glutamine, 450 µM monothioglycerol, 50 U/mL penicillin, and 50 µg/mL streptomycin, supplemented with 20 ng/mL bFGF and 50 ng/mL VEGF.
3. Seed the cells in 100-mm ultralow attachment dish to form embryoid body.
4. Culture the cells in suspension in an incubator at 37°C, with 5% CO_2.
5. Change medium every other day with differentiation medium until 12 days. Whole-mount immunostaining shows that CD31$^+$ and CD144$^+$ positive cells are organized into tube-like structure in 12-day-old EBs (Fig. 1b).

3.2.2. Sprouting Differentiation (Stage 2)

1. Harvest 12-day-old EBs by spinning at $200 \times g$ for 5 min at room temperature.
2. Aspirate the supernatant carefully and resuspend the EBs with 1.5 mg/mL rat tail collagen type I solution by diluting the collagen stock in differentiation medium described as in Sect. 3.2.1.
3. Mix EBs and collagen type I medium completely. Transfer the 1.5 mL mixture to a 6-well plate.
4. Incubate the dish at 37°C for 45 min to allow polymerization.
5. Add EGM-2 medium and incubate at 37°C with 5% CO2, for 3 days. Figure 1c shows sprouting EB after 3 days of culturing in collagen CD31 and CD144 double positive. FACS analysis indicates CD31$^+$/CD144$^+$ double positive cells at ~11% after subculture in collagen (Fig. 1d).

3.2.3. Dissociation of EB

1. Aspirate the medium carefully. Take care not to destroy the gel.
2. Add 2 mL 0.25% collagenase I by dissolving collagenase I powder in EBM-2 medium at 37°C for 30 min.
3. Then add 2 mL 0.56 units/mL librerase blendzyme IV by dissolving its powder in EBM-2 medium at 37°C for 20 min.
4. After becoming single cell suspension, transfer the mixture into a 15-mL conical tube.
5. Add another 6 mL EGM-2 medium into the conical tube.
6. The mixture medium including single cell suspension is passed through 40-μm cell strainer. Harvest the filtrate including single cells.

3.2.4. Purification of EC with CD31+/CD144+ Double Positive from Adherent EB (Stage 3)

1. Centrifuge the cells at 300 ×g for 5 min. Aspirate supernatant completely and carefully.
2. Rinse the cell with 10 mL DPBS. Spin down the cell at 300 ×g for 5 min and aspirate the supernatant.
3. Resuspend cells to maximum concentration of 1×10^6 cells per 60 μL of DPBS.
4. Add 20 μL of FITC Mouse anti-human CD31 antibody reagent per 1×10^6 cells. Mix briefly. Then add 20 μL of PE Mouse anti-human CD144 antibody reagent to the mixture. Mix thoroughly.
5. Incubate for 30 min on ice, in dark room.
6. Add 1 mL of cold DPBS per 1×10^6 cells and centrifuge the cells at 300 ×g for 5 min.
7. Resuspend the cell pellet in 1 mL of DPBS, centrifuge the cells at 300 ×g for 5 min.
8. Sort CD31+/CD144+ double positive cells with FACS machine from BD Bioscience.
9. Seed the isolated CD31+/CD144+ human ESC-ECs or iPSC-ECs on 4 μg/cm² fibronectin-coated dish in EGM-2 medium. Figure 2a shows the typical morphology of derived ECs.

3.3. Biological Characterization of Human ESC-ECs and iPSC-ECs

3.3.1. Flow Cytometry Analysis

1. Aspirate the supernatant completely.
2. Wash the cells with DPBS and suck out the supernatant.
3. Add 2 mL 0.05% Trypsin/EDTA for one 100-mm dish, and incubate 3 min at room temperature.
4. Add Trypsin Neutralizer Solution to neutralize Trypsin solution.
5. Transfer the liquid mixture into a 15-mL conical tube and spin the cells at 300 ×g for 5 min.

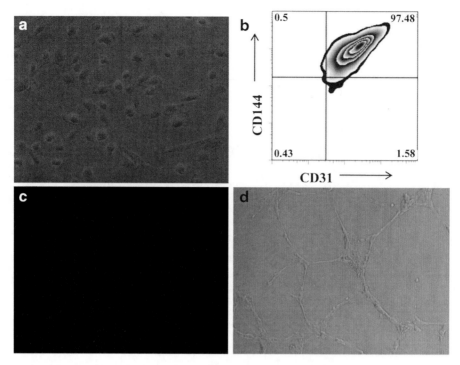

Fig. 2. Characterization of human ESC-ECs. (**a**) Typical cobblestone-like morphology of human ESC-ECs. (**b**) FACS analysis indicates both CD31 and CD144 are expressed on the cell membranes. (**c**) Human ESC-ECs show low-density lipoprotein uptake (*red*). (**d**) In vitro angiogenesis assay indicates ESC-ECs can form tube-like structure on Matrigel.

6. Aspirate the supernatant carefully, and resuspend the cells with DPBS.
7. Centrifuge the cells at 300 ×*g* for 5 min.
8. Resuspend the cell pellets to maximum concentration of 1×10^6 cells per 60 μL of DPBS.
9. Add 20 μL of FITC Mouse anti-human CD31 antibody reagent per 1×10^6 cells. Mix briefly. Then add 20 μL of PE Mouse anti-human CD144 antibody reagent to the mixture. Mix thoroughly.
10. Incubate 30 min on ice, in dark room.
11. Wash the cells with DPBS twice.
12. Analyze the stained cells with flow cytometry from BD Bioscience. Figure 2b shows the derived ECs being >97% $CD31^+/CD144^+$ double positive.

3.3.2. Low-Density Lipoprotein (LDL) Uptake

1. Human ESC- or iPSC-derived ECs are plated in 35-mm fibronectin-coated dish in EGM-2 medium. Incubate the cells overnight at 37°C with 5% CO_2.
2. Aspirate the medium and add 10 μg/mL of Alexa Fluor® 594 AcLDL in EGM-2 medium. Culture the ECs at 37°C for 4 h.

3. Wash the cells with DPBS twice.
4. Fix the cells with 4% Formaldehyde in DPBS for 5 min at room temperature.
5. Wash the cells with DPBS twice.
6. Add DAPI at a concentration of 0.5 µg/mL and incubate for 5 min.
7. Aspirate the DAPI medium, wash the cells with DPBS three times.
8. Check and take pictures under fluorescence microscope. Figure 2c shows Alexa Fluor® 594 AcLDL uptake by ECs (red color).

3.3.3. In Vitro Matrigel Angiogenesis Assay

1. Allow BD Matrigel to thaw on ice. Keep the thawed Matrigel on ice always or it will solidify at room temperature.
2. Coat 24-well plates with 250 µL per well of Matrigel. Let the plates stay at room temperature for at least 30 min to allow gelling.
3. Seed 250 µL of the cell mixture with 50,000 cells into each Matrigel-coated 24-well plate.
4. Incubate the plates overnight at 37°C with 5% CO_2.
5. The next day, gently aspirate the medium from each well, add new medium. Observe tube structures under a microscope. Figure 2d shows the formed vascular tube-like structures on Matrigel.

4. Notes

1. For human ESC digestion, long incubation time will damage the cells. Monitor the cells under a light microscope to avoid over digestion.
2. For medium change during EB formation, both medium and EBs are suspended in 15-mL conical tube. Allow pooled EBs to settle by gravity (about 10 min), then aspirate the supernatant without disturbing EBs. Gently resuspend EBs in 10-mL differentiation medium and transfer to 10-cm ultralow dishes.
3. If the EBs grow in too concentrated conditions, they will tend to form large clumps.
4. During the preparation of collagen I, keep the medium on ice, otherwise collagen will be gelling.
5. The digestion processes of step 2 are harsh. To get high viability cells, shake plate gently during digestion processes, and observe under a light microscope to monitor the processes.

References

1. Thomson, J.A., et al., Embryonic stem cell lines derived from human blastocysts. Science, 1998. 282(5391): p. 1145–7.
2. Takahashi, K., et al., Induction of pluripotent stem cells from adult human fibroblasts by defined factors. Cell, 2007. 131(5): p. 861–72.
3. Yu, J., et al., Induced pluripotent stem cell lines derived from human somatic cells. Science, 2007. 318(5858): p. 1917–20.
4. Sun, N., M.T. Longaker, and J.C. Wu, Human iPS cell-based therapy: considerations before clinical applications. Cell Cycle, 2010. 9(5): p. 880–5.
5. Kocher, A.A., et al., Neovascularization of ischemic myocardium by human bone-marrow-derived angioblasts prevents cardiomyocyte apoptosis, reduces remodeling and improves cardiac function. Nat Med, 2001. 7(4): p. 430–6.
6. Li, Z., et al., Differentiation, survival, and function of embryonic stem cell derived endothelial cells for ischemic heart disease. Circulation, 2007. 116(11 Suppl): p. I46-54.
7. Yu, J., et al., nAChRs mediate human embryonic stem cell-derived endothelial cells: proliferation, apoptosis, and angiogenesis. PLoS One, 2009. 4(9): p. e7040.
8. Kooreman, N.G. and J.C. Wu, Tumorigenicity of pluripotent stem cells: biological insights from molecular imaging. J R Soc Interface, 2010.
9. Li, Z., Z. Han, and J.C. Wu, Transplantation of human embryonic stem cell-derived endothelial cells for vascular diseases. J Cell Biochem, 2009. 106(2): p. 194–9.
10. Wang, Z.Z., et al., Endothelial cells derived from human embryonic stem cells form durable blood vessels in vivo. Nat Biotechnol, 2007. 25(3): p. 317–8.
11. Yamahara, K., et al., Augmentation of neovascularization [corrected] in hindlimb ischemia by combined transplantation of human embryonic stem cells-derived endothelial and mural cells. PLoS One, 2008. 3(2): p. e1666.
12. Levenberg, S., et al., Endothelial cells derived from human embryonic stem cells. Proc Natl Acad Sci USA, 2002. 99(7): p. 4391–6.
13. Chen, T., et al., Stromal cell-derived factor-1/CXCR4 signaling modifies the capillary-like organization of human embryonic stem cell-derived endothelium in vitro. Stem Cells, 2007. 25(2): p. 392–401.
14. Li, Z., et al., Comparison of reporter gene and iron particle labeling for tracking fate of human embryonic stem cells and differentiated endothelial cells in living subjects. Stem Cells, 2008. 26(4): p. 864–73.
15. Feraud, O., Y. Cao, and D. Vittet, Embryonic stem cell-derived embryoid bodies development in collagen gels recapitulates sprouting angiogenesis. Lab Invest, 2001. 81(12): p. 1669–81.
16. Li, Z., et al., Functional and transcriptional characterization of human embryonic stem cell-derived endothelial cells for treatment of myocardial infarction. PLoS One, 2009. 4(12): p. e8443.
17. Li, Z., et al., Functional characterization and expression profiling of human induced pluripotent stem cell- and embryonic stem cell-derived endothelial cells. Stem Cells Dev, 2011. [Epub ahead of print]

Chapter 23

Differentiation of Human Embryonic and Induced Pluripotent Stem Cells into Blood Cells in Coculture with Murine Stromal Cells

Feng Ma, Yanzheng Gu, Natsumi Nishihama, Wenyu Yang, Ebihara Yasuhiro, and Kohichiro Tsuji

Abstract

The establishment of human embryonic stem cell (hESC) lines, as well as the recent induced pluripotent stem cells (hiPSC), has greatly expanded our knowledge about the early development in human ontogeny. In the past decade, hESCs and hiPSCs have been proven excellent tools in characterization of molecular and cellular mechanisms underlying the normal and diseased differentiation of hematopoietic progenitors and mature, functional blood cells. Most of the types of hematopoietic cells (HCs) derived from hESCs have recently been shown with functionally mature properties, including erythrocytes, neutrophils, platelets, megakaryocytes, eosinophils, monocytes, dendritic cells (DC), nature killer (NK) cells, mast cells (MCs), and B- and T-lineage lymphoid cells. Along with the advances in research, a clinical translation of hESC/hiPSC-derived HCs as novel therapies is foreseen in the near future. However, different efficiencies in blood cell production have been reported when using different culture systems. Because of the restriction to use living human embryos, most of the hematopoiesis-inducing cultures are based on murine stromal cells. In our laboratory, we established efficient blood cell-inducing systems by coculturing hESC/hiPSCs with murine fetal stromal cells derived from aorta-gonad-mesonephros (AGM) region and fetal livers. These fetal hematopoietic tissue-derived cells showed strong supporting effects on hESC/hiPSCs, gradually inducing them to terminally mature blood cells if given proper conditions. The murine fetal hematopoietic tissue-derived stromal cells, AGM stromal cells [Xu M et al. Blood 92:2032–2040, 1998], and mid-gestation fetal liver stromal cells [Ma F et al. Blood 97:3755–3762, 2001; Ma F et al. Proc Natl Acad Sci USA 105:13087–13092, 2008] are maintained in our laboratory and radiated right before coculture. When undifferentiated hESC/hiPSC colonies are plated on these stromal cells, they grow up and differentiate to, firstly, a mesoderm-like structure. On days 10–14 in cocultures, some floating cells free themselves from the adherent layer, and they are characterized as hematopoietic progenitor cells. In the second culture system, these hematopoietic progenitors are further induced along to a specific blood cell lineage, such as erythrocytes, MCs, eosinophils, etc. At certain time points, these hESC/hiPSC-derived blood cells are examined with maturity and function. In this chapter, we will describe a coculture protocol developed in our laboratory for differentiating hESC/hiPSCs into hematopoietic cells.

Key words: Human ES cells, Human iPS cells, Murine fetal stromal cells, Hematopoiesis, AGM, Fetal liver, Differentiation

1. Introduction

In a living body, functionally mature blood cells are originated from hematopoietic stem cells (HSCs) residing in the bone marrow (BM). However, during the embryo/fetus stage, the early development of hematopoiesis is a complex progression over time and space. It is believed that the first blood cells appear in the blood islands of the yolk sac, where large nucleated erythroblasts are generated. These yolk sac-derived blood cells represent a primitive wave of the initial hematopoiesis. A second wave of the blood cell generation, also termed as definitive hematopoiesis, has been originated at aorta/gonad/mesonephros (AGM) region, where the HSCs that can reconstitute lethally radiated mice generate. At mid-gestation, fetal liver becomes the predominant location for blood cell production. Finally, the hematopoietic center shifts to BM where HSCs inhabit lifelong. So far, most of our knowledge about early development of hematopoiesis has been accumulated from experiments on mice. Because of the restriction to use living human embryos as the experimental tools, the early genesis of the human embryonic/fetal hematopoietic system is largely unknown.

In 1998, Thomson first established human embryonic stem cell (hESC) lines (4–6). The ESCs derived from the inner cell mass of the human balstocyst are capable of growing indefinitely while maintaining the potential to differentiate into all cell types of the body, including blood cells. This finding greatly expanded our view to elucidate the events in early human ontogeny. The characteristics of hESCs provide two main expectations on hESCs in basic research as well as in clinical applications. On one side, hESCs are an ideal cell source for studying mechanisms responsible for disease development and screening new drugs for treatments of diseases such as diabetes, spinal cord injury, Parkinson's disease, myocardial infarction, and cancers. On the other hand, the pluripotency and embryonic property of hESCs provide a unique model for exploring basic mechanisms underlying early development of human beings. Recently, human induced pluripotent stem cells (hiPSCs) have also been established by reprogramming human somatic cells using defined stemness factors (7–9). The hiPSCs share common features with hESCs in morphology, proliferation manner, feeder cell dependence, gene expression, surface marker property, telomerase activities, differentiation, and teratoma formation in vivo. Although the question whether hiPSCs can fully function as hESCs has still remained elusive, their developmental and differentiation potency comparable to hESCs has already been well defined. Because of the simpleness in establishing cell lines and little ethical concerns, hiPSCs can become a major cell source for regenerative medicine. Since hiPSCs derived from individual patients are disease-tailored stem cells, they provide a new hope for patient-specific therapies for the cure of various diseases. In fact, many disease models have been established using hiPSCs (10–12).

In the past decade, hESCs have been utilized as good models for characterizing molecular and cellular mechanisms responsible for hematopoietic cell development. Various hESC-derived blood cells with functional maturation have been reported by many research groups, including ours. These cells include functionally matured erythrocytes and neutrophils, platelets, megakaryocytes, eosinophils, monocytes, dendritic cells (DCs), nature killer (NK) cells, mast cells (MCs), and B- and T-lineage lymphoid cells.

The methods applied to produce hematopoietic cells from hESC/hiPSCs can be mainly categorized in the formation of embryoid bodies (EBs, three-dimensional colonies of differentiated ESCs) and the coculturing of ESCs with stromal cells. In in vitro culture, hESCs develop into EBs, a sac-like structure mimicking early development of embryonic microenvironments, providing a suitable condition for spontaneous differentiation. Because the hESC-derived EBs mimic the early yolk sac structure, when hematopoiesis-directing factors are added, the primitive blood cells can be generated from hESCs. However, because the accessibility of external factors in this complex structure is limited, EBs may be disadvantageous in regulating the development of hESCs toward definitive hematopoiesis to generate functionally matured blood cells (13). In addition, a more subtle and efficient way to generate mature blood cells from ESCs can be achieved by coculturing with stromal cells derived from fetal/newborn hematopoietic niches. There are a variety of cell lines employed in coculture systems with mouse and human ESCs, among them the OP9 being most widely used (14–20).

We have reported efficient methods to direct human or non-human primate ESCs differentiating into blood cells by coculture with mouse AGM region-derived and fetal liver-derived stromal cells (AGMS and mFLSC, respectively) (1–3, 21, 22). Coculture with these mouse fetal hematopoiesis-breeding tissue stromal cells supports hematopoietic development from hESC/hiPSCs to generate functionally mature blood cells. In this chapter, we introduce the coculture method along with the procedure for establishing the mFLSC. The derivation of AGMS cells has been described elsewhere (1). The characterization of hESC/hiPSC-derived hematopoietic cells is also discussed.

2. Materials

2.1. hESC and hiPSC Lines

1. hESC line, H1 can be acquired from the WiCell Research Institute (Medison, WI). It has been maintained in our laboratory for more than 9 years.

2. hiPSC lines (253G1, 253G4, 201B6, 201B7; kindly provided by Professor S Yamanaka at Center for iPS Cell Research and Application, Kyoto University, Japan).
3. H1 maintaining medium:
   - DMEM/F12 (Invitrogen, Cat. No. 11330)
   - KSR (Invitrogen, Cat. No. 10828)
   - 2-ME (Wako, Cat. No. 137-06862)
   - L-Glutamine (Invitrogen, Cat. No. 25030)
   - Nonessential Amino Acid Solution (Invitrogen, Cat. No. 11140)
   - Basic FGF (Wako, Cat. No. 068-04544)
4. hiPSCs maintaining medium
   - DMEM/F12 (Sigma, Cat. No. D6421)
   - KSR (Invitrogen, Cat. No. 10828)
   - 2-mercaptoethnol (2-ME. Wako, Cat. No. 137-06862)
   - L-Glutamine (Invitrogen, Cat. No. 25030)
   - Nonessential Amino Acid Solution (Invitrogen, Cat. No. 11140)
   - Basic FGF (Wako, Cat. No. 068-04544)
5. Gelatin-coated culture dishes and plates (Sumilon, MS-0390G/10 cm dish; MS—0006G/6-well plate).
6. Trypsin/EDTA solution (WAKO, 202–16931/0.05%, 209–16941/0.25%).
7. Scraper (IWAKI, Asahi Glass Co. TDT, Cat. No. 9000-220).

2.2. Murine AGMS-3 Cell Line

1. The murine AGM region-derived stromal cell lines, AGMS-3, had been established and maintained in our laboratory since 1998, and their hematopoiesis-supporting potential has not been decreased (1) (Fig. 1a).
2. Gelatin-coated culture dishes (Sumilon, Cat. No. MS-0390 G).
3. AGMS maintaining medium:
   - α-Minimum essential medium (α-MEM) (Invitrogen, Cat. No. 12571).
   - Fetal Bovine Serum (FBS) (Hyclone, Cat. No. SH-30396) 10% in volume.

2.3. mFLS Cells

1. Mice (Pregnant day 14–15, Strain: C57/Black 6) (Fig. 1b).
2. Ophthalmology surgery scissors and forceps (Autoclaved before use).
3. Stereomicroscope (Olympus Optical Co. LTD. Model: SZX-ILLK100).
4. Sterile culture dish (Corning Incorporated, Cat. No. 430167).

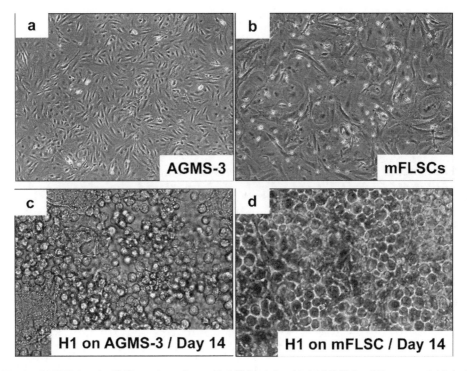

Fig. 1. Photos of AGMS-3 and mFLSCs and coculture with hESCs at day 14. (**a**) AGMS-3 cell line was established in our laboratory in 1998 and has been maintained without loss in hematopoiesis-supporting potential. (**b**) The mFLSCs were derived from E14.5 C57/Black 6 mouse fetal livers and maintained at 1P. (**c**) Coculture of hESCs (H1) with AGMS-3 cells at day 14, showing a robust proliferation of cobblestone-like hematopoietic cells. (**d**) Coculture of hESCs (H1) with mFLSCs at day 14, showing a robust proliferation of cobblestone-like hematopoietic cells.

5. 6-well culture plates (Sumilon, Cat. No. MS-0006G).
6. D-PBS(−) (Wako, Cat. No. 045-29795).
7. 0.05% Trypsin/EDTA solution (WAKO, 202–16931).
8. mFLSC culture medium:
   - Dulbecco's Modified Eagle's Medium (DMEM, Sigma, Cat. No. D5796).
   - FBS (Hyclone) 10% in volume.

2.4. Induction of Hematopoietic Progenitor Cells

1. Radiation system (Gamma Cell 40, Canadian Nuclear Co. LTD.)
2. Gelatin-coated 6-well culture plates (Sumilon, Cat. No. MS-0006G)
3. Undifferentiated hESCs, hiPSCs
4. Radiated AGMS cells or mFLSCs (Radiation dose: 15Gy for AGMS and 25Gy for mFLSCs)
5. Hematopoiesis-inducing medium in coculture
6. Iscove Modified Dulbecco's Medium (IMDM; Sigma, Cat. No. 13390)

- FBS (10% in volume, Hyclone)
- Nonessential Amino Acid Solution (Invitrogen, Cat. No. 11140)
- 2-ME (Wako, Cat. No. 137-06862)
- Glutamine (200 mM, WAKO, Cat. No. 073-05391)
- Vascular endothelial growth factor (VEGF, PeproTech EC LTD, Cat. No. 100-20)

2.5. Flow Cytometric (FCM) Analysis of hESC/hiPSC-Derived Hematopoietic Cells

1. 0.05% Trypsin/EDTA solution (Wako, Cat. No. 202-16931)
2. Sorting medium (SM):
   - D-PBS(-) (Wako, Cat. No. 045-29795)
   - 5% FBS (Hyclone)
3. Antibodies:
   - Mouse antihuman CD45, phycoerythrin (PE)-conjugated (DAKO Cytomation, Cat. No. Nr. R7087)
   - Mouse antihuman Glycophorin A (GPA), PE-conjugated (DAKO Cytomation, Cat. No. Nr. R 7078)
   - Mouse antihuman CD34, Fluorescein isothiocyanate (FITC)-conjugated (BD Bioscience, Cat. No. 348053)
   - Mouse antihuman CD31, FITC-conjugated (BD Bioscience, Cat. No. 555445)
   - Flow cytometry system (FACS Calibur, Becton Dickinson Company)

2.6. Hematopoietic Colony Formation Assay of hESC/hiPSC-Derived Hematopoietic Cells

1. Methylcellulose (SM-4000, Shin-etsu Chemistry, Tokyo)
2. α-MEM (Invitrogen, Cat. No. 12571)
3. FBS (Hyclone, Cat. No. SH-30071; 56°C/30 min heat inactivated)
4. Bovine serum albumin (BSA, Sigma, Cat. No. A-4161)
5. 2-ME (Wako, Cat. No. 137-06862)
6. Hematopoietic growth factors:
   - Stem Cell Factor (SCF, Wako, Osaka, Japan. Cat. No. 199-12813)
   - Interleukin 3 (IL-3, provided by Kirin Brewery Company, Tokyo, Japan)
   - IL-6 (PeproTech, NJ, USA, Cat. No. 200-06)
   - Flt-3 ligand (FL, Wako, Osaka, Japan, Cat. No. 061-04051)
   - Trombopoietin (TPO, provided by Kirin Brewery Company, Tokyo, Japan)
   - Erythropoietin (EPO, provided by Kirin Brewery Company, Tokyo, Japan)

- Granulocyte colony stimulating factor (G-CSF, provided by Kirin Brewery Company, Tokyo, Japan)

7. hESC/hiPSC-derived hematopoietic cells (routinely, after 0.05% trypsin/EDTA treatment; total harvested cells were used)
8. 35 mm Petri dish (FALCON, Becton Dickinson Labware, Cat. No. 35-1008)
9. 2-mL plastic syringe (Top Surgical Taiwan Corporation)
10. 9, 18 gauge syringe needle (TERUMO, Cat. No. 1838S)

3. Methods

Although the differentiation potential toward hematopoiesis varies among cell lines of hESCs and hiPSCs, a successful induction of hematopoietic cells from hESC/hiPSCs is mostly depended on the quality of the mouse fetal stromal cells. The stromal cell lines derived from the microenvironment of hematopoietic centers at midgestation, AGMS-3 and mFLSC, should be maintained to the best to keep their primary property as in the fetus circumstances. Cautions need to be taken when maintaining these fetal stromal cells. First, the fetal stromal cells should be at low passage and stored in liquid nitrogen. Second, do not let the cells be cultured too long. We commonly culture the cells within 1 week to maintain or to passage them again. Third, to control the growth rate, FBS concentration needs to be adjusted. A routine maintaining medium of AGMS-3 and mFLSC contains 10% of FBS. However, to ensure the viability and activity, a higher concentration of FBS may be used when preparing stromal cells for coculture with undifferentiated hESC/hiPSCs. Fourth, apparently, mFLSCs derived from different mouse strains and different fetal time points show diverse potentials in supporting hematopoiesis. We routinely use C57/Black 6 strain to prepare fetal livers at embryonic day 14 (E14) to embryonic day 15 (E15).

Since the AGMS-3 cells had already been established in our laboratory some decades ago, we omit the description of its establishment (please refer to 1, Fig. 1). The methodology that is described below has been proved efficient to isolate and maintain a good stromal cell line from E14~E15 mFLSC (Fig. 1). The established mFLSCs can strongly induce hematopoietic differentiation when cocultured with undifferentiated hESC/hiPSCs (3, 22).

Because there is some difference in lots and stocks of mFLSCs, it is difficult to compare the hematopoiesis supporting potential between AGMS-3 and mFLSCs. Although both stromal cell lines give rise to a good induction of hematopoietic cells from hESC/hiPSCs, to our experiences, coculture with AGMS-3 generates more stable blood cell production, while coculture with mFLSCs shows a more diverse blood cell production. However, the latter

328   F. Ma et al.

generate huge erythrocyte burst forming units (BFU-E) compared to the former, providing an experimental model for investigating erythrocyte development (3).

3.1. Maintenance of hESC and hiPSC Lines

1. The hESC line (H1) can be maintained and passaged weekly on irradiated mouse embryonic fibroblast (MEF) feeder cells, as described in (4).
2. The hiPSCs (Lines 253G1, 253G3, 201B6, 201B7) can be maintained and passaged every 5 days on irradiated MEF cells, as described in (7). Lines 253G1 and 253G4 can be established by transfecting with three factors (Oct4, Sox 2, and EKLF4), while lines 201B6 and 201B7 can be established with one more factor (c-Myc).

3.2. Establishment of mFLSC Line

The procedure has largely based on our previous work with some modifications (2). The procedure for preparation of the mFLCs is depicted in Fig. 2.

1. Sacrifice pregnant mouse at E14-15 by cervical dislocation using a forceps.
2. Spray enough 75% ethanol to cover the whole body.
3. Lay the mouse on an autoclaved aluminum sheet in a clean bench, wait for 2–4 min to let the ethanol evaporated.
4. Use a scissor and a forceps to cut a skin section in the middle of the abdomen.

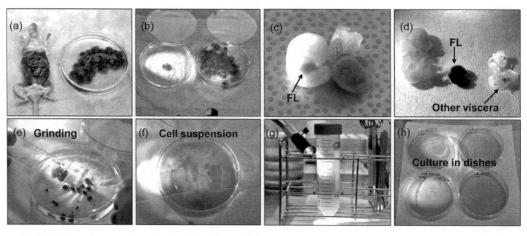

Fig. 2. Establishing mFLSC lines from E14 ICR fetus. (a) After sacrifice, uterus was taken out from the pregnant mouse. (b) Separating individual fetus with amnion and placenta. This ensures the viability of the fetuses. (c) Using forceps to clear away the amnion and placenta. Free the fetus. (d) Take out fetal liver by fine forceps (under stereomicroscope). (e) Grinding the fetal liver lobes between two sterile glass slides (using the ground surface). (f) The fetal liver cell suspension after grinding and pipetting. (g) Collect fetal liver cells into a 50-mL tube and centrifuge at 1500 rpm for 8 min. (h) Culture the fetal liver cells in gelatin-coated culture dishes.

5. Strip off the skin to both opposite terminals over to the neck and rear legs and open the abdomen. Be careful not to let the skin and fur touch on the abdomen membrane.
6. Use another sterile scissor to cut open the abdomen membrane at the middle.
7. Carefully pick up the uterus and cut along the uterus mesentery to free the uterus. Cut off the uterus at the cervix.
8. Move the uterus to a 10 cm culture dish with 20 mL PBS (−), wash once and move to a new dish.
9. Use a fine ophthalmologic scissor and forceps to cut open the uterus. Be careful not to damage the fetus.
10. Use a pair of forceps to separate the whole fetus entirely with placenta and amnion and then move the fetus to a new dish.
11. Use a pair of fine forceps to tear open the amniotic membrane and free the fetus.
12. Cut the umbilical cord with forceps, wash the fetus by gently moving the fetus in PBS (−) to let the umbilical cord blood drain out.
13. Move the fetus to a new dish with PBS (−). The fetus older than E14 can be lifted by gently gripping the upper leg with a fine forceps.
14. Use a pair of fine forceps to tear off the abdomen of the fetus and take out the viscera. Be careful not to damage the fetal liver, which can be easily recognized by its red-colored mass at the middle of the right side.
15. Under a stereomicroscope, remove the mesenteric tissue of the other organs and collect the fetal liver masses in another new dish.
16. Use two glass slides (with one end ground) to triturate the fetal liver masses gently to and fro beneath the ground surfaces until they become melted in the suspension.
17. Mix the fetal liver cell suspension by pipetting and harvest the whole suspension to a 50-mL Falcon tube. Centrifuge at 1,500 rpm for 8 min.
18. Remove the supernatant and add in culture medium (10% FBS-DMEM). Culture the fetal liver cells in 10-cm culture dishes that are coated with gelatin. Routinely, 2–3 whole fetal livers from E14 fetuses are cultured in one 10-cm culture dish (60 cm^2).
19. Do not touch the culture dish within the first 24 h and add 5 mL fresh medium after then. On the third day of the culture, remove the supernatant with floating cells (mostly consisted of erythrocytes and other myeloid cells). Wash once with PBS (−) and add in fresh medium.

20. When the fetal liver-derived stromal cells reach a confluent condition, harvest the cells by treating with 0.25% trypsin/EDTA solution and replate into a new culture dish.

21. Routinely, from primary culture to passage 1 (P-1), the mFLSCs will continually proliferate, and the ratio for the passage is about 1.5–2 in area. However, after P-1, the proliferation of the mFLSCs is drastically decreased over the passaging. This is especially dominant when frozen mFLSCs (after P-1) are thawed and re-cultured. In this situation, a reduction in area ration should be accounted.

22. In culture, a good P-1 mFLSC line consists of a loosely distributed stromal cell layer, even in cell size at large (Fig. 1b). They are not easily to be compacted or occupied by narrow and paralleling fibroblastic cells.

3.3. Coculture of Undifferentiated hESC/hiPSCs with AGMS-3 or mFLSCs

This coculture method has been continuously used in our laboratory to achieve an efficient hematopoietic differentiation from various human and nonhuman primate ESCs (3, 21, 22) or iPSCs. It provides a stable and balanced development of hESC/hiPSCs toward maturation of blood cells, mostly myeloid and erythroid progenitor cells in serum-containing culture system that are comparable to those derived from human cord blood (CB) CD34$^+$ cells. The hESC-derived erythrocytes developed in the coculture with mFLSCs show typical patterns in the maturation pathway, from a primitive wave that mostly expresses embryonic globin to a later mature stage that expresses definitive β-globins. The hESC-derived erythrocytes not only phenotypically mimic the mature erythrocytes derived from CB CD34$^+$ cells but also exert functions as release of oxygen (3). This demonstrates that this coculture system mostly mimic the developmental processes happening in human early embryonic hematopoiesis and may be an excellent experimental model for research on human embryonic/fetal development.

1. Prepare AGMS-3 cells or mFLSCs in gelatin-coated 6-well culture plates. Routinely, after thawing, $1-2 \times 10^5$ cells per one 6-well are cultured for 2 days to reach a good confluence.

2. Radiate the confluent cells at 15Gy for AGMS-3 and 25Gy for mFLSCs. After radiation, the stromal cells can be used at anytime within 5 days. Right before coculture with hESC/hiPSCs, the culture medium should be exchanged to ESC medium before 1–2 h. This step ensures a smooth adaptation of the ESCs to the new matrix.

3. Pick up undifferentiated hESC/hiPSC colonies by poking the colony free with a pipet tip under a reverse microscope in a clean bench. For harvesting big colonies, it is practical to cut the colony first into several small pieces then harvest them.

4. Gently collect the masses of lifted hESC/hiPSC colonies and wash once with fresh hESC medium.

5. Plate the undifferentiated hESC/hiPSCs onto radiated AGMS-3 cells or mFLSCs. For the first 2–3 days, change the culture medium by adding fresh hESC medium everyday. When the hESC/hiPSC colonies keep growing bigger and still appear as undifferentiated, exchange the medium to 10% FBS-IMDM inducing medium. After then, change culture medium every 2 days.

6. During the first 4–6 days, the hESC/hiPSC colonies grow larger and begin to differentiate, with the outskirt looking like a mesoderm differentiation. The hematopoietic cells appear at around day 8–10, commonly at the outskirt area of an expanding colony. These hESC/hiPSC-derived hematopoietic cells are having typical phenotype of hematopoietic progenitors, small, round in shape, and with no cytoplasm granules. Our primitive experiments showed that they mostly consist of hemangioblastic progenitors that can both give rise to mature blood cells and endothelial cells (22).

7. After culture day 10, these hematopoietic cells proliferate rapidly and begin to differentiate to myeloid progenitor cells, some of them showing granules in their cytoplasm. According to our data, on days 12–14 the production of the hematopoietic cells reaches a peak and gradually decreases in number after then.

8. Routinely, we harvest the cells on day 14 of the culture to investigate their maturation in various lineage-specific culture systems. After wash once with PBS (−), 0.3 mL of 0.05% trypsin/EDTA solution is added in one 6-well and then incubated in 37°C for 5 min. Total coculture cells are then collected by pipetting and moved to a 15-mL tube through a 70-μm cell strainer to collect the single cells in suspension. Cells are then put on ice for further analysis.

9. Figure 1c and d shows the generation of cobblestone-like hematopoietic cells derived from hESCs (H1) that cocultured with AGMS-3 or mFLSCs for 14 days, mostly consisting of CD34+ progenitor cells when checked by immunostaining (22).

3.4. Characterization of hESC/hiPSC-Derived Hematopoietic Cells by FACS Analysis

A representative hESC/hiPSC-AGMS-3 coculture result is shown in Fig. 3.

1. Divide the total harvested coculture cells into $0.5–1 \times 10^6$ per sample in 0.1 mL volume.

2. Pre-incubate the cells with 10 μL normal rabbit serum in each sample to block non-specific binding.

3. Wash once with SM and refresh with 0.1 mL SM.

4. Add monoclonal antibodies that conjugated with FTIC, PE, or APC. Stain the cells on ice in a dark place for 30 min.

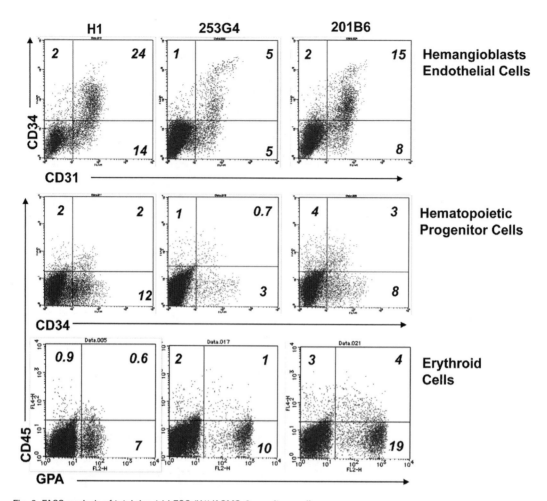

Fig. 3. FACS analysis of total day 14 hESC (H1)/AGMS-3 coculture cells.

5. Wash once with SM, refresh with 0.5–1 mL SM per sample and pass the cell suspension through a 40-μm cell strainer. Put the stained samples on ice. Before analysis, add PI (propidium iodide, finally 1 μg/mL) to stain the dead cells.

6. Analyze the stained samples on a Calibur II by using a CellQuest software (BD). The living cells are gated by PI negative fraction. Record of the data can be further analyzed by either CellQuest or FlowJo software (Tomy Digital Biology Co., LTD).

3.5. Colony Formation Analysis of hESC/hiPSC-Derived Hematopoietic Progenitors

The colony assay is based on the method established by M. Ogawa and T. Nakahata (23–25). This method provides a useful approach to subtly evaluate the potential of hematopoiesis. By addition of a cytokine cocktail favoring hematopoietic cell development, most of the myeloid and erythroid cells can be generated in the colony assay system. These individually developed hematopoietic colonies are derived from single cells and thus specify a lineage potential at clone level. Table 1 gives a result of a representative experiment

Table 1
Hematopoietic colony assay of hESC/hiPSC-AGMS-3 coculture cells

G	Mφ	GM	Mix	CFU-E	BFU-E	Total
hiPSC(253 G1)-derived colonies						
24.0 ± 3.6	15.3 ± 3.1	68.8 ± 10.3	5.3 ± 0.5	17.8 ± 7.4	(C+B)	131.0 ± 18.5
hiPSC(253 G4)-derived colonies						
27.0 ± 8.3	16.3 ± 2.9	43.3 ± 6.8	4.8 ± 1.3	13.8 ± 6.5	(C+B)	105.2 ± 20.5
hESC(H1)-derived colonies						
18.5 ± 3.1	14.5 ± 4.0	21.8 ± 2.9	5.5 ± 1.3	50.3 ± 5.6	10.8 ± 3.0	121.3 ± 5.1

hESCs (H1) or hiPSCs (253 G1 & 253 G4) were cocultured with AGMS-3 for 13 days. Total coculture cells were harvested by 0.05% trypsin/EDTA solution, and 1/4 of a 6-well coculture was planted in a 35-mm Petri dish ($\approx$ from 2,500 undifferentiated hESCs or hiPSCs, or 2.5×10^5 total coculture cells) for colony assay

Hematopoietic colony formation was stimulated by a cocktail of seven factors: SCF, IL-3, Flt3-ligand, IL-6, TPO, EPO, and G-CSF

from day 13 hESC/hiPSC-AGMS-3 cocultures. The recognition of the colony types is based on the criteria established by T. Nakahata and M. Ogawa (23–25).

1. Prepare a mixture of semisolid cell culture. The components for a 5-mL semisolid culture are listed as below:
    - 2 mL α-methylcellulose (Final concentration: around 1%)
    - 0.5 mL BSA solution (10% solution) (Final 1%)
    - 1.5 mL heat-inactivated FBS (Final 30%)
    - 0.05 mL 2-ME/α-MEM (3 μL original 2-ME solution in 4.3 mL α-MEM) (Final 10^{-4} mM in culture)
    - rhSCF 500 ng (Final 100 ng/mL)
    - rhFL 50 ng (Final 10 ng/mL)
    - rhIL-3 50 ng (Final 10 ng/mL)
    - rhIL-6 250 ng (Final 50 ng/mL)
    - rhTPO 50 ng (Final 10 ng/mL)
    - rhEPO 20U (Final 4 U/mL)
    - rhG-CSF 50 ng (Final 10 ng/mL)
    - Harvested total coculture cells
    - Add α-MEM to 5 mL
2. Mix well using a 2-mL syringe and an 18-G needle.
3. Drastically shake the tube to further mix the culture and put away for 20 min to let the bulbs diminish.
4. Use an 18-G needle and 2-mL syringe to tranfer1 mL of the culture to a 35-mm Petri dish. From 5 mL preparation, 4 mL aliquots can be made.

5. Put two culture dishes and another with distilled water into a 10-cm dish. Culture at 37°C in a 5% CO_2 incubator.

6. Observation and calculation of colonies can be applied around days 10–15.

References

1. Xu M, Tsuji K, Ueda T. et al. (1998) Stimulation of mouse and human primitive hematopoiesis by murine embryonic aorta-gonad-mesonephros-derived stromal cell lines. Blood 92, 2032–2040.
2. Ma F, Wada M, Yoshino H. et al. (2001) Development of human lymphohematopoietic stem and progenitor cells defined by expression of CD34 and CD81, Blood 97, 3755–3762.
3. Ma F, Ebihara Y, Umeda K. et al. (2008) Generation of functional erythrocytes from human embryonic stem cell-derived definitive hematopoiesis. Proc Natl Acad Sci USA 105, 13087–13092.
4. Thomson JA, Itskovitz-Eldor J, Shapiro SS. et al. (1998) Embryonic stem cell lines derived from human blastocysts. Science 282, 1145–1147.
5. Bongso A, Fong CY, Ng SC, et al. (1994) Isolation and culture of inner cell mass cells from human blastocysts. Hum Reprod 9, 2110–2117.
6. Reubinoff BE, Pera MF, Fong CY, et al. (2000) Embryonic stem cell lines from human blastocysts: somatic differentiation in vitro. Nat Biotechnol 18,399–404.
7. Takahashi K, Tanabe K, Ohnuki M.et al. (2007) Induction of pluripotent stem cells from adult human fibroblasts by defined factors. Cell 131, 861–872.
8. Yu J, Vodyanik MA, Smuga-Otto K, et al. (2007) Induced pluripotent stem cell lines derived from human somatic cells. Science 318,1917–1920.
9. Park IH, Zhao R, West JA, et al. (2008) Reprogramming of human somatic cells to pluripotency with defined factors. Nature 451,141–146.
10. Hanna J, Wernig M, Markoulaki S, et al. (2007) Treatment of sickle cell anemia mouse model with iPS cells generated from autologous skin. Science 318, 1920–1923.
11. Wernig M, Zhao JP, Pruszak J, et al. (2008) Neurons derived from reprogrammed fibroblasts functionally integrate into the fetal brain and improve symptoms of rats with Parkinson's disease. Proc Natl Acad Sci USA 105, 5856–5861.
12. Dimos JT, Rodolfa KT, Niakan KK, et al. (2008) Induced pluripotent stem cells generated from patients with ALS can be differentiated into motor neurons. Science 321, 1218–1221.
13. Sakamoto H, Tsuji-Tamura K, Ogawa M, et al. (2010) Hematopoiesis from pluripotent stem cell lines. Int J Hematol 91, 384–391.
14. Nakano T, Kodama H, Honjo T. (1994) Generation of Lymphohematopoietic Cells from Embryonic Stem Cells in Culture. Science 265, 1098–1101.
15. Mukouyama Y, Hara T, Xu M. et al. (1998) In Vitro Expansion of Murine Multipotential Hematopoietic Progenitors from the Embryonic Aorta–Gonad–Mesonephros Region. Immunity 8, 105–114.
16. Slukvin II, Vodyanik MA, Thomson JA, et al. (2006) Directed differentiation of human embryonic stem cells into functional dendritic cells through the myeloid pathway. J Immunol 176, 2924–32.
17. Umeda K, Heike T, Yoshimoto M, et al. (2004) Development of primitive and definitive hematopoiesis from non-human primate embryonic stem cells in vitro. Development 131, 1869–1879.
18. Gaur M, Kamata T, Wang S, et al. (2006) Megakaryocytes derived from human embryonic stem cells: a genetically tractable system to study megakaryocytopoiesis and integrin function. J Thromb Haemost 4, 436–442.
19. Takayama N, Nishikii H, Usui J.et al. (2008) Generation of functional platelets from human embryonic stem cells in vitro via ES-sacs. VEGF-promoted structures that concentrate hematopoietic progenitors. Blood 111, 5298–5306.
20. Timmermans F, Velghe I, Vanwalleghem L.et al. (2009) Generation of T Cells from Human Embryonic Stem Cell-Derived Hematopoietic Zones. J Immunol 182, 6879–6888.
21. Ma F, Kambe N, Wang D, et al. (2008) Direct development of functionally mature tryptase/chymase double-positive connective tissue-type mast cells from primate embryonic stem cells. Stem Cells 26, 706–714.

22. Ma F, Wang D, Hanada S, et al. (2007) Novel Method for Efficient Production of Multipotential Hematopoietic Progenitors from Human Embryonic Stem Cells. Int J Hematol 85, 371–379.
23. Nakahata T, Ogawa M. (1982) Hemopoietic colony-forming cells in umbilical cord blood with extensive capability to generate mono- and multipotential hemopoietic progenitors. J Clin Invest 70, 1324–1328.
24. Nakahata T, Spicer SS, Cantey JR, Ogawa M. (1982) Clonal assay of mouse mast cell colonies in methylcellulose culture. Blood 60, 352–361.
25. Nakahata T, Ogawa M. (1982) Identification in culture of a class of hemopoietic colony-forming units with extensive capability to self-renew and generate multipotential hemopoietic colonies. Proc Natl Acad Sci USA 79, 3843–3847.

Chapter 24

Generation of Multipotent CD34⁺CD45⁺ Hematopoietic Progenitors from Human Induced Pluripotent Stem Cells

Tea Soon Park, Paul W. Burridge, and Elias T. Zambidis

Abstract

Human embryonic stem cells (hESCs) and patient-specific human induced pluripotent stem cells (hiPSCs) are valuable reagents for studying the earliest stages of hematopoietic genesis and for modeling the developmental basis of hematologic disorders, and they may also have the potential for generating an unlimited supply of autologous transplantable hematopoietic stem cells. We have previously described the development phases of hematopoiesis that arise from differentiated hESCs using a human embryoid body (hEB) culture system. This hEB system produces hematopoietic and endothelial progenitors through a hemangioblast intermediate. In this chapter, we describe a modified and optimized version of this hEB system that generates high frequencies of multipotent CD34⁺CD45⁺ hematopoietic progenitors *via* employing hEB hematopoietic differentiation followed by adherent hemogenic endothelium generation. Multipotent hematopoietic progenitors can be reproducibly generated in robust amounts with this differentiation system from a wide variety of hiPSC or hESC for biochemical, transcriptomic, epigenetic, and transplantation studies.

Key words: Induced pluripotent stem cells, Human embryonic stem cells, Embryoid body, Hematopoiesis, Hemogenic endothelium, HSC

1. Introduction

The efficient generation of hematopoietic progenitors from patient-specific induced pluripotent stem cells (hiPSCs) (1, 2) has enormous potential for treating a multitude of hematologic and vascular disorders. Moreover, directed hematopoietic differentiation of human pluripotent stem cells provides an unprecedented tool for studying normal human developmental hematopoiesis as well as for the modeling of hematologic disorders (3). However, significant technical challenges remain for efficiently differentiating transplantable hematopoietic stem cells from human pluripotent stem cells. For example, we previously demonstrated that human embryonic stem cell (hESC) differentiation produces blood cells similar to those found during embryonic development and has limited engraftment potential (4, 5). A sophisticated developmental biologic approach is necessary for elucidating effective strategies that can differentiate hiPSC into adult-type transplantable HSC comparable to those normally found in adult bone marrow. Interestingly, recent studies have suggested that iPSCs generated from a hematopoietic origin may retain an epigenetic memory of their origins and possess a natural propensity for hematopoietic differentiation (6). Thus, an efficient and reproducible protocol for generating large numbers of hematopoietic progenitors from a wide variety of human pluripotent stem cells will serve as a useful tool for probing the epigenetic mechanisms that regulate hiPSC pluripotency and hematopoietic differentiation. In this chapter, we provide an optimized protocol developed in our laboratory for the generation of large numbers of multipotent $CD34^+CD45^+$ hematopoietic progenitors from both hESCs and hiPSCs. This method is modified from our previously described human embryoid (hEB) body hematopoietic differentiation system in which hematopoietic lineages emerge from a hemangioblast intermediate (5). We have further evolved this original hEB system into a protocol where robust amounts of multipotent CD34+CD45+ hematopoietic progenitors arise from hemogenic endothelial precursors expanding in endothelial culture conditions. Multipotent hematopoietic progenitors generated with this system can be quantitated by FACS analysis and colony-forming cell (CFC) assays.

2. Materials

2.1. Materials for Generation of Human Embryoid Bodies That Are Competent for Hematopoietic Differentiation

2.1.1. Cell Culture Reagents

1. Dispase I (Invitrogen, cat. no. 17105-041), 2 mg/mL in DMEM/F12. Filter sterilize and store at 4°C for up to 2 weeks.
2. Phosphate-buffered saline (PBS), pH 7.4, without $CaCl_2$ and $MgCl_2$ (Invitrogen, Carlsbad, CA, cat. no. 10010-023). Store at room temperature (RT)
3. DMEM/F-12 with 2.5 mM GlutaMAX™-I with sodium pyruvate and no HEPES buffer (Invitrogen, cat. no. 10565-018). Store at 4°C
4. Cell scraper (Sarstedt, cat. no. 83-1830)
5. StemSpan® SFEM (STEMCELL Technologies, cat. no. 09650). Store at 4°C
6. Fetal bovine serum (FBS) (HyClone, Thermo Scientific, cat. no. SH30071.03). Do not heat to inactivate it. Thaw overnight at 4°C, aliquot, and store at –20°C.

7. ES-Cult® fetal bovine serum (FBS) (STEMCELL Technologies, cat. no. 06950) Do not heat to inactivate it. Thaw overnight at 4°C, aliquot, and store at −20°C.

8. L-ascorbic acid (Sigma-Aldrich, cat. no. A4403, 50 mg/mL in PBS). Aliquot and store at −20°C.

9. EX-CYTE (Millipore, cat. no. 81-129-1). Aliquot and store at 4°C, light-sensitive

10. Penicillin-streptomycin (Invitrogen, cat. no. 15140-122). Aliquot and store at −20°C; once thawed, store at 4°C for up to 2 weeks

11. Insulin-transferrin-selenium-X (ITS-X) (Invitrogen, cat. no. 51500-056). Store at 4°C

12. MethoCult® SF H4236 (STEMCELL Technologies, cat. no. 04236). Thaw in a 100-mL bottle overnight at 4°C, aliquot, and store at −20°C for single use; do not freeze-thaw multiple times)

13. Protein-free hybridoma mix II (PFHM II) (Invitrogen, cat. no. 12040-077). Store at 4°C

14. StemPro34 (Invitrogen, cat. no. 16039)

15. L-glutamine (Invitrogen, cat. no. 25030)

16. 1-Thioglycerol (Sigma-Aldrich, cat. no. M1753)

17. Ultralow adherent, 6-well plates (Corning, cat. no. 3471)

18. Recombinant human bone morphogenic protein 4 (BMP4) (R&D Systems, cat. no. 314-BP-010). Reconstitute to 100 ng/µL in SFEM, make 50 µL aliquots, and store at −20°C. Once thawed, store at 4°C for up to 2 weeks.

19. Recombinant human vascular endothelial growth factor A_{165} ($VEGFA_{165}$) (R&D Systems, cat. no. 293-VE-50). Reconstitute to 100 ng/µL in SFEM, make 50 µL aliquots, and store at −20°C. Once thawed, store at 4°C for up to 1 week.

20. Recombinant human FGF2 (FGF basic) (R&D Systems, Minneapolis, MN, cat. no. 233-FB-025). Reconstitute to 100 ng/µL in DMEM/F-12, make 50 µL aliquots, and store at −20°C. Once thawed, store at 4°C for up to 1 week

21. Heparan sulfate sodium salt from bovine kidney (Sigma-Aldrich, cat. no. H7640). Reconstitute to 0.2 mg/mL in PBS, make 250 µL aliquot, and store at −20°C. Acting concentration is 5 µg/mL.

22. 16G needle (BD Biosciences, cat. no. 305198)

23. 5-mL syringe (BD Biosciences, cat. no. 309603)

24. Saran Wrap (SC Johnson)

2.1.2. Characterization of Hematopoietic Cells by FACS Analysis

1. 0.05% Trypsin-EDTA (Invitrogen, cat. no. 25300-054). Make 5 mL aliquots of 0.5% trypsin and store at –20°C. Before use, thaw and add 45 mL of PBS. Store 0.05% trypsin at 4°C and immediately prior to application warm to 37°C in water bath for 5 min.

2. Accutase® (Sigma-Aldrich, cat. no. A6964). Store 5 mL aliquots at –20°C. Once thawed, store aliquots at 4°C; warm to 37°C in water bath for 5 min before use.

3. Tissue culture-treated 6-well plates (Greiner Bio-One, Monroe, NC, cat. no. 657160, through ISC BioExpress, cat. no. T-3026-3)

4. 21G needle (BD Biosciences, cat. no. 305165)

5. 70-µm cell strainers (Fisher Scientific, cat. no. 22363548)

6. 5-mL flow cytometry tubes (BD Biosciences, cat. no. 352052)

7. FIX & PERM® Fixation and Permeabilization Kit (Invitrogen, cat. no. GAS-003)

8. Mouse antihuman CD34-PE (BD Biosciences, cat. no. 555822), 5 µL per 1×10^6 cells

9. Mouse antihuman CD45-APC (BD Biosciences, cat. no. 555485), 5 µL per 1×10^6 cells

10. Mouse antihuman BB9-APC (BD Biosciences, cat. no. 557929), 5 µL per 1×10^6 cells

2.2. Materials for Generation of Hemogenic Endothelium

1. Human plasma fibronectin (Invitrogen, 33016-015). Dilute to 1 mg/mL in PBS, make 100 µL aliquots, and store at –20°C. To use, thaw aliquots and dilute to 10 µg/mL in PBS.

2. Collagenase IV (Invitrogen, cat. no. 17104-019). Dilute to 1 mg/mL in DMEM–F12, filter sterilize, and store at 4°C up to 1 month.

3. Endothelial Cell Growth Medium-2 (EGM®-2 BulletKit,t Lonza, cat. no. CC-3162). EGM-2 is a complete optimized low-FBS medium for endothelial cell expansion and contains 2% FBS, hydrocortisone, ascorbic acid, and heparin sulfate and is supplemented with the following human growth factors: hEGF, hVEGF, FGF-2, and hIGF-2.

2.3. Materials for Hematopoietic CFC Assays

2.3.1. Cell Culture

1. Hemocytometer (Hausser Scientific, Horsham, PA, cat. no. 3200)

2. Trypan blue (Invitrogen, cat. no. 15250)

3. 35-mm petri dish (NUNC, cat. no. 171099)

4. 35-mm cell culture dish with grid (NUNC, cat. no. 174926)

5. 10-cm cell culture dish (Corning, cat. no. 430167)

6. MethoCult® SF H4436 with growth factors (STEMCELL Technologies, cat. no. 04436). Thaw overnight at 4°C, aliquot,

and store at −20°C. Thawed aliquots can be thawed once only at 4°C overnight for single use, then stored for up to 1 week.

7. Blunt-end needles (STEMCELL Technologies, cat. no. 28110)
8. 1-mL syringe (BD Biosciences, cat. no. 309602)

2.3.2. Materials for Characterization of Hematopoietic Cells by FACS Analysis

1. FIX & PERM® Fixation and Permeabilization Kit (cell permeabilization reagents, Invitrogen, cat. no. GAS-003)
2. Mouse antihuman CD235a (Glycophrin A)-PE (BD Biosciences, cat. no. 555570)
3. Mouse antihuman CD71-PE-Cy5 (BD Biosciences, cat. no. 551143)
4. Mouse antihuman embryonic hemoglobin-ε-FITC (Fitzgerald Industries International, cat. no. 61C-CR8008M1F). Use 0.3 µL per 1×10^6 cells.
5. Mouse antihuman fetal hemoglobin-FITC (BD Biosciences, cat. no. 552829), 3 µL per 1×10^6 cells
6. Mouse antihuman purified hemoglobin-β (Santa Cruz Biotechnology, cat. no. sc-21757), 2 µL for ~1×10^6 cells
7. Goat antimouse IgG_1-PE (SouthernBiotech, cat. no. 1707-09), 0.5 µL per 1×10^6 cells

3. Methods

hEBs are clusters of differentiating human pluripotent stem cells in three-dimensional liquid suspension cultures. In our laboratory, we have optimized medium conditions specifically for hEB mesodermal differentiation and subsequent hematopoietic, endothelial, and mesenchymal lineages. The following modified protocol (summarized in Fig. 1) is divided into two sequential phases: phase 1 includes (1) adaptation culture, (2) methylcellulose culture, and (3) serum-free liquid hEB suspension culture, and phase 2 consists of a sequential adherent hemogenic endothelium culture phase. Robust amounts of BB9+ (ACE+/CD143+) hemangioblast-containing populations (5) expand during endothelial EGM2 medium culture. After 3–4 days of EGM2 culture, a wave of multipotent CD34+CD45+ hematopoietic progenitors comprising 20–50% of the culture begins to emerge both as cobblestone areas as well as floating cells from an adherent endothelial-like hEB cell population (Fig. 2). In general, ~ 1×10^6 hematopoietic cells can be harvested from ~ one plate of initial hEB culture, thus making this two-phase protocol highly ideal for generating high yield amounts of multipotent progenitors for subsequent molecular and hematologic analyses.

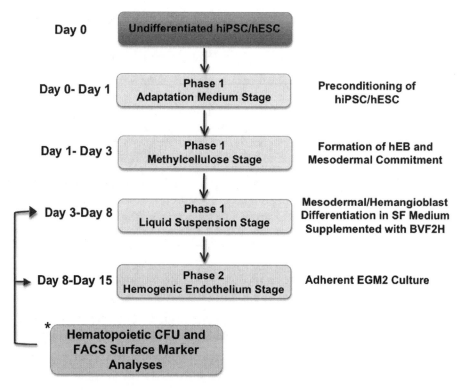

Fig. 1. Experimental summary of methodologic stages of hematopoietic differentiation of hESC and hiPSC. Replicate hematopoietic colony assays or FACS analyses of hEB cells can be performed on day 3–15 of phase 1 or phase 2 (adherent EGM2) cultures*.

3.1. Generation of Human Embryoid Bodies (hEBs): Phase 1

3.1.1. Adaptation Step (Day 0–Day 1)

1. hESC should be grown to approximately 80% confluent on irradiated MEF (6–7 days following last passage).
2. Aspirate hESC medium.
3. Wash with 2 mL PBS per well (optional).
4. Add 2.5 mL adaptation medium (Table 1) and culture for 24 h.

3.1.2. Methylcellulose Culture (Day 1–Day 3: Phase 1)

1. Aspirate adaptation medium.
2. Wash hESC with 2 mL PBS per well.
3. Add 1mL of 2 mg/mL dispase into each well
4. Incubate at 37°C for 5 min.
5. Aspirate dispase.
6. Add 1 mL of DMEM–F12 and combine cells into a 15-mL tube.
7. Harvest hESC cell clumps with cell scraper.
8. Wash wells with DMEM–F12 and combine cells into a 15-mL tube; centrifuge them at $300 \times g$ for 5 min (see Note 1).

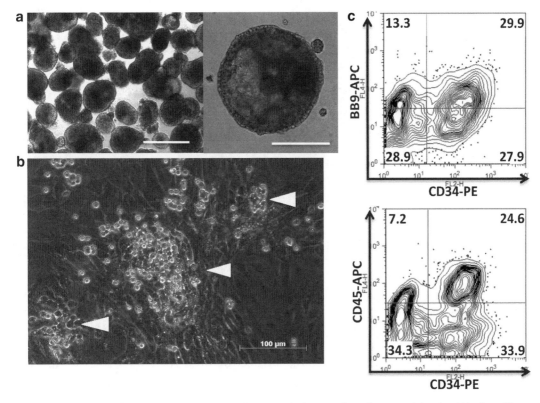

Fig. 2. Differentiation of hiPSC into hemangioblast and hematopoietic progenitors via sequential embryoid body and hemogenic endothelial cultures. (**a**) Phase-contrast images of hEB generated from hiPSC derived from CD34+ cord blood cells (CB-iPSC) that were reprogrammed with expression of seven factors (*SOX2, OCT4, KLF4, MYC, NANOG, LIN28,* and *SV40 T antigen*) expressed with EBNA1-based episomal plasmids (7). Scale bars are 500 μm (*left*) and 200 μm (*right*). (**b**) Phase-contrast image of emerging hematopoietic progenitors after culturing iPS(IMR90)-4 fibroblast-derived hiPSC hEB cells that were differentiated for 8 days as suspension hEB and then further cultured for an additional 5 days in adherent EGM2 culture conditions. Shown are hematopoietic "cobblestone areas" that emerge directly from adherent stromal cells. Floating hematopoietic cells bud off from adherent endothelial layers (*arrow heads*). (**c**) Flow cytometry analysis of nonadherent cells from iPS(IMR90)-4 fibroblast iPSC that emerge from cultures from day 8 hEB cells that were further cultured for 5 days in EGM2 conditions. These cells are enriched in populations that contain hemangioblasts (BB9+/ACE+/CD143+ cells) as well as multipotent hematopoietic progenitors (CD34+CD45+ cells).

Table 1
Composition of the adaptation medium

Component	Volume (mL)/100 mL
StemSpan SFEM	82
FCS	15
Ascorbic acid	1
EX-CYTE	1
Penicillin/streptomycin	1
ITS	0.5

Table 2
Composition of the methylcellulose medium

Component	Volume (mL)/100 mL
MethoCult SF H4236	80
FCS	15
Ascorbic acid	1
PFHM II	3.5
EX-CYTE	0.5

9. Resuspend hESC clumps in DMEM–F12 and centrifuge at 200×g for 5 min (see Note 2).
10. Resuspend hESC clumps in 0.5 mL of adaptation medium per well.
11. Add 2–2.5 mL of methylcellulose medium (Table 2) per well of an ultralow adherent 6-well plate.
12. Distribute 0.5 mL of the hESC clump suspension into each well (see Note 3).
13. Rock the plate back and forth to evenly distribute hESC clumps.
14. Wrap the plate in Saran Wrap to create semihypoxic conditions. Incubate the plate at 37°C for 48 h.

3.1.3. Serum-Free Liquid Suspension Culture (Day 3: Phase 1)

1. Add 2 mL PBS per well.
2. Transfer PBS/methylcellulose/hEB suspension into 50-mL tubes.
3. Wash the wells with 3 mL PBS per well to collect any remaining hEB; gather all in to one 50-mL tube.
4. Repeat step 3 (from three wells of a 6-well plate, total volume will be approximately 30–35 mL) (see Note 4).
5. Centrifuge at 300×g for 5 min (see Note 5).
6. Aspirate methycellulose/PBS and wash centrifuged hEB in 2.5 mL PBS per well and centrifuge at 200×g for 5 min
7. Aspirate PBS and replace with fresh PBS; allow hEB to settle out by gravity for approximately 5–10 min.
8. Using a pipette, remove top 2/3 of medium to remove any remaining dead MEF.
9. Centrifuge at 200×g for 5 min.
10. Resuspend in liquid differentiation medium (LDM; Table 3) that includes fresh supplementation with BMP4, VEGFA, FGF2, and heparan sulfate (BVF2H)

Table 3
Composition of the liquid differentiation medium (LDM)

Component	Weight (mL)/100 mL
StemPro34	98
Ascorbic acid	1
L-Glutamine	1
1-Thioglycerol	0.0035
BMP4[a]	0.05
VEGF[a]	0.05
Basic-FGF[a]	0.05
Heparan sulfate[a]	0.25

[a]Add fresh growth factors right before using LDM only for the amount being used

11. Distribute hEB in to one well of an ultralow adherent 6-well plate with 2–2.5 mL volume.
12. Cover with Saran Wrap to create semihypoxic conditions.
13. Change medium every 2–3 days.
14. To change medium, collect hEB/medium in 15-mL tube, allow hEBs to settle by gravity for 5 min, remove supernatant, and resuspend in fresh medium.

3.1.4. Characterization of hEB Cells by FACS Analysis

1. Harvest hEB in PBS into a 15-mL tube.
2. Allow hEB to settle by gravity for 5 min.
3. Aspirate PBS/medium/dead MEF.
4. Add 5 mL of PBS to wash.
5. Repeat steps 2 and 3.
6. Add 2 mL of Accutase and transfer to one well of 6-well plate.
7. Incubate at 37°C for 5 min.
8. Gently pass hEB clumps through a 21G needle and 5-mL syringe 3 times.
9. Incubate for an additional 5 min at 37°C.
10. Repeat step 8 and add 3 mL of hESC medium to stop Accutase.
11. Pass cells through a 70-μm cell strainer to remove nonsingle cells.
12. Centrifuge at $200 \times g$ for 5 min.

13. Resuspend in 5% FBS/PBS.
14. Stain in hematopoietic cell markers (e.g., CD34, CD45, BB9) for 20 min (see Sec. 3.3.2 and Fig. 2c).

3.2. Preparation of the Hemogenic Endothelium Differentiation Stage (Phase 2)

1. Harvest hEB at day 8 in PBS into a 15-mL tube (see Note 6).
2. Allow hEB to settle by gravity for 3–5 min.
3. Aspirate PBS/medium/dead MEF.
4. Add 5 mL of PBS to wash.
5. Repeat steps 2–3.
6. Add 2 mL of collagenase IV and transfer to one well of a 6-well plate.
7. Incubate at 37°C for 5 min.
8. Use a P1000 pipet tip to break hEB into small clumps.
9. Move clumps into 15-mL tube in hESC medium.
10. Centrifuge at $200 \times g$ for 5 min.
11. Resuspend clumps in EGM2 supplemented with an additional 25 ng/mL VEGFA.
12. Plate clumps into a fibronectin-coated 6-well plate (see Note 7).
13. After 1 day of culture, replace medium with fresh EGM2 plus additional VEGFA.
14. 2 to 4 days later (day 3–5 in EGM2), harvest floating cells for characterization by flow cytometry and hematopoietic colony assay.

3.3. Hematopoietic CFC Assays

3.3.1. Cell Culture

1. Prepare cell-strained hEB cells (optimal at days 10–12) or cells harvested from hemogenic endothelium stage (optimal at days 3–5).
2. Resuspend cells in SFEM medium and count using a hemocytometer.
3. For 3 of 35-mm dishes, take 3×10^5 hEB cells (1×10^5 per dish) or 1.5×10^5 EGM2-cultured cells (5×10^4 cells per dish) into a 15-mL tube (see Note 8).
4. Centrifuge at $200 \times g$ for 5 min.
5. Resuspend cells in 450 µL of SFEM medium.
6. Add methylcellulose medium (H4436) up to 4.5 mL.
7. Vortex and let it sit for 5 min to allow bubbles to resolve.
8. Distribute cells into 2 mm × 35 mm gridded dishes equally using 16G needle or blunt needle with 1-mL syringe.
9. Place 2 mm × 35 mm gridded dishes into a 10-cm dish and add one 35-mm nongrid dish with sterile water without lid to provide humidity.
10. Incubate in 37°C for 2 weeks (see Note 9).

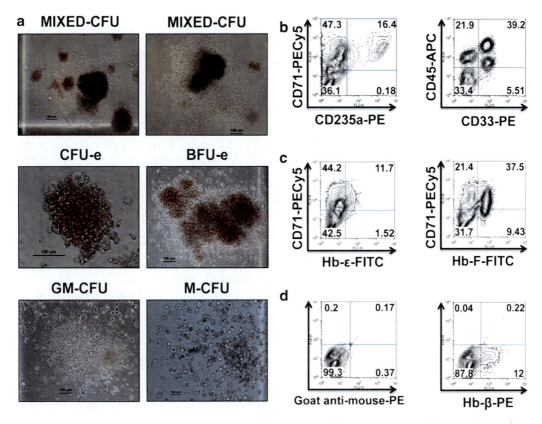

Fig. 3. Morphologies and surface marker expressions of hematopoietic CFU generated from hiPSC. (**a**) Phase-contrast images of hematopoietic colonies generated from CB-iPSC differentiation in EGM2 culture conditions, as described in the text. Colonies which are generated in methylcellulose include mixed erythromyeloid CFU (MIXED-CFU), colony-forming unit-erythroid CFU (CFU-e), burst-forming unit-erythroid CFU (BFU-e), granucocyte CFU (G-CFU), and macrophage CFU (M-CFU). Scale bars are 100 μm. (**b**, **c**) Flow cytometry analysis of surface markers of pooled hematopoietic cells from fibroblast-derived iPS(IMR90)-1 that were obtained from methylcellulose colony assays of floating hematopoietic progenitors generated after 5 days of EGM2 differentiation cultures. (**b**) Shown are expressions of erythroid (CD71$^+$CD235a$^+$) and myeloid (CD33$^+$CD45$^+$) progenitors, as well as intracytoplasmic expressions of erythrocyte hemoglobins: (**c**) *Hb-ε* epsilon chain (embryonic), *Hb-F* gamma chain (fetal) hemoglobin, (**d**) *Hb-β* beta chain (adult) hemoglobin.

11. Enumerate hematopoietic colony-forming units (CFUs) by morphology primitive erythroid (EryP), adult-type erythroid-CFU (CFU-e, BFU-e), mixed erythromyeloid (Mixed-CFU), and myeloid/machrophage/granulocyte (GM-CFU, G-CFU, M-CFU) (see Fig. 3, and Note 10).

3.3.2. Characterization of Hematopoietic Cells by Flow Cytometry Analysis

1. 14–20 days after hematopoietic colony assay, add 2 mL of PBS to each 35-mm dish.
2. Harvest all cells/medium into a 50-mL tube and add 5 mL of PBS into the dish to wash out any remaining cells (see Note 11).
3. Repeat step 2.
4. Centrifuge at 300×*g* for 5 min.

5. Aspirate media and wash with 20 mL PBS.
6. Centrifuge at $200 \times g$ for 5 min.
7. Resuspend cells in 5% FBS/PBS.
8. Count cells using a hemocytometer.
9. Distribute ~1×10^5 cells per flow cytometry tube.
10. For cell surface antigens (e.g., CD71, CD235a (glycophorin A), CD33, or CD45), add antibody to cell solution and incubate at RT for 20 min in the dark.
11. For hemoglobin analysis, add Reagent A from the FIX and PERM Kit and incubate for 20 min.
12. Wash in PBS.
13. Add Reagent B and antihemoglobin-ϵ, hemoglobin-F, and hemoglobin-β antibodies.
14. Incubate at RT for 20 min in the dark.
15. Wash in PBS.
16. Only for hemoglobin-β staining, resuspend cells in 100 µL of 5% FBS/PBS and add goat antimouse PE (see Notes 12–14)
17. Incubate at RT for 20 min in the dark.
18. Wash in PBS.
19. Resuspend in 5% FBS/PBS for flow cytometry or FACS analysis.

4. Notes

1. Pipette vigorously using a 10-mL serological pipette to ensure adequate trituration into small hiPSC/hESC clumps.
2. Washing twice in DMEM–F12 medium is critical for disaggregating hiPSC/hESC and washing away debris. Ensure strong pipetting for the second wash as well.
3. For forming high-quality formed hEB (Fig. 2a), the ratio of hiPSC or hESC to hEB should be ~2:1 (i.e., two wells of a 6-well plates transferred into one well of an ultralow adherent 6-well plate).
4. Maintain the washed plates in sterile conditions. Place washed hEB back into the same ultralow adherent plate after resuspending with liquid differentiation medium.
5. Centrifugation at $300 \times g$ or 1200 rpm minimizes loss of human cells from the viscous washing solutions.
6. Optimization experiments in our lab have found that day 8 hEBs are most effective at establishing the hemogenic endothelium differentiation stage.

7. We generally avoid adding additional hematopoietic expansion growth factors (e.g., TPO, FLT3L) during the endothelial EGM2 culture (phase 2) since we have found that this decreases the frequency of multipotent $CD34^+CD45^+$ progenitors and favors lineage-specific differentiation (e.g., erythropoiesis).

8. It is critical to plate collagenase-treated hEB with enough space to expand and differentiate during the adherent hemogenic endothelium stage (Fig. 2b, c). One estimate is to plate ~500 EBs (equivalent to ~1/2 well of hEB in suspension cultures) into one well of a 6-well plate.

9. For the hematopoietic colony-forming cell (CFC) methylcellulose assays of EGM2-cultured cells, we recommend assaying fewer numbers of cells (e.g., 10,000–30,000 viable cells per plate) than those typically used from day 10 hEB cells (e.g., 100,000–150,000 viable cells per plate) since they contain significantly higher frequencies of progenitors.

10. Since maturation of hematopoietic CFU is slower in serum-free H4436 methylcellulose medium and may take up to 21 days for complete maturation of colonies, we recommend *gently* feeding with ~1 mL of fresh, prewarmed H4436 methylcellulose medium on the top of culture in 35-mm dishes using a 16G needle/1-mL syringe after 7–10 days of the culture.

11. Hematopoietic CFUs that develop from hEB cells are morphologically different from those that arise from human cord blood or bone marrow cells (5).

12. Use at least two pooled replicate dishes of hematopoietic cells from methylcellulose cultures for flow cytometry analysis. These pooled cells can also be used for Western blot or RT-PCR.

13. Stain hematopoietic cell surface markers (e.g., CD71-PE/Cy7) prior to incubating with Reagent A and proceeding to intracytoplasmic staining since fixation often destroys antigenicity.

14. Assure that proper controls for flow cytometry analysis are prepared which include unstained cells and secondary antibody-only stained negative controls.

References

1. Takahashi, K., Tanabe, K., Ohnuki, M., Narita, M., Ichisaka, T., Tomoda, K., and Yamanaka, S. (2007) Induction of pluripotent stem cells from adult human fibroblasts by defined factors, *Cell* 131, 861–872.

2. Yu, J., Vodyanik, M. A., Smuga-Otto, K., Antosiewicz-Bourget, J., Frane, J. L., Tian, S., Nie, J., Jonsdottir, G. A., Ruotti, V., Stewart, R., Slukvin, II, and Thomson, J. A. (2007) Induced pluripotent stem cell lines derived from human somatic cells, *Science* 318, 1917–1920.

3. Park, I. H., Arora, N., Huo, H., Maherali, N., Ahfeldt, T., Shimamura, A., Lensch, M. W., Cowan, C., Hochedlinger, K., and Daley, G. Q. (2008) Disease-specific induced pluripotent stem cells, *Cell* 134, 877–886.

4. Zambidis, E. T., Peault, B., Park, T. S., Bunz, F., and Civin, C. I. (2005) Hematopoietic differentiation of human embryonic stem cells progresses through sequential hematoendothelial, primitive, and definitive stages resembling human yolk sac development, *Blood 106*, 860–870.

5. Zambidis, E. T., Park, T. S., Yu, W., Tam, A., Levine, M., Yuan, X., Pryzhkova, M., and Peault, B. (2008) Expression of angiotensin-converting enzyme (CD143) identifies and regulates primitive hemangioblasts derived from human pluripotent stem cells, *Blood 112*, 3601–3614.

6. Kim, K., Doi, A., Wen, B., Ng, K., Zhao, R., Cahan, P., Kim, J., Aryee, M. J., Ji, H., Ehrlich, L. I., Yabuuchi, A., Takeuchi, A., Cunniff, K. C., Hongguang, H., McKinney-Freeman, S., Naveiras, O., Yoon, T. J., Irizarry, R. A., Jung, N., Seita, J., Hanna, J., Murakami, P., Jaenisch, R., Weissleder, R., Orkin, S. H., Weissman, I. L., Feinberg, A. P., and Daley, G. Q. (2010) Epigenetic memory in induced pluripotent stem cells, *Nature*.

7. Yu, J., Hu, K., Smuga-Otto, K., Tian, S., Stewart, R., Slukvin, I. I., and Thomson, J. A. (2009) Human induced pluripotent stem cells free of vector and transgene sequences. *Science 324*, 797–801.

Chapter 25

Adipogenic Differentiation of Human Induced Pluripotent Stem Cells

Michio Noguchi, Masakatsu Sone, Daisuke Taura, Ken Ebihara, Kiminori Hosoda, and Kazuwa Nakao

Abstract

Human induced pluripotent stem (iPS) cells are promising sources of disease modeling and regenerative medicine. However, differentiation properties of human iPS cells to specific lineages still remain to be fully clarified. Here, we describe a protocol for differentiating human iPS cells into adipocytes via embryoid body formation. We found that human iPS cells differentiated using this protocol exhibit lipid accumulation and the expression of adipogenic genes, suggesting that human iPS cells have adipogenic properties in vitro.

Key words: Human iPS cells, Adipocytes, Lineage-specific differentiation of human iPS cells, Stem cell differentiation, Adipogenic differentiation

1. Introduction

Adipogenesis is an area of intense interest because of the large and growing prevalence of obesity and type 2 diabetes. Studies with preadipocyte models, such as 3T3-L1 and 3T3-F442A cell lines, have clarified the mechanisms of adipocyte terminal differentiation (1, 2). However, little is known about the development of adipocytes from human embryonic stem (ES) cells and induced pluripotent stem (iPS) cells.

Generally, adipocyte is considered to have a mesodermal origin. The formation of the mesoderm begins with the migration of a layer of cells between the primitive endoderm and ectoderm. The mesoderm generates all the organs between the ectodermal wall and the endodermal tissues. In addition, the mesoderm helps the ectoderm and endoderm to form their own tissues. The trunk mesoderm of an embryo can be divided into four regions: chordamesoderm, paraxial mesoderm, intermediate mesoderm, and lateral plate mesoderm. Given that adipocytes are derived from multipotent mesenchymal progenitors, the origin of adipocytes may be paraxial mesoderm. However, a part of adipocytes and their progenitors from ES cells are derived from neuroepithelial cells or neural crest cells (3–5). Regarding adipogenesis from human ES and iPS cells, further studies will be needed.

2. Materials

2.1. Cells

1. Human iPS cells (201 B7 and 253 G1). These cells were kindly provided by Prof. Yamanaka (Center for iPS Cells Research and Application, Kyoto University).
2. Mitomycin-treated SNL 76/7 cells (DS Pharma Biomedical)

2.2. Maintenance Medium for Human iPS Cells

1. Dulbeco's modified Eagle's medium: Nutrient mixture F-12 (DMEM-F12) (Sigma)
2. Knockout serum replacement (KSR) (Invitrogen)
3. L-Glutamine (Invitrogen)
4. Nonessential amino acids solution (Invitrogen)
5. 2-Mercaptoethanol (Sigma)
6. Basic fibroblast growth factors (bFGF), human recombinant (WAKO)
7. Penicillin/streptomycin (Invitrogen)

Prepare a complete medium (DMEM/F12 containing 20% KSR, 2 mM L-glutamine, 1×10^{-4} M nonessential amino acids, 1×10^{-4} M 2-mercaptoethanol, 50 units and 50 mg/mL penicillin/streptomycin, and 4 ng/mL bFGF) as follows:

- DMEM/F12 390 mL
- KSR 100 mL
- L-glutamine (200 mM) 5 mL
- NEAA 5 mL
- 2-Mercaptoethanol 5 μL
- Penicillin/streptomycin (10,000 units and 10,000 mg/mL) 2.5 mL
- bFGF (2 μg/mL) 1 mL
- Store at 4°C up to 1 week.

2.3. Dissociation Solution

1. Collagenase IV (Invitrogen)
2. 2.5% Trypsin (Invitrogen)
3. KSR (Invitrogen)

4. CaCl$_2$ (Kanto Chemical)

5. Phosphate-buffered saline, calcium and magnesium free (Sigma)

Prepare the dissociation solution as follows:

- Collagenase IV (1 mg/mL) 10 mL
- 2.5% Trypsin 10 mL
- KSR 20 mL
- CaCl$_2$ (1 M) 100 µL
- PBS 60 mL
- Store at –20°C. Avoid repeated freezing and thawing.

2.4. Maintenance Medium for Mitomycin C-Treated SNL Cells

1. Dulbeco's modified Eagle's medium (DMEM) (Invitrogen)
2. L-glutamine (Invitrogen)
3. Fetal bovine serum (FBS) (Invitrogen)
4. Penicillin/streptomycin (Invitrogen)

Prepare a complete medium (DMEM/F12 containing 7% FBS, 2 mM L-glutamine, and 50 units and 50 mg/mL penicillin/streptomycin) as follows:

- DMEM (Invitrogen 11960) 915 mL
- L-Glutamine (200 mM) 10 mL
- FBS 70 mL
- Penicillin/streptomycin (10,000 units and 10,000 mg/mL) 5 mL
- Store at 4°C.

2.5. EB Formation Medium

1. DMEM-F12 (Sigma)
2. KSR (Invitrogen)
3. L-glutamine (Invitrogen)
4. Nonessential amino acids solution (NEAA) (Invitrogen)
5. 2-Mercaptoethanol (Sigma)
6. Retinoic acid (RA) (Sigma)

Prepare the EB formation medium (DMEM/F12 containing 20% KSR, 2 mM L-glutamine, 1×10^{-4} M nonessential amino acids, 1×10^{-4} M 2-mercaptoethanol, and 50 units and 50 mg/mL penicillin and streptomycin, respectively) as follows:

- DMEM/F12 390 mL
- KSR 100 mL
- L-Glutamine (200 mM) 5 mL
- NEAA 5 mL

- 2-Mercaptoethanol 5 μL
- Penicillin/streptomycin (10,000 units and 10,000 mg/mL) 2.5 mL
- Store at 4°C up to 1 week.

2.6. Adipogenic Induction Medium

1. α–Minimum Essential Medium (MEM) (Invitrogen)
2. FBS (Invitrogen)
3. Insulin, human recombinant (Roche)
4. 3-Isobutyl-1-methylxanthine (IBMX) (Nacalai)
5. Dexamethasone (Nacalai)
6. Pioglitazone (Takeda)
7. Penicillin/streptomycin (Invitrogen)

Prepare adipogenic differentiation medium as follows:

Basal medium (α–MEM containing 10% FBS and 100 units and 100 mg/mL penicillin/streptomycin)

- α–MEM (Invitrogen 12571) 890 mL
- FBS 100 mL
- Penicillin/streptomycin (10,000 units and 10,000 mg/mL) 10 mL
- Store at 4°C.

Adipogenic induction medium

- 1 μg/mL insulin, 0.5 mM 3-Isobutyl-1-methylxanthine (IBMX), 0.25 μM dexamethasone, and 1 μM pioglitazone
- Basal medium 20 mL
- Insulin (100 μg/mL) 200 μL
- IBMX (50 mM) 200 μL
- Dexamethasone (25 μM) 200 μL
- Pioglitazone (10 mM) 2 μL

The differentiation medium must be prepared just before use. Therefore, prepare only the volume of medium required for experiments.

3. Methods

3.1. Adipogenic Differentiation of Human iPS Cells

Adipocyte differentiation of hiPS cells is initiated by aggregation of iPS cells to form embryoid bodies (EBs). We routinely use the floating culture for the formation of EBs and subsequently the attachment culture for adipogenic differentiation (6) based on the adipocyte differentiation protocol from mouse and human ES cells (7–9).

In adipogenesis from mouse ES cells, two phases can be distinguished. The first phase of adipogenic induction is considered to the period for the commitment of pluripotent stem cells to the adipocyte lineage. The second phase is considered to the period for the terminal differentiation. In the second phase, human iPS cells are subjected to the adipogenic induction cocktails including insulin, dexamethasone, IBMX, and pioglitazone.

3.1.1. Maintenance of Human iPS Cells

1. Plate human iPS cells at $1–2 \times 10^6$ cells in 100-mm dish
2. Culture for 5–7 days until they reach 90% confluency.
3. For passage, wash human iPS cells once with PBS and then incubate in the *dissociation solution* for 5 min at 37°C.
4. As soon as colonies start dissociating from the dish, remove the solution.
5. After washing with PBS, add *maintenance medium for human iPS cells* to the dish.
6. Scrape cells and transfer into a conical tube.
7. Add an appropriate volume of the medium and then plate the cells onto mitomycin C-treated SNL cells or mouse embryonic fibroblasts (MEFs) (CF-1 strain).

3.1.2. EB Formation

To induce adipocyte differentiation from iPS cells, the following procedure can be used. The procedure has been previously used for adipogenic differentiation of mouse ES cells (Fig. 1). For EB formation, hiPS and hES colonies are treated with *dissociation solution* and plated onto non-adherent bacterial culture dishes, where they are allowed to aggregate in an *EB formation medium*. Retinoic acid is added to the medium with the concentration of 100 nM from day 2 to day 5.

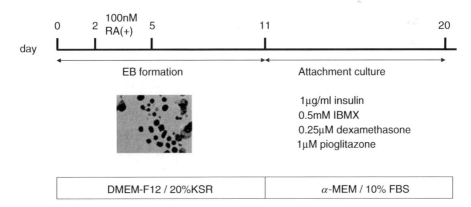

Fig. 1. Schematic diagram of experimental protocol used for adipogenic differentiation of human iPS cells. *EB* embryoid body; *RA* retinoic acid; *IBMX* 3-Isobutyl-1-methylxanthine.

1. Aspirate *maintenance medium for human iPS cells* in a 100-mm dish and wash cells twice with PBS.
2. Add 1 mL of *dissociation solution* and incubate for 2–5 min.
3. Aspirate the *dissociation solution* and wash the cells with 5 mL of *EB formation medium* and scrape the cells. Transfer the cells to a non-adherent bacterial culture dish with 20 mL of *EB formation medium*.
4. At day 2, aspirate the medium and let stand for 5 min to allow the aggregates (EBs) to sediment. Aspirate the supernatant and resuspend the pellet in 10 mL of *EB formation medium*. Transfer the EBs to a non-adherent bacterial culture dish with 20 mL of *EB formation medium* supplemented with a concentration of 100 nM RA. Incubate for 3 days in the presence of RA changing the medium every day.
5. At day 5 and 8, change the medium to *EB formation medium* (without RA).

3.2. Attachment Culture with Adipogenic Cocktails

After EB formation, EBs are transferred to a plate coated with gelatin or other matrix. Differentiation can be induced for approximately 10 days using *adipogenic differentiation medium*. Differentiated iPS cells with lipid droplets are observed spreading outward from the attached EBs.

1. At day 11, transfer EBs to 6-well plates coated with gelatin or other matrix (see Note 1).
2. After plating, exchange the adipogenic differentiation medium every 3 days.
3. Approximately, at day 20, observe differentiated cells with lipid droplets.

3.3. Oil Red O Staining

1. Wash cells with PBS twice after adipogenic induction.
2. Fix the cells in 3.7% formaldehyde for 1 h.
3. Stain the cells with 0.6% (w/v) Oil Red O solution (60% isopropanol, 40% water) for 30 min–2 h at room temperature.
4. Wash cells with water to remove unbound dyes. Differentiated iPS cells with lipid droplets can be stained with Oil Red O (Fig. 2).

3.4. Analysis of Adipocyte-Related Gene Expression

1. Examine the expression of adipocyte marker genes, including peroxisome proliferator-activated receptor γ2 (PPARγ2) and aP2 genes.

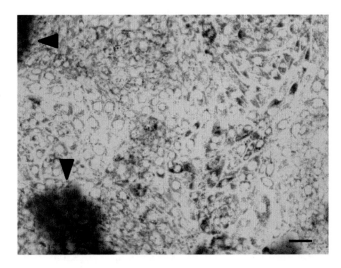

Fig. 2. Lipid accumulation of differentiated iPS cells. Human iPS cells were subjected to adipogenic differentiation. Lipid accumulation was examined by Oil Red O staining. Arrowheads indicate attached EBs. Scale bar, 50 μm.

2. At 20 days after differentiation, detect gene expression of PPARγ2 and aP2 through polymerase chain reaction (PCR) analysis (6). Primers used for PCR analyses are as follows:

PPARγ2
- Forward: 5′-TTCCATGCTGTTATGGGTGA-3′
- Reverse: 5′-ACCCTTGCATCCTTCACAAG-3′

aP2
- Forward: 5′-TGCAGCTTCCTTCTCACCTT-3
- Reverse: 5′-TGGTTGATTTTCCATCCAT-3′

GAPDH
- Forward: 5′-AGCCGCATCTTCTTTTGCGTC-3′
- Reverse: 5′-TCATATTTGGCAGGTTTTTCT-3′

Total RNA is extracted using TRIzol Reagent. Semi-quantitative PCR was carried out using a thermal cycler as instructed by the manufacturer. PCR analyses are performed at following annealing temperatures (PPARγ2: 58°C, aP2: 56°C, and GAPDH: 55°C). Expected size of each PCR product is 496 bp (PPARγ2), 357 bp (aP2), and 816 bp (GAPDH), respectively.

4. Note

1. In cases of clones whose capacities of attachment are low, use dishes coated with a combination of 30 μg/mL Poly-L-ornithine and 2 μg/mL fibronectin, type I collagen or type IV collagen.

References

1. Lane, M.D., Q.Q. Tang, and M.S. Jiang, *Role of the CCAAT enhancer binding proteins (C/EBPs) in adipocyte differentiation.* Biochem Biophys Res Commun, 1999. **266**(3): p. 677–83.
2. Rosen, E.D. and O.A. MacDougald, *Adipocyte differentiation from the inside out.* Nat Rev Mol Cell Biol, 2006. 7(12): p. 885–96.
3. Takashima, Y., et al., *Neuroepithelial cells supply an initial transient wave of MSC differentiation.* Cell, 2007. **129**(7): p. 1377–88.
4. Billon, N., et al., *The generation of adipocytes by the neural crest.* Development, 2007. **134**(12): p. 2283–92.
5. Lee, G., et al., *Isolation and directed differentiation of neural crest stem cells derived from human embryonic stem cells.* Nat Biotechnol, 2007. **25**(12): p. 1468–75.
6. Taura, D., et al., *Adipogenic differentiation of human induced pluripotent stem cells: comparison with that of human embryonic stem cells.* FEBS Lett, 2009. **583**(6): p. 1029–33.
7. Dani, C., et al., *Differentiation of embryonic stem cells into adipocytes in vitro.* J Cell Sci, 1997. **110 (Pt 11)**: p. 1279–85.
8. van Harmelen, V., et al., *Differential lipolytic regulation in human embryonic stem cell-derived adipocytes.* Obesity (Silver Spring), 2007. **15**(4): p. 846–52.
9. Xiong, C., et al., *Derivation of adipocytes from human embryonic stem cells.* Stem Cells Dev, 2005. **14**(6): p. 671–5.

Chapter 26

Chondrogenic Differentiation of hESC in Micromass Culture

Deborah Ferrari, Guochun Gong, Robert A. Kosher, and Caroline N. Dealy

Abstract

Human adult cartilage has limited capacity for self-renewal. Accordingly, repair of cartilage tissue damaged as a result of acute traumatic injury or via chronic wear or degenerative disease, such as arthritis, is a major clinical problem. Human embryonic stem cells (hESCs) could provide an unlimited source of chondrogenic progenitors for cartilage repair. In order to realize this potential, it is necessary to develop methodologies for the directed differentiation of hESCs into chondrocytes. In this chapter, we describe culture systems and conditions, which we have developed for direct, progressive, and substantially uniform differentiation of pluripotent hESCs into the chondrogenic lineage.

Key words: Human embryonic stem cells, Cartilage, Cartilage repair, Chondrogenic differentiation, Micromass culture, Bone morphogenetic protein-2, Transforming growth factor-β1

1. Introduction

Because of their unlimited potential for self-renewal, human embryonic stem cells (hESCs), when differentiated into the chondrogenic lineage, offer a potentially powerful tool for cell-based repair of cartilage damaged as a result of injury or degenerative disease, such as osteoarthritis (1–4). We have developed culture systems and conditions, which direct the progressive and substantially uniform differentiation of pluripotent hESC into the chondrogenic lineage (5). The methodology is based on the high density micromass culture conditions that we have extensively used to direct the progressive and uniform differentiation of embryonic limb bud mesenchymal cells into chondrocytes (6–12). The high density micromass culture system recapitulates the close juxtaposition and subsequent cell–cell interactions required for chondrogenic differentiation. Chondrogenic differentiation in micromass culture is promoted by exogenous supplementation with pro-chondrogenic factors including BMP2 and TGFβ1, which are also utilized by embryonic limb mesenchymal cells during chondrogenic differentiation in vivo (13). Our method is simpler and is more efficient than others that have been reported for chondrogenic differentiation of hESCs, which typically involve the use of embryoid bodies which introduce non-mesodermal and non-chondrogenic cell types (14–18) and/or require FACS sorting to purify the mesenchymal or chondrogenic progenitors (14–20). Other protocols also typically rely on pellet culture (14–19, 21) in which chondrogenic differentiation of human stem cells has been shown to be less optimal than in micromass culture (22).

Briefly, our method involves trypsin dissociation of pluripotent hESCs to single cells followed by plating of the cells in 10 µL spots at high density on tissue culture dishes. Dishes are incubated for 2 h to allow attachment of the cells prior to addition of serum-free hESC media, which is replaced 1 day later by chondrogenic media, to which BMP2 or a combination of BMP2 and TGFβ1 is subsequently added. Cultures are maintained with media changes every other day, and the progression of chondrogenic differentiation assayed by whole-mount Alcian blue staining (pH 1.0). Micromass cultures of pluripotent hESC undergo direct, rapid, progressive, and substantially uniform chondrogenic differentiation under these conditions (5). Dissociation of pluripotent hESCs into single cells is also achieved using Accutase or TrypLE Select (23–26), which offers more gentle alternatives to trypsin and is better aligned with the current direction of human stem cell differentiation protocols toward more defined and/or xenogeneic conditions (27, 28). We have also found that the optimal density for cell spotting is 2×10^5 cells/10 µL spot (2×10^7 cells/mL), and that the optimal time for growth factor supplementation is 24 h after the switch to chondrogenic media.

2. Materials

2.1. Culture of hESC on MEFs

1. DMEM/F12 (1:1): Invitrogen, cat. no. 11330-032 (057); store at 4°C.

2. Knockout serum replacement (KSR): Invitrogen, cat. no. 10828-028; store at –20°C.

3. MEM nonessential amino acids (NEAA), 100×: Invitrogen, cat. no. 11140-050; store 4°C.

4. L-Glutamine, 100×: Invitrogen, cat. no. 25030-081; store at –20°C.

5. EmbryoMax ES cell qualified 2-mercaptoethanol, 100×: Millipore, cat. no. ES-007-E, 20 mLs; store at 4°C for up to 18 months.

6. FGF2 (fibroblast growth factor 2), human, recombinant: Invitrogen, cat. no. 13256–029; store at −20°C. Working stock: 2 µg/mL bFGF in 0.1% BSA; store at −80°C.

7. Bovine serum albumin (BSA), fraction V, 7.5%: Invitrogen, cat. no. 15260. Working stock: 0.1% BSA in Dulbecco's PBS; store at 4°C.

8. PES filter, 0.22 µm: Millipore Stericup sterile vacuum filter unit w/ flask, 250 mLs, Fisher, cat. no. SCGP U02 RE.

9. Collagenase type IV, lyophilized: Invitrogen, cat. no. 17104-019; store at 4°C; 1 mg/mL in DMEM/F12: filter sterilize with 0.22 µm PES filter and store at 4°C for up to 1 month.

10. Tissue culture plates, 6-well: Nunc no. 140675, Fisher, cat. no. 14-832-11.

11. Serum-free hESC medium: DMEM/F12 supplemented with 20% knockout serum replacement (KSR), 1% vol/vol nonessential amino acids, 1 mM L-glutamine, 4 ng/mL fibroblast growth factor-2 (FGF2), and 0.1 mM b-mercaptoethanol. Filter sterilize with 0.22 µm PES filter and store at 4°C for up to 1 week.

2.2. Dissociation of Undifferentiated hESC to Single Cells

1. Dispase, lyophilized: Invitrogen, cat. no. 17105-041; store at 4°C

2. 0.05% Trypsin-EDTA, 1×: Invitrogen, cat. no. 25300; store at 4°C

3. TrypLE Select: Invitrogen, cat. no. 12563; store at 4°C

4. Fetal bovine serum (FBS), characterized, HI: Hyclone, cat. no. SH30071.03; store at −20°C

5. Accutase, activity 500–720 units/mL: Innovative Cell Technologies, cat. no. AT104; store at −20°C for up to 2 years or at 4°C for up to 2 months

6. PBS, Dulbecco's 1×: Invitrogen, cat. no. 14190; store at 4°C

7. Cell strainer, 40 µm: BD Biosciences, cat. no. 352340

8. Trypan blue: Sigma, cat. no. T6146, powder, bioreagent; store at room temperature

9. ROCK inhibitor, Y27632: Calbiochem, cat. no. 688000, 10 mM stock in sterile mH$_2$O; store at −20°C for up to 6 months

2.3. Establishment and Maintenance of Micromass Cultures

1. Tissue culture plates, 24-well: Nunc, cat. no. 142475, Fisher, cat. no. 12-565-163.

2. DMEM (high glucose): Invitrogen, cat. no. 11965; store at 4°C.

3. ITS+ Universal Culture Supplement Premix: BD Biosciences, cat. no. 354352; store at 4°C.

4. L-proline: Sigma, cat. no. P5607, 40 mg/mL in mH_2O stock; store at 4°C.

5. Sodium Pyruvate Solution, 100×, 100 mM: Invitrogen, cat. no. 11,360-070; store at 4°C.

6. MEM nonessential amino acids, 10 mM, 100×: Invitrogen, cat. no. 11140-050; store at 4°C.

7. Penicillin/streptomycin (10,000 units/10,000 μg/mL, respectively): Invitrogen, cat. no. 15140-122; store at −20°C.

8. Ascorbic acid: Sigma, cat. no. A8960, 50 mg/mL stock in mH_2O, made fresh before use.

9. Dexamethasone: Sigma, cat. no. D2915, 10^{-4} M stock in mH_2O; store at −20°C for up to 6 months.

10. Chondrogenic medium: DMEM, 1% ITS, 40 μg/mL L-proline, 1% sodium pyruvate, 1% nonessential amino acids + PenStrep (100 units/100 μg), filter sterilize and store at 4°C for up to 1 week. Add ascorbic acid (50 μg/mL) and dexamethasone (10^{-7}M) on day of media change.

11. BMP2 (bone morphogenetic protein-2), human, recombinant, lyophilized: R&D Systems, cat. no. 355-BM. Reconstitute in sterile 4 mM HCL +0.1% BSA, aliquot, and store at −20°C for up to 3 months without detectable loss of activity.

12. BSA (bovine serum albumin), Probumin Media Grade: Millipore, cat. no. 81-068-3.

13. TGFβ1 (transforming growth factor beta-1), human, recombinant, lyophilized: R&D Systems, cat. no. 240-B. Reconstitute in sterile 4 mM HCL +0.1% BSA, aliquot, and store at −20°C and for up to 3 months without detectable loss of activity.

2.4. Alcian Blue Staining

1. 10% formalin, 0.5% CPC in mH_2O; store at room temperature

2. CPC (cetylpyridinium chloride): Sigma, cat. no. C-5460

3. Glacial acetic acid, 3% in mH_2O (pH 1.0); store at room temperature

4. 0.5% Alcian blue 8GX in 3% glacial acetic acid, pH 1.0 (filter through no. 1 whatman paper); store at room temperature

5. Alcian blue 8GX certified: Sigma cat. no. A3157

6. Glacial acetic acid, 3% mH_2O (pH 2.5); store at room temperature

3. Methods

3.1. Culture of hESCs on MEFs

We have utilized the H9 hESC line generated at the WiCell Research Institute (29) and maintained and provided by the University of Connecticut Stem Cell Core. The H9 cells were used at passage numbers 36–53. hESCs are cultured in serum-free hESC medium (30) on a feeder layer of irradiated CF1 mouse embryonic fibroblasts (MEFs) in 6-well plates. MEFs are prepared using standard protocols (31, 32) and used within 7 days of first plating. Cryopreserved hESCs are plated onto the MEFs, and hESC cultures are maintained with daily 2-mL media changes. Cultures are viewed microscopically daily for quality and size of the colonies, differentiation, and contamination (see Note 1). Undifferentiated hESC colonies are passaged every 4–7 days following enzymatic treatment with 1 mg/mL collagenase IV using standard stem cell protocols (31, 32).

3.2. Dissociation of Undifferentiated hESCs to Single Cells for Micromass Culture

The hESC colonies should be healthy and moderate in size (typically 4–6 days after splitting) with all differentiated cells removed (see Notes 2 and 31, 32). Undifferentiated colonies are first detached from the MEF feeder layer by treatment with Dispase and then dissociated to single cells using enzymatic dissociation with Trypsin-EDTA or TypLE Select or Accutase (see Note 3).

1. *Dispase treatment of MEF-hESC*: Aspirate media from hESC/MEF plate and add 1 mL of Dispase to each well (1 mg/mL Dispase in DMEM/F12, filter sterilized and stored at 4°C). Incubate at 37°C for 30–45 min (see Note 4). When most of the hESC colonies peel off the MEF layer, transfer the colonies with serological pipette to 50 mL conical tubes. Rinse well with DMEM/F12 and transfer to the same 50 mL tubes (see Note 5). Allow the colonies to settle, aspirate Dispase/media, and wash three times with fresh DMEM/F12, allowing cells to settle each time. Aspirate media.

2. *Enzymatic dissociation to single cells*: Add 2 mL of Trypsin-EDTA, TypLE Select, or Accutase (pre-warmed to 37°C) and transfer colonies to a 60-mm Petri dish. Add an additional 4 mL Trypsin-EDTA, TrypLE Select, or Accutase and incubate at 37°C for up to 5 min. Transfer the colonies to a 50-mL conical tube and triturate with a 10-mL serological pipette to determine when colonies are dissociating and then quickly add 6 mL DMEM + 10%FBS, 10%KSR (see Note 6).

3. Triturate the cells gently with a serological pipette to completely dissociate the cells (see Note 7). Pass the dissociated cells through a 40-μm cell strainer into a 50-mL conical tube.

4. Aliquot 50 µL of the cell suspension for counting and add 50 µL Trypan blue (0.4% in sterile mQ water; store at room temperature) to the cells to identify dead cells. Mix well and transfer a 10-µL aliquot to a hemacytometer for counting.

5. Centrifuge remaining cell suspension to pellet cells (200 × g for 5 min). Aspirate media from pelleted cells and resuspend pellet in DMEM +10%FBS, 10%KSR at 2×10^7 cells/mL (see Note 8).

3.3. Establishment and Maintenance of Micromass Cultures

1. *Spotting plates*: Prepare 24-well plates by adding a moat of 65 µL of DMEM +10% FBS, 10%KSR around the edge of each well (see Note 9). Pipette a 10 µL of spot (2×10^5 cells) into the middle of each well. Gently transfer the plates to incubator for 2 h at 37°C to allow attachment of cells to the plate (see Note 10).

2. Very gently pipette 500 µL DMEM + 10%FBS + 10% KSR into each well (see Note 11).

3. Incubate plates for the remainder of the culture period (5% CO_2, 37°C). Day of plating is considered day 0.

4. After 24 h in culture (at day 1), change the culture media to chondrogenic media (500 µL/well). Ascorbic acid (50 µg/mL) and dexamethasone (10^{-7}M) are added to the media immediately before use.

5. After an additional 24 h (day 2), exchange the culture media and add growth factors (100 ng/mL BMP2 or 100 ng/mL BMP2 + 10 ng/mL TGFβ1) to the cell culture medium (see Note 12). Exchange the chondrogenic media + growth factors every 2 days.

3.4. Alcian Blue Staining

To monitor progressive chondrogenic differentiation, micromass cultures can be stained whole mount with Alcian blue, pH 1.0 (8) (Fig. 1). This method stains sulfated glycosaminoglycans at a low pH.

1. Aspirate media and in the fume hood, add 500 µL 10% formalin + 0.5% CPC for 10 min at room temperature.

2. Briefly rinse dishes two times with 3% glacial acetic acid, pH 1.0 and add 500 µL 0.5% Alcian blue 8GX in 3% glacial acetic acid, pH = 1.0. Let sit overnight at room temperature.

3. Briefly rinse dishes in 3% glacial acetic acid, pH 1.0, followed by two rinses in 3% acetic acid, pH 2.5. Airdry.

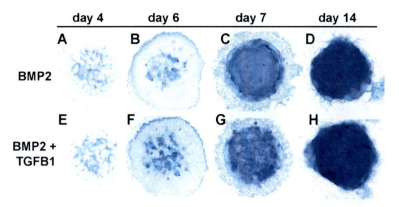

Fig. 1. Progression of chondrogenic differentiation by hESC-derived micromass cultures as assayed by whole-mount Alcian blue staining. (**a–d**) Micromass cultures established as described and maintained in micromass culture in the presence of BMP2. (**e–h**) Micromass cultures established as described and maintained in micromass culture in the presence of BMP2 and TGFβ1. (**a, e**) After 4 days of micromass culture, numerous Alcian blue-positive nodules are present in the central region of the culture. (**b, f**) After 6 days of micromass culture, the nodules are larger and stain more intensely with Alcian blue and occupy much of the central region of the cultures. (**c, g**) After 7 days of micromass culture, relatively uniform Alcian blue staining is detected throughout the majority of the cultures. (**d, h**) After 14 days of micromass culture, Alcian blue-positive staining is intense, substantially uniform and present throughout the entire culture.

4. Notes

1. Spontaneous differentiation commonly occurs at any time during hESC culture. Differentiated cells have variable morphologies compared to the uniform even morphology and distinct colony borders of undifferentiated hESC. Spontaneous differentiation is undesirable, and differentiated cells can overrun the culture and therefore become unusable.

2. If only a few differentiated cells are present, they may be removed by scraping them off the MEF feeder layer using a sterile, pulled and fire-polished glass pipette tip. Aspirate the media containing the differentiated scraped clumps, wash the dish with media, and replace with fresh media.

3. We have not observed major differences in the appearance of chondrogenic differentiation of micromass cultures established following dissociation with trypsin-EDTA, TrypLE Select, or Accutase.

4. Do not tap the plate to dislodge the hESCs, as this will disturb the MEF layer.

5. If there are colonies still attached to the dish following the first Dispase treatment, add additional Dispase and re-incubate. When dislodged, transfer to same 50 mL tube.

6. A xenofree KSR is also commercially available (Invitrogen).

7. If dissociation is incomplete, triturate with a fire-polished sterile Pasteur pipette to complete dissociation.

8. We have compared different stem cell micromass densities ranging from normal density (2×10^7 cells/mL) to one-half normal density (1×10^7 cells/mL) or to twice normal density (4×10^7 cells/mL) or to four times normal density (8×10^7 cells/mL) and found that normal density (2×10^7 cells/mL or 2×10^5 cells/10 µL spot) is optimal for stem cell chondrogenic differentiation.

9. It is important that the incubator humidity be adequate to prevent drying of the spots during attachment.

10. In some experiments, we have used ROCK inhibitor, which promotes survival of dissociated hESC (33, 34), during spotting and during the first 24 h of culture in serum-free hESC medium (10 µM). ROCK inhibitor is light sensitive.

11. When feeding the cells, it is important to direct the flow of the media toward the edge of the culture dish in order not to disturb the newly attached cells.

12. We have found that addition of BMP2 is most effective in promoting chondrogenesis when added at 24 h.

References

1. Khan W.S., Johnson D.S., Hardingham T.E. (2010) The potential of stem cells in the treatment of knee cartilage defects. Knee. Epub ahead of print
2. Nelson L., Fairclough J., Archer C.W. (2010) Use of stem cells in the biological repair of articular cartilage. Expert Opin Biol Ther 10, 43–55
3. Oldfield R.S.N., Archer C.W. (2005) Current strategies for articular cartilage repair. Eur Cells Mat 9, 23–32
4. Vinatier C., Bouffi C., Merceron C., et al. (2009) Cartilage tissue engineering: towards a biomaterial assisted mesenchymal cell stem cell therapy. Curr Stem Cell Res Ther 4, 318–329
5. Gong G., Ferrari D., Dealy C.N. et al.(2010) Direct and progressive differentiation of human embryonic stem cells into the chondrogenic lineage. J Cell Physiol 224, 664–671
6. Ahrens P.B., Solursh M., Reiter R.S. (1977) Stage-related capacity for limb chondrogenesis in cell culture. Dev Biol 60, 69–82
7. Solursh M., Reiter R.S. (1980) Evidence for histogenic interactions during in vitro limb chondrogenesis. Dev Biol 78, 141–150
8. Gay S.W., Kosher R.A. (1984) Uniform cartilage differentiation in micromass cultures prepared from a relatively homogeneous population of chondrogenic progenitor cells of the chick limb bud: effect of prostaglandins. J Exp Zool 232, 317–326
9. Kosher R.A., Gay S.W., Kamanitz J.R. et al. (1986) Cartilage proteoglycan core protein gene expression during limb cartilage differentiation. Dev Biol 118, 112–117
10. Kosher R.A., Kulyk W.M., Gay S.W. (1986) Collagen gene expression during limb cartilage differentiation. J Cell Biol 102, 1151–1156
11. Kulyk W.M., Coelho C.N., Kosher R.A. (1991) Type IX collagen gene expression during limb cartilage differentiation. Matrix 11, 282–288
12. Daniels K., Reiter R., Solursh M. (1996) Micromass cultures of limb and other mesenchyme. Methods Cell Biol 51, 237–247
13. Kulyk W.M., Rodgers B.J., Greer K. et al. (1989) Promotion of embryonic chick limb cartilage differentiation by transforming growth factor-beta. Dev Biol 135, 424–430
14. Koay E.J., Hoben G.M., Athanasiou K.A. (2007) Tissue engineering with chondrogenically differentiated human embryonic stem cells. Stem Cells 25, 2183–2190
15. Lee E.U., Lee H-N., Kang H-J. et al. (2010) Novel embryoid body-based method to derive mesenchymal stem cells from human embryonic stem cells. Tisse Eng A 16, 705–715

16. Brown S.E., Tong W., Krebsbach P.H. (2008) The derivation of mesenchymal stem cells from human embryonic stem cells. Cells Tiss Organs 189, 256–260
17. Hwang N.S., Varghese S., Lee H.J. et al. (2008) In vivo commitment and functional tissue regeneration using human embryonic stem cell-derived mesenchymal cells. Proc Natl Acad Sci 105, 20641–20646
18. Arpornmaeklong P., Brown S.E., Wang Z., et al. (2009) Phenotypic characterization, osteoblastic differentiation, and bone regeneration capacity of human embryonic stem cell-derived mesenchymal stem cells. Stem Cells Dev 18, 955–968
19. Stavropoulos, M.E., Mengarelli I., Barberi T. (2009) Differentiation of multipotent mesenchymal precursors and skeletal myoblasts from human embryonic stem cells. Curr Protocols in Stem Cell Biology 9, 1 F8.1–10
20. Toh W.S., Guo X.M., Choo A.B. et al. (2009) Differentiation and enrichment of expandable chondrogenic cells from human embryonic stem cells. J Cell Mol Med 13, 3570–3590
21. Nakagawa T., Lee S.Y., Reddi A.H. (2009) Induction of chondrogenesis from human embryonic stem cells without embryoid body formation by bone morphogenetic protein 7 and transforming growth factor β1. Arthritis Rheum 60, 3686–3692
22. Zhang L., Su P., Xu C. et al. (2010) Chondrogenic differentiation of human mesenchymal stem cells: a comparison between micromass and pellet culture systems. Biotech Lett. Epub May 13
23. Englund M.C.O., Caisander G., Noaksson K. et al (2010) The establishment of 20 different human embryonic stem cell lines and subclones: a report on derivation culture characterisation and banking. In vitro Cell Dev Biol 46, 217–230
24. Gong G., Roach M.L., Jiang L. et al. (2010) Culture conditions and enzymatic passaging of bovine ESC-like cells. Cell Reprogram 12,151–160
25. Phillips B.W., Horne R., Lay T.S. et al (2008) Attachment and growth of human embryonic stem cells on microcarriers. Biotechnol 138, 24–32
26. Bajpai R., Lesperance J., Kim M, et al. (2008) Efficient propagation of single cells Accutase dissociated human embryonic stem cells. Mol Reprod Dev 75, 818–827
27. Venuri M.C., Schimmel T., Colls P. et al. (2007) Derivation of human embryonic stem cells in xeno-free conditions. Methods Mol Biol 407, 1–10
28. Rajala K., Lindroos B., Hussein S.M. et al. (2010) A defined and xeno-free culture method enabling the establishment of clinical-grade human embryonic, induced pluripotent and adipose stem cells. PLOS One 5, e10246
29. Thomson J.A., Itskovitz-Eldor J., Shaprio S.S. et al. (1998) Embryonic stem cells lines derived from human blastocysts. Science 282, 1145–1147
30. Toh W.S., Yang Z., Liu H. et al (2007) Effects of culture conditions and bone morphogenetic protein 2 on extent of chondrogenesis from human embryonic stem cells. Stem Cells 25, 950–960
31. Carpenter M.K., Xu C., Daigh C.A. et al. (2003) Protocols for the isolation and maintenance of human embryonic stem cells. Humana Press, New Jersey, USA
32. WiCell Research Institute (2008) Introduction to human embryonic stem cell culture methods, version 4. Madison, Wisconsin, USA.
33. Li X., Krawetz R., Liu S. et al. (2009) ROCK inhibitor improves survival of cryopreserved serum/feeder-free single human embryonic stem cells. Hum Reprod 24, 580–589
34. Watanabe K., Ueno M., Kamiya D. et al (2007) A ROCK inhibitor permits survival of dissociated human embryonic stem cells. Nat Biotechnol 25, 681–686

Chapter 27

Deriving Metabolically Active Hepatic Endoderm from Pluripotent Stem Cells

Claire N. Medine, Zara Hannoun, Sebastian Greenhough, Catherine M. Payne, Judy Fletcher, and David C. Hay

Abstract

The human liver is a vital organ within the body and plays a major role in normal homeostasis. The "work horse" of the liver, termed the "hepatocyte," is estimated to make up approximately 70–80% of the liver's mass. Therefore, the study of hepatocyte biology has an important role to play in medicine and the drug discovery process. At present the routine use of human primary hepatocytes is limited due to poor supply and their loss of function upon isolation. Therefore, additional and renewable sources of hepatocytes are being sought. Rodent hepatocytes have been utilised for many years, and although informative, they possess significant limitations and do not accurately extrapolate to human liver. To overcome the issue of cell viability, several groups have tried to generate immortalised hepatocytes; however, the derivative cells exhibit dramatic decreases in function and karyotypic instability over prolonged culture. It has therefore been necessary to find an alternative source of hepatocytes and efficient methods for deriving hepatic endoderm from stem cells in vitro. We have employed human embryonic stem cells (hESCs) and induced pluripotent stem cells (iPSCs) to derive human hepatic endoderm (HE). hESCs and iPSCs represent scalable and highly efficient resources with which to generate human HE in vitro, and hESC-derived HE will be the focus of this chapter.

Key words: Stem cell, Human embryonic stem cell, Human induced pluripotent stem cell, Hepatic endoderm, Hepatocyte, Liver, Hepatic differentiation, Hepatocyte-like cell

1. Introduction

The liver is the largest organ in the body and performs a variety of important endocrine and exocrine functions essential for bodily homeostasis. Hepatocytes represent the main cell type of the liver parenchyma accounting for ~70–80% of cells. Therefore, the study of primary hepatocyte biology has an important role to play in many aspects of modern medicine from in vitro modelling to cell-based therapies. At present, the routine use of human primary hepatocytes is limited due to poor supply and their loss of function and proliferation upon isolation. Therefore, additional and renewable sources of hepatocytes are being sought.

To overcome the limitations associated with primary hepatocytes, several groups have immortalised hepatocytes by the introduction of telomerase (1), SV40 T antigen (2) and viral transfection (2, 3). Although routine cell growth was achieved, genetically modified cells exhibited poor liver function and karyotypic instability (4). Other groups have focused on the use of transformed cells from human liver cancers, including HepaRG (5) and HepG2 (6). Although these cell lines demonstrate some of the attributes of primary hepatocytes, they still possess limited function which restricts their utility in vitro. Given these limitations, hepatocytes of animal origin have also been employed for many years, and although informative, they possess significant limitations and do not accurately extrapolate to human liver.

More recently, advances in human liver progenitor cell biology have provided hope for reliable sources of human hepatocytes. A number of studies have proposed the use of foetal liver stem cells (hepatoblasts) and adult liver stem cells (oval cells) to generate primary hepatocytes and cholangiocytes (7–9). Whilst these studies are highly encouraging, hepatic progenitor cells are detected in low numbers within tissue which has made their isolation, purification and expansion *ex vivo* extremely difficult. Although the utility of tissue-specific stem cells is currently limited, other stem cell populations represent attractive candidates. The ability of human embryonic stem cells (hESCs) and induced pluripotent stem cells (iPSCs) to self-replicate and differentiate to all cell types in the human body (pluripotency) means that they are currently being explored as a resource for hepatocyte generation in vitro. hESCs are derived from the inner cell mass of pre-implantation embryos (10, 11) and iPSCs are generated by the reprogramming of somatic cells through the introduction of a defined set of transcription factors (12). Both populations exhibit similar characteristics and, in theory, offer the potential to produce unlimited supplies of somatic cells, making them an ideal resource for human drug development and regenerative medicine (13–16).

A variety of approaches have been utilised to differentiate hESCs and iPSCs to functional hepatic endoderm (17–25). These studies have adopted both spontaneous differentiation through the formation of embryoid bodies (EBs) and directed differentiation to drive hepatocyte differentiation from stem cells. Although successful, differentiation through EBs is spontaneous and occurs with limited efficiency, requiring downstream purification of cell populations. Therefore, we and several groups have employed direct differentiation models to drive stem cell differentiation towards a hepatic lineage in a 2-dimensional environment (17, 19–24).

Our initial attempts to derive hESCs to HE resulted in poor yields (~10%) (22). Therefore, in order to overcome this limitation, we used human liver development as a reference system to improve our in vitro model. Subsequently, we developed a highly efficient and scalable differentiation protocol to produce HE (24). We have since successfully translated this technology to iPSCs (25). In this chapter we describe the differentiation of hESCs to HE using our validated model (26–28).

2. Materials

2.1. Matrigel Coating Plates and Flasks

1. Matrigel (10 mL, BD Biosciences, UK); store at −20°C.
2. KO-DMEM (500 mL, Gibco, Invitrogen, UK); store at 4°C.
3. Tissue culture plates (6-well, 12-well, Corning, UK).
4. Tissue culture flask (25 cm^2, vented, Corning, UK).

2.2. hESC Maintenance

1. Mouse embryonic fibroblast conditioned medium (MEF-CM) (100 mL, R&D Systems, USA); store at −20°C.
2. BSA solution (50 mL, Sigma-Aldrich, UK); store at 4°C.
3. Human basic fibroblast growth factor (100 µg, PeproTech, USA); store at −20°C.

2.3. Passaging hESCs with Collagenase

1. Confluent well or flask of hESCs.
2. Matrigel-coated wells or flasks as appropriate.
3. Phospate-buffered saline (−MgCl$_2$, −CaCl$_2$) (500 mL, Gibco, Invitrogen, UK); store at room temperature.
4. Collagenase IV (1 g, Gibco, Invitrogen, UK); store at 4°C.
5. Mouse embryonic fibroblast conditioned medium (MEF-CM) (100 mL, R&D Systems, USA).
6. Human basic fibroblast growth factor (100 µg, PeproTech, USA).

2.4. Embryoid Body Formation from hESCs

1. Confluent well of hESCs.
2. Phosphate-buffered saline (−MgCl$_2$, −CaCl$_2$) (500 mL, Gibco, Invitrogen, UK); store at room temperature.
3. Collagenase IV (1 g, Gibco, Invitrogen, UK); store at 4°C.
4. DMEM (500 mL, Gibco, Invitrogen, UK); store at 4°C.
5. Foetal Bovine Serum (500 mL, Gibco, Invitrogen, UK); store at −20°C.
6. L-Glutamine (100 mL, Gibco, Invitrogen, UK); store at −20°C.
7. Low cluster microplates (Costar, UK).
8. Chamber slides (BD Biosciences, UK).

2.5. Differentiation of hESCs to Hepatic Endoderm

1. RPMI 1640 (500 mL, Gibco, Invitrogen, UK); store at 4°C.
2. B27 supplement (10 mL, Gibco, Invitrogen, UK); store at −20°C.
3. Activin A (2 µg, PeproTech, USA); store at −20°C.
4. Recombinant mouse Wnt3a (2 µg, R&D Systems, USA); store at −20°C.
5. KO-DMEM (500 mL, Gibco, Invitrogen, UK); store at 4°C.
6. KO-SR (500 mL, Gibco, Invitrogen, UK); store at −20°C.
7. Non-essential amino acids (100 mL, Gibco, Invitrogen, UK); store at 4°C.
8. β-Mercaptoethanol (10 mL, Gibco, Invitrogen, UK); store at 4°C.
9. DMSO (Sigma-Aldrich, UK); store at room temperature.
10. Leibovitz L-15 culture medium (500 mL, Sigma-Aldrich, UK); store at 4°C.
11. Tryptose phosphate broth (100 mL, Sigma-Aldrich, UK); store at 4°C.
12. Foetal bovine serum, heat-inactivated (500 mL, Gibco, Invitrogen, UK); store at −20°C.
13. Hydrocortisone 21-hemisuccinate (100 mg, Sigma-Aldrich, UK); store at −20°C.
14. Insulin (bovine pancreas) (100 mg, Sigma-Aldrich, UK); store at −20°C.
15. L-Glutamine (100 mL, Gibco, Invitrogen, UK); store at −20°C.
16. Ascorbic acid (25 g, Sigma-Aldrich, UK); store at −20°C.
17. Human HGF (10 µg, PeproTech, USA); store at −20°C.
18. Recombinant human oncostatin M (OSM) (50 µg, R&D Systems, USA); store at −20°C.
19. Syringe-driven filter unit 0.22 µm (Millipore, UK)

2.6. Characterisation of hESC-Derived Hepatic Endoderm

2.6.1. Immunostaining

1. Phosphate-buffered saline (−$MgCl_2$, −$CaCl_2$) (500 mL, Gibco, Invitrogen, UK); store at room temperature.
2. PBST, PBS made up with 0.1% TWEEN 20 (Sigma-Aldrich, UK).
3. Paraformaldehyde (PFA) (Sigma-Aldrich, UK) is made up in PBS; store −20°C.
4. Glycerol (Sigma-Aldrich, UK); store at room temperature.
5. Tris base (Sigma-Aldrich, UK); store at room temperature.
6. Ethanol.
7. Serum (AbD Serotech, UK); store at −20°C.
8. Primary antibodies (Table 1).

Table 1
The antibodies used for hESC-derived HE immunostaining (Fig. 2c), the concentrations used, the species developed in and the supplier are detailed below

Primary antibodies

Antigen*	Type	Supplier	Dilution
AFP	Mouse monoclonal	Sigma-Aldrich	1/500
ALB	Mouse monoclonal	Sigma-Aldrich	1/500
β-Tubulin III	Mouse monoclonal	Sigma-Aldrich	1/1,000
Cyp 3A	Sheep polyclonal	University of Dundee	1/1,000
E-CAD	Mouse monoclonal	Dako	1/500
HNF4α	Rabbit polyclonal	Santa Cruz	1/100
SMA	Mouse monoclonal	Santa Cruz	1/50

9. Secondary antibody, Alexa fluorophores (Molecular Probes, Invitrogen, UK).
10. MOWIOL 4–88 (Polysciences Inc, USA) is made up in tris HCL and glycerol as per manufacturer's instructions. DAPI (Pierce, Thermo Fisher Scientific, UK) is added to the MOWIOL solution at a 1:1000 dilution.

2.6.2. RNA Isolation and Extraction

1. TRIZOL (Invitrogen, UK); store at 4°C.
2. Chloroform (Sigma-Aldrich, UK); store at room temperature.
3. Isopropanol (Sigma-Aldrich, UK); store at room temperature.
4. Ethanol.

2.6.3. Reverse Transcription PCR

1. RT kit – random hexamers, nucleotides, reverse transcriptase and buffer (Promega, USA); store at –20°C.
2. 0.5 mL thin-walled eppendorf (Eppendorf, Fisher Scientific, UK).

2.6.4. TAQMAN Quantitative Reverse Transcriptase Polymerase Chain Reaction

1. Primers (Applied Biosystems, UK) (Fig. 2a, b).
2. PBS (–$MgCl_2$, –$CaCl_2$) (500 mL, Gibco, Invitrogen, UK).
3. TRIZOL (Invitrogen, UK).
4. Ethanol.
5. RNA extraction kit (Invitrogen, UK).
6. SuperScript III RT kit (Invitrogen, UK).

2.7. Functional Analysis of Hepatic Endoderm and Normalisation (per mg Protein)

1. P450-GLO CYP3A, CYP1A2 (Promega, USA) (Fig. 3).
2. White flat bottom 96-well assay plate (BD Biosciences, UK).
3. BCA assay kit (Pierce, Thermo Fisher Scientific, UK).

2.7.1. Cytochrome P450 Assays

2.7.2. Albumin Secretion

1. Albumin ELISA kit (Alpha Diagnostic, USA).

2.7.3. Ureagenesis

1. PBS (Gibco, Invitrogen, UK).
2. Ammonium chloride (Sigma-Aldrich, UK).
3. Cuvette (Innovative Lab Supply, USA).
4. Glutamate dehydrogenase (Sigma-Aldrich, UK).
5. Urea (Sigma-Aldrich, UK).
6. NADH (Sigma-Aldrich, UK).
7. ADP (Sigma-Aldrich, UK).
8. α-Ketoglutarate (Sigma-Aldrich, UK).

3. Methods

3.1. Coating of Culture Dishes and Flasks with Matrigel

1. Thaw the 10 mL stock bottle of Matrigel overnight at 4°C on ice and then add 10 mL of KO-DMEM. Mix well using a pipette and store 1 mL aliquots at −20°C.
2. Thaw an aliquot of Matrigel at 4°C for at least 2 h or overnight to avoid the formation of a gel (can be stored at 4°C for 2 weeks).
3. Add 5 mL of cold KO-DMEM to the Matrigel; mix well with a pipette.
4. Make up to 15 mL with cold KO-DMEM and mix using a pipette.
5. Add Matrigel to the plate or flask to be coated (Table 2).

Table 2
Recommended volumes of Matrigel for coating typical plasticware for hESC culture

Plate/flask	Volume/well or flask
12-well plate	0.5 mL per well
6-well plate	1 mL per well
25-cm^2 flask	2 mL per flask

6. Incubate the coated plate or flask overnight at 4°C.
7. Plates or flasks which have been coated with Matrigel can be stored at 4°C for up to 1 week. They should be clearly labelled with the date they were coated. Discard any plates or flasks not used within 1 week.
    (a) Before use, allow the coated culture container to come up to room temperature inside a tissue culture hood.
    (b) Immediately prior to use, aspirate the Matrigel and add the cell suspension to the well or flask.

3.2. Routine hESC Maintenance

3.2.1. Preparation of bFGF

1. All steps are to be carried out in a tissue culture hood under aseptic conditions.
2. Prepare 10% BSA solution in PBS and filter through a 0.22 μm filter.
3. From the 10% BSA solution, prepare a 0.2% BSA solution.
4. Add 10 mL 0.2% BSA solution/100 μg hbFGF.
5. Pre-wet a 0.22-μm filter by filtering 5 mL 10% BSA solution through the filter. Discard the 10 mL of BSA wash.
6. Filter the hbFGF through the pre-washed filter.
7. Aliquot the hbFGF in sterile eppendorfs and store at −20°C.

3.2.2. The Cells Need to Be Examined and Fed Daily

1. Examine under the microscope for contamination, cell morphology and confluence.
2. Aspirate the spent medium.
3. Add an appropriate volume of fresh MEF-CM+human bFGF (final concentration 4 ng/mL) or other serum free media (26).

3.3. Passaging Cells with Collagenase

Normal hESC lines will reach confluence every 5–7 days following passaging at a 1:3 split ratio. Early passage hESCs in the presence of stroma grow slower, and the time on the Matrigel can become an important factor. As a rule, hESCs should not be left longer than 14 days on the same Matrigel due to matrix degradation.

1. All steps are to be carried out in a tissue culture hood under aseptic conditions.
2. Ensure there is a new Matrigel-coated TC dish or flask prepared as per Sect. 3.1.
3. Decide on the desired split ratio for the cells. A number of factors are involved in deciding the split ratio:
    (a) High level of stroma with low numbers of colonies can be passaged back to a smaller well size or can be passaged 1:1, which will get rid of some stroma and thus increase the colony to stroma ratio, promoting hESC growth.

(b) High level of stroma with large colonies, depending on the number of colonies, can be passaged 1:1 or 1:2 if there are enough large hESC colonies.

(c) Typical growing hESCs with a little stroma or no stroma and/or some differentiation can be passaged 1:2 or 3.

4. Aspirate the media from the well or flask.
5. Wash once with 2 mL PBS ($-MgCl_2$, $-CaCl_2$).
6. Add an appropriate volume of collagenase (200 U/mL diluted in KO-DMEM) and incubate at 37°C for 2–5 min. From 2 min onwards examine regularly under the microscope at 1 minute intervals. At the point the differentiated cells in the culture begin to lift off and the colonies begin to lift at the edge, the cells are ready to be passaged.
7. Aspirate the collagenase.
8. Wash once with 2 mL PBS ($-MgCl_2$, $-CaCl_2$).
9. Add an appropriate volume of MEF-CM depending on the split ratio, and using a cell scraper, physically remove the cells from the surface of the well or flask, then triturate gently by pipetting up and down 2–3 times using a 10 mL pipette. It is important that the hESCs are kept in clumps of cells and are not broken up into single cells.
10. Replate the resulting cell suspension onto the new Matrigel-coated flasks or wells.
11. Make the volume of MEF-CM up to 4 mL for a well of a 6-well plate.
12. When placing the cells in the incubator, agitate the tissue culture container to ensure as even as possible a distribution of colonies as the colonies tend to settle in the centre of the tissue culture plate/flask affecting cell replating and subsequently growth and differentiation.

3.4. Embryoid Body Formation from hESCs

3.4.1. Preparation of Embryoid Body Medium

1. All steps are to be carried out in a tissue culture hood under aseptic conditions.
2. For EB medium, mix 80% DMEM, 20% FCS and 1% L-glutamine.
3. Add all components to a 500 mL filter unit and filter under vacuum; store at 4°C.
4. Aspirate the medium from the cells.
5. Add 2 mL of PBS.
6. Aspirate the PBS.
7. Add an appropriate volume of collagenase and incubate at 37°C for 2–5 min.

8. From 2 min onwards examine the cells under the microscope. At the point the differentiated cells in the culture begin to lift off and the colonies begin to lift at the edge, the cells are ready to be passaged.
9. Aspirate the collagenase.
10. Add 2 mL of PBS.
11. Aspirate the PBS.
12. Add 4 mL of EB medium.
13. Gently scrape the colonies off the well.
14. Gently pipette the cells using a 5 mL pipette to dissociate the cells into smaller clumps.
15. Transfer to a low cluster plate, label with cell line information and incubate at 37°C.
16. 48 h later transfer the EBs to a 15 mL conical bottom tube and leave on the bench for 15 min to allow the clumps to settle to the bottom of the tube.
17. Aspirate the supernatant.
18. Add 5 mL of fresh EB media to the tube.
19. Transfer the EBs back into the low cluster plate and incubate at 37°C.
20. Feed in this way every 2 days.
21. After 7 days in suspension, transfer the aggregates to the 15 mL tube and leave for 15 min for the aggregates to settle.
22. Plating of EBs for analysis. Transfer the aggregates to gelatin-coated chamber slides in 1 mL of EB medium and incubate at 37°C. Aim to have approximately 4 aggregates per chamber.
23. Feed the slides every 2–3 days for an additional 14 days by aspirating the medium and adding fresh EB medium.
24. Examine the slides daily to check for cells starting to differentiate and note the presence of beating bodies.
25. After 14 days in culture on the slides, aspirate the medium.
26. Add 0.5–1 mL of PBS.
27. Aspirate the PBS.
28. Add 0.5 mL of 4% PFA in PBS.
29. Leave at room temperature for 20 min.
30. Aspirate the PFA.
31. Wash 3 times with PBS.
32. After the final wash, slides can be stored in PBS at 4°C, until ready to do the staining assay.

3.5. Preparation of Media for Differentiation of hESCs to Hepatic Endoderm

All media preparation should be carried out in a tissue culture hood under aseptic conditions.

3.5.1. Preparation of RPMI: B27 Priming Medium for Hepatocyte Differentiation

1. For RPMI-B27 medium, mix RPMI 1640 (500 mL) and B27 (50x, 10 mL).
2. Swirl to mix components.
3. Add all components to a filter unit and filter under vacuum; store at 4°C.

Preparation of Human Activin A Stock Solution

1. Add 1 mL of 0.2% BSA into a syringe and pre-wet the filter.
2. Dilute activin A in 0.2% BSA to a stock concentration of 100 μg/mL.
3. Filter the activin A solution and aliquot in sterile eppendorfs; store at −20°C.

Preparation of Mouse Wnt3a Stock Solution

1. Add 200 μL of PBS to a 2 μg vial of Wnt3a to a stock concentration of 10 μg/mL.
2. Aliquot in sterile eppendorfs and store at −20°C.

3.5.2. Preparation of SR-DMSO Medium for Hepatocyte Differentiation

1. For SR-DMSO medium, mix 80% KO-DMEM, 20% KO-SR, 0.5% L-glutamine, 1% non-essential amino acids, 0.1 mM β-mercaptoethanol and 1% DMSO.
2. Filter the solution under vacuum, store at 4°C and aliquot and store at −20°C if required.
3. Use 4 mL per well of a 6-well plate and 6 mL per T25 flask.

3.5.3. Preparation of L15 Maturation Medium for Hepatocyte Differentiation

1. For L-15 medium, mix 500 mL Leibovitz L-15 medium, tryptose phosphate broth (final concentration 8.3%), heat-inactivated foetal bovine serum (final concentration 8.3%), 10 μM hydrocortisone 21-hemisuccinate, 1 μM insulin (bovine pancreas), 1% L-glutamine and 0.2% ascorbic acid.
2. Filter the solution under vacuum, store at 4°C and aliquot and store at −20°C if required.

Preparation of Human HGF Stock Solution (1,000×)

1. Dilute the HGF in PBS to a stock concentration of 10 μg/mL.
2. Filter the HGF solution and aliquot in sterile eppendorfs; store at −20°C.

Preparation of Oncostatin M Stock Solution (1,000×)

1. Dilute the OSM in PBS to a stock concentration of 20 μg/mL.
2. Filter the OSM solution and aliquot in sterile eppendorfs; store at −20°C.

3.5.4. Preparation of Final RPMI 1640-B27 Priming Medium

1. Dispense the required volume of priming medium for the experiment (Sect. 3.5.1) (1 mL per well of a 6-well plate and 2 mL per T25 flask).
2. Add activin A to a final concentration of 100 ng/mL.
3. Add recombinant Wnt3a to a final concentration of 50 ng/mL.
4. Mix well and the media is now ready for use.
5. This final media should be made up fresh each day.

3.5.5. Preparation of Final L-15 Maturation Medium

1. Dispense the required volume of L-15 medium for the experiment (Sect. 3.5.3) (4 mL per well of a 6-well plate and 6 mL per T25 flask).
2. Add HGF to a final concentration of 10 ng/mL.
3. Add OSM to a final concentration of 20 ng/mL.
4. Mix well and the media is now ready for use.
5. This final media should be made up fresh each day.

3.6. Differentiation of hESCs to Hepatic Endoderm

3.6.1. Priming hESCs to Definitive Endoderm

1. Culture hESCs and propagate on Matrigel-coated plates with mouse embryonic fibroblast (MEF)-CM supplemented with bFGF.
2. Initiate hepatic differentiation (Fig. 1) when hESCs reach a confluency level of approximately 30–60% (depending on the hESC line) by replacing the MEF-CM with priming medium (RPMI 1640-B27 (Sect. 3.5.1) supplemented with 100 ng/mL activin A (section "Preparation of human activin A stock solution") and 50 ng/mL Wnt3a (section "Preparation of mouse Wnt3a stock solution").
3. The cells are cultured in priming medium for 3 days (changing the medium every 24 h), and activin A and Wnt3a are added fresh daily (Fig. 1).

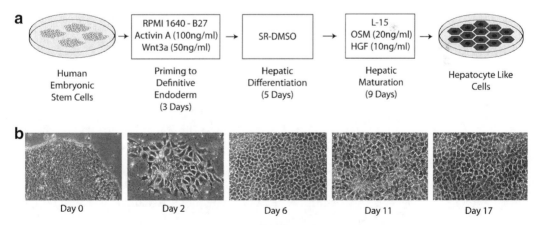

Fig. 1. Hepatic endoderm differentiation from hESCs. (**a**) Flow diagram of the differentiation protocol as described by Hay et al. (2008). (**b**) Phase contrast microscopy images (x10) demonstrating changes in morphology over the 17-day differentiation protocol. At day 0, cells display typical hESC morphology and grow as colonies of small cells with large nuclei. By day 2, cells are differentiated into definitive endoderm, and by day 6 they have taken on a hepatic shape. By day 11 they display mature hepatocyte morphology, and at day 17 we observe large hexagonal cells with multiple nuclei and visible cannaliculi.

4. After 72 h in priming medium, change the medium to differentiation medium (SR-DMSO (Sect. 3.5.2)) for 5 days (changing the medium every 48 h, Fig. 1).

5. Culture the cells in maturation and maintenance medium (L-15 (Sect. 3.5.3)) supplemented with 10 ng/mL hHGF (section "Preparation of Human HGF Stock Solution (1,000×)") and 20 ng/mL OSM (section "Preparation of Oncostatin M Stock Solution (1,000×)") for 9 days (changing medium every 48 h). hHGF and OSM are added fresh each day.

6. The cells gradually exhibit morphological changes from a spiky/triangular shape to a characteristic liver morphology displaying a polygonal appearance (Fig. 1b).

3.7. Characterisation of hESC-Derived Hepatic Endoderm

3.7.1. Immunostaining

1. Wash hESC-derived HE with PBS twice, 5 min each wash (Fig. 2c).

2. Fix the HE with 4% PFA for 20 min at room temperature (the cells can be stored in PBS at 4°C and stained at a later date). The cells can also be fixed with ice-cold methanol for 10 min.

3. Wash the cells twice with PBS, 5 min each wash.

4. Incubate the cells for 2 min at room temperature with 100% ethanol for nuclear staining (this step is not required if methanol fixation is used).

5. Wash the cells twice with PBS, 5 min each wash.

6. Block the cells with PBS/T (0.1% TWEEN)/10% serum for 1 h at room temperature.

7. Remove the serum and add the respective primary antibody diluted in 1% serum (made up in PBST) and incubate for 2 h at room temperature, or overnight at 4°C with agitation. (For primary antibody details, see Table 1).

8. Wash the cells 3 times with PBS at room temperature, 5 min each wash.

9. Add the appropriate secondary Alexa Fluor antibody (1:400) diluted in PBS to the cells and incubate at room temperature for 1 h in the dark with agitation.

10. Wash the cells 3 times with PBS, 5 min each wash in the dark.

11. Mount each well with MOWIOL 4–88 and DAPI (1:1,000). Cover the well with a cover slip and store at 4°C in the dark until analysis.

3.7.2. RNA Isolation and Extraction

1. Wash the hESC-derived HE with PBS and aspirate.

2. Add 1 mL of TRIZOL reagent and incubate at room temperature for 5 min.

3. Scrape the cells and place in a 1.5 mL eppendorf (store at −80°C for later use if required).

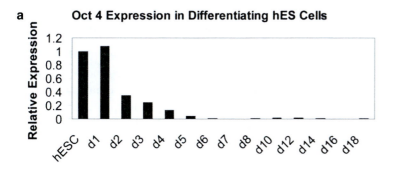

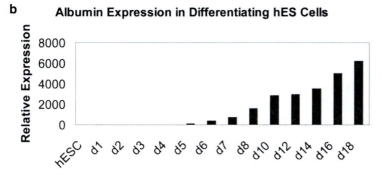

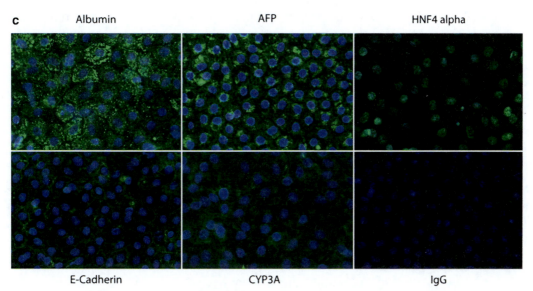

Fig. 2. Characterization of hESC-derived hepatic endoderm by qPCR and immunostaining. (a) qPCR analysis demonstrates a decrease in expression of the hESC marker Oct4 over the course of the hepatic differentiation protocol. (b) qPCR analysis demonstrates an increase in expression of the hepatocyte marker albumin over the course of the hepatic differentiation protocol. (c) Fluorescence microscopy of hESC-derived hepatic endoderm on day 17 of differentiation confirms the homogeneity of our process (~90%) and the presence of several hepatocyte markers: albumin, AFP, CYP3A, E-cadherin and HNF4α. Results are compared to an IgG primary antibody negative control.

4. Add 0.5 mL of chloroform to the eppendorf and mix by inverting; make sure this is done in a fume hood.
5. Centrifuge the solution at 13,000 rpm for 15 min at 4°C.
6. Collect the aqueous layer and place into a clean eppendorf; make sure there is no contamination from the interface.
7. Add 1 mL of isopropanol and mix by inverting; leave at room temperature for 10 min to precipitate the RNA.
8. Centrifuge at 13,000 rpm for 10 min at 4°C.
9. Aspirate the supernatant and make sure not to disturb the RNA pellet. Wash with 0.5 mL of 70% ethanol and leave at room temperature for 5 min.
10. Centrifuge at 8,000 rpm for 5 min at 4°C.
11. Aspirate the ethanol and leave to dry at room temperature for 5–10 min.
12. Once all the ethanol has evaporated, resuspend the pellet in 30 µL of deionised water. Store the RNA at −80°C for later use.
13. Quantify the RNA concentration using a nanodrop.

3.7.3. Reverse Transcription PCR

1. Set up a reaction using previously isolated RNA (200 ng), random hexamers, nucleotides (10 mM) reverse transcriptase and the respective buffer in a thin-walled 0.5 mL eppendorf.
2. Set up a negative RT, the above reaction without the reverse transcriptase.
3. Place the tubes into a thermocycler, PCR machine, and set up the appropriate program.
4. Store the cDNA at −20°C for later use if required.

For details regarding semi-quantitative PCR, please refer to Note 7.

3.7.4. TAQMAN Quantitative Reverse Transcriptase Polymerase Chain Reaction

Harvest cells at different time points throughout the differentiation protocol. Extract the RNA and carry out reverse transcription qPCR (Fig. 2a, b) using the primers below from Assay-on-Demand, Applied Biosystems:

1. Oct 4 Hs03005111_g1
2. Albumin Hs00910225 m1

For RNA isolation and extraction, please refer to Sect. 3.7.2.

1. Reverse transcription and TAQMAN qPCR:
   (a) Take 1 µg of RNA and reverse-transcribe to cDNA using Invitrogen's SuperScript III reverse transcription kit, as per manufacturer's instructions.
   (b) 1 µL of the cDNA used in a 25-µL TAQMAN reaction consisting of the appropriate primers from Applied Biosystems,

18 S ribosomal control primers and Invitrogen's 2x platinum qPCR supermix UDG with rox and an appropriate volume of water.

(c) Mix well and place 10 µL of each sample into 2 wells of either a 96- or 384-well qPCR plate.

(d) Once all samples are loaded (plus the appropriate controls), seal the plate and analyse on the Applied Biosystems 7900HT TAQMAN machine.

(e) Results are expressed as relative expression over a control sample.

3.8. Functional Analysis of Hepatic Endoderm

3.8.1. Cytochrome P450 Assays

1. Incubate day 17 hESC-derived HE with the specific substrate for 5 h at 37°C ($n=3$). Use tissue culture media, supplemented with the appropriate substrate, as a negative control and incubate at 37°C for 5 h (Fig. 3).

2. Collect the supernatants and carry out the assay as per manufacturer's instructions.

3. Measure the relative levels of basal activity and normalise to per mg protein as determined by the BCA assay.

3.8.2. Albumin Secretion

1. Add 1 mL of L-15 medium on days 14, 15 and 16 of the differentiation protocol to the hESC-derived HE and leave for 24 h at 37°C ($n=3$).

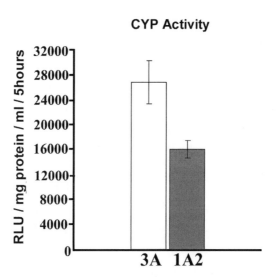

Fig. 3. PGlo system (Promega) was used to assess cytochrome P450 activity (CYP3A and CYP1A2). hESC-derived hepatic endoderm was incubated with the appropriate substrate for 5 h. Following this tissue culture, supernatants were harvested and activities measured using a luminometer. Relative luminescence units (*RLU*) are stated with background removed. Values are normalised to protein (mg/well) and media volume (mL), and are shown as the mean ± standard deviation ($n=4$).

3.8.3. Ureagenesis

2. Collect the supernatants (store at −80°C for later use if required).
3. Carry out the albumin assay as per manufacturer's instructions and normalise to per mg protein as determined by the BCA assay.

1. Wash cells with PBS.
2. Add 1 mL of PBS to each well.
3. Add ammonium chloride made up in PBS to a final concentration of 16 µM and use the PBS-only wells as a control.
4. Incubate for 4 h at 37°C.
5. Collect the supernatant (store at −20°C for later use).
6. Add 100 µL of the sample to a cuvette and add 1 mL of buffer (150 mL PBS, 300 µL 0.1 M ADP, 50 mg NADH and 750 µL of 0.5 M alpha ketoglutarate, pH 7.4) and read at 340 nm on a spectrophotometer.
7. Dilute glutamate dehydrogenase 1:5 with water and add 10 µL into the cuvette. Mix by inverting and incubate at room temperature for 2 h.
8. Read the absorbance at 340 nm.
9. Dilute urease 1:3.5 with water and add 10 µL into the cuvette. Mix by inverting and incubate at room temperature for 2 h.
10. Read the absorbance at 340 nm.
11. Calculate the urea concentration using the following formula: Urea (mmol/L) = Abs/6.22/2 (originated from the following formula: Abs = $\varepsilon \times l \times c$, where ε is the molar extinction coefficient, which in this case is 6.22 mmol^{-1} cm^{-1} for NADH, l is the length of the cuvette and c is the concentration).

4. Notes

1. All volumes are based on a 6-well plate format. Adjust the volumes accordingly for the required plate or flask.
2. All priming, differentiation and maturation media are filtered under vacuum before use.
3. Priming, differentiation and maturation media are stored at 4°C for no longer than 2 weeks. Assess how much medium is required for the experiment and aliquot the remaining media and store at −20°C for future use.
4. Matrigel is made up as per manufacturer's instructions; 1 mL aliquots can be stored at −20°C until use.

5. Growth factors, once made up and aliquoted, can be stored at −20°C, and when thawed, can be stored at 4°C for no longer than 2 weeks.

6. Primary antibodies usually used to characterise HE are albumin (1:500, Sigma-Aldrich, Saint Louis, MO), HNF4α (1:100, Santa Cruz Biotechnology Inc, CA), AFP (1:500, Sigma-Aldrich, Saint Louis, MO) and cytochrome P450 3A.

7. For details regarding the PCR primers and conditions required for generating the semi-quantitative expression of hepatic specific genes, please refer to the recent publication (26).

References

1. Wege H, Chui MS, Le HT, et al. (2003). In vitro expansion of human hepatocytes is restricted by telomere-dependent replicative aging. Cell Transplant 12: 897–906.
2. Cai J, Ito M, Westerman KA, et al. (2000). Construction of a non-tumorigenic rat hepatocyte cell line for transplantation: reversal of hepatocyte immortalization by site-specific excision of the SV40 T antigen. J Hepatol 33: 701–708.
3. Allain JE, Dagher I, Mahieu-Caputo D, et al. (2002). Immortalization of a primate bipotent epithelial liver stem cell. Proc Natl Acad Sci USA 99: 3639–3644.
4. Delgado JP, Parouchev A, Allain JE, et al. (2005). Long-term controlled immortalization of a primate hepatic progenitor cell line after Simian virus 40 T-Antigen gene transfer. Oncogene 24: 541–551.
5. Turpeinen M, Tolonen A, Chesne C, et al. (2009). Functional expression, inhibition and induction of CYP enzymes in HepaRG cells. Toxicol In Vitro 23: 748–753.
6. Yoshitomi S, Ikemoto K, Takahashi J, et al. (2001). Establishment of the transformants expressing human cytochrome P450 subtypes in HepG2, and their applications on drug metabolism and toxicology. Toxicol In Vitro 15: 245–256.
7. Herrera MB, Bruno S, Buttiglieri S, et al. (2006). Isolation and characterization of a stem cell population from adult human liver. Stem Cells 24: 2840–2850.
8. Lazaro CA, Rhim JA, Yamada Y, and Fausto N. (1998). Generation of hepatocytes from oval cell precursors in culture. Cancer Res 58: 5514–5522.
9. Rogler LE. (1997) Selective bipotential differentiation of mouse embryonic hepatoblasts in vitro. Am J Pathol 150: 591–602.
10. Reubinoff BE, Pera MF, Fong CY, et al. (2000). Embryonic stem cell lines from human blastocysts: somatic differentiation in vitro. Nat Biotechnol 18: 399–404.
11. Thomson JA, Itskovitz-Eldor J, Shapiro SS, et al. (1998). Embryonic stem cell lines derived from human blastocysts. Science 282: 1145–1147.
12. Takahashi K, and Yamanaka S. (2006). Induction of pluripotent stem cells from mouse embryonic and adult fibroblast cultures by defined factors. Cell 126: 663–676.
13. Dalgetty DM, Medine C, Iredale JP, et al. (2009). Progress and Future Challenges in Stem Cell-Derived Liver Technologies. American Journal of Physiology – Gastrointestinal and Liver Physiology. 2009. 297(2):G241–248.
14. Greenhough S, Medine CN, and Hay DC. (2010). Pluripotent Stem Cell Derived Hepatocyte Like Cells and their Potential in ToxicityScreening. Toxicology 278: 250–255
15. Hay DC. (2010) Cadaveric Hepatocytes Repopulate Diseased Livers: Life After Death. Gastroenterology 139: 715–716
16. Medine CN, Greenhough S, and Hay DC. (2010). Role of stem-cell-derived hepatic endoderm in human drug discovery. Biochem. Soc. Trans 38: 1033–1036
17. Agarwal S, Holton KL, Lanza R, et al. (2008). Efficient differentiation of functional hepatocytes from human embryonic stem cells. Stem Cells 26: 1117–1127.
18. Basma H, Soto-Gutierrez A, Yannam G, et al. (2009). Differentiation and Transplantation of Human Embryonic Stem Cell-Derived Hepatocytes. Gastroenterology 136: 990–999.
19. Cai J, Zhao Y, Liu Y, et al. (2007). Directed differentiation of human embryonic stem cells into functional hepatic cells. Hepatology 45: 1229–1239.

20. Duan Y, Catana A, Meng Y, et al. (2007). Differentiation and enrichment of hepatocyte-like cells from human embryonic stem cells in vitro and in vivo. Stem Cells 25: 3058–3068.
21. Fletcher J, Cui W, Samuel K, et al. (2008). The inhibitory role of stromal cell mesenchyme on human embryonic stem cell hepatocyte differentiation is overcome by Wnt3a treatment. Cloning Stem Cells 10: 331–339.
22. Hay DC, Zhao D, Ross A, et al. (2007). Direct differentiation of human embryonic stem cells to hepatocyte-like cells exhibiting functional activities. Cloning Stem Cells 9: 51–62.
23. Hay DC, Zhao D, Fletcher J, et al. (2008). Efficient differentiation of hepatocytes from human embryonic stem cells exhibiting markers recapitulating liver development in vivo. Stem Cells 26: 894–902.
24. Hay DC, Fletcher J, Payne C, et al. (2008). Highly efficient differentiation of hESCs to functional hepatic endoderm requires ActivinA and Wnt3a signaling. Proc Natl Acad Sci USA 105: 12301–12306.
25. Sullivan GJ, Hay DC, Park IH, et al. (2010). Generation of functional human hepatic endoderm from human induced pluripotent stem cells. Hepatology 51: 329–335.
26. Hannoun Z, Fletcher J, Greenhough S, et al. (2010). The Comparison between Conditioned Media and Serum-Free Media in Human Embryonic Stem Cell Culture and Differentiation, Cellular Reprogramming, 12:2 133–140.
27. Hay DC, Pernagallo S, Diaz-Mochon JJ, et al (2011). Unbiased Screening of Polymer Libraries to Define Novel Substrates for Functional Hepatocytes with Inducible Drug Metabolism. Stem Cell Research, 6: 92–101.
28. Payne C.M, Samuel K, Pryde A, et al (2011). Persistence of Functional Hepatocyte Like Cells in Immune Compromised Mice. Liver International, 31(2):254–62.

Chapter 28

Multistage Hepatic Differentiation from Human Induced Pluripotent Stem Cells

Su Mi Choi, Hua Liu, Yonghak Kim, and Yoon-Young Jang

Abstract

Efficient hepatic differentiation technologies designed for patient specific induced pluripotent stem (iPS) cells may provide an unlimited hepatocyte source which can be utilized in drug screening, disease modeling, and cell therapy. This chapter describes the methods we use to differentiate human iPS cells into functional hepatic cells. Our hepatic differentiation protocol has been applicable for many human iPS cell types of various sources and conditions (i.e., iPS cells derived from healthy donor or patients, as well as iPS cells generated using virus free or viral methods). The protocol is composed of three stages: definitive endoderm induction, hepatic progenitor induction and expansion, and hepatocyte maturation. We also describe the protocols for detecting iPS cell-derived functional hepatocytes both in vitro and in vivo.

Key words: Induced pluripotent stem (iPS) cells, Hepatic differentiation, Endoderm, Hepatic progenitors, Hepatocytes, Cytochrome P450, Patient specific iPS cells, Liver disease

1. Introduction

In the US, 20,000 patients actively await liver transplantation, whereas only 7,000 are performed annually (1). Clearly, the need for liver replacement far outstrips current supply, and thus, we are forced to critically evaluate alternative approaches to traditional solid organ transplantation. The use of ex vivo adult human hepatocytes is a desirable option for cellular therapies or drug testing. However, these cells have limited proliferation potential and lose function and viability upon isolation. Although there have been great advances in liver stem cell biology (2–4), hepatic stem cells are infrequent within tissue, making their isolation and expansion unfavorable for large-scale applications (5). Attempts to immortalize hepatocytes by introducing telomerase constructs and viral transfection also suffer the shortcomings of phenotypic changes, poor liver function, and karyotypic abnormalities (6, 7). Therefore, recently, there has been a focus on deriving human hepatocytes from other sources, in particular, embryonic stem (ES) cells and induced pluripotent stem (iPS) cells (8–11). These pluripotent stem cells have advantages over their adult tissue-specific counterparts because they can be expanded in culture indefinitely while maintaining normal karyotypes and differentiation capacity.

Human ES and iPS cell differentiation to hepatocyte-like cells has been improved with more efficient and functional methods including our protocol (8–11). This holds a great promise as an unlimited hepatocyte source which can be utilized in drug screening, disease modeling, and cell therapy, although some of their functions remain to be improved.

Our hepatic differentiation protocol is composed of three stages: definitive endoderm (DE) induction for 5–7 days, hepatic progenitor (HP) induction and expansion for another 4–5 days, and hepatic maturation for another 10 days (11). Cells from these three stages (approximately at days 5, 10, and 20) were chosen because we observed that these time frames were most distinct in terms of their specific marker expression profile for each stage cells (i.e., CXCR4 for the DE stage, AFP for the HP stage, or ALB positivity for the mature hepatocyte stage). The functionality of iPS cell-derived mature hepatocytes can be analyzed by multiple in vitro methods including analyses for cytochrome P450 (CYP) activities and glycogen storage ability with periodic acid-Schiff (PAS) assay. Although these in vitro methods are highly informative and convenient, the most definitive proof for the functionality of human iPS cell-derived hepatic cells would be the demonstration of (1) hepatic engraftment in vivo using animal models and (2) detection of secreted human liver proteins in animal serum/plasma. Here, we describe our hepatic differentiation protocol that is applicable for a variety of human iPS cell types (i.e., iPS cells generated using either viral or non-viral methods), as well as protocols for detecting functional hepatocytes both in vitro and in vivo.

2. Materials

2.1. Human iPS Cell Culture

1. ES culture medium–knockout Dulbecco's modified Eagle's medium (DMEM) supplemented with 20% knockout serum replacement (KOSR), 0.1 mM nonessential amino acids (NEAA), 0.1 mM 2-mercaptoethanol, 2 mM GlutaMAX, and 6 ng/mL basic fibroblast growth factor (FGF); all from Invitrogen (Carlsbad, CA).

2. Mouse embryonic fibroblast (MEF) inactivated with mitomycin C (Millipore, Billerica, MA).

3. MEF medium–DMEM (high glucose) supplemented with 10% defined fetal bovine serum (FBSd), 0.1 mM nonessential amino acids (NEAA); all from Invitrogen.
4. DMEM–F12 medium (Invitrogen).
5. Gelatin-coated 6-well plates – add 1 mL of 0.1% gelatin (Sigma-Aldrich, Saint Louis, MO) in water to 6-well plate and let sit for at least 1 h. Before use, aspirate the gelatin solution from the plate.

2.2. Differentiation of Human iPS Cells into Multistage Hepatocytes

1. *Stage 1* (Differentiation of human iPS cells to DE cells) – Roswell Park Memorial Institute 1640 medium containing 2 mM GlutaMAX, 0.5% defined FBS (or KOSR), and 100 ng/mL activin A (R&D Systems, Minneapolis, MN).
2. *Stage 2* (Differentiation of DE to HP cells) – minimal Madin–Darby bovine kidney maintenance medium (Sigma-Aldrich) supplemented with GlutaMAX and 0.5 mg/mL bovine serum albumin, 10 ng/mL human FGF4, and 10 ng/mL human HGF (R&D Systems).
3. *Stage 3* (Differentiation of HP cells to mature hepatocytes) – Williams' E medium (WEM), 5% heat-inactivated FBS, 1×GlutaMAX, 15 mM 4-(2-hydroxyethyl)-1-piperazine ethanesulfonic acid (Hepes), 2 μg/mL insulin, 10 ng/mL FGF4, 10 ng/mL HGF, 10 ng/mL oncostatin M (R&D Systems), and 10^{-7} M dexamethasone (Sigma-Aldrich).
4. 0.05% trypsin/0.53 mM ethylene diamine tetraacetic acid (EDTA) (Invitrogen).
5. Collagen I-coated dishes – add 1 mL of 50–60 μg/mL collagen I solution (Invitrogen, dissolved in 0.02 M acetic acid, Sigma-Aldrich) to each well of a 6-well cell culture plate and let it sit for ~1 h. Rinse the plate three times with PBS to remove the acid and use the plate immediately.

2.3. Immunofluorescence and Flow Cytometry

1. Permeabilizing solutions – 0.1% Triton X-100 and 0.3% bovine serum albumin (BSA) in PBS.
2. Blocking solution for immunofluorescence – 0.1% Triton X-100 and 1% BSA in PBS.
3. 1 mg/mL of 4′,6-diamidino-2-phenylindole (DAPI).
4. Rat (IgM) anti-human stage-specific embryonic antigen-3 (SSEA-3) (1:200 dilution, Millipore, Temecula, CA).
5. Rabbit (IgG) anti-human alpha-fetoprotein (AFP, 1:200 dilution, Dako, Carpinteria, CA).
6. Rabbit polyclonal anti-human alpha 1-antitrypsin (AAT; 1:200 dilution, Thermo Scientific, Rockford, IL).
7. Rabbit anti-human cytochrome P450 3A4 (CYP3A4; 1:200 dilution, Enzo Life Sciences, Farmingdale, NY).

8. The secondary antibodies all belonged to the Alexa Fluor series from Invitrogen.
9. FITC-conjugated polyclonal rabbit anti-human albumin (1:200 dilution, Dako).
10. PE-conjugated mouse anti-human chemokine (C-X-C motif) receptor 4 (CXCR4, 1:100 dilution, BioLegend, San Diego, CA).

2.4. Periodic Acid-Schiff (PAS) Assay

1. 4% paraformaldehyde (MP Biomedicals, Aurora, OH) in phosphate-buffered saline (PBS).
2. 0.5% periodic acid (Cat. No. AR165, Dako); store at room temperature.
3. Schiff's reagent (Cat. No. AR165, Dako); store at 4°C (see Note 1).
4. Mayer's hematoxylin (Cat. No. AR165, Dako); store at room temperature.
5. Bluing reagent (Cat. No. AR165, Dako); store at room temperature.
6. VectaMount AQ (Vector Laboratories, Burlingame, CA); store at room temperature.

2.5. Cytochrome P450 (CYP450) Assay

1. P450-Glo™ CYP1A2 assay (Cat. No. V8771, Promega, Madison, WI); store at −20°C, except luminogenic substrate which must be stored at −70°C.
2. P450-Glo™ CYP3A4 assay (Cat. No. V8901, Promega); store at −20°C, except luminogenic substrate which must be stored at −70°C.
3. The abovementioned kits include a luminogenic CYP substrate, a luciferin detection reagent (lysophilized), and a reconstitution buffer.
4. 96-well opaque white luminometer plates (Costar Corning, Lowell, NY).
5. A luminometer (GloMax, Promega).

2.6. Liver Engraftment Assay

1. 0.05% trypsin/0.53 mM EDTA (Invitrogen).
2. Phosphate-buffered saline (PBS) Ca^{2+} and Mg^{2+} free (Invitrogen).
3. 40-μm cell strainer (BD Falcon™, Franklin Lakes, NJ).
4. 0.3- to 1-mL syringe (BD, Franklin Lakes, NJ).
5. NOD.Cg-$Prkdc^{scid}$ $Il2rg^{tm1Wjl}$/SzJ (NSG) mice (Jackson, Bar Harbor, Maine).

3. Methods

3.1. Cell Culture

1. Prepare MEF feeder plates (or matrigel plates) 1 day before passaging iPS cells.
2. Passage human iPS cells onto either the MEF feeder layers with ES medium or matrigel-coated culture plates with MEF-conditioned media (CM).
3. Culture these cells at 37°C in a humidified atmosphere of 5% CO_2 by changing medium everyday for 3–5 days.

3.2. Differentiation of Human iPS Cells into Multistage Hepatocytes

Stage 1. Definitive endoderm induction

1. When the confluence of iPS cells reaches 50–60%, replace the medium with DE medium (described in materials) which contains activin A (see Note 2).
2. Change the DE medium everyday or every other day for 5–7 days.
3. (Optional) After the endoderm induction, we recommend testing the efficiency of DE differentiation. Cell surface marker analysis for the expression of both undifferentiated pluripotent stem cells (i.e., either SSEA3 or TRA-1-60) and DE cells (i.e., CXCR4) using flow cytometry provides the most sensitive and quantitative way of endoderm differentiation (see the corresponding section "Flow cytometry" and Fig. 1). The efficiency of DE induction is usually ~80% at day 4 and over 90% at day 5 after activin A treatment, as we have previously shown for various types of human iPS cells generated using either viral or non-viral methods (11, 12).

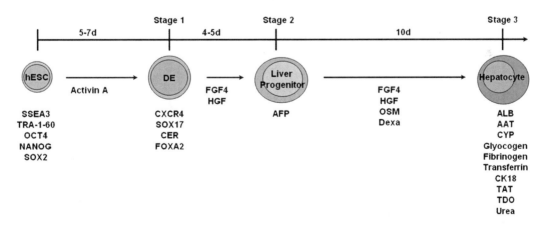

Fig. 1. A schematic representation of differentiation steps of human iPS cells into hepatocytes. The key factors that are responsible for each stage differentiation and the time course of hepatic commitment are shown in this diagram. Stage specific markers including genes and surface phenotypes are indicated under each stage cell types.

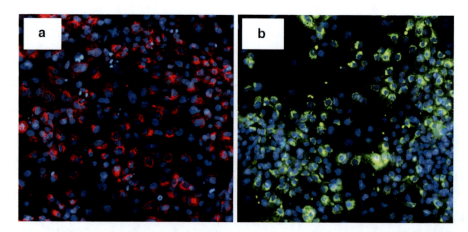

Fig. 2. Hepatic differentiation from virus-free and integration-free patient iPS cells. Human iPS cells were generated with a non-viral method (13) from primary fibroblasts derived from a non-tumor portion of a hepatocellular carcinoma (named as iNF1). These integration-free iNF1 cells were able to differentiate into (**a**) Hepatic progenitors and (**b**) Mature hepatocytes with a comparable efficiency shown with the iPS cells generated with a viral method (11).

Stage 2. Hepatic progenitor induction and expansion

4. On day 5–7 after DE induction, dissociate the cells with 0.05% trypsin/0.53 mM EDTA and plate the DE cells onto collagen I-coated culture dishes in the presence of HP stage medium and reagents (described in materials). Change medium every other day.

5. (Optional) After HP cell induction, we recommend testing the expression of early hepatic cell markers such as AFP by immunofluorescence analysis (see section "Immunofluorescence"). The differentiation efficiency for HP cells is usually over 90% for all of our human iPS cell types including the virus-free iPS cells from liver disease patients (Fig. 2) and those from healthy donor tissues of different origins (11).

Stage 3. Hepatic maturation

6. On day 10 after the initiation of hepatic differentiation, switch the medium with complete hepatocyte culture medium and reagents (described in materials). Change medium every other day for another 10 days.

7. (Optional) At the end of the final differentiation stage, we recommend testing the expression of mature hepatocyte markers such as ALB, AAT, and CYP3A4, as well as analyses for functional hepatocytes (see sections "Immunofluorescence to Cytochrome P450 (CYP450) Assay – CYP3A4 and CYP1A2 Activity").

3.3. Assessment of Hepatic Differentiation

3.3.1. Flow Cytometry

1. Dissociate cells with 0.05% trypsin/0.53 mM EDTA and spin the cells at $200 \times g$ for 5 min.

2. Incubate the cells with each fluorescence-conjugated specific antibodies (such as CXCR4 and SSEA3) in PBS (if needed, use a permeabilizing solution) and isotype control antibodies as

negative controls for 1 h at room temperature in the dark, and wash the cells three times with PBS.

3. Analyze the expression with a flow cytometry using the software CellQuest™.

3.3.2. Immunofluorescence

1. Fix cells with 4% paraformaldehyde in PBS for 20 min at room temperature.

2. To permeabilize the cells and inhibit nonspecific antibody binding, incubate cells with 0.1% Triton X-100 and 1% bovine serum albumin for 30 min at room temperature, and wash the cells two times with PBS.

3. Incubate the cells with a primary antibody (i.e., AFP, ALB, AAT, and CYP3A4) in the permeabilizing solution for 1–4 h (depending upon antibodies) at room temperature in the dark, and wash the cells three times with PBS.

4. Incubate cells with an appropriate fluorescence-conjugated secondary antibody for 1 h at room temperature in the dark and counterstain with 1 mg/mL of DAPI for 5 min at room temperature in the dark.

5. Wash the cells three times with PBS in the dark.

6. Image the labeled cells with an immunofluorescence microscope.

3.3.3. Periodic Acid-Schiff (PAS) Assay (see Note 3)

1. Fix human iPS-derived mature hepatocytes with 4% paraformaldehyde in PBS for 20 min at room temperature.

2. Wash the cells three times with PBS (see Note 4).

3. Add 0.5% periodic acid solution for 10 min and wash the cells three times with distilled water for 3 min.

4. Stain the cells in Schiff's reagent for 10 min and wash them three times with distilled water for 3 min (see Note 5).

5. Counterstain the cells with Mayer's hematoxylin solution for 30 s to 2 min, and wash three times with distilled water for 3 min.

6. Dehydrate the cells with 95% alcohol and 100% alcohol sequentially, and mount them with a mounting solution (VectaMount AQ).

7. Image the stained cells with microscope; glycogen is pink or red color in cytoplasm, and nuclei are blue.

3.3.4. Cytochrome P450 (CYP450) Assay – CYP3A4 and CYP1A2 Activity

To measure CYP activities in mature heptocytes, we use non-lytic assays which are useful for cell-based experiments. Since the cell numbers in each hepatic culture dish may vary, we normalize the CYP activities using the percentages of ALB-expressing hepatocyte-like cells (11).

1. Incubate iPS cell-derived mature hepatocytes with fresh hepatocyte culture medium(described in materials) containing 50 μM of luminogenic CYP3A4 substrate or 100 μM of CYP1A2 substrate for 4 h at 37°C. To determine background luminescence, add luminogenic substrate in medium to set of empty wells (no cells).
2. Transfer 50 μL of medium from each well to a 96-well opaque white luminometer plate (or appropriate luminometer tubes) at room temperature, and add 50 μL of Luciferin Detection Reagent.
3. Incubate the plate at room temperature for 20 min and read luminescence using a luminometer (see Note 6).
4. For normalization, dissociate the cells with 0.05% trypsin/0.53 mM EDTA and centrifuge at $200 \times g$ for 5 min.
5. Fix the cells with 4% paraformaldehyde for 20 min at room temperature, and wash them with PBS.
6. Incubate the cells with FITC-conjugated polyclonal rabbit anti-human albumin in permeabilizing solution (0.1% Triton X-100 and 0.3% bovine serum albumin in PBS) for 1 h at room temperature in the dark, and wash the cells three times with PBS.
7. Analyze the albumin positivity using flow cytometry.
8. To normalize CYP activities, calculate the measured luminescence values by the percentage of albumin-expressing hepatocytes (11).

3.3.5. Liver Engraftment and In Vivo Functional Assays

1. Trypsinize human iPS-derived hepatic cells (i.e., DE, HP, and mature hepatocytes) to generate single cells for transplantation (see Note 7).
2. Wash and re-suspend the cells using sterile PBS.
3. Filter the cells using 40-μm sterile filter.
4. Suspend 1 to 2×10^6 cells in 100 μL PBS (see Note 8), and transfer the cells to a 0.3–1-mL syringe before injection.
5. Transplant these cells intra-splenically, or intravenously (i.e., via portal vein or tail vein) into immunodeficient NSG mice (14) (see Note 9).
6. Measure the efficiency of engraftment periodically by immunohistochemical quantitation for human liver-specific markers including human ALB (Fig. 3) or measuring donor hepatocyte secreted proteins in mouse serum/plasma (14).

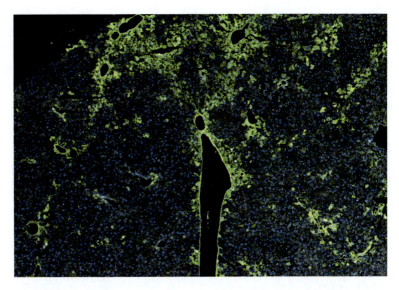

Fig. 3. In vivo engraftment of human iPS cell-derived hepatic cells. Human ALB staining of NSG mouse liver 8 week post-transplantation of 2 million DE cells derived from hHiPS cells (11). Human hepatocytes that are producing the albumin protein (shown in green) are identified by an antibody specifically recognizing human but not mouse albumin. Liver images were taken using the motorized Nikon Ti-E microscope with a Nikon encoded motorized XY stage and a function in NIS-Elements – advanced research software called "Scan Large Image" to generate these montaged images. The camera used is a Coolsnap HQ2 = (Photometrics).

4. Notes

1. When a pinkish discoloration appears in Schiff's reagent, discard the reagent.
2. Too high or too low cell densities seem to affect the efficiency of the hepatic differentiation; more spontaneous differentiation into other lineage cells rather than endoderm can occur. The activities of activin A may vary upon different lots or manufacturers.
3. The experiments should be performed under the hood. All reagents should be brought to room temperature prior to use.
4. The fixed plate can be stored at 4°C in the presence of PBS.
5. The washing step not only removes any excess reagent from cells but also promotes the development of the rich pink-red color.
6. Use an integration time of 0.25–1 s per well.
7. Purify iPS-derived hepatic cells before transplantation by flow cytometry or immunomagnetic sorting to remove ES/iPS marker (such as SSEA3) positive cells, if you suspect that there are any, because it might cause tumors in the mice.

8. Too high numbers of cells per each injection ($> 3 \times 10^6$ cells) tend to decrease the immediate survival rate of the recipient mice.
9. To enhance donor cell engraftment, the hepatic cells can be transplanted into liver-injured (i.e., partial hepatectomy) mice (14).

References

1. Locke J.E., Sun Z., Warren D.S., et al. (2008) Generation of humanized animal livers using embryoid body-derived stem cell transplant. Ann Surg. 248, 487–493
2. Herrera M.B., Bruno S., Buttiglieri S., et al. (2006) Isolation and characterization of a stem cell population from adult human liver. Stem Cells 24, 2840–2850
3. Lazaro C.A., Rhim J.A., Yamada Y., et al. (1998) Generation of hepatocytes from oval cell precursors in culture. Cancer Res 58, 5514–5522
4. Rogler L.E. (1997) Selective bipotential differentiation of mouse embryonic hepatoblasts in vitro. Am J Pathol 150, 591–602
5. Czyz J., Wiese C., Rolletschek A., et al. (2003) Potential of embryonic and adult stem cells in vitro. Biol Chem 384, 1391–1409
6. Dalgetty D.M., Medine C.N., Iredale J.P., et. al. (2009) Progress and future challenges in stem cell-derived liver technologies Am J Physiol Gastrointest Liver Physiol 297, G241–G248
7. Delgado J.P., Parouchev A., Allain J.E., et al. (2005) Long-term controlled immortalization of a primate hepatic progenitor cell line after Simian virus 40 T-Antigen gene transfer. Oncogene 24, 541–551
8. Song Z., Cai J., Liu Y., et al. (2009) Efficient generation of hepatocyte-like cells from human induced pluripotent stem cells. Cell Res.19:1233–1242
9. Agarwal S., Holton K.L., and Lanza R. (2008) Efficient differentiation of functional hepatocytes from human embryonic stem cells. Stem Cells 26, 1117–1127
10. Sullivan G.J., Hay D.C., Park I.H., et al. (2010) Generation of functional human hepatic endoderm from human induced pluripotent stem cells. Hepatology 51, 329–335
11. Liu H., Ye Z., Kim Y., et al. (2010) Generation of endoderm-derived human induced pluripotent stem cells from primary hepatocytes. Hepatology 51, 1810–1819
12. Liu H., Kim Y., Sharkis S., et al. (2010) Hepatic differentiation from virus-free and integration-free human induced pluripotent stem cells. Hepatology 52, 1169–1170
13. Yu J., Hu K., Smuga-Otto K., et al. (2009) Human induced pluripotent stem cells free of vector and transgene sequences. Science 324, 797–801
14. Jang, Y.Y., Collector, M.I., Baylin, S.B., et al. (2004) Hematopoietic stem cells convert into liver cells within days without fusion. Nat Cell Biol. 6, 532–539

Chapter 29

Hepatic Maturation of hES Cells by Using a Murine Mesenchymal Cell Line Derived from Fetal Livers

Takamichi Ishii and Kentaro Yasuchika

Abstract

Hepatocytes derived from human embryonic stem (hES) cells are a potential cell source for regenerative medicine. However, the successful differentiation of hES cells into mature hepatocytes has been difficult to achieve because the definitive mechanisms governing hepatocyte differentiation have not yet been well defined. The $CD45^-CD49f^{\pm}Thy1^+gp38^+$ mesenchymal cells that reside in murine fetal livers induce hepatic progenitor cells to differentiate into mature hepatocytes by direct cell–cell contact. A cell line named MLSgt20 was also successively established from these mesenchymal cells. The MLSgt20 cells possess the ability to promote the hepatic maturation of not only murine ES cells but also hES cell-derived endodermal cells. hES cells were treated with a two-step procedure for hepatic maturation; first, hES cells were differentiated into endodermal cells or hepatic progenitor cells, and second, hES cell-derived endodermal cells were matured into functional hepatocytes by co-culture with MLSgt20 cells, forming cell aggregates. The hES cell-derived hepatocyte-like cells possess hepatic functions. In this chapter, we describe a two-step protocol for the hepatic maturation of hES cells utilizing the MLSgt20 cells.

Key words: Embryonic stem cell, Fetal liver, Hepatocyte, AFP, Mesenchymal cell, Thy1, MLSgt20, Cell aggregate

1. Introduction

Embryonic stem (ES) cells are established from inner cell masses and possess a pluripotency to differentiate into cells from all three germ layers. Hepatocytes derived from human embryonic stem (hES) cells could be a potential cell source for cell transplantation, bio-artificial livers, and drug discovery systems (1). However, it is difficult to obtain fully functional hepatocytes from hES cells because there is limited knowledge of molecular mechanisms for the hepatic maturation.

Mesenchymal cells residing in murine fetal livers (CD45⁻CD49f±Thy1⁺gp38⁺ cells) promote hepatic maturation of hepatic progenitor cells (HPCs) and murine ES cell-derived alpha-fetoprotein (AFP)-producing cells (2–4). Moreover, the MLSgt20 cell line was successfully established from the CD45⁻CD49f±Thy1⁺gp38⁺ murine fetal liver mesenchymal cells by transfecting the immortalizing SV40 large T antigen gene (5). The MLSgt20 cells promoted the maturation of murine HPCs and murine ES cell-derived AFP-producing cells as well as hES cell-derived AFP-producing cells into functional hepatocyte-like cells (6). The MLSgt20 cells do not have the ability to directly mature undifferentiated ES cells into functional hepatocytes. Therefore, a hES cell line expressing enhanced green fluorescent protein (EGFP) under the control of the human AFP enhancer/promoter was generated, and the AFP-producing cells derived from hES cells were isolated using flow cytometry, followed by co-culture with the MLSgt20 cells.

In this chapter, we describe a two-step procedure for the hepatic maturation of hES cells utilizing the MLSgt20 cells (6). First, undifferentiated hES cells are differentiated into AFP-producing endodermal cells using sequential addition of activin A and hepatocyte growth factor (HGF) and Matrigel. Then, the AFP-producing cells are isolated as the EGFP-positive cells using flow cytometry. Second, the hES cell-derived AFP-producing cells are matured into functional hepatocyte-like cells by co-culture and forming cell aggregates with the MLSgt20 cells.

2. Materials

2.1. Culture of hES Cells

1. *A hES cell line*: A transgenic hES cell line that expresses EGFP under the control of human AFP enhancer/promoter was employed (7).
2. *ESC medium*: 1:1 mixture of Dulbecco's modified Eagle's medium (DMEM) and Ham's nutrient mixture F12 (Sigma-Aldrich) supplemented with 20% Knockout SR (KSR, Gibco, Grand Island, NY), 0.1 mM 2-mercaptoethanol (Sigma-Aldrich), and MEM non-essential amino acids (Gibco; see Note 1).
3. A solution of 0.05% collagenase IV (Gibco), 0.25% trypsin (Gibco), and 20% KSR (CTK solution). The CTK solution is stored in aliquots at –80°C and thawed before use.
4. Sixty-millimeter plastic culture dishes (BD Biosciences, Franklin Lakes, NJ) with a mouse embryonic fibroblast (MEF) feeder layer treated with mitomycin C (Wako Pure Chemical Co., Osaka, Japan; see Note 2).

2.2. Differentiation of hES Cells into Endoderm

1. Low-serum endoderm differentiation medium (ED medium): RPMI1640 (Gibco) with 0.5% fetal bovine serum (FBS) (HyClone, Logan, UT), 1 mM sodium pyruvate (Sigma-Aldrich), 10 mM nicotinamide (Sigma-Aldrich), 2 mM L-ascorbic acid phosphate (Wako Pure Chemical, Osaka, Japan), insulin-transferrin-selenium supplement (Gibco), and 0.1 µM dexamethasone (Sigma-Aldrich).

2. Activin A (R&D Systems, Inc., Minneapolis, MN) is dissolved to a concentration of 10 µg/mL, and HGF (R&D Systems, Inc.) is of 20 µg/mL in phosphate-buffered saline (PBS, Ca^{2+}-free) supplemented with 0.5% bovine serum albumin (BSA, Sigma), and stored in aliquots at −80°C. These growth factors are added to culture dishes as required (see Note 3).

3. Sixty-millimeter culture dishes coated with Matrigel (BD Biosciences). Matrigel is thawed at 4°C and diluted at 1:80 by cold RPMI1640 (see Note 4). The Matrigel solution is added onto 60-mm culture plastic dishes, which are incubated for 1 day at 37°C.

2.3. Culture of the MLSgt20 Cells

1. The MLSgt20 cell line.

2. HD medium; DMEM supplemented with 10% FBS, 1 mM sodium pyruvate (Sigma-Aldrich), 10 mM nicotinamide, 2 mM L-ascorbic acid phosphate, insulin-transferrin-selenium supplement, 0.1 µM dexamethasone, and 20 ng/mL HGF.

3. Sixty-millimeter collagen type I-coated dishes (AGC Techno Glass, Chiba, Japan).

4. A solution of 0.25% trypsin (Gibco) and 1 mM ethylenediaminetetraacetic acid (EDTA; Dojindo laboratories, Kumamoto, Japan; 0.25% trypsin–EDTA solution).

2.4. Isolation of hES Cell-Derived AFP-Producing Cells Using Flow Cytometry

1. A solution of 0.05% trypsin and 1 mM EDTA (0.05% trypsin–EDTA solution; see Note 5).

2. Hank's balanced salt solution (HBSS; 30 mL) with 3% FBS (1 mL; 3% FBS/HBSS).

3. 5 mL round-bottom tubes with 35 µm nylon meshes (BD Biosciences).

4. FACSVantage SE (BD Biosciences).

5. HD medium (see Sect. 2.3.2).

2.5. Co-culture of the hES Cell-Derived AFP-Producing Cells and MLSgt20 Cells

1. Sumilon Celltight Spheroid 96-well plates (Sumitomo Bakelite Co., Ltd., Tokyo, Japan).

2. Oncostatin M (R&D Systems, Inc.) is dissolved to a concentration of 10 µg/mL in PBS supplemented with 0.5% BSA, and stored in aliquots at −80°C.

3. HD medium (see Sect. 2.3.2) supplemented with 10 ng/mL oncostatin M.

4. Matrigel-coated 24-well culture plates. The plates are coated by 1:40 diluted Matrigel as described in Sect. 2.2.3.

3. Methods

hES cells are cultured in the undifferentiated state on MEF feeder layers. They are subcultured using CTK solution (8). MEFs are prepared according to a standard protocol. This section will focus on the two-step procedure for hepatic maturation of hES cells utilizing the MLSgt20 cell line. A key step in this procedure is to obtain a viable cell fraction of the isolated AFP-producing cells from hES cells, and to form cell aggregates with the MLSgt20 cells. The procedures are summarized in Fig. 1.

3.1. Step 1: Differentiation of hES Cells into Endodermal Cells

1. Dissociate confluent, undifferentiated hES cells cultured on a 60-mm dish using the CTK solution. The dissociated cells are transferred into a 15-mL plastic tube in the ESC medium and left for 5 min at room temperature to deplete MEFs (see Note 6). The supernatant fluid is discarded by decantation.

2. Centrifuge at 1,000 rpm for 3 min to harvest the cells. Resuspend the cell pellet mildly and replate onto a 60-mm culture dish coated with Matrigel in the ESC medium (day 0; see Note 7).

3. Change the culture medium from ESC medium to ED medium at day 1. Add 100 ng/mL of activin A to the ED medium for the first 4 days (days 1–4), and add 20 ng/mL of GF for the next 5 days (days 5–9).

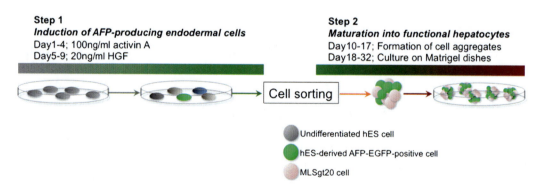

Fig. 1. Summary of the culture protocol.

4. Add 2 mL of 0.05% trypsin–EDTA solution to every 60-mm culture dish at day 10 to dissociate the differentiated cells, and incubate at 37°C for 5 min. The cultured cells are well dissociated into a single cell suspension and resuspended in cold 3% FBS/HBSS solution for flow cytometric sorting (see Note 8).

5. Isolate the EGFP-positive cell fraction by flow cytometry (see Note 9). The harvested cells can be resuspended in HD medium supplemented with oncostatin M at a concentration of 1×10^5 cells/mL for further experimentation (see Sect. 3.3).

3.2. Preparation of the MLSgt20 Cells

1. Grow the MLSgt20 cells in HD medium on collagen type I-coated dishes at 33°C (Fig. 2f; see Note 10). The cells are subcultured using 0.25% trypsin–EDTA solution.

2. Dissociate the MLSgt20 cells using 0.25%-EDTA solution into a single cell suspension, centrifuge at 1,000 rpm for 3 min, and then resuspend in HD medium supplemented with oncostatin M at a concentration of 1×10^5 cells/mL for further experimentation (see Sect. 3.3).

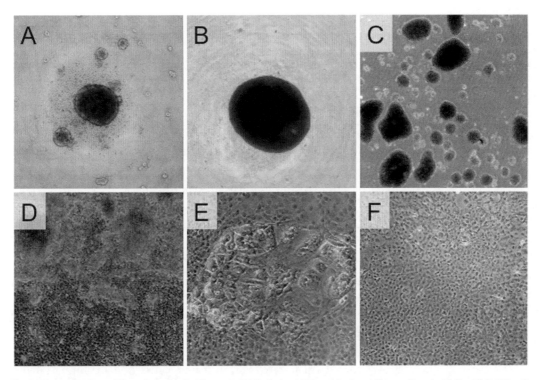

Fig. 2. The morphology of the cultured cells. The sorted hES-derived AFP-producing cells cocultured with the MLSgt20 cells form one cell aggregate in each well 1 day (day 11; **a**), and 7 days after cell sorting (day 17; **b**). The collected cell aggregates are transferred onto Matrigel-coated dishes (day 18; **c**) and expanded on the dishes (day 32, **d**). High-magnification microscopy shows that the hES-derived cells have a polygonal appearance like mature hepatocytes (**e**). Figure (**f**) shows the original morphology of the MLSgt20 cells.

3.3. Step 2: Maturation of the hES Cell-Derived Endodermal Cells

1. Combine (1:1) the isolated AFP-producing EGFP-positive endodermal cell suspension with the dissociated MLSgt20 cell suspension so that the cell mixture contains the same number of the hES-derived AFP-producing cells and MLSgt20 cells.
2. Add 100 μL of the cell mixture to each well of the Sumilon Celltight Spheroid 96-well plates, allowing forming cell aggregates (see Note 11). Incubate the culture plates at 37°C.
3. Add 10–20 μL of the fresh HD medium with oncostatin M into each well every 2 days. The culture can be continued for 7 days (day 10–17; Fig. 2a, b).
4. Harvest the cell aggregates after the 7-day period using a 500-μL micropipette from individual wells. Approximately 12–16 cell aggregates are transferred onto one well of 24-well Matrigel-coated culture plates (Fig. 2c).
5. The culture is maintained for an additional 14 days, and the medium is changed with fresh HD medium with oncostatin M each day (day 18–32; Fig. 2d, e; see Note 12).

4. Notes

1. All culture media should be used within a month.
2. Culture dishes with an MEF-feeder layer can be used for 1 week after mitomycin C treatment.
3. Unless stated otherwise, growth factors are added from stock solutions to culture medium as required.
4. All instruments including pipette tips and culture dishes are kept cold using a freezer until just before use to prevent the Matrigel from demonstrating gel formation.
5. A 0.05% trypsin–EDTA solution is preferable to a 0.25% trypsin–EDTA solution because hES cells are vulnerable to enzymatic damage.
6. The CTK solution dissociates hES cell colonies into cell clusters that contain several dozen hES cells so that hES clusters settle to the bottom of tubes.
7. Human ES cells can be damaged by mechanical manipulation; therefore, all processes dealing with hES cells have to be performed carefully and mildly. Undifferentiated hES cells on one 60-mm dish can be transferred onto one 60-mm Matrigel-coated dish.
8. Again, hES cells are vulnerable to enzymatic and mechanical damage. Therefore, the trypsin reaction time should be minimal, and the flow cytometric procedure should be done as quickly as possible.

9. Approximately 20% of the differentiated hES cells are positive for EGFP (7).

10. Theoretically, the MLSgt20 cells can proliferate only at 33°C because they are transfected with temperature-sensitive SV40 large T antigen. However, the temperature-sensitive element does not work appropriately. The MLSgt20 cells can be subcultured at least 50 times even at 37°C (5). However, the morphology of the MLSgt20 cells with multiple subcultures at 37°C is somehow different from that of the original MLSgt20 cells cultured at 33°C.

11. One well should contain 5,000 MLSgt20 cells and 5,000 hES-derived AFP-producing cells. One spheroid is formed in one well only by using Sumilon Celltight Spheroid 96-well plates. Other devices did not allow the formation of a single spheroid in one well. Flat cultures utilizing the MLSgt20 cells as a feeder layer do not work in this system.

12. The differentiated cells possess ammonia removal activity, glycogen storage and synthesis ability, and cytochrome P450 enzyme activity at day 32.

References

1. Ishii, T., Yasuchika, K., Machimoto, T., Kamo, N., Komori, J., Konishi, S., Suemori, H., Nakatsuji, N., Saito, M., Kohno, K., Uemoto, S., and Ikai, I. (2007) Transplantation of embryonic stem cell-derived endodermal cells into mice with induced lethal liver damage., *Stem Cells 25*, 3252–3260.

2. Hoppo, T., Fujii, H., Hirose, T., Yasuchika, K., Azuma, H., Baba, S., Naito, M., Machimoto, T., and Ikai, I. (2004) Thy1-positive mesenchymal cells promote the maturation of CD49f-positive hepatic progenitor cells in the mouse fetal liver., *Hepatology 39*, 1362–1370.

3. Kamo, N., Yasuchika, K., Fujii, H., Hoppo, T., Machimoto, T., Ishii, T., Fujita, N., Tsuruo, T., Yamashita, J. K., Kubo, H., and Ikai, I. (2007) Two populations of Thy1-positive mesenchymal cells regulate in vitro maturation of hepatic progenitor cells, *Am J Physiol Gastrointest Liver Physiol 292*, G526–534.

4. Ishii, T., Yasuchika, K., Fujii, H., Hoppo, T., Baba, S., Naito, M., Machimoto, T., Kamo, N., Suemori, H., Nakatsuji, N., and Ikai, I. (2005) In vitro differentiation and maturation of mouse embryonic stem cells into hepatocytes., *Exp Cell Res 309*, 68–77.

5. Fukumitsu, K., Ishii, T., Yasuchika, K., Amagai, Y., Kawamura-Saito, M., Kawamoto, T., Kawase, E., Suemori, H., Nakatsuji, N., Ikai, I., and Uemoto, S. (2009) Establishment of a cell line derived from a mouse fetal liver that has the characteristic to promote the hepatic maturation of mouse embryonic stem cells by a coculture method., *Tissue Eng Part A 15*, 3847–3856.

6. Ishii, T., Yasuchika, K., Fukumitsu, K., Kawamoto, T., Kawamura-Saitoh, M., Amagai, Y., Ikai, I., Uemoto, S., Kawase, E., Suemori, H., and Nakatsuji, N. (2010) In vitro hepatic maturation of human embryonic stem cells by using a mesenchymal cell line derived from murine fetal livers., *Cell Tissue Res 339*, 505–512.

7. Ishii, T., Fukumitsu, K., Yasuchika, K., Adachi, K., Kawase, E., Suemori, H., Nakatsuji, N., Ikai, I., and Uemoto, S. (2008) Effects of extracellular matrixes and growth factors on the hepatic differentiation of human embryonic stem cells., *Am J Physiol Gastrointest Liver Physiol 295*, G313–321.

8. Suemori, H., Yasuchika, K., Hasegawa, K., Fujioka, T., Tsuneyoshi, N., and Nakatsuji, N. (2006) Efficient establishment of human embryonic stem cell lines and long-term maintenance with stable karyotype by enzymatic bulk passage, *Biochem Biophys Res Commun 345*, 926–932.

Chapter 30

Generation of Lung Epithelial-Like Tissue from hESC by Air–Liquid Interface Culture

Lindsey Van Haute, Gert De Block, Inge Liebaers,
Karen Sermon, and Martine De Rycke

Abstract

Human embryonic stem cells (hESCs) have the capacity to differentiate in vivo and in vitro into cells from all three germ lineages. The in vitro generation of lung cells and tissues from hESCs creates opportunities for fundamental research, drug development or cell-replacement therapy.

In this chapter, we describe a reliable and simple protocol to differentiate hESCs into the major cell types of lung epithelial tissue. In this protocol, undifferentiated hESCs, grown on a porous membrane in hESC medium for 4 days, are switched to a differentiation medium and cultured for 4 days, followed by differentiation in air–liquid interface conditions for 20 days. The expression of several lung markers is analysed using quantitative real-time PCR and confirmed by immunohistochemistry. The functionality is determined through an enzyme immunoassay. For a more detailed discussion regarding the obtained cells and their gene and protein expression, refer to Van Haute et al. (Generation of lung epithelial-like tissue from human embryonic stem cells Respiratory Research 10: 105, 2009).

Key words: Air–liquid interface, hESC, Lung epithelium, Clara cells, Differentiation

1. Introduction

The in vitro generation of lung cells and tissues from hESCs creates opportunities for fundamental research, drug development or cell-replacement therapy. The availability of lung epithelial lineages and tissues facilitates research on lung development and pharmaceutical studies as it diminishes the need for laboratory animals and accelerates testing procedures. Cell-replacement therapy might provide an alternative to lung transplantation in patients with lung injury (due to chronic pulmonary disease or inherited genetic diseases such as cystic fibrosis).

The respiratory system originates from the foregut endoderm that differentiates into many kinds of specialised epithelial cells. These include ciliated, secretory and neuroendocrine cells of the proximal bronchi and the alveolar cells. The latter can be divided into two types of cells: type I cells, with a highly flattened morphology ideally suited for gas exchange, and the less differentiated, more cuboidal type II cells serving as progenitor cells for type I cells. Furthermore, these alveolar type II cells are important in synthesising and secreting pulmonary surfactant proteins, a complex mixture of proteins and phospholipids that lower surface tension (2).

Clara Cells are non-ciliated, non-mucous cells lining the bronchioles of the lung. A variety of proteins, such as the surfactant proteins SP-A, SP-B and SP-D, and most importantly CC16, a Clara cell diffusible 16 kDa protein, are secreted by these cells (3).

Various protocols that stimulate the differentiation of ESCs into lung cells have been reported. Many studies rely on growth factors to direct lung cell differentiation. These growth factors are directly supplemented to the medium (4–6) or indirectly via the use of conditioned medium or coculture systems (7). For a more detailed overview of the different protocols, refer to a review of Rippon et al. (8). It was recently shown that extracellular matrices can upregulate lung cell differentiation (9, 10). A transgenic approach to obtain relatively pure alveolar type II cells was followed by Wang et al. (11). Coraux et al. (12) developed an air–liquid interface (ALI) culture system for murine ESC, which primarily relied on the physical forces of air instead of growth factors to support lung cell differentiation. We have adapted this approach and established a convenient protocol to differentiate hESC into a lung epithelial-like tissue.

With this method, hESCs can differentiate into lung epithelial-like cells without specific growth factors. The protocol relies on an ALI system that mimics the conditions of an adult trachea. The ALI on a porous membrane in combination with low serum is sufficient to prime the cells to form an airway epithelial-like tissue. Details on the major cell types and data on mRNA expression as well as protein expression and localisation have been described in Van Haute et al. (1). The differentiation protocol may be further optimised by applying selected growth factors.

2. Materials

2.1. Coating of the 12-Well Plates with MEFs (Mouse Embryo Fibroblasts)

1. Porcine gelatine (Sigma-Aldrich, St. Louis, MO, USA)
2. Stericup-GP, 0.22 μm filter unit (Millipore Corporation, Billerica, MA, USA)
3. MEF Embryomax® Primary Mouse Embryo Fibroblasts, strain CF1, Mytomycin C treated, passage 3 (Millipore) (see Note 1)

4. 12-well plates (Nunc, Thermo Fisher Scientific)
5. MEF medium (see Note 2):
   (a) 90% Dulbecco's modified Eagle's medium (Invitrogen Life technologies, Paisley, UK)
   (b) 10% Fetal calf serum (Invitrogen)
   (c) 2 mM L-glutamine (Invitrogen)
   (d) 1% nonessential amino acids (Invitrogen)
6. 15 mL sterile centrifuge tubes (Falcon)

2.2. Differentiation

1. Human embryonic stem cells (see Note 3)
2. Differentiation medium (see Note 4)
   (a) 80% knockout Dulbecco's modified Eagle's medium (Invitrogen)
   (b) 20% knockout-serum replacement (KO-SR) (Invitrogen)
   (c) 2 mM L-glutamine (Invitrogen)
   (d) 1% nonessential amino acids (Invitrogen)
   (e) Penicillin/streptomycin (100 U/mL) (Invitrogen) (see Note 5)
3. Collagenase type IV (Invitrogen)
4. Millicell Cell Culture Insert 12 mm HA mixed cellulose esters (Millipore) (see Note 6)
5. Stericup-GP, 0.22 µm filter unit (Millipore)
6. 15 mL or 50 mL sterile centrifuge tubes (Falcon)

3. Protocol

Undifferentiated hESCs are maintained in culture using mechanical passaging. To initiate differentiation, cells are detached and grown on a porous membrane in hESC medium for 4 days before switching to a differentiation medium for 4 days. This can be followed by culture in air–liquid interface conditions during another 20 days. For all steps, one should work in a sterile flow cabinet.

3.1. Coating of 12-Well Plates with MEF Feeders (see Note 7)

1. Prepare MEF medium and filter sterilise.
2. Pour approximately 1 mL of 0.1% gelatine into each well of a 12-well plate.
3. Wait at least half an hour before removing the excess of gelatine solution.
4. Seed MEFs at a density of $4 \times 10^4/cm^2$ and maintain in MEF medium in a 37°C, 5% CO_2 incubator.

3.2. Differentiation of hESC

Day 1

1. Remove hESC medium and harvest cells using collagenase IV (incubation at 37°C till colonies start to detach) (see Note 8).
2. Collect colonies and dilute by adding and excess of hESC medium, centrifuge 3 min 800 rpm and remove the supernatans.
3. Resuspend the cells in hESC medium and pipet up and down to break the clumps into smaller pieces.
4. Plate ± 30 clumps on each porous membranes of millicell-HA culture inserts, placed in 12-well plates coated with MEF feeders (see Sect. 3.1 and Note 9).
5. Grow the cells in 1 mL hESC medium in liquid-liquid conditions (400 μL medium in upper and 600 μL medium in lower compartment).

Day 2–4

6. Change hESC medium daily.

Day 5 Initiate differentiation in liquid-liquid conditions (Fig. 1)

7. Remove the hESC medium.
8. Switch to 1 mL differentiation medium, again in liquid–liquid conditions.

Day 6–9:

9. Change differentiation medium daily.

Day 10: Start air–liquid interface: (Fig. 2)

10. Remove medium.
11. Add 560 μL differentiation medium to the lower compartment (see Note 10).

Day 11–30 (see Notes 11–13)

12. Change differentiation medium daily (see Note 14).

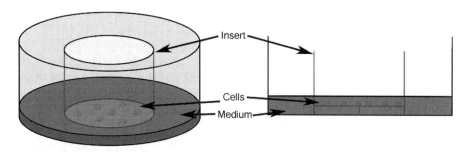

Fig. 1. Schematic overview of liquid-liquid conditions. Medium covers the cells.

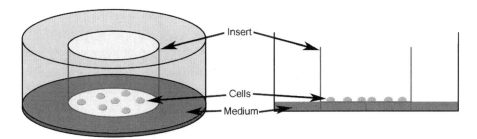

Fig. 2. Schematic overview of an ALI setting. Cells grow on top of the membrane in open air. Medium is only added in the lower compartment.

3.3. Controls That Might Be Included

1. hESCs plated on 0.1% gelatine-coated culture dishes on MEF feeders that had differentiated spontaneously in hESC medium.
2. hESC plated on porous membranes for 4 days in hESC medium followed by 24 days in differentiation medium in liquid–liquid conditions.

4. Notes

1. This is the type of MEF used in our experiments. However, there is no reason to assume that other types of MEF that support hESC growth would not be suitable.
2. MEF medium is stable at 4°C for up to 1 week.
3. hESC lines were derived in our lab from human preimplantation embryos after IVF or preimplantation genetic diagnosis. Lines were fully characterised according to the International Stem Cell Initiative guidelines and karyotyped by array-based comparative genomic hybridization. Five different cell lines were used to check for possible variations in their differentiation potential: VUB03_DM1, VUB04_CF carrying one mutation in the CFTR gene, VUB09_FSHD, VUB07 and VUB14.
4. Differentiation medium is stable at 4°C for up to 1 week.
5. Penicillin/streptomycin might not be necessary for the protocol itself; however, it was used in our experiments.
6. In our experiments, we tested other types of (translucent) membranes, but none of them gave similar hESC attachment.
7. Perform this step preferable exactly 1 day before Step 3.2 (1).
8. If large parts of the hESC colonies show differentiation, mechanical slicing rather than collagenase IV treatment should be used to harvest only the undifferentiated cells. If this method is used, proceed immediately to Step 3.2. Day 1 Step 4.

9. The membrane itself should not be coated.
10. The exact volume might depend upon the type of 12-well plate that is used. The correct level of medium allows a little bit of medium at the edge of the membrane but keeps the middle part dry.
11. The highest CC16 expression can be measured around day 10 of ALI culture. If one is interested in cells expressing CC16, the differentiation protocol should be stopped at this point, which is day 21 (± 10 days ALI).
12. The porous membranes that are used in this protocol are non-translucent, so you cannot study the cells under a light microscope, yet after a few days, the clumps can be seen by bare eye. Morphological evaluation of the cells by light microscopy can only be performed on paraffin-embedded sections. Cells can be fixed on the membrane to perform immunostaining for fluorescence microscopy. It should be taken into account that the membrane itself is autofluorescent. In our opinion, this was not a problem when evaluating the cells. Prior to embedding in paraffin, cut loose the membrane to which the cells are attached.
13. To perform RNA extraction, it is possible to lyse the cells directly on the membrane. Wash the cells with PBS and add RLT buffer (lysis buffer Qiagen kit) to the membrane. The lysate was transferred to an eppendorf tube and could be used for further treatment.
14. It is important to change the medium daily in order to maintain the same liquid level and to avoid drying out of the cells due to evaporation.

References

1. Van Haute, L., De Block, G., Liebaers, I. et al. (2009) Generation of lung epithelial-like tissue from human embryonic stem cells Respiratory Research 10, 105
2. Creuwels, L. A. J. M., vanGolde, L. M. G., and Haagsman, H. P. (1997) The pulmonary surfactant system: Biochemical and clinical aspects Lung 175, 1–39
3. Singh, G. and Katyal, S. L. (2000) Clara cell proteins Uteroglobin/Clara Cell Protein Family 923, 43–58
4. Samadikuchaksaraei, A., Cohen, S., Isaac, K. et al. (2006) Derivation of distal airway epithelium from human embryonic stem cells Tissue Engineering 12, 867–875
5. Ali, N. N., Edgar, A. J., Samadikuchaksaraei, A. et al. (2002) Derivation of type II alveolar epithelial cells from murine embryonic stem cells Tissue Engineering 8, 541–550
6. Roszell, B., Mondrinos, M. J., Seaton, A. et al. (2009) Efficient Derivation of Alveolar Type II Cells from Embryonic Stem Cells for In Vivo Application Tissue Engineering Part A15, 3351–3365
7. van Vranken, B. E., Romanska, H. M., Polak, J. M. et al. (2005) Coculture of embryonic stem cells with pulmonary mesenchyme: A microenvironment that promotes differentiation of pulmonary epithelium Tissue Engineering 11, 1177–1187

8. Rippon, H. J., Lane S, Qin M et al. (2008) Embryonic stem cells as a source of pulmonary epithelium in vitro and in vivo Proc Am Thorac Soc 5, 717–722
9. Lin, Y. M., Zhang, A., Rippon, H. J. et al. (2010) Tissue Engineering of Lung: The Effect of Extracellular Matrix on the Differentiation of Embryonic Stem Cells to Pneumocytes Tissue Engineering Part A16, 1515–1526
10. Cortiella, J., Niles, J., Cantu, A. et al. (2010) Influence of Acellular Natural Lung Matrix on Murine Embryonic Stem Cell Differentiation and Tissue Formation Tissue Engineering Part A16, 2565–2580
11. Wang, D., Haviland, D. L., Burns, A. R. et al. (2007) A pure population of lung alveolar epithelial type II cells derived from human embryonic stem cells Proceedings of the National Academy of Sciences of the United States of America 104, 4449–4454
12. Coraux, C., Nawrocki-Raby, A., Hinnrasky, J. et al. (2005) Embryonic stem cells generate airway epithelial tissue American Journal of Respiratory Cell and Molecular Biology 32, 87–92

Chapter 31

Direct Differentiation of Human Embryonic Stem Cells into Selective Neurons on Nanoscale Ridge/Groove Pattern Arrays

Kye-Seong Kim, Hosup Jung, and Keesung Kim

Abstract

Human embryonic stem cells (hESCs) are pluripotent cells that have the potential to be used for tissue engineering and regenerative medicine. Biochemical and biological agents are widely used to induce hESC differentiation. However, it would be better if we could induce the differentiation of hESCs without using such agents because these factors are expensive. It is also difficult to determine optimal concentrations of agents for efficient differentiation. Moreover, the mechanism of differentiation induced by these factors is still not fully understood. Using UV-assisted capillary force lithography, we constructed nanoscale ridge/groove pattern arrays with a dimension and alignment that were finely controlled over a large area. Human embryonic stem cells seeded onto the 350-nm ridge/groove pattern arrays differentiated into neuronal lineage after 5 days, in the absence of differentiation-inducing agents. This nanoscale technique could be used for a new neuronal differentiation protocol of hESCs and may also be useful for nanostructured scaffolding for nerve injury repair. In this chapter, we describe this method in detail. This protocol can be used to create nanoscale ridge/groove pattern arrays for effective and rapid directing of the differentiation of hESCs into a neuronal lineage without the use of any differentiation-inducing agents.

Key words: Human embryonic stem cells, Capillary force lithography, Neuronal differentiation, Nanotopography

1. Introduction

Human embryonic stem cells (hESCs) have the properties of self-renewal and pluripotency and can give rise to over 200 different cell types of the human body. Because of their unlimited capacity for self-renewal and differentiation, hESCs have been proposed for use in a wide range of applications, including toxicology testing, tissue engineering, cellular therapies, and basic biological stem cell research. Of particular interest to the medical community is the potential for using hESCs to heal tissues that have a naturally limited capacity of renewal, such as the human heart, liver, and brain.

To induce the differentiation of hESCs into specific lineages, biochemical and biological agents are widely used (1). However, recent reports have shown that mechanical factors such as stress (2–4), adhesion area (5, 6), substrate elasticity (7), and topography on micro- and nanoscale (8–11) can also induce the differentiation of stem cells. While most studies report on the effects of both mechanical and biochemical factors on the differentiation of stem cells, few have suggested the possibility that mechanical factors alone, such as micro- and nanostructured substrates, could induce the differentiation of stem cells (11).

It has been reported that micro- and nanostructures (12–17), especially aligned micro- and nanostructures (12), can effectively induce neuronal differentiation of stem cells. The dimensions of these structures can also affect the neural differentiation of stem cells (13, 14, 17). There are a few studies on neuronal differentiation of ESCs (12), especially hESCs (18), on micro- and nanostructures, in comparison with those of adult stem cells (7, 13, 14, 17, 19–22). It is necessary to study the neuronal differentiation of hESCs on micro- and nanostructures with precisely controlled dimensions and alignment. We examined whether hESCs would differentiate into three germ layers, including a neuronal lineage, on nanoscale ridge/groove pattern arrays. To construct precisely controlled nanoscale ridge/groove pattern arrays on a large surface area, we used UV-assisted capillary force lithography, a versatile tool for studying cell-substrate topography interactions (23–25) and for making nanostructured scaffolds that can support and induce the direct differentiation of stem cells into specific lineage cells (25–27). We established the protocol that nanoscale ridge/groove pattern arrays alone can effectively and rapidly induce differentiation of hESCs into neuronal lineage cells without the addition of any biochemical or biological agents.

2. Materials

2.1. Silicon Master

1. 4″ Silicon wafer (Silicon Technology Corp., Japan, Cat #05-0530-04)
2. Photoresist (PR) (AZ Electronic Material, Korea, AZ®9200)
3. PR Developer (AZ Electronic Material, Korea, AZ®300MIF)
4. Mask alignment & exposure with the Karl Suss Mask Aligner and Quintel Mask Aligner
5. Developing exposed wafers

2.2. Polyurethane Acrylate Replica Mold

1. Polyurethane acrylate (PUA) (Minuta Technology, Korea, MINS-311RM)
2. Polyethylene terephthalate (PET) film (SK Chemical Co. Korea)
3. UVO surface treatment system (Minuta Technology, Korea, Ozone Cure 16)

2.3. Treatment of Nanoscale Ridge/Groove-Patterned Surface

1. Oxygen plasma (60 W, PDC-32 G, Harrick Scientific, Ossining, NY)
2. 0.1% Gelatin solution
3. Dulbecco's modified Eagle's medium (DMEM)/F12 (GIBCO, USA, Cat #11320-033)
4. 15% Fetal bovine serum (FBS) (Hyclone, USA, Cat #SH30070.03)

2.4. Maintenance of hESCs

1. Mouse embryonic fibroblast (CF1, 13.5 dpc)
2. Mitomycin C (10 μg/mL, Sigma, USA, Cat #G1393)
3. Dulbecco's modified Eagle's medium (DMEM)/F12 (GIBCO, USA, Cat #11320-033), 20% serum replacement (GIBCO, USA, Cat #10828-028), 1 mM glutamine (Invitrogen, USA, Cat #25030-081), 0.1% nonessential amino acids (GIBCO, USA, Cat #11140-050), 0.1% penicillin/streptomycin (GIBCO, USA, Cat #15140), 0.1 mM beta-mercaptoethanol (Sigma, USA, Cat #M-7522), and 4 ng/mL recombinant human FGF-2 (GIBCO, USA, Cat #13256-029)

2.5. Scanning Electron Microscopy

1. 2.5% Glutaraldehyde in sodium cacodylate buffer
2. Japan Electron Optics Laboratory model (JSM 5600, Tokyo, Japan)

2.6. Immunocytochemistry

1. 4% (w/v) Paraformaldehyde
2. 0.1% (v/v) Triton X-100 in PBS
3. Mouse anti-Nestin (1:1,000, R and D system, USA), goat anti-PDX-1 (1:500, R and D system, USA), goat anti-Brachyury (1:500, R and D system, USA), mouse anti-Tuj-1 (1:1,000, Chemicon, USA), rabbit anti-GFAP (1:500, Chemicon, USA), mouse anti-HuC/D (1:500, Invitrogen, USA), rabbit anti-Calbindin (1:200,Chemicon, USA), rabbit anti-MAP2 (1:500, Santa Cruz Biotechnology, USA)
4. FITC- and rhodamine-conjugated secondary antibodies (Molecular Probe, USA)
5. Confocal microscope (LSM 510; Zeiss)

3. Methods

3.1. Fabrication of Silicon Master

This part requires specialized equipment and training; thus, it is best to fabricate the master in collaboration with a nanofabrication facility. The desired photoresist master can also be purchased from companies or nanofabrication laboratories that fabricate custom-made microstructured masters on demand.

This protocol provides guidelines for the various parameters needed to fabricate a photoresist master. Details of this photolithography protocol, however, must be adapted to the specific needs of each laboratory (e.g., size of the microfeatures, distance between features, incubation times, and temperatures).

1. Clean the silicon wafer in an ultrasonic bath for 10 min in acetone. Dry it with nitrogen gas.
2. Place the silicon wafer on a spin coater.
3. Cover the central half of the wafer's surface with PR AZ® 9200.
4. Spin coat the PR onto the wafer at 500 rpm for 10 s.
5. Spin coat the PR onto the wafer at 2,500 rpm for 30 s.
6. Dry the silicon wafer under a hood for 10 min.
7. Bake the PR layer onto the wafer using either a hot plate or an oven for 1 min at 100°C. Cool the wafer on the bench for 15 min until it returns to room temperature.
8. Place the PR layer of silicon wafer and the optical mask into contact with one another on the mask aligner or on the custom-made vacuum mask holder.
9. Illuminate with the UV lamp (UV power 45 mW/cm^2 at 365 nm) for 10 s.
10. Dilute the stock developer AZ® 300 MIF using one part developer and four parts distilled water.
11. Develop the PR in diluted developer for 90–120 s.
12. Rinse the wafer in a distilled water bath, which stops the development process.
13. Dry the resist master with a flow of nitrogen gas.

The UV illumination time, the developer concentration, and the development time in Step 11 depend upon the resist used and on its thickness. It may be necessary to adjust these.

3.2. Fabrication of PUA Replica Mold and Nanoscale Ridge/Groove-Patterned Surface

Fabrication of the elastomeric stamp and UV-assisted capillary force lithography of the nanoscale pattern can be performed in any biology laboratory without the need for specialized equipment. Polyurethane acrylate (PUA) micro- and nanoscale ridge/groove pattern arrays were fabricated on glass coverslips using UV-assisted

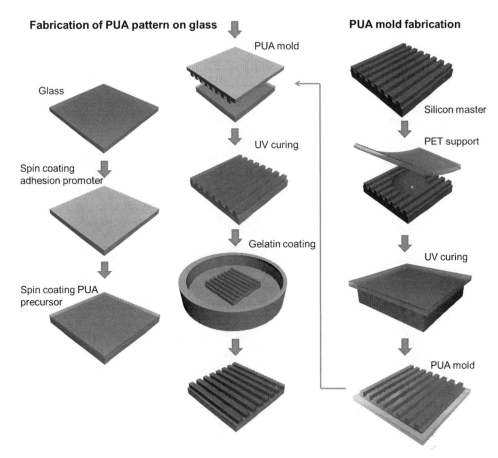

Fig. 1. Fabrication of nanoscale ridge/groove pattern arrays. Schematic diagram of the UV-assisted capillary force lithography to fabricate PUA nanoscale ridge/groove pattern arrays.

capillary force lithography, as shown in Fig. 1. The PUA mold for nanoscale patterning was fabricated by curing PUA pre-polymer (311RM, Minuta) on silicon master molds prepared by photolithography. The ultraviolet (UV)-curable PUA mold material consists of a functionalized precursor with an acrylate group for cross-linking, a monomeric modulator, a photoinitiator, and a radiation-curable releasing agent for surface activity.

1. Dispense liquid precursor dropwise onto a silicon master mold.
2. Bring the polyethylene terephthalate (PET) film into contact with the precursor surface.
3. Expose silicon master mold to UV light ($\lambda = 200$–400 nm) for 20 s through the transparent backplane (dose = 100 mJ/cm^2).
4. After UV curing, peel the mold from the master and additionally cure overnight to terminate the remaining active acrylate groups prior to use as a first replica.
5. Rinse the glass coverslip with ethanol in an ultrasonic bath for 30 min.

6. Wash in a flow of distilled water and dry in a drying oven.

7. To increase the adhesion, coat adhesion promoter (phosphoric acrylate: isopropyl alcohol = 1:10, volume ratio) onto the glass substrate.

8. Dispense a small amount of the PUA precursor (~0.1–0.5 mL) dropwise onto the substrate.

9. Place the first-replicated PUA mold (same material but without active acrylate groups) directly onto the surface.

10. Move the PUA precursor spontaneously into the cavity of the mold by means of capillary action.

11. Expose to UV light ($\lambda = 250$–400 nm) for ~30 s through the transparent backplane (dose = 100 mJ/cm^2).

12. After curing, peel the PUA mold from the substrate using sharp tweezers.

3.3. Gelatin Coating and Seeding hESCs

Nanoscale ridge/groove-patterned surfaces need to be coated with 0.1% gelatin to enable adherence of the hESCs. Although there are reports that FN and laminin enhance the differentiation of hESCs and neural progenitor generation, gelatin is more generally used for the attachment of undifferentiated hESCs onto surfaces (28).

1. Treat nanoscale ridge/groove-patterned surface in oxygen plasma for 60 s (60W, PDC-32G, Harrick Scientific, Ossining, NY).

2. Immerse in a 0.1% gelatin solution for 12 h and rinsed with deionized water.

3. Seed hESCs in DMEM medium containing 15% FBS until the hESCs adhere to the substrate.

4. After 24 h, count the attached hESC colonies under a microscope.

3.4. Imaging Cell Morphology with SEM

Scanning electron microscopy (SEM) was employed to examine the surface topography of the PUA nanoscale ridge/groove pattern structures and the cultured hESC colonies, as shown in Fig. 2.

1. Wash samples and fix in 2.5% glutaraldehyde in sodium cacodylate buffer at 4°C for 5 h.

2. After fixing, wash the cells with cacodylate buffer, place through a series of graded ethanol dehydrations, and allow to air-dry.

3. Sputter-coat the samples with gold and analyze under a Japan Electron Optics Laboratory model (JSM 5600) scanning electron microscope at 20 kV.

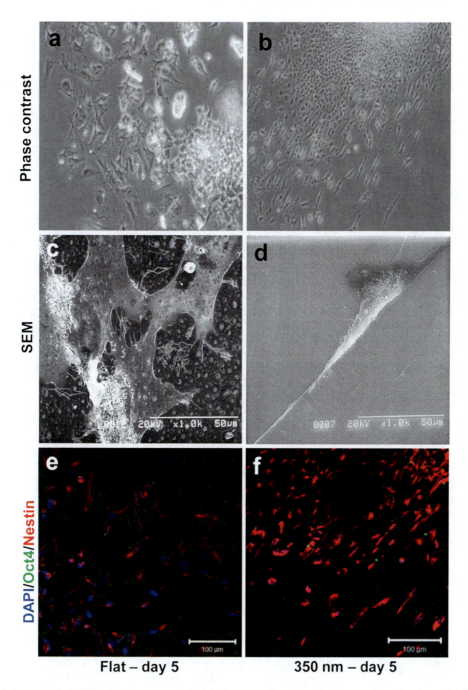

Fig. 2. Enhancement of hESC adhesion on nanoscale ridge/groove pattern arrays by oxygen plasma treatment. (**a**) An SEM image of gelatin-coated 350-nm ridge/groove pattern arrays. (**b**) A phase contrast image of hESCs on gelatin-coated 350-nm ridge/groove pattern arrays. The white circle indicates the cell-detached area. (**c**) An SEM image of gelatin-coated 350-nm ridge/groove pattern arrays treated by oxygen plasma for 60 s prior to gelatin coating. (**d**) A phase contrast image of hESCs on gelatin-coated 350-nm ridge/groove pattern arrays treated by oxygen plasma for 60 s prior to gelatin coating. (**e**) A graph showing the fraction of adherent hESCs on the gelatin-coated 350-nm ridge/groove pattern arrays and on the gelatin-coated 350-nm ridge/groove pattern arrays treated by oxygen plasma prior to gelatin coating.

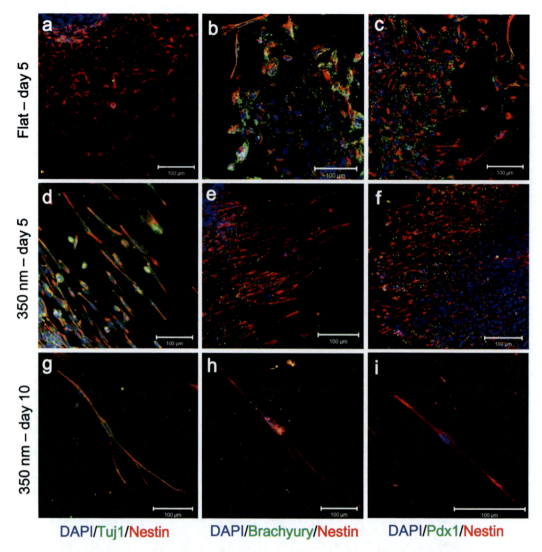

Fig. 3. Early differentiation of hESCs cultured on the PUA flat surface and on the 350-nm nanoscale ridge/groove pattern arrays for five days. (**a**) Phase contrast images, (**c**) SEMimages, and (**e**) fluorescence images of hESCs on the flat surface and (**b**), (**d**), and (**f**) show the corresponding images for hESCs on the 350-nm nanoscale ridge/groove pattern arrays. Cells were triple-immunolabeled for DAPI, Oct4 and nestin.

3.5. Immuno-cytochemistry

1. Fix cells with 4% (w/v) paraformaldehyde for 30 min and permeabilize with 0.1% (v/v) Triton X-100 in PBS for 5 min. After treatment with a blocking solution of 10% (v/v) goat serum for 30 min, incubate the cells with primary antibodies at 4°C overnight.

2. Use the following primary antibodies for immunocytochemistry: mouse anti-Nestin (1:1,000), goat anti-PDX-1 (1:500), goat anti-Brachyury (1:500), mouse anti-Tuj-1 (1:1,000), rabbit anti-GFAP (1:500), mouse anti-HuC/D (1:500), rabbit anti-Calbindin (1:200), rabbit anti-MAP2 (1:500).

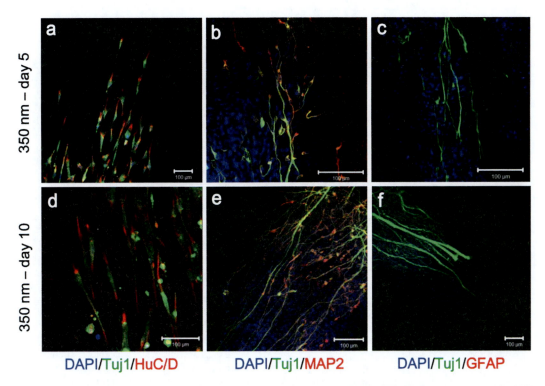

Fig. 4. Immunofluorescence staining of hESCs with neural and glial markers. (**a**, **d**) hESCs were immunolabeled for DAPI, Tuj1, and HuC/D. (**b**, **e**) hESCs were immunolabeled for DAPI, Tuj1, and MAP2. (**c**, **f**) hESCs were immunolabeled for DAPI, Tuj1, and GFAP. hESCs cultured for 5 days (**a–c**) and 10 days (**d–f**) on the 350-nm ridge/groove pattern arrays.

3. After washing with PBS, visualize the stained cells using a confocal microscope (LSM 510; Zeiss) using FITC- and rhodamine-conjugated secondary antibodies.

Neural differentiation of hESCs on the PUA flat surface and on the 350-nm ridge/groove pattern arrays can be analyzed through immunofluorescence microscopy, as shown in Figs. 3 and 4.

4. Notes

1. Steps 3–6: PR is corrosive and toxic, avoid direct contact and always handle it in the fume hood.
2. Step 4: Delamination is another problem contributing to unsuccessful demolding. For clean demolding, it is necessary that the cured resin adheres to the polyester film substrate more strongly than it does to the mold such as treat with adhesion promoter (phosphoric acrylate).
3. Step 9: Avoid trapping air bubbles between the stamp and the substrate by initiating physical contact between these two surfaces from the edge with an angle.

References

1. Dhara, S. K., and Stice, S. L. (2008) Neural Differentiation of Human Embryonic Stem Cells, *Journal of Cellular Biochemistry 105*, 633–640.
2. Kurpinski, K., Chu, J., Hashi, C., and Li, S. (2006) Anisotropic mechanosensing by mesenchymal stem cells, *P Natl Acad Sci USA 103*, 16095–16100.
3. O'Cearbhaill, E. D., Punchard, M. A., Murphy, M., Barry, F. P., McHugh, P. E., and Barron, V. (2008) Response of mesenchymal stem cells to the biomechanical environment of the endothelium on a flexible tubular silicone substrate, *Biomaterials 29*, 1610–1619.
4. Ruiz, S. A., and Chen, C. S. (2008) Emergence of Patterned Stem Cell Differentiation Within Multicellular Structures, *Stem Cells 26*, 2921–2927.
5. McBeath, R., Pirone, D. M., Nelson, C. M., Bhadriraju, K., and Chen, C. S. (2004) Cell shape, cytoskeletal tension, and RhoA regulate stem cell lineage commitment, *Dev Cell 6*, 483–495.
6. Park, J., Cho, C. H., Parashurama, N., Li, Y. W., Berthiaume, F., Toner, M., Tilles, A. W., and Yarmush, M. L. (2007) Microfabrication-based modulation of embryonic stem cell differentiation, *Lab Chip 7*, 1018–1028.
7. Engler, A. J., Sen, S., Sweeney, H. L., and Discher, D. E. (2006) Matrix elasticity directs stem cell lineage specification, *Cell 126*, 677–689.
8. Hashi, C. K., Zhu, Y. Q., Yang, G. Y., Young, W. L., Hsiao, B. S., Wang, K., Chu, B., and Li, S. (2007) Antithrombogenic property of bone marrow mesenchymal stem cells in nanofibrous vascular grafts, *P Natl Acad Sci USA 104*, 11915–11920.
9. Xin, X. J., Hussain, M., and Mao, J. J. (2007) Continuing differentiation of human mesenchymal stem cells and induced chondrogenic and osteogenic lineages in electrospun PLGA nanofiber scaffold, *Biomaterials 28*, 316–325.
10. Dalby, M. J., Gadegaard, N., Tare, R., Andar, A., Riehle, M. O., Herzyk, P., Wilkinson, C. D. W., and Oreffo, R. O. C. (2007) The control of human mesenchymal cell differentiation using nanoscale symmetry and disorder, *Nat Mater 6*, 997–1003.
11. Oh, S., Brammer, K. S., Li, Y. S. J., Teng, D., Engler, A. J., Chien, S., and Jin, S. (2009) Stem cell fate dictated solely by altered nanotube dimension, *P Natl Acad Sci USA 106*, 2130–2135.
12. Xie, J. W., Willerth, S. M., Li, X. R., Macewan, M. R., Rader, A., Sakiyama-Elbert, S. E., and Xia, Y. N. (2009) The differentiation of embryonic stem cells seeded on electrospun nanofibers into neural lineages, *Biomaterials 30*, 354–362.
13. Yang, F., Murugan, R., Wang, S., and Ramakrishna, S. (2005) Electrospinning of nano/micro scale poly(L-lactic acid) aligned fibers and their potential in neural tissue engineering, *Biomaterials 26*, 2603–2610.
14. Christopherson, G. T., Song, H., and Mao, H. Q. (2009) The influence of fiber diameter of electrospun substrates on neural stem cell differentiation and proliferation, *Biomaterials 30*, 556–564.
15. Ellis-Behnke, R. G., Liang, Y. X., You, S. W., Tay, D. K. C., Zhang, S. G., So, K. F., and Schneider, G. E. (2006) Nano neuro knitting: Peptide nanofiber scaffold for brain repair and axon regeneration with functional return of vision, *P Natl Acad Sci USA 103*, 5054–5059.
16. Silva, G. A., Czeisler, C., Niece, K. L., Beniash, E., Harrington, D. A., Kessler, J. A., and Stupp, S. I. (2004) Selective differentiation of neural progenitor cells by high-epitope density nanofibers, *Science 303*, 1352–1355.
17. Yim, E. K. F., Pang, S. W., and Leong, K. W. (2007) Synthetic nanostructures inducing differentiation of human mesenchymal stem cells into neuronal lineage, *Exp Cell Res 313*, 1820–1829.
18. Sridharan, I., Kim, T., and Wang, R. (2009) Adapting collagen/CNT matrix in directing hESC differentiation, *Biochemical and Biophysical Research Communications 381*, 508–512.
19. Shih, Y. R. V., Chen, C. N., Tsai, S. W., Wang, Y. J., and Lee, O. K. (2006) Growth of mesenchymal stem cells on electrospun type I collagen nanofibers, *Stem Cells 24*, 2391–2397.
20. Jan, E., and Kotov, N. A. (2007) Successful differentiation of mouse neural stem cells on layer-by-layer assembled single-walled carbon nanotube composite, *Nano Letters 7*, 1123–1128.
21. Kim, S. J., Lee, J. K., Kim, J. W., Jung, J. W., Seo, K., Park, S. B., Roh, K. H., Lee, S. R., Hong, Y. H., Kim, S. J., Lee, Y. S., Kim, S. J., and Kang, K. S. (2008) Surface modification of polydimethylsiloxane (PDMS) induced proliferation and neural-like cells differentiation of umbilical cord blood-derived mesenchymal stem cells, *Journal of Materials Science-Materials in Medicine 19*, 2953–2962.
22. Recknor, J. B., Sakaguchi, D. S., and Mallapragada, S. K. (2006) Directed growth and selective differentiation of neural progenitor

cells on micropatterned polymer substrates, *Biomaterials 27*, 4098–4108.

23. Kim, D. H., Seo, C. H., Han, K., Kwon, K. W., Levchenko, A., and Suh, K. Y. (2009) Guided Cell Migration on Microtextured Substrates with Variable Local Density and Anisotropy, *Advanced Functional Materials 19*, 1579–1586.

24. Kwon, K. W., Choi, S. S., Lee, S. H., Kim, B., Lee, S. N., Park, M. C., Kim, P., Hwang, S. Y., and Suh, K. Y. (2007) Label-free, microfluidic separation and enrichment of human breast cancer cells by adhesion difference, *Lab Chip 7*, 1461–1468.

25. Seidlits, S. K., Lee, J. Y., and Schmidt, C. E. (2008) Nanostructured scaffolds for neural applications, *Nanomedicine 3*, 183–199.

26. Chai, C., and Leong, K. W. (2007) Biomaterials approach to expand and direct differentiation of stem cells, *Mol Ther 15*, 467–480.

27. Engel, E., Michiardi, A., Navarro, M., Lacroix, D., and Planell, J. A. (2008) Nanotechnology in regenerative medicine: the materials side, *Trends in Biotechnology 26*, 39–47.

28. Ilic, D., Genbacev, O., and Krtolica, A. (2007) Derivation of hESC from intact blastocysts, *Curr Protoc Stem Cell Biol Chapter 1*, Unit 1A 2.

Part V

Directed Differentiation of hES and iPS Cells in 3D Environments

Chapter 32

Neural Differentiation of Human ES and iPS Cells in Three-Dimensional Collagen and Martigel™ Gels

Eric Derby, Dezhong Yin, Wei-Qiang Gao, and Wu Ma

Abstract

In this chapter, we describe an effective and reproducible protocol for neural differentiation of human pluripotent stem cells in three dimensional (3D) collagen and Martigel™ gels. We have used this protocol to generate embryoid bodies (EBs) from dissociated suspension cultures of human embryonic stem cells (hESCs) and induced pluripotent stem cells (hiPSCs) which are immobilized in 3D collagen or Martigel™ gels. The gel-entrapped EBs differentiated into robust Pax6⁺ neuroectodermal cells of neural rosettes in N2 neural-inducing medium, then into nestin⁺ neural progenitors and further differentiated into TuJ1⁺ neurons, GFAP⁺ astrocytes, and O4⁺ oligodendrocytes. This 3D neural differentiation model, derived from human ES and iPS cells, has potential applications in the study of early human embryonic development, drug screening, and cell therapy.

Key words: Human embryonic stem cell, Induced pluripotent stem cell, 3D hydrogels, Embryoid body, Neural rosettes, Neural progenitors

1. Introduction

The combination of living cells with polymer scaffolds has potential for generating functional three-dimensional (3D) constructs for cell and tissue replacement. There is a particular need for engineered neural tissues, as the mammalian central nervous system has little capacity of self-repairing after injury. Functional recovery following brain and spinal cord injuries and neurodegenerative diseases is likely to require the transplantation of exogenous neural cells or tissues. This remains a significant challenge for neural tissue engineering since neurons are not capable of proliferating and are short-lived in culture. Recent advances in stem cell biology suggest that both human embryonic stem (hES) and induced pluripotent (iPS) cells hold tremendous promise for the development of novel therapies for many diseases and/or injuries because of their ability to self-renew and to differentiate into all cell types in human body. Both hESCs and hiPSCs have shown the potential to be valuable sources of specific neural cell types. Various strategies have been developed to direct the *in vitro* differentiation of pluripotent stem cells into neurons. Most neural differentiations have been performed on 2D plates coated with laminin or other extracellular matrix (ECM) proteins (1–4). In the present protocol, we describe a reliable method to achieve neural differentiation of human ES and iPS cells in 3D matrices.

In this protocol, we choose two biological hydrogels, collagen and Martigel™, for cell immobilization. The hydrogels are attractive polymer scaffolds because of their highly porous and hydrated structure, allowing cells to assemble spontaneously and become organized into a recognized tissue and permitting the infusion of nutrients and oxygen and transfer of waste products and CO_2 out of the cells. Collagen is the major class of insoluble fibrous proteins in the mammalian extracellular matrix, and Martigel™ is a solubilized basement membrane extracted from EHS mouse sarcoma (a tumor rich in ECM proteins). Major components of Martigel™ are laminin (56%), followed by collagen IV (31%), heparan sulfate proteoglycans, and entactin (8%). Both collagen and Martigel™ provide a physiologically relevant environment for stimulation of cell proliferation and differentiation.

In previous studies, the first step of hESCs or hiPSCs neural differentiation is to form embryoid bodies (EBs) from a dissociated suspension culture. A number of studies have shown that the differentiation of ES cell through EBs can mimic the embryonic development process, in which EBs recapitulate early embryonic developmental phases. Therefore EBs can be used to study early human development (5). We developed a protocol and showed that the pluripotent stem cell-derived EBs are neuro-inducible and can be expanded in N2 medium in 3D collagen and Martigel™ gels. These EBs are enriched with Pax6+ neuroectodermal cells in neural rosettes. The hydrogel-entrapped Pax6+ neuroectodermal cells further differentiate into nestin+ neural progenitors and then TuJ1+ neurons, GFAP+ astrocytes, and O4+ oligodendrocytes. The Pax6+ neural rosettes are immobilized within the 3D collagen or Martigel™ gels. Immunocytochemical analysis reveals the generation of a high percentage of nestin+, TuJ1+, GFAP+, and O4+ neural cells from the N2 medium-induced Pax6+ EBs.

This protocol uses a chemically defined neural-inducing N2 medium to direct neural commitment of pluripotent stem cells. We found that precise spatial and temporal exposure of cells to the N2 medium and 3D matrices is crucial to achieve homogeneous and efficient neural differentiation. In the past 10 years, we have developed various stem cell-based 3D neural tissue models (6–13). These 3D systems have been used to study stem cell proliferation, differentiation, gene expression, signaling transduction, synaptic transmission, and toxicology in a mimic body environment. The 3D human tissue models derived from human ES and iPS cells have potential applications not only for the study of early human development but also for toxicology, drug screening, and cell therapy (14, 15).

2. Materials

2.1. Preparation of MEF Feeder Layers

1. Dulbecco's minimum essential medium (DMEM) 1× (ATCC, Manassas, VA); store at 4°C
2. Embryonic stem cell grade fetal bovine serum (FBS) (Invitrogen); store as frozen aliquots at −20°C
3. DMEM/F12 1× (ATCC); store at 4°C
4. DMSO (Sigma, St. Louis, MO)
5. L-Alanyl/L-glutamine, 100× (ATCC); store as aliquots at −20°C
6. Nonessential amino acid (NEAA) solution, 100× (ATCC); store at 4°C
7. Mouse embryonic fibroblast (MEF) irradiated, SCRC-1040.1 (ATCC)
8. Mouse embryonic fibroblast culture medium
   Prepare the following solution sterilely in a 500 mL filter unit:
   - 449.5 mL DMEM (ATCC)
   - 50 mL FBS (Invitrogen, Carlsbad, CA)
   - 500 µL 2-mercapoethanol (BME) (Invitrogen)

2.2. Maintenance and Expansion of Human ESCs

1. Human ES cell line TE-06 (NIH stem cell registry)
2. Flasks T75, Vent cap (Corning or equivalent)
3. Flasks T225, Vent cap (Corning or equivalent)
4. 15-mL conical tubes (Corning or equivalent)
5. 50-mL polypropylene conical tubes (Falcon or equivalent)
6. Embryonic stem cell grade fetal bovine serum (ES-FBS) (ATCC)
7. Collagenase IV (400 U/mL) (Invitrogen). Dissolve collagenase IV powder in DMEM 1× to final concentration of 400 U/mL and filter sterilize. Store as frozen aliquots at −20°C
8. DMEM/F12 1× (ATCC)
9. L-Alanyl/L-glutamine, 100× (ATCC)

10. Non-essential amino acid (NEAA) solution, 100× (ATCC)
11. Mouse embryonic fibroblast (MEF) irradiated, SCRC-1040.1 (ATCC)
12. Dimethyl sulfoxide (DMSO) (ATCC)
13. Bovine serum albumin (BSA) (SIGMA)
14. Knockout serum replacer (KOSR) (Invitrogen)
15. Basic fibroblastic growth factor (bFGF, R&D Systems)
16. Human ES culture medium
    Prepare the following solution sterilely in a 500 mL filter unit:
    - 384 mL DMEM/F12
    - 25 mL KOSR
    - 75 mL ES-FBS
    - 5 mL NEAA
    - 5 mL L-alanyl/L-glutamine
    - 909 µL BME (55 mM)
    - 200 µL bFGF (10 µg/mL)
    - 5 mL Pen/strep

2.3. Maintenance and Expansion of hiPSC on Martigel™

1. Human iPSC lines
2. mTeSR™1 5× supplement (StemCell Technologies)
3. mTeSR™1 basal medium (StemCell Technologies)
4. KO-DMEM (Invitrogen)
5. KOSR Knockout serum replacer (Invitrogen)
6. Martigel™, growth factor reduced (BD Biosciences)
7. Collagenase IV (400 U/mL) (Invitrogen). Dissolve Collagenase IV powder in DMEM 1× to final concentration of 400 U/mL and filter sterilize. Store as frozen aliquots at −20°C.
8. DMEM/F12 1× (ATCC)
9. Dimethyl sulfoxide (DMSO) (Sigma)
10. Isopropyl alcohol, 70% (EMD)
11. PBS w/o Ca^{2+} and Mg^{2+} (Mediatech)

2.3.1. Preparation of Martigel™ Aliquots

1. Thaw Martigel™ in refrigerator at 2–8°C to prevent gelling.
2. Aliquots must be prepared while stock is at 2–8°C. Place stock bottle in ice bath in a pipette tip box while aliquoting.
3. Aliquot 120 µL into a 15 mL conical tube. Place aliquots immediately on ice in an ice bucket.
4. Store aliquots in a freezer at −20°C.

2.3.2. Preparation of Martigel™-Coated Dishes

1. Resuspend 120 µL of Martigel™ aliquot in 12 mL cold KO-DMEM medium and plate to dishes at 1 mL of diluted

Martigel™ per 10 cm² of surface area (6 mL is appropriate for a 100-mm dish).

2. Spread it to make Martigel™ covers entire surface. It will begin gelling before it reaches room temperature.

3. Place coated culture dishes in a 37°C incubator for at least 1 h prior to use.

2.4. Preparation of Embryonic Bodies from hESCs and hiPSCs

1. Human ES cell line (see Sect. 2.2)
2. Collagenase IV (400 U/mL) (Invitrogen). Dissolve collagenase IV in DMEM 1× to 400 U/mL and filter sterilize. Store as frozen aliquots at −20°C
3. Ultra-low attachment dish 100 mm (Corning, Corning, NY)
4. Tissue culture dishes 100 mm (Corning)
5. DMEM (ATCC)
6. F12 (ATCC)
7. N-2 supplement, 100× (Invitrogen)
8. Nonessential amino acids, NEAA 100× (ATCC)
9. Penicillin (10,000 IU/mL)/streptomycin (10,000 µg/mL) 100× (ATCC)
10. Basic fibroblastic growth factor (bFGF, R&D Systems). Stock solution at 10 µg/mL in 1× PBS
11. MEF (ATCC) as feeder cells. (see Sect. 2.1)
12. 6-well plates (Nunc, Fisher Scientific, Rochester, NY)
13. Sterilized Pasteur pipettes (Fisher Scientific)
14. L-Glutamine (ATCC)
15. PBS 1× without Ca^{2+} and Mg^{2+} (ATCC)
16. 2-Mercaptoethanol (Sigma)
17. Human embryoid body media preparation.
    Prepare the following solution sterilely in a 500 mL filter unit:
    - 390 mL KO DMEM (Invitrogen)
    - 50 mL Knockout serum replacement (Invitrogen)
    - 50 mL Plasmanate (Talecris)
    - 5 mL Glutamax (Invitrogen)
    - 5 mL Nonessential amino acids (Invitrogen)
      Store at 2–8°C up to 2 weeks

2.5. Martigel™ Gel Preparation for 3D Culture

1. Human ESC qualified Martigel™ matrix (BD Biosciences, Bedford, MA)
2. DMEM/F12 medium (ATCC, Manassas, VA)
3. 24-well culture cluster plates (Corning, Corning, NY)
4. bFGF (see Sect. 2.2)

2.6. Collagen Preparation for 3D Culture

1. Rat tail tendon, type I (Roche, Indianapolis, IN)
2. 0.2% v/v acetic acid, sterile, diluted in molecular grade sterile water
3. 1N NaOH
4. Phosphate-buffered saline (ATCC, Manassas, VA), made 2× by 1:5 dilution of 10X PBS in sterile molecular biological grade water.
5. 24-well tissue culture plate

2.7. Neural Differentiation-Inducing N2 Medium Preparation

Prepare the following solution sterilely in a 500 mL filter unit:

- 163 mL of F-12 (ATCC)
- 326 mL of DMEM (ATCC)
- 5 mL N-2 supplement (Gibco, Invitrogen)
- 5 mL Nonessential amino acids solution (NEAA) (ATCC)
- 254 µL bFGF stock (10 µg/mL, R&D systems)
  Store media at 2–8°C up to 2 weeks

3. Methods

This step-by-step protocol comprises the following major steps (Fig. 1). The differentiation begins with EB formation from dissociated suspension cultures of hESCs or iPSCs. The EBs are preinduced to Pax6+ neuroectodermal cells in neural differentiating-inducing N2 medium. The next step is to immobilize neural rosettes in pre-induced EBs into collagen or Martigel™ gels. The gel-entrapped neuroectodermal cells differentiate into nestin+ neural progenitors and further TuJ1+ neurons, GFAP+ astrocytes, and O4+ oligodendrocytes.

3.1. Maintenance and Expansion of hESCs

1. Seed 100-mm culture dishes (55 cm^2) with 55,000 MEF per cm^2 in MEF medium (DMEM 1×, 10% ES-FBS, L-alanyl/L-glutamine 1×, NEAA 1×, and 0.1 mM 2-mercapoethanol. This will be 3 × 10^6 MEF cells per 100-mm tissue culture dish (see Note 1).
2. Thaw a vial of hESCs (TE-06) rapidly in a 37°C water bath.
3. Transfer vial contents to a 15-mL tube and add 12 mL hESC culture medium (DMEM/F12 1×, 15% ES-FBS, 5% knockout serum replacement, pen-strep 1×, NEAA 1×, L-alanyl/-L-glutamine 1×, 2-ME (0.1 mM), and bFGF (4 ng/mL) (see Note 2).
4. Remove medium from previously prepared MEF monolayers (see Sect. 3.1.1)

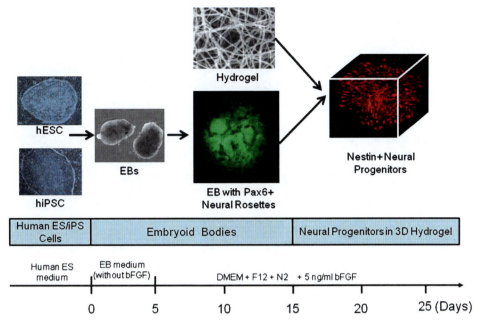

Fig. 1. Schematic diagram of the protocol for directed neural differentiation of hESCs and hiPSCs in a 3D collagen or Martigel™ gels.

5. Dispense hESCs evenly into two 100-mm tissue culture dishes containing the MEF monolayers.
6. Add 12 mL of hESC culture medium to each dish.
7. Add an additional 4 ng/mL of bFGF to the medium at 18–24 h post seeding,
8. Incubate at 37°C, 5% CO_2, exchange the hESC medium at day 2 and then exchange the medium daily thereafter.
9. Observe colony formation on MEFs under a microscope (see Note 3).
10. Passage cells every 6–8 days by removing medium from dishes and adding 4 mL of 37°C prewarmed collagenase IV solution per dish.
11. Incubate at 37°C for 30–45 min. Observe the dishes for rounding and lifting of hESC colonies. Gently tap the dishes to dislodge the colonies. If colonies do not dislodge, place dishes at 37°C for an additional 10–15 min. Observe dishes for rounding colonies.
12. Collect hESC suspension to a 50-mL conical tube and centrifuge at $200 \times g$ for 5 min. Discard the supernatant carefully.
13. Resuspend the cell pellets in hESC culture medium by pipetting up and down gently to break the colonies to small aggregates (see Note 4). Distribute hESC aggregates evenly to six 100-mm tissue culture dishes prepared with MEF feeders as in Sect. 3.1.

Add 12 mL medium to each dish. Add an additional 4 ng/mL of bFGF to the medium after 18–24 h.

14. When colonies reach desired size and density, split again at 1:3 split ratio following steps 9–13. Use 4 mL of collagenase solution (400 U/mL) per 100-mm dish for passaging.

15. This provides enough number of hESCs for EB production.

3.2. Maintenance and Expansion of hiPSCs

3.2.1. Thawing Human iPS Cells

1. Thaw a vial of hiPSC by immersing the vial in a 37°C water bath without submerging the cap. Swirl the vial gently.

2. Spray the vial with 70% ethanol or isopropyl alcohol to sterilize the outside of the tube. Briefly (30–60 s) air-dry the vial in the sterile biosafety cabinet.

3. Transfer the cells gently into a sterile 15-mL conical tube using a 5-mL pipette with 4 mL warm mTeSR. Dispense 3 mL of collected cells into labeled 15-mL conical tube. Use the remaining 2 mL to rinse cryovial and add to the same 15-mL conical tube.

4. Centrifuge the cells in 15 mL conical tube at $200 \times g$ for 5 min.

5. Aspirate and discard the supernatant and gently resuspend pellet in 2 mL mTeSR. Take care not to break cell clumps into single cells.

3.2.2. Plating and Passaging iPS Cells on MEFs

1. Seed 100-mm tissue culture dishes with 55,000 MEF per cm^2 in MEF medium (DMEM 1×, 10% ES-FBS, L-alanyl/-L-glutamine 1×, NEAA 1×, and 0.1 mM 2-mercapoethanol). This will be 3 × 10^6 MEF cells per 100-mm dish.

2. Remove the MEF medium from 100-mm dishes, and add 5 mL hES media to the dish.

3. Aspirate the hES media from the MEF dish prior to plating.

4. Add the iPS cell suspension drop-wise on fresh MEF plates.

5. Place dishes in incubator and do not disturb for 24 h.

6. On day 2, feed hiPSC by aspirating off media and replacing with 12–15 mL fresh hES media. Repeat the procedure daily until cells require passaging.

7. Prepare MEFs the day before you plan to passage cells.

8. Aspirate the hESC media from the culture to be split. Wash the wells with 10 mL of 1× PBS per dish.

9. Add 4 mL prewarmed collagenase IV solution (400 U/mL) to each dish. Incubate for 10–20 min at 37°C. Examine cells under microscope to confirm colony separation from the plate. Tap dish gently to help cells detach.

10. Pool cells in 10× the volume of collagenase IV solution of hESC media prepared in a 50-mL conical tube. Wash plate with 3–8 mL of fresh hES media and add to a conical tube.

11. Centrifuge cells at $200 \times g$ for 5 min.

12. Resuspend cell pellet in fresh hESC media by pipetting up and down gently to break the colonies to small aggregates. How much additional medium required to cell suspension is dependent on the split ratio and the number of dishes used. There should be a total of 12 mL of medium and cells in each of the new dish.

13. Aspirate MEF medium off the MEF dish. Plate cells on fresh MEF dishes. Label with date, cell line, and passage number.

14. Return the dish to incubator. Do not disturb for 24 h.

3.2.3. Plating and Passaging iPS Cells on Martigel™

1. Prepare the new Martigel™ dishes for culture by aspirating the Martigel™ from precoated 100-mm dish and add 8 mL mTeSRTM1 medium. Label the coated dish with the cell line, the passage number from the vial, the date and your initials.

2. Transfer iPS cell suspension dropwise into the 100-mm dish and place into the 37°C incubator overnight to allow colonies to attach.

3. The next day, add 4 mL of mTeSRTM1 medium to the first original dish. Place the dish back to the incubator overnight.

4. Then feed cells with full fluid changes with 10–15 mL mTeSRTM1 medium daily. If feeding more than one dish, use a different pipet for each to reduce risk of contamination.

5. Examine cells closely under the inverted microscope. Colonies may not be visible for up to 3–5 days.

6. To passage cells, warm mTeSRTM1 medium and collagenase IV to 37°C prior to operation.

7. Remove iPS cell dish from incubator and place it in the biosafety cabinet. Aspirate the medium from the wells with a pipette.

8. Rinse with 3 mL PBS without Ca^{2+} or Mg^{2+}. Then aspirate rinse.

9. Add 4 mL room temperature collagenase IV solution to each passaged dish. Incubate for 10–20 min at 37°C.

10. Examine cells closely under the microscope. When the edges of the colonies begin to lift, tap the dish to loosen colonies.

11. Add 8 mL mTeSRTM1 medium to dilute enzyme. If the colonies are not coming completely off the plate, a pipette can be used to dispense media to rinse colonies off culture surface. Minimize bubbles by pipetting gently.

12. Pool the media/cell suspension into the 15-mL conical tube.
13. Centrifuge the tube at $200 \times g$ for 5 min.
14. Aspirate supernatant from the cell pellet and gently re-suspend the pellet again with 4 mL of mTeSRTM1 medium. Add sufficient cells to each new Martigel™-coated dish based on the split ratio and the number of dishes used. There should be a total of 10–12 mL of medium and cells in each of the new dish (see Note 5).
15. Add cells to new dishes and place into 37°C incubator.
16. Conduct full fluid changes daily to include day after passage (see Note 6).

3.3. hESC and hiPSC-Derived EB Formation and Expansion

1. On day 0, when human ES/iPS cell cultures are confluent on irradiated MEFs or on Martigel™, remove media and wash cells with PBS, then add 4 mL prewarmed collagenase IV solution per 10 cm dish. Incubate at 37°C for 10–20 min, checking dishes every 10 min thereafter, until colonies lift off the MEF monolayer or Martigel™. Dissociate colonies by gently agitating and triturating.
2. Pool cells with collagenase IV solution into 50 mL conical tubes with 5× volume EB media. Wash plate two times with EB media and collect cells in the same tube.
3. Centrifuge cell suspension at $200 \times g$ for 5 min.
4. Remove collagenase IV/media and gently resuspend cell pellet in a small volume of EB medium (for example, 5 mL per 100-mm dish of starting cells) for cell number evaluation.
5. Seed ES/iPS cells in 100-mm ultra-low attachment dishes at a density of about 4.5×10^6 cells per dish with a total volume of 12 mL EB media per dish.
6. Incubate the cells in a 37°C incubator. Culture the EBs in suspension for another 2 days without medium change.
7. At day 3, add an additional 10 mL of EB media per dish.

3.4. Neural Induction of EBs

1. At day 5, pool EBs in a 50 mL falcon tube and allow settling by gravity for 10 min. Aspirate the media carefully, and resuspend EBs with N2 neural differentiation-inducing medium (see Note 7) in a volume to allow re-plating of 10 mL EB suspension per dish.
2. Aliquot 10 mL EB suspension per dish in N2 medium to new 100-mm ultra-low attachment dishes.

3.5. Maintenance and Expansion of Neuro-Induced EBs

1. N2 medium containing bFGF (20 ng/mL) is used for maintenance and expansion of neuro-induced EBs. Feed EB cultures every 2–3 days with N2 medium. Place 100-mm dishes on the

flat surface of the biological safety cabinet. Swirl dishes in a circular motion (see Note 8). Carefully remove about 50% of N2 medium per dish without removing floating EBs. Replace the N2 medium with 50% fresh N2 medium.

2. Passaging and expanding neuro-induced EBs involve the breaking up of individual EB clusters and the expanding of the broken clusters to new 100-mm ultra-low attachment dishes. The EBs should be broken up three times every 10–14 days using a P200 tip and P200 pipetteman. This is executed by aspirating EBs into the bore of the tip then expelling the EBs until the individual EB break into 2–3 sections. Expand half of the newly broken EBs into a new 100-mm low-attachment dish with fresh N2 medium when there are approximately a 100 EB pieces in a dish (see Note 9).

3. EB cultures can be maintained in the N2 medium for up to 10 months. Neuro-induced EBs aged 1–10 months are ready for immobilization in 3D gels (see Note 10).

3.6. Immobilization of hESC and hiPSC-Derived EBs in 3D Collagen Gels

3.6.1. Preparation of Collagen Gels

1. Add 3.3 mL of sterile 0.2% (v/v) acetic acid to 10 mg rat tail collagen I powder to make a stock collagen at 3 mg/mL and incubate it at 4°C overnight.

2. Prepare 6 mL of collagen working solution at 0.5 mg collagen/mL by adding the following components:
   (a) Take 1 mL collagen stock at 3 mg/mL by using a 5 mL pipette and add it to a sterile 15 mL tube.
   (b) Add 1 mL sterilely 2× PBS by using the same pipette.
   (c) Add 4 mL N2 medium by using the same pipette and mix well.
   (d) Adjust pH to 7.4 by using sterile 1 N NaOH and add in increments of 20 µL until color of medium is corrected to approximately 7.3–7.5 by using a pH paper
   (e) Place diluted collagen solution (0.5 mg/mL) on ice.

3.6.2. Immobilization of Neuro-Induced EBs into 3D Collagen Gels

1. Mix 3.2 mL collagen working solution (0.5 mg/mL) with 0.8 mL EBs.
2. Load 0.4 mL of EB–collagen mixture per well for 24-well plates.
3. Allow gel to form for 1–2 h at 37°C.
4. Slowly add 0.5 mL of N2 medium per well for 24-well plates (see Note 11).
5. Change the N2 medium twice per week.
6. Observe development of cell–collagen constructs over day 1 through day 28.

3.7. Immobilization of hESC and hiPSC-Derived EBs in 3D Martigel™ Gels

3.7.1. Prepare Martigel™ Gels

1. Thaw human ESC qualified Martigel™ matrix overnight at 4°C (see Note 12).
2. Dilute Martigel™ 1:6 (approximately 16% Martigel™) in cold N2 medium. This is sufficient to coat a 24-well plate.
3. Place a 24-well culture plate on a level ice surface so as not to gel the Martigel™ prematurely.
4. Dispense 0.25 mL of Martigel™–NP medium mixture per well of a 24-well plate.
5. Leave coated plates for 30–60 min on ice.
6. Then aspirate excess Martigel™ and place plates to room temperature for 30–60 min prior to addition of EB suspension.
7. Dilute cold Martigel™ (16%) with cold N2 medium to approximately 10% Martigel™ for suspension of cells. (e.g., approximately, mixing 1 mL N2 medium and 1.6 mL of 16% Martigel™).

3.7.2. Immobilization of Neuro-Induced EBs into 3D Martigel™ Gels

1. Add EBs in N2 medium to 10% Martigel™ in equal volumes. This produces EB–Martigel™ solution containing 5% Martigel™.
2. Add bFGF to EB–Martigel™ solution to make a final concentration of 10 ng/mL.
3. Load 0.4 mL of EB–Martigel™ solution per well of a 24-well plate.
4. Place plates in a 37°C incubator.
5. Observe 3D cell-Martigel™ constructs for expression of pluripotency and neural markers.

3.8. Characterization of Hydrogel-Entrapped Cells for Expression of Pluripotency and Neural Makers Using Immunocytochemistry

During the neural specialization of human pluripotent stem cells-derived EBs in suspension and 3D hydrogel cultures, expression of pluripotency markers is down-regulated, whereas neural markers is up-regulated. To characterize this transition, NANOG, Oct3/4, SSEA4, TRA-1-60, and TRA-1-81 are used as puripotency markers. Human pluripotent stem cell-derived neural stem/progenitor cells and progeny are identified based upon the presence of molecular markers that are correlated with the precursor state along with the absence of a more differentiated phenotype as assessed through immunocytochemical analysis. Pax6, nestin, and Sox1 are used to identify neuroectodermal cells and neural stem/progenitor cells along with more differentiated lineage markers, including TuJ1 for neurons, GFAP for astrocytes, and O4 for oligodendrocytes. All of the antibodies used have been tested and optimized for use in immunocytochemistry on human neural progenitors and progeny.

Table 1
Primary antibodies for immunocytochemistry

Primary antibodies	Host	Dilution	Source
Pax6	Mouse IgG1	1:100	*Abcam*
Sox1	Chicken Ig G(Yolk sac derived)	1:300	Millipore/Chemicon
Nestin	Mouse IgG1	1:200	Millipore/Chemicon
Nestin	Rabbit IgG	1:200	Millipore/Chemicon
TuJ1	Mouse IgG2a	1:300	Covance
GFAP	Rabbit IgG	1:500	Millipore/Chemicon
O4	Mouse IgM	1:100	Millipore/Chemicon
NANOG	Mouse IgG2a	1:100	Millipore/Chemicon
TRA-1-60	Mouse IgM	1:50	Millipore/Chemicon
OCT3/4	Mouse IgG1	1:250	BD Bioscieces
TRA-1-81	Mouse IgM	1:50	Millipore/Chemicon
SSEA4	Mouse IgM	1:100	Millipore/Chemicon

1. Aspirate the media and rinse cell–hydrogel constructs with 2 mL of 1X PBS in each well of the 24-well plates. Fix cells with 1 mL of 4% paraformaldehyde (PFA) (mix 10 mL of 20% PFA, 35 mL of dH_2O, and 5 mL of 10× PBS) to each well, for 30 min at RT.
2. Carefully aspirate the fixative and rinse three times with 1× PBS.
3. Add 0.4–0.5 mL of saponin solution (0.5% w/v saponin in PBS 1×) to inside wells for detection of intracellular markers. Allow to sit at RT for 30 min. Remove saponin, rinse well with 1× PBS twice.
4. Apply the appropriate blocking solutions for 2 h at room temperature. For optimal results, use the blocking solution (5% normal donkey serum, and 0.3% Triton X-100 in 1×PBS) with antibodies directed against Pax6, nestin, Sox2, TuJ1, or GFAP. Use the non-permeable blocking solution (5% normal donkey serum in 1× PBS) with the antibody directed against the oligodendrocyte marker O4.
5. Dilute the primary antibodies to working concentrations (Table 1).
6. Aspirate PBS from wells and apply 0.5 mL of primary antibody to designated wells.

Table 2
Secondary antibodies for immunocytochemistry

Secondary antibodies	Dilution	Source
Donkey anti mouse IgG-FITC	1:50	Jackson
Donkey anti rabbit IgG-Rhodamine	1:50	Jackson
Goat anti mouse IgG (H + L) Alexa Fluor 488 nm	1:50	Invitrogen
Goat anti mouse IgG1 Alex Fluor 488 nm	1:50	Invitrogen
Goat anti mouse IgM-Rhodamine	1:50	Millipore/Chemicon
Donkey anti-mouse IgG2a FITC-conjugated	1:50	Southern Biotech, Birmingham

7. Incubate cell–hydrogel constructs with primary antibody overnight at 4°C.
8. Aspirate primary antibody and rinse the constructs three times with 1× PBS.
9. Dilute secondary antibodies (Table 2) in the appropriate blocking solution just before use.
10. Aspirate PBS from wells and apply 0.5 mL of the appropriate secondary antibody to designated wells for 2 h at room temperature.
12. Wash constructs three times (5–10 min each) with 1× PBS.
13. Counterstain the cell nuclei with DAPI solution.
14. Visualize the cell staining with a fluorescent microscope.

4. Notes

1. MEF feeders may be used for hESC culture within a week.
2. All bFGF containing culture media should be used within a week.
3. Observe the colony formation in 3–4 days of culture by microscopy and passage hESCs every 6–8 days.
4. hESCs will not recover well after subculture if they are broken into single cell suspensions.
5. Take care not to break cell clumps too much. 15–30 cells per clump is ideal. Be careful not to break up the colonies too much or cause bubbles in the media.
6. Split ratio can vary based on confluency of culture prior to split, and can be adjusted to offer quicker or delayed passage

time. In general, a 1:6 on a 60% confluent, 1:7 on a 70% confluent, and 1:8 on an 80% confluent culture will set up for following split in 4–6 days.

7. No bFGF should be in the N2 medium for the first 5 days of initiation of culture.

8. EBs are very lightly attached to the plates. Swirling the 100-mm low attachment plates moves the floating EBs to the center of the plate. Care must be taken to avoid aspirating EB when removing old N2 medium. Alternatively, EB can be removed to 50 mL-tubes and replaced into the 100 mm plates with fresh N2 medium.

9. Breaking up EBs may also be transferred to new ultra-low attachment dishes at 1:2–3 split ratio according to the density of EBs.

10. EBs may also be cryopreserved and thawed according to standard procedures of hESC cryopreservation.

11. Take care not to disturb the collagen–EB construct by slowly adding the N2 medium when the gel is formed.

12. Martigel™ and materials associated with formation of the 3D culture must be kept below 8°C to maintain liquid state. Above this critical temperature the Martigel™ will gel. (e.g., prechill 5 mL pipettes, P1000 tips, and culture tubes). Work on ice for all initial Martigel™ coating.

References

1. Zhang SC, Wernig M, Duncan ID, Brüstle O, Thomson JA. (2001) In vitro differentiation of transplantable neural precursors from human embryonic stem cells. Nat Biotechnol. 19:1129–1133.
2. Ma W., Tavakoli T., Derby E., Serebryakova Y., Rao MS., Mattson MP. (2008a) Cell-extracellular matrix interactions regulate neural differentiation of human embryonic stem cells. *BMC Developmental Biology*, 8:90.
3. Hu BY, Zhang SC. (2009) Differentiation of spinal motor neurons from pluripotent human stem cells. Nat Protoc. 4:1295–304.
4. Tavakoli T., Derby E., Serebryakova Y., Rao MS., Mattson MP., Ma W. (2009) Self-Renewal and differentiation capabilities are variable between human embryonic stem cell lines I3, I6 and BG01V. *BMC Cell Biology*, 10:44.
5. Hwang NS, Varghese S, Elisseeff J. (2008) Controlled differentiation of stem cells. Adv Drug Deliv Rev. 60:199–214.
6. O'Connor SM., Stenger DA., Shaffer KM., Maric D., Barker JL, Ma W. (2000a) Primary neural precursor O cell expansion, differentiation and cytosolic Ca2+ response in three-dimentional collagen gel. *Journal of Neuroscience Methods*, 102:187–195.
7. O'Connor SM., Andreadis JD., Shaffer KM., Ma W., Pancrazio J., Stenger DA. (2000b) Immobilization of neural cells in three-dimensional matrices for biosensor applications. *Biosensors & Bioelectronics*, 14:871–881.
8. O'Connor SM., Stenger DA., Shaffer KM., Ma W. (2001) Survival and neurite outgrowth of rat cortical neurons in three-dimensional agarose and collagen gel matrices. *Neuroscience Letter*, 304:189–193.
9. O'Shaughnessy TJ., Lin HJ., Ma W. (2003) Functional synapse formation among rat cortical neurons grown on three-dimensional collagen gels. *Neuroscience Letter*, 340:169–172.
10. Lin JH., O'Shaughnessy TJ., Kelly J., Ma W. (2004) Neural stem and progenitor cell differentiation in a cell-collagen-bioreactors culture system. *Developmental Brain Research*, 153:163–173.

11. Ma W., Fitzgerald W., O'Shaughnessy TJ., Lin HJ., Liu QL., Maric D., Alkon DL., Barker JL. (2004) CNS stem and progenitor cell differentiation into functional neuronal circuits in three-dimensional collagen gels. *Experimental Neurology*, 190: 276–288.
12. Ma W., Chen S., Fitzgerald W., Maric D., Lin HJ., O'Shaughnessy TJ., Kelly J., Liu X-H., Barker JL. (2005) Three-dimensional collagen gel networks for neural stem cell-based neural tissue engineering. *Macromolecular Symposia*, 227:327–333.
13. Ma W., Tavakoli T, Chen S., Maric M., Lin HJ., O'Shaughnessy TJ., Barker JL. (2008b) Reconstruction of functional cortical-like tissues from neural stem and progenitor cells. *Tissue Engineering*, 10: 1673–1685.
14. Stenger DA., Gross GW., Keefer EW., Shaffer KM., Andreadis JD., Ma W., Pancrazio JJ. (2001) Detection of physiologically active compounds using cell-based biosensors. *Trends in Biotechnology*, 19:304–309.
15. Kunlin Jin, et al. (2010) Transplantation of Human Precursor Cells in Martigel Scaffolding Improves Outcome from Focal Cerebral Ischemia after Delayed Postischemic Treatment in Rats. *J Cereb Blood Flow Metab.* 30(3): 534–544.

Part VI

Qualitative and Quantitative Analysis of Dynamics of hES Cell Differentiation

Chapter 33

Single-Cell Transcript Profiling of Differentiating Embryonic Stem Cells

Jason D. Gibson, Caroline M. Jakuba, Craig E. Nelson, and Mark G. Carter

Abstract

Heterogeneity of stem cell populations is a well-known but poorly characterized phenomenon. Here, we demonstrate the qualitative and quantitative power of single-cell transcript analysis to characterize transcriptome dynamics in embryonic stem cells (ESC). In this chapter, we describe a method for isolation, characterization, and analysis of single-cell transcript profiles of individual human ESC that clearly identifies cellular heterogeneity within undifferentiated populations and identifies novel cell types in differentiating cultures. This analysis is presented at a level of depth and resolution not attainable by other methods. None of the insights in this study would have been possible with standard population-level transcript analysis or single-cell FACS analysis of known cell-surface markers. Only by developing robust single-cell transcript profiling techniques and applying these to established stem cell differentiation paradigms were we able to deconstruct complex populations of cells into their component parts. Single-cell analysis can systematically determine unique cellular profiles for use in cell sorting and identification, show the potential to augment standing models of cellular differentiation, and elucidate the behavior of stem cells exiting pluripotency.

Key words: Single-cell transcript analysis, Embryonic stem (ES) cells, Pluripotency, Differentiation, Gene expression profiling

1. Introduction

1.1. Embryonic Stem Cells and Single-Cell Resolution

Embryonic stem cells (ESC) are derived from the inner cell mass of the developing embryo and retain the capacity to form derivatives of each of the three embryonic germ layers: ectoderm, mesoderm, and endoderm (1). Effectively mapping the developmental trajectory of stem cells undergoing differentiation is critical for the optimization of protocols to generate clinically relevant cell types. From determining the range of cellular responses to inductive cues to identifying markers for cell type profiling and purification, optimization of directed differentiation protocols requires studying stem cells as individual entities. The direct assessment of transcript profiles from individual cells has been important for the elucidation of olfactory receptor expression (2) and cell type determination in the retina (3) and pancreas (4). These and other studies demonstrate how single-cell transcript profiling has already contributed to the understanding of complex developmental decisions in vivo and suggest that this approach could provide similar insights into stem cell biology (5) through the identification of distinct, but often cryptic, cell types in complex populations. Here, we detail a method that enables analysis of transcript levels in individual cells and present results of a study characterizing transcriptome dynamics in human embryonic stem cells (hESC) based on single-cell analysis to illustrate the power, utility, and innovation of the technique.

1.2. Cell Type Heterogeneity Within Differentiating Cultures

An accurate description of the genetic activities driving the differentiation of individual cells within complex populations is essential for optimizing stem cell protocols to generate clinically relevant cell types. Through a comprehensive study of transcription in undifferentiated and differentially induced human embryonic stem cells at single-cell resolution, we demonstrate the power of this technique as a qualitative and quantitative method to characterize transcriptome dynamics in individual cells. By comparing transcriptional profiles of individual undifferentiated and differentiated stem cells, we were able to show that undifferentiated hESC colonies displayed consistent gene expression profiles characterized by uniform transcripts and sporadically detected transcripts. We identified new stages of hESC differentiation and markers of intermediate progenitor populations and discovered the extensive co-expression of pluripotency and differentiation markers in individual cells over several days or weeks of differentiation.

2. Materials

2.1. Reagents

1. 0.05% trypsin–EDTA (Invitrogen #25300)
2. Accutase (Chemicon #SCR005)
3. Accumax (Chemicon #SCR006)
4. DMEM-F12 media (Invitrogen #11330)
5. Knockout serum replacer (Invitrogen #10828)
6. L-Glutamine (Invitrogen #25030)
7. Non-essential amino acids (Invitrogen #11140)
8. b-Mercaptoethanol (Invitrogen #21985)
9. bFGF (Invitrogen #AA10-155)

10. Collagenase type IV (Gibco #17104)
11. Dispase (Gibco #17105)
12. HyClone characterized fetal bovine serum (Thermo Scientific #SH30071)
13. NP-40 detergent solution (Thermo Scientific #28324)
14. RNasin® Plus RNase inhibitor (Promega #N261)
15. TaqMan® Gene Expression Assays 20× (Applied Biosystems #4331182)
16. dNTP Mix, 100 mM (25 mM of each dNTP) (Thermo Scientific #AB-1124)
17. M-MLV Reverse Transcriptase with 5× Reaction Buffer (Promega #M170)
18. $MgCl_2$ (Applied Biosystems #AM9530G)
19. TaqMan® PreAmp Master Mix (Applied Biosystems #4391128)
20. TaqMan® Universal PCR Master Mix (Applied Biosystems #4304437)
21. MicroAmp Fast Optical 96-well Reaction Plates (Applied Biosystems #4346906)
22. TaqMan® PreAmp Pool for Human Stem Cell Pluripotency Array (Applied Biosystems #4405625)
23. TaqMan® Stem Cell Pluripotency Low Density Array Cards (Applied Biosystems: Human #4385344; Mouse #4385363)
24. Tissue culture plastics
25. Pasteur pipettes
26. Serological pipettes
27. PCR tubes
28. Micropipette tips

2.2. Equipment

1. Class II biosafety cabinet (tissue culture hood)
2. Humidified cell culture incubator at 5% CO_2 and 37°C
3. Microliter pipettes
4. Serological pipette filler
5. Inverted microscope with phase contrast optics
6. Centrifuge
7. Thermal cycler
8. ABI 7500 and 7900HT Fast Real-time Systems (Applied Biosystems)
9. JMP (SAS, Cary, NC) or similar statistical analysis software

3. Methods

3.1. Single-Cell Isolation Methods

1. Incubate hESC colonies or embryoid bodies (EBs) for 5 min in 0.05% trypsin–EDTA or longer in other enzymatic reagents (see Note 1, i.e., accutase, accumax, etc.) with gentle trituration.
2. Add serum-containing media to inactivate trypsin, pellet cells, and wash once with PBS, dilute throughout a 6-well plate to space single cells for picking and allow cells to settle prior to isolation.
3. Pick single cells using a 2 µL pipette and note the presence of any neighboring cells to ensure that they remain in place after picking.
4. Transfer single cells isolated in 2.0 µL PBS to 3.0 µL of lysis solution containing 0.5 µL 5% NP-40, 0.5 µL RNAse inhibitor, and 2.0 µL 0.25× pooled TaqMan® assays for immediate lysis (see Note 9). For increased throughput, single cells can also be isolated *via* fluorescent activated cell sorting (FACS). They can be sorted directly into wells containing 2 µL PBS and 3 µL lysis solution as described above.

3.2. Direct Cell Denaturation and Reverse Transcription of Single-Cell mRNA

1. Denature single cells immediately by incubating at 70°C for 10 min.
2. Cool in a cycler to 4°C to allow binding of TaqMan® primers (see Note 2).
3. Immediately following denaturation, add 5.0 µL reverse transcription (RT) mix consisting of 0.2 µL 25 mM dNTPs, 0.5 µL 200 U/mL MMLV, 2.0 µL 5× M-MLV RT Buffer, 1.2 mL 25 mM $MgCl_2$, 0.5 µL RNAse inhibitor, and 1.0 µL ddH_2O to each sample.
4. Incubate single-cell reverse transcription reactions in a thermal cycler using the program 40 cycles (37°C 20 s, 42°C 10 s, 50°C 100 s), 1 cycle (85°C 50 s) (see Note 3).

3.3. Multiplex Pre-amplification

1. Pre-amplify single-cell cDNA samples in multiplex prior to singleplex qPCR. Pre-amplification reactions consist of 4.0 µL single-cell cDNA, 5.0 µL 0.25× TaqMan® pooled assays (see Note 4), 1.0 µL ddH_2O, and 10 µL TaqMan® PreAmp Master Mix. Single-cell pre-amplification reactions are amplified using the program 1 cycle (95°C 30 s, 55°C 20 s, 72°C 20 s), 16 cycles (95°C 15 s, 60°C 20 s, 72°C 20 s), and diluted 1:5 with ddH_2O (see Note 5).

3.4. Real-Time Quantitative Polymerase Chain Reaction

1. Use diluted cDNA products for individual TaqMan® Gene Expression Assay qPCR reactions performed using an ABI 7500 Fast Real-time PCR System with the same TaqMan® assays (Table 1) included in the original 0.25× pool. TaqMan® qPCR

Table 1
Applied Biosystems TaqMan® Assay ID references for genes examined. Genes selected for transcript detection are on the left, assay ID numbers on the right. (16) (Reproduced by permission of The Royal Society of Chemistry (RSC))

Gene	TaqMan® Assay ID
ACTB	Hs99999903_m1
AFP	Hs00173490_m1
CDX2	Hs01078080_m1
CGA	Hs00174938_m1
EOMES	Hs01015629_m1
FGFR1	Hs00241111_m1
FOXA2	Hs00232764_m1
GAPDH	Hs99999905_m1
GATA4	Hs01034629_m1
GATA6	Hs00232018_m1
HAND1	Hs00231848_m1
KDR	Hs00911700_m1
LEFTY2	Hs00745761_s1
LHX1	Hs00232144_m1
MIXL1	Hs00430824_g1
NANOG	Hs02387400_g1
POU5F1	Hs00999632_g1
SOX1	Hs01057642_s1
SOX2	Hs01053049_s1
T	Hs00610080_m1
TBX6	HS00365539_m1

reactions consist of 0.5 µL diluted pre-amplified cDNA, 5.0 µL 2× TaqMan® Universal PCR Master Mix, 2.0 µL 5× TaqMan® gene expression assay, and 2.5 µL ddH$_2$O.

2. Perform reactions with a standard cycling program, 1 cycle (50°C 20 s, 95°C 10 s), 40 cycles (95°C 15 s, 60°C 10 s).

3. Use diluted pre-amplified cDNA at the same concentrations for analysis with TaqMan® Low Density Array (TLDA) cards using the ABI 7900HT Fast Real-Time System (see Note 6).

3.5. Data Processing and Analysis

Single-cell average cycle threshold (C_T) values obtained from qPCR runs are normalized, and this normalized data is used to cluster the gene expression profiles. Normalization is achieved *via* ΔC_T calculations with ACTB and GAPDH values for each cell using the following equation: $\Delta C_T = C_T - \left[\dfrac{C_{T,ACTB} + C_{T,GAPDH}}{2}\right]$. Undetected transcripts are set to $\Delta C_T = 40$. All ΔC_T values are reversed (ΔC_T^*) to obtain positive values using the equation: $\Delta C_T^* = 40 - \Delta C_T$ for each sample. JMP analysis software is used for unsupervised hierarchical clustering and statistical analysis.

3.6. Embryonic Stem Cell Culture Maintenance

Maintain H9 hESC (WA09 (H9)) (6) on mouse embryonic fibroblasts (MEFs) in DMEM-F12 media supplemented with 20% knockout serum replacer, 2 mM (0.5%) L-glutamine, 0.1 mM (1%) non-essential amino acids, 0.55 mM b-mercaptoethanol and 4 ng/mL bFGF. Human ESC are mechanically passaged using 1.0 mg/mL (1.0%) collagenase.

3.7. Uniformity and Precocious Differentiation in hESC Colonies

To examine the uniformity of hESC colonies, the aforementioned single-cell isolation, cDNA synthesis, and pre-amplification protocol platform are used upstream of the TaqMan® Low Density Array (TLDA®) cards to profile the expression of individual cells from phenotypically undifferentiated (d0) hESC cultures (Fig. 1). hESC are notorious for precocious differentiation, but the degree to which this differentiation is reflected in phenotypically uniform and well-defined hESC colonies is largely unknown. However, all seven markers of the undifferentiated pluripotent ES cell state (POU5F1, NANOG, SOX2, TDGF1, DNMT3B, GABRB3, and GDF3) should be detected in every day 0 cell tested. Uniform expression of these factors and 18 other stemness-associated transcripts (as per ABI TaqMan® Human Stem Cell Pluripotency Array) in all day 0 single cells analyzed is consistent with an undifferentiated pluripotent state. In addition, these cells should be negative for a large set of differentiation markers.

The analysis on unsupervised clustering of these transcript profiles from day 0 cells reveal three classes of transcripts: (1) absent: not detected in any cells; (2) uniform: detected in all cells; and (3) sporadic: detected in some, but not all cells. The confirmation of these characteristics suggests that the colonies surveyed reside in an undifferentiated state. However, a subset of other genes could also be expressed sporadically in these cells.

3.8. False-Negative and False-Positive Rates

In order to estimate a false-negative rate for this assay (or the rate at which genes expressed at levels typical of the core members of the pluripotency network would go undetected), sample the 96 undifferentiated hESC and analyze each for POU5F1, NANOG, GAPDH, and ACTB, expecting positive signals for each assay in all 96 wells. In our experiment, nearly all cells (95/96) tested positive

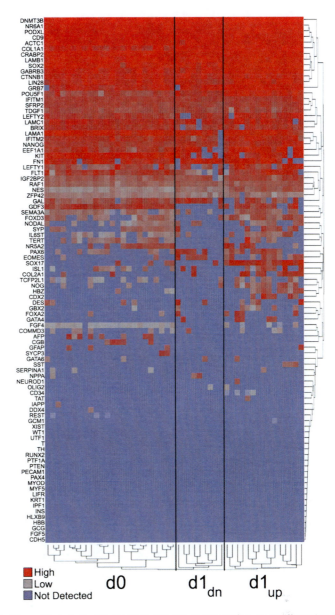

Fig. 1. *Hierarchical clustering of single-cell expression profiles in undifferentiated hESC (d0), and after 24 h of differentiation in serum (d1).* Undifferentiated cells display uniformly present and uniformly absent markers, as well as a set of markers that are sporadically expressed. After 24 h in serum, two populations of cells are observed: one that has up-regulated a suite of sporadic markers (d1$_{up}$), and one which has down-regulated several markers (d1$_{dn}$). Each column represents an individual cell; each row, a transcript. Normalized expression values (delta-Ct) range from blue (not detected) to red (highly expressed). (16) (Reproduced by permission of The Royal Society of Chemistry (RSC)).

for POU5F1, NANOG, GAPDH, and ACTB (data not shown). The one cell that was negative for all markers and, as the human TaqMan® probes used are species-specific, was likely a contaminated MEF feeder cell. This test suggests that the false-negative rate for these assays is 0.5%. Due to the extreme sensitivity of our PCR protocol, false-positive results represent a potentially serious hazard. To detect possible false-positive results, we routinely run no-RT, no-probe, and no-cell negative control reactions. These assays should be all consistently negative.

3.9. Single-Cell Transcript Analysis of Directed Differentiation

In directed differentiation experiments, it is specifically interesting to determine how the earliest lineage decisions are made by ESC as they exit from pluripotency and how they adopt early lineage fates. For this analysis, you could assemble a small panel of TaqMan® assays for genes routinely used to mark the three main germ layers and the extra-embryonic endoderm (Table 1). You could use this panel of markers to profile single cells differentiating in three culture conditions: exposure to activin, BMP4, or high serum, each of which have been shown to give rise to endoderm, mesoderm, or all three germ layers, respectively (7–9) (and references therein). In order to ensure that the culture conditions can replicate prior findings, you could compare the single-cell transcript results with the population-level gene expression profiles previously published for these hESC culture protocols by constructing cell-averaged expression graphs from expression levels measured over the first 7 days of differentiation (Fig. 2a). Up to 12 cells isolated at each selected time point of differentiation could be analyzed, and the corresponding C_T values at each of these time points could be averaged to generate the spline curves in Fig. 2a (activin = d1,3,5,7; BMP4 = d1,2,6; serum = d1,3,5,7). These splines reveal several common features of all three differentiation protocols: early downregulation of POU5F1 and NANOG, transient expression of mesendoderm markers EOMES and T, and gradual upregulation of endoderm markers GATA6 and FOXA2. These graphs also allow direct comparison of the three culture conditions and confirm several expected differences between conditions.

Deconstructing the data to single-cell resolution, Fig. 2b plots the expression of 19 genes over time in each of the three differentiation protocols. Following the trend of gene expression (red) from day 0 to day 21 confirms the expected differential effects of culture conditions, as seen in the splines, as well as common features of all three protocols: slow downregulation of pluripotency genes, transient expression of mesendoderm markers, and gradual upregulation of endoderm markers. A striking common feature that could be revealed only in the single-cell data is the extensive simultaneous detection of pluripotency transcripts with transcripts widely associated with lineage commitment and differentiation. We observed this phenomenon in the majority of cells sampled, persisting through day 21 (Fig. 3a–c).

The extended expression of pluripotency markers in EBs is frequently interpreted as the persistence of undifferentiated cells within the EB (10). However, this interpretation is called into question by studies that fail to find any indication of undifferentiated proliferating cells within teratomas (11, 12). Our data suggest that,

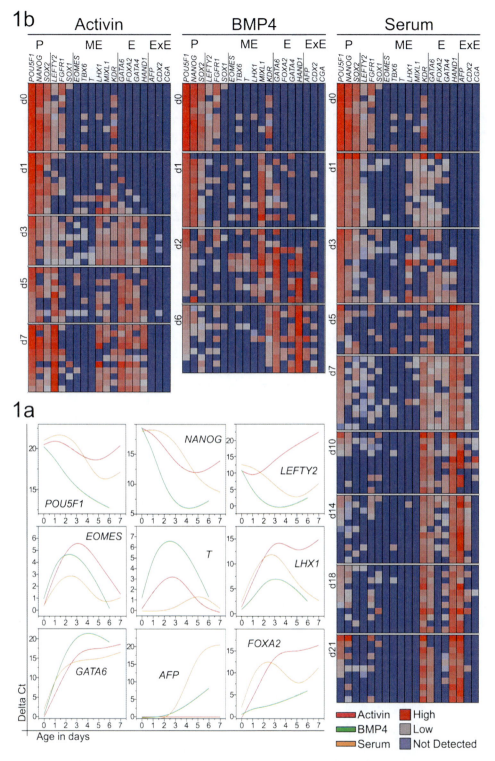

Fig. 2. *Expression profiles of single hESC differentiating in three culture conditions.* (**1a**) Cell-averaged expression plots of nine genes over the first 7 days of differentiation. Lines are splines fit to the average of all cells collected at each time point and are color-coded to represent different culture conditions (*Orange*=serum, *Red*=activin, *Green*=BMP4). Relative expression intensity (ΔCT^*) is on the X-axis, age of cells (in days) is on the Y-axis. (**1b**) Relative expression (ΔCT^*) of 19 genes in single cells under three culture conditions. Rows represent individual cells, columns are gene markers. Markers are grouped as: *P* pluripotency, *ME* mesendoderm, *E* endoderm, *ExE* extra-embryonic endoderm. Expression levels range from *blue* (not detected) to *red* (highly expressed) (16) (Reproduced by permission of The Royal Society of Chemistry (RSC)).

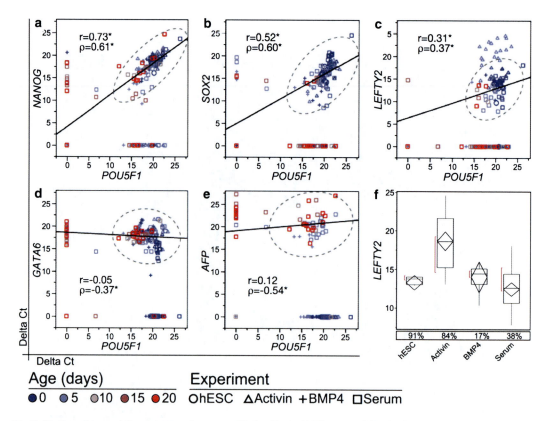

Fig. 3. *Single-cell transcript measurements are quantitative.* (**2a** and **b**) The per-cell levels of pluripotency markers co-vary and diminish over time. (**2d** and **e**) Despite the extensive co-expression of both pluripotency and endoderm markers (as illustrated in Fig. 1), no significant quantitative relationship is observed between the pluripotency marker *POU5F1* and endoderm markers. (**2c** and **f**) Activin up-regulates the per-cell levels of LEFTY2 transcripts. Age of cells is a range from *dark blue* (d0) to *red* (d21). Culture conditions are encoded by shape: open circles = undifferentiated hESC, triangles = activin, squares = serum, crosses = BMP4. Asterisks (*) mark statistically significant correlation coefficients. In the box and whisker plot in 2f, the diamond is centered on the mean response and ends at the 95% confidence intervals. The box is split by the median and bounded by the 25th and 75th percentiles. The whiskers extend from the ends of the box to the upper quartile + (1.5 × interquartile range) and the lower quartile + (1.5 × interquartile range). *Dots* beyond each whisker represent outliers. The *bracket* along the edge of the box identifies the shortest half. Box percentages in 2f indicate the number of cells positive for LEFTY2 (16) (Reproduced by permission of The Royal Society of Chemistry (RSC)).

rather than indicating the presence of undifferentiated cells within the EB, the persistent expression of pluripotency markers is an aspect of normal differentiation in these cells. Thus, what was originally seen as overlapping expression of POU5F1 with mesendodermal and endodermal markers at the population level, we now show to be part of a larger pattern, where multiple markers of pluripotency, mesendoderm, and/or endoderm are co-expressed at the single-cell level (Figs. 2b and 5), a fact obscured in population assays.

3.10. Quantitative Analysis and Modulation of Transcript Levels

Many previous attempts at single-cell transcript profiling have restricted analysis of the data to qualitative, presence/absence detection. However, this approach will allow you to quantitatively analyze the gene expression in differentiating hESC. Using this approach, we found that bivariate analysis of ΔC_T values between pluripotency markers (Fig. 3a–c) could give highly significant correlation coefficients between POU5F1, NANOG, LEFTY2, and SOX2 across all experimental conditions (undifferentiated, serum, activin, and BMP4, p-values ranging from 0.03 to 0.0001) with r-values ranging from 0.73 for NANOG and POU5F1 to 0.52 between SOX2 and POU5F1. These single-cell analysis correlations are approximately equivalent to those observed at the population level between these same genes across many hESC lines and culture conditions (10). When compared to genes whose expression patterns are uncorrelated or negatively correlated, we observed no significant quantitative correlation (Fig. 3d and e), suggesting that quantitative correlations between transcripts detected in single cells are reliable reflections of cellular transcript levels. Furthermore, we observed that as pluripotency markers are down-regulated upon differentiation, their diminishing levels of expression are clearly correlated with those of other pluripotency markers (Fig. 3a–c). Another advantage of this approach is the clear detection of LEFTY2 upregulation upon activin treatment (Fig. 3c, f). Comparison of LEFTY2 levels in day 0 hESC and hESC treated with activin for 7 days reveals a marked (>27-fold) upregulation of per-cell LEFTY2 transcript levels in response to activin treatment, but no increase in the percentage of LEFTY2-positive cells. This observation is consistent with previous studies describing the upregulation of SMAD2/3 targets, including LEFTY2, in response to activin treatment (13).

3.11. Qualitative Analysis and Non-parametric Correlations

In addition to the quantitative correlations described above, many non-quantitative (presence/absence) positive and negative correlations between genes can be found in the aforementioned panel. Because this single-cell assay has low false-positive and false-negative rates, non-parametric, qualitative relationships between genes could be readily detected (Fig. 4). Indeed, we found many such correlations, including positive correlations between markers of different cell states (pluripotent, mesendoderm, and embryonic and extra-embryonic endoderm) and negative correlations between genes marking different cell lineages. The 19 positive correlations and 13 negative correlations we identify here are supported by other investigations of individual genes (Table 2). The extensive support of these relationships by the literature suggests that the novel relationships identified here (e.g., positive correlations between KDR and AFP and negative correlations between LHX1 and KDR) may suggest regulatory relationships between these genes that bear further investigation. Because of the heterogeneity of population samples, many of these transcriptional relationships may only be readily detected using single-cell analysis, thus illustrating the utility of this approach for augmenting currently mapped regulatory networks and discovering novel cell types and lineages.

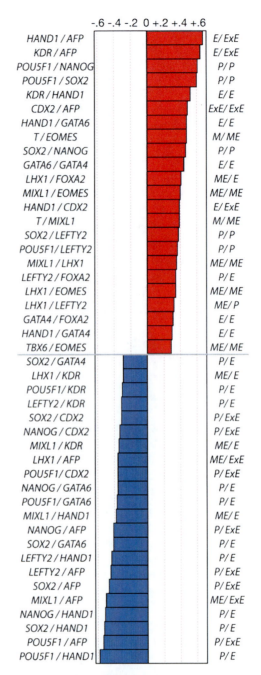

Fig. 4. Non-parametric correlations between all markers. Combined data from undifferentiated hESC and serum, activin, and BMP4 cultures were screened for significant correlations between markers. Novel correlations are in *bold*, previously observed correlations are in plain text. Correlations are expressed as Spearman rho values with a Bonferroni corrected p-value cutoff of 0.0001. *Red* = positive correlation, *blue* = negative correlation. Cell type designations: P (pluripotency), ME (mesendoderm), E (endoderm), ExE (extra-embryonic endoderm) (16) (Reproduced by permission of The Royal Society of Chemistry (RSC)).

Table 2
Non-parametric correlations detected by single-cell analysis supported by current literature. Pairwise correlations are on the left, supporting references on the right (16) (Reproduced by permission of The Royal Society of Chemistry (RSC))

Gene correlation	Cell types	References
Positive		
HAND1/AFP	E/ExE	Peiffer et al. (17)
KDR/AFP	ME/ExE	**NOVEL**
POU5F1/NANOG	P/P	Boyer et al. (18), Kim et al. (19), Kuroda et al. (20)
POU5F1/SOX2	P/P	Card et al. (21), Li et al. (22), Boyer et al. (18), Kim et al. (19), Kuroda et al. (20)
KDR/HAND1	ME/E	**NOVEL**
CDX2/AFP	ExE/ExE	Hyslop et al. (23)
HAND1/GATA6	E/E	Niu et al. (24)
T/EOMES	M/ME	Izumi et al. (25)
SOX2/NANOG	P/P	Boyer et al. (18), Kim et al. (19), Kuroda et al. (20)
GATA6/GATA4	E/E	Fujikura et al. (26), Zhang et al. (27), Xin et al. (28)
LHX1/FOXA2	ME/E	Perea-Gómez et al. (29), Foucher et al. (30)
MIXL1/EOMES	ME/ME	Izumi et al. (25)
HAND1/CDX2	E/ExE	Coucouvanis et al. (31), Peiffer et al. (17), Hough et al. (32)
T/MIXL1	M/ME	Izumi et al. (25)
SOX2/LEFTY2	P/P	Westfall et al. (33)
POU5F1/LEFTY2	P/P	Westfall et al. (33)
MIXL1/LHX1	ME/ME	**NOVEL**
LEFTY2/FOXA2	P/E	Merrill et al. (34)
LHX1/EOMES	ME/ME	**NOVEL**
LHX1/LEFTY2	ME/P	Tsang et al. (35)
GATA4/FOXA2	E/E	Matsuura et al. (36), Denson et al. (37)
HAND1/GATA4	E/E	Nagao et al. (38)
TBX6/EOMES	ME/ME	Yi et al. (39)
Negative		
SOX2/GATA4	P/E	Murakami et al. (40)
LHX1/KDR	ME/ME	**NOVEL**
POU5F1/KDR	P/ME	Nelson et al. (41)
LEFTY2/KDR	P/ME	**NOVEL**
SOX2/CDX2	P/ExE	Sherwood et al. (42), Chickarmane and Peterson (43)
NANOG/CDX2	P/ExE	Ralston and Rossant (44), Yasuda et al. (45), Hyslop et al. (23), Strumpf et al. (46)
MIXL1/KDR	ME/ME	Lim et al. (47)
LHX1/AFP	ME/ExE	**NOVEL**
POU5F1/CDX2	P/ExE	Hough et al. (32), Ralston and Rossant (44), Babaie et al. (48)
NANOG/GATA6	P/E	Hough et al. (32), Chickarmane and Peterson (43), Singh et al. (49)
POU5F1/GATA6	P/E	Babaie et al. (48)
MIXL1/HAND1	ME/E	**NOVEL**
NANOG/AFP	P/ExE	Hyslop et al. (23)
SOX2/GATA6	P/E	Kobayashi et al. (50)
LEFTY2/HAND1	P/E	**NOVEL**
LEFTY2/AFP	P/ExE	**NOVEL**

(continued)

**Table 2
(continued)**

Gene correlation	Cell types	References
SOX2/AFP	P/ExE	NOVEL
MIXL1/AFP	ME/ExE	NOVEL
NANOG/HAND1	P/E	Degrelle et al. (51)
SOX2/HAND1	P/E	NOVEL
POU5F1/AFP	P/ExE	Bettiol et al. (52), Chen et al. (53)
POU5F1/HAND1	P/E	Hough et al. (32), Takada et al. (54)

3.12. Hierarchical Clustering and Inference of Developmental Trajectory

In order to directly compare transcript profiles from all conditions (day 0, serum, activin, and BMP4), all expression profiles could be clustered by similarity using JMP analysis software, as shown in Fig. 5. In our study, unsupervised hierarchical clustering revealed five clusters of transcriptionally similar cells (labeled as cell types 1–5, Fig. 5) that roughly correspond to the following cell type designations: (1) pluripotency, (2) exit from pluripotency, (3) mesendoderm, (4) endoderm, and (5) extra-embryonic endoderm. Cells from all conditions differentiated towards an endoderm-like profile as cultures aged. While activin-treated cells do not express extra-embryonic endoderm markers, cells isolated from BMP4- and serum-induced cultures assume an extra-embryonic endoderm-like fate (cell type 5) marked by the expression of both the endoderm markers and the extra-embryonic markers. Across all culture conditions, we found very few non-day 0 cells with undifferentiated gene expression profiles. However, we observed occasional young cell profiles clustering within older profile groups, infrequent older cell profiles clustering within younger groups, and occasional cell profiles from each protocol intermixing with those from other protocols, indicating that these cultures display both temporal asynchrony and cell type heterogeneity (Fig. 5).

3.13. Determination of Unique Cellular Profiles

Possibly the greatest use of single-cell analysis will be the ability to systematically uncover unique cellular subtypes within heterogeneous populations. By analyzing the gene expression profiles of single cells, you will be able to discriminate individual cell types within complex mixtures. Because there is no *a priori* assumption regarding cell type or lineage, this method is ideal for the identification of novel cell types and cell type markers. Our data show that in hESC exposed to serum for 24 h (Fig. 1), two subpopulations of cells arise, one which up-regulates many sporadic genes ($d1_{up}$) and the other one which down-regulates several transcripts ($d1_{dn}$). These subpopulations differ in the expression of many transcripts, including cell-surface markers such as SEMA3A and IL6ST (Fig. 1). The identification of cell-surface markers by single-cell transcript

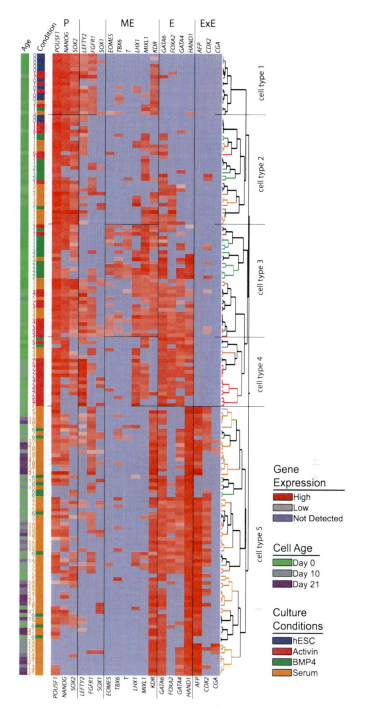

Fig. 5. *Clustering of all cells across culture conditions.* Combined data from undifferentiated hESC and serum, activin, and BMP4 cultures were analyzed using unsupervised hierarchical clustering. Cells with similar expression profiles are grouped together. Expression level is indicated by the color of each square, while age and culture condition are indicated to the left. Single-cell expression values (ΔC_T^*) range from *blue* (not expressed) to *red* (highly expressed). Age of individual cells is indicated by the color map in the far left column (*green* = d0, *magenta* = d21) and listed in the second column from the *left*. Culture conditions are in the third column from the *left* (*blue* = d0, *red* = activin, *orange* = serum, *green* = BMP4. Five major cells types discussed in the text are indicated on the far *right* (16) (Reproduced by permission of The Royal Society of Chemistry (RSC)).

analysis could be quite useful for FACS purification and subsequent characterization by microarrays, differentiation assays, and chromatin immunoprecipitation. Since roughly equal numbers of cells adopt each fate after 24 h and regulate these transcripts in opposite directions, this phenomenon would have been difficult to detect using population-level assays.

Also, the expression pattern of KDR (Fig. 5) illustrates how important the ability to unravel temporal asynchrony and cell type heterogeneity is to the identification of subtypes and markers in differentiating ESC populations. KDR has been proposed as a marker of mesoderm in human ESC cultures (14), but our single-cell data shows a more complex story, with KDR expressed widely in cell types 1, 2, 3, and 5, but almost completely absent from cell type 4 (endoderm), consistent with the utility of a KDR-negative profile to identify definitive endoderm-like cells as used in mouse ESC (15). Standard expression profiling of differentiating cultures would be confounded by the mixture of cell types at the population level. Similarly, many of the qualitative correlations detected by single-cell profiling and reported in Fig. 4 have not been previously reported in population-level studies. The mixing of cell types and temporal stages within populations has likely obscured some such relationships in previous work. While the single-cell profiling data we present here is based on limited numbers of transcripts and cells and, in the case of Fig. 2, limited time points, it has uncovered several novel transcriptional relationships and cell types. Single-cell analysis gives researchers previously unattainable depth to their experiments and serves as a powerful tool for the characterization of stem cell differentiation dynamics.

4. Notes

1. Although 0.05% trypsin–EDTA can be used effectively to dissociate single cells from ESC colonies, we have found that the trituration necessary to produce a homogenous single-cell suspension results in the lysis of many cells. As the presence of cell lysate may interfere with accurate single-cell transcript profiling, it is suggested that accutase be used to dissociate ESC colonies and accumax be used for the dissociation of embryoid bodies or primary tissues. These enzymatic reagents may be incubated with adherent and/or suspension cell cultures for extended periods of time without imposing the risk of lysing cell membranes.

2. Degradation of cellular mRNA has been shown to occur when isolated single cell samples are left in lysis solution for more than 1 h on ice. It is highly recommended that single-cell samples are denatured as soon as possible and processed through the RT step of this protocol. However, snap-freezing single-cell

lysis samples on dry ice and storing at −80°C have provided consistent results up to at least 1 week.

3. Processed single-cell RT reactions may be snap-frozen and stored at −80°C for at least 1 week and likely longer without degradation of the output cDNA.

4. The low cycle number for single-cell pre-amplification is known not to skew transcript level relationships. The 16 cycles of pre-amplification provide enough cDNA to run 200 individual qPCR assays per reaction. Processed single-cell pre-amplification reactions should be stored at −80°C.

5. Pre-amplification of single-cell RT reactions for use on TaqMan® Low Density Array (TLDA) cards is performed with the corresponding TaqMan® PreAmp Pool (i.e., TaqMan® PreAmp Pool for Human Stem Cell Pluripotency Array).

6. TLDA cards require a tenfold lower volume per reaction (1 μL) and are useful for increasing the number of reactions per single-cell sample. For a 96-assay card, a 1:5 diluted single-cell pre-amplification reaction is further diluted to 1:20 in a 200 μL reaction (10 μL of 1:5 diluted pre-amp reaction, 100 μL 2× TaqMan® Universal PCR Master Mix, and 90 μL ddH$_2$0) for loading into two 48-well ports of a TLDA card.

7. Besides using species-appropriate TaqMan assays and pre-amplification pools, this protocol can be used without modification for mouse ESC as well as human ESC. We have found, however, that trypsin is more effective than accutase or accumax for dissociating mESC and does not result in the high level of cell lysis seen in hESC dissociated with trypsin.

8. When picking single cells, we found that mESC adhere quickly to a 6-well dish in PBS and, therefore, prefer to pick mESC in a dilution of TBS.

9. To make the TaqMan pool, mix equal parts of all of the 20× assays you wish to test by qPCR and dilute with water to a final 0.25× concentration. This pool is the same in the pre-amplification step so we suggest making enough at once for both steps of the protocol.

References

1. Gadue P., Huber T.L., Nostro M.C. et al. (2005) Germ layer induction from embryonic stem cells, *Exp Hematol* **33**, 955–964.
2. Tietjen I., Rihel J.M., Cao Y. et al. (2003) Single-cell transcriptional analysis of neuronal progenitors, *Neuron* **38**, 161–175.
3. Trimarchi J.M., Stadler M.B., Roska B. et al. (2007) Molecular heterogeneity of developing retinal ganglion and amacrine cells revealed through single cell gene expression profiling, *J Comp Neurol* **502**, 1047–1065.
4. Chiang M.K., Melton D.A. (2003) Single-cell transcript analysis of pancreas development, *Dev Cell* **4**, 383–393.
5. Kurimoto K., Yabuta Y., Ohinata Y. et al. (2006) An improved single-cell cDNA amplification method for efficient high-density oligonucleotide microarray analysis, *Nucleic Acids Res* **34**, e42.

6. Thomson J.A., Itskovitz-Eldor J., Shapiro S.S. et al. (1998) Embryonic stem cell lines derived from human blastocysts, *Science* **282**, 1145–1147.
7. Cerdan C., Hong S.H., Bhatia M. (2007) Formation and hematopoietic differentiation of human embryoid bodies by suspension and hanging drop cultures, *Curr Protoc Stem Cell Biol* **Chapter 1**, Unit 1D 2.
8. D'Amour K.A., Agulnick A.D., Eliazer S. et al. (2005) Efficient differentiation of human embryonic stem cells to definitive endoderm, *Nat Biotechnol* **23**, 1534–1541.
9. Zhang P., Li J., Tan Z. et al. (2008) Short-term BMP-4 treatment initiates mesoderm induction in human embryonic stem cells, *Blood* **111**, 1933–1941.
10. Adewumi O., Aflatoonian B., Ahrlund-Richter L. et al. (2007) Characterization of human embryonic stem cell lines by the International Stem Cell Initiative, *Nat Biotechnol* **25**, 803–816.
11. Blum B., Benvenisty N. (2007) Clonal analysis of human embryonic stem cell differentiation into teratomas, *Stem Cells* **25**, 1924–1930.
12. Gertow K., Wolbank S., Rozell B. et al. (2004) Organized development from human embryonic stem cells after injection into immunodeficient mice, *Stem Cells Dev* **13**, 421–435.
13. Besser D. (2004) Expression of nodal, lefty-a, and lefty-B in undifferentiated human embryonic stem cells requires activation of Smad2/3, *J Biol Chem* **279**, 45076–45084.
14. Yang L., Soonpaa M.H., Adler E.D. et al. (2008) Human cardiovascular progenitor cells develop from a KDR+ embryonic-stem-cell-derived population, *Nature* **453**, 524–528.
15. Nostro M.C., Cheng X., Keller G.M. et al. (2008) Wnt, activin, and BMP signaling regulate distinct stages in the developmental pathway from embryonic stem cells to blood, *Cell Stem Cell* **2**, 60–71.
16. Gibson J.D., Jakuba C.M., Boucher N. et al. (2009) Single-cell transcript analysis of human embryonic stem cells, *Integr Biol (Camb)* **1**, 540–551.
17. Peiffer I., Belhomme D., Barbet R. et al. (2007) Simultaneous differentiation of endothelial and trophoblastic cells derived from human embryonic stem cells, *Stem Cells Dev* **16**, 393–402.
18. Boyer L.A., Lee T.I., Cole M.F. et al. (2005) Core transcriptional regulatory circuitry in human embryonic stem cells, *Cell* **122**, 947–956.
19. Kim J., Chu J., Shen X. et al. (2008) An extended transcriptional network for pluripotency of embryonic stem cells, *Cell* **132**, 1049–1061.
20. Kuroda T., Tada M., Kubota H. et al. (2005) Octamer and Sox elements are required for transcriptional cis regulation of Nanog gene expression, *Molecular and Cellular Biology* **25**, 2475–2485.
21. Card D.A.G., Hebbar P.B., Li L. et al. (2008) Oct4/Sox2-regulated miR-302 targets cyclin D1 in human embryonic stem cells, *Molecular and Cellular Biology* **28**, 6426–6438.
22. Li J., Pan G., Cui K. et al. (2007) A dominant-negative form of mouse SOX2 induces trophectoderm differentiation and progressive polyploidy in mouse embryonic stem cells, *J Biol Chem* **282**, 19481–19492.
23. Hyslop L., Stojkovic M., Armstrong L. et al. (2005) Downregulation of NANOG induces differentiation of human embryonic stem cells to extraembryonic lineages, *Stem Cells* **23**, 1035–1043.
24. Niu Z., Iyer D., Conway S.J. et al. (2008) Serum response factor orchestrates nascent sarcomerogenesis and silences the biomineralization gene program in the heart, *Proc Natl Acad Sci USA* **105**, 17824–17829.
25. Izumi N., Era T., Akimaru H. et al. (2007) Dissecting the molecular hierarchy for mesendoderm differentiation through a combination of embryonic stem cell culture and RNA interference, *Stem Cells* **25**, 1664–1674.
26. Fujikura J., Yamato E., Yonemura S. et al. (2002) Differentiation of embryonic stem cells is induced by GATA factors, *Genes & Development* **16**, 784–789.
27. Zhang C., Ye X., Zhang H. et al. (2007) GATA factors induce mouse embryonic stem cell differentiation toward extraembryonic endoderm, *Stem Cells Dev* **16**, 605–613.
28. Xin M., Davis C.A., Molkentin J.D. et al. (2006) A threshold of GATA4 and GATA6 expression is required for cardiovascular development, *Proc Natl Acad Sci USA* **103**, 11189–11194.
29. Perea-Gómez A., Shawlot W., Sasaki H. et al. (1999) HNF3beta and Lim1 interact in the visceral endoderm to regulate primitive streak formation and anterior-posterior polarity in the mouse embryo, *Development* **126**, 4499–4511.
30. Foucher I., Montesinos M.L., Volovitch M. et al. (2003) Joint regulation of the MAP1B promoter by HNF3beta/Foxa2 and Engrailed is the result of a highly conserved mechanism for direct interaction of homeoproteins and Fox transcription factors, *Development* **130**, 1867–1876.
31. Coucouvanis E., Martin G.R. (1999) BMP signaling plays a role in visceral endoderm differentiation and cavitation in the early mouse embryo, *Development* **126**, 535–546.

32. Hough S.R., Clements I., Welch P.J. et al. (2006) Differentiation of mouse embryonic stem cells after RNA interference-mediated silencing of OCT4 and Nanog, *Stem Cells* **24**, 1467–1475.
33. Westfall S.D., Sachdev S., Das P. et al. (2008) Identification of oxygen-sensitive transcriptional programs in human embryonic stem cells, *Stem Cells Dev* **17**, 869–881.
34. Merrill B.J., Pasolli H.A., Polak L. et al. (2004) Tcf3: a transcriptional regulator of axis induction in the early embryo, *Development* **131**, 263–274.
35. Tsang T.E., Kinder S.J., Tam P.P. (1999) Experimental analysis of the emergence of left-right asymmetry of the body axis in early post-implantation mouse embryos, *Cell Mol Biol (Noisy-le-grand)* **45**, 493–503.
36. Matsuura R., Kogo H., Ogaeri T. et al. (2006) Crucial transcription factors in endoderm and embryonic gut development are expressed in gut-like structures from mouse ES cells, *Stem Cells* **24**, 624–630.
37. Denson L.A., McClure M.H., Bogue C.W. et al. (2000) HNF3beta and GATA-4 transactivate the liver-enriched homeobox gene, Hex, *Gene* **246**, 311–320.
38. Nagao K., Taniyama Y., Kietzmann T. et al. (2008) HIF-1alpha signaling upstream of NKX2.5 is required for cardiac development in Xenopus, *J Biol Chem* **283**, 11841–11849.
39. Yi C.H., Terrett J.A., Li Q.Y. et al. (1999) Identification, mapping, and phylogenomic analysis of four new human members of the T-box gene family: EOMES, TBX6, TBX18, and TBX19, *Genomics* **55**, 10–20.
40. Murakami A., Shen H., Ishida S. et al. (2004) SOX7 and GATA-4 are competitive activators of Fgf-3 transcription, *J Biol Chem* **279**, 28564–28573.
41. Nelson T.J., Faustino R.S., Chiriac A. et al. (2008) CXCR4+/FLK-1+ biomarkers select a cardiopoietic lineage from embryonic stem cells, *Stem Cells* **26**, 1464–1473.
42. Sherwood R.I., Chen T.-Y.A., Melton D.A. (2009) Transcriptional dynamics of endodermal organ formation, *Dev Dyn* **238**, 29–42.
43. Chickarmane V., Peterson C. (2008) A computational model for understanding stem cell, trophectoderm and endoderm lineage determination, *PLoS ONE* **3**, e3478.
44. Ralston A., Rossant J. (2008) Cdx2 acts downstream of cell polarization to cell-autonomously promote trophectoderm fate in the early mouse embryo, *Developmental Biology* **313**, 614–629.
45. Yasuda S.-y., Tsuneyoshi N., Sumi T. et al. (2006) NANOG maintains self-renewal of primate ES cells in the absence of a feeder layer, *Genes Cells* **11**, 1115–1123.
46. Strumpf D., Mao C.-A., Yamanaka Y. et al. (2005) Cdx2 is required for correct cell fate specification and differentiation of trophectoderm in the mouse blastocyst, *Development* **132**, 2093–2102.
47. Lim S., Pereira L., Wong M. et al. (2008) Enforced Expression of Mixl1 During Mouse ES Cell Differentiation Suppresses Hematopoietic Mesoderm and Promotes Endoderm Formation, *Stem Cells*.
48. Babaie Y., Herwig R., Greber B. et al. (2007) Analysis of Oct4-dependent transcriptional networks regulating self-renewal and pluripotency in human embryonic stem cells, *Stem Cells* **25**, 500–510.
49. Singh A.M., Hamazaki T., Hankowski K.E. et al. (2007) A heterogeneous expression pattern for Nanog in embryonic stem cells, *Stem Cells* **25**, 2534–2542.
50. Kobayashi M., Takada T., Takahashi K. et al. (2008) BMP4 induces primitive endoderm but not trophectoderm in monkey embryonic stem cells, *Cloning Stem Cells* **10**, 495–502.
51. Degrelle S.A., Campion E., Cabau C. et al. (2005) Molecular evidence for a critical period in mural trophoblast development in bovine blastocysts, *Developmental Biology* **288**, 448–460.
52. Bettiol E., Sartiani L., Chicha L. et al. (2007) Fetal bovine serum enables cardiac differentiation of human embryonic stem cells, *Differentiation* **75**, 669–681.
53. Chen Y., Soto-Gutierrez A., Navarro-Alvarez N. et al. (2006) Instant hepatic differentiation of human embryonic stem cells using activin A and a deleted variant of HGF, *Cell Transplant* **15**, 865–871.
54. Takada T., Nemoto K.-i., Yamashita A. et al. (2005) Efficient gene silencing and cell differentiation using siRNA in mouse and monkey ES cells, *Biochemical and Biophysical Research Communications* **331**, 1039–1044.

Chapter 34

Using Endogenous MicroRNA Expression Patterns to Visualize Neural Differentiation of Human Pluripotent Stem Cells

Agnete Kirkeby, Malin Parmar, and Johan Jakobsson

Abstract

Many existing protocols for neuronal differentiation of human pluripotent cells result in heterogeneous cell populations and unsynchronized differentiation, necessitating the development of methods for labeling specific cell populations. Here we describe how microRNA-regulated lentiviral vectors can be used to visualize specific cell populations by exploiting endogenous microRNA expression patterns. This strategy provides a useful tool for visualization and identification of neural progeny derived from human pluripotent stem cells. We provide detailed protocols for lentiviral transduction, neural differentiation, and subsequent analysis of human embryonic stem cells.

Key words: Embryonic stem cell, Induced pluripotent stem cell, MicroRNA, Neuronal differentiation, FACS, Immunohistochemistry, Lentiviral vector

1. Introduction

During the last decade, a number of protocols that enable efficient differentiation of human pluripotent stem cells into neurons have been developed (1–3). Given this, human pluripotent stem cells, such as human embryonic stem cells (hESCs) and induced pluripotent stem cells (hiPSCs), have become a unique cell source for studying early human brain development. They may also serve as an unlimited source of therapeutically active cells for regenerative medicine. However, current protocols for deriving neurons from human pluripotent stem cells give rise to heterogeneous cell populations both in regard to the temporal aspects and the cellular composition. This complicates molecular analysis of specific cell populations and may be a large hurdle when using cells for transplantation since

Kaiming Ye and Sha Jin (eds.), *Human Embryonic and Induced Pluripotent Stem Cells: Lineage-Specific Differentiation Protocols*, Springer Protocols Handbooks, DOI 10.1007/978-1-61779-267-0_34, © Springer Science+Business Media, LLC 2011

contaminating undifferentiated cells may give rise to unwanted proliferation and tumors upon transplantation (4).

To track differentiating cell populations, reporter cell lines generated by homologous recombination (knock-in) or *via* BAC transgenes have been widely used in mouse cells (5–8). These strategies provide a robust way to visualize and isolate specific cell populations of differentiated pluripotent stem cells. Although possible, these strategies are often complicated to transfer to human cells due to technical issues (9), and only a few successful cases have been described (10, 11). To circumvent these difficulties, we have employed an alternative strategy that exploits the cells' endogenous microRNA (miRNA) machinery (12). Our approach is simple to use and offers a reporter system that can be as reliable as BAC transgenesis or knock-in technology.

MicroRNAs are small noncoding RNAs of 21–23 nucleotides that negatively regulate gene expression by binding complementary mRNAs and inhibit subsequent protein expression (13). Our system is based on a lentiviral vector that encodes a fluorescent reporter gene (in this case GFP) and tandem target sites for an miRNA (14, 15). When a microRNA is present in the cell, it binds to the target sites and downregulates GFP expression, while in cells that do not express the microRNA GFP, it is expressed. Based on this strategy, we used miR-292. It is specifically expressed in pluripotent cells, therefore only allowing GFP expression in differentiated cells (12). However, the versatility of the system allows the use of any microRNA of choice, including neuron-specific microRNAs (14).

Here we describe how this system can be used for human embryonic stem cells. We provide detailed protocols for the generation of lentiviral vectors, the transduction of hES-cells, and their differentiation into neurons. Finally, we provide protocols for analyzing these cells through flow cytometry and immunocytochemistry.

2. Materials

All catalog numbers are European.

2.1. Lentiviral Production and Titering

1. Cell culture lab certified for lentiviral vector production and use.
2. Virkon and tissue culture hood with UV-light.
3. 15-cm tissue culture-treated plastic dishes.
4. Ultracentrifugation tubes (Beckman L 60).
5. 293T cells – must never be grown to confluency.
6. 0.05% Trypsin (diluted from Invitrogen, cat. no. 15090-046).

7. *293T growth medium*:

   DMEM (Invitrogen, cat. no. 61965)

   10% FBS (Invitrogen, cat. no. 26140-095)

   1% Penicillin/streptomycin (Invitrogen, cat. no. 15140-122)

8. *IMDM virus production medium*:

   IMDM (Invitrogen, cat. no. 31980)

   10% FBS (Invitrogen, cat. no. 26140-095)

   1% Penicillin/streptomycin (Invitrogen, cat. no. 15140-122)

9. 0.1× TE buffer (1 mM Tris, pH 8.0, 1 mM EDTA, pH 8.0). Sterile filter and store at 4°C.

10. 2.5 M $CaCl_2$ solution (Sigma-Aldrich, cat. no. C7902. Sterile filter and store at –20°C).

11. 2× HBS buffer (281 mM NaCl, 100 mM HEPES, 1.5 mM Na_2HPO_4, pH 7.12). Sterile filter and store in aliquots at –20°C.

12. Plasmids for packaging of third-generation viral vectors (16):

    (a) pMD2.G envelope plasmid.

    (b) pMDL Gag-Pol plasmid.

    (c) pRSV-Rev reverse transcriptase plasmid.

    (d) Transfer vector plasmid with reporter gene (i.e., GFP) under control of a constitutively active promoter (i.e., PGK) containing microRNA target sequences in the 3′UTR (see Note 11 and Fig. 1).

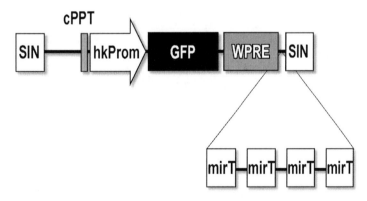

Fig. 1. *Design of microRNA-regulated reporter construct.* Schematic drawing of the integrated form of a typical microRNA-regulated lentiviral vector. A housekeeping promoter drives expression of GFP. The transgene expression is then regulated by four tandem copies of exact complementary sequences of a mature microRNA of choice. *SIN* self-inactivating LTR, *cPPT* central polypurine tract; *hkProm* housekeeping promoter; *GFP* enhanced green fluorescent protein; *WPRE* woodchuck hepatitis post regulatory element; *miRT* microRNA target sequence.

13. Ultracentrifuge (Beckman L60).
14. Rotor for ultracentrifuge (Beckman Coulter, SW32).
15. FACS cytometer instrument for flow cytometric analysis of GFP expression.

2.2. Culturing and Transduction of hESCs

1. Human ESC line (i.e., H9, HuES10).
2. Tissue culture-treated plasticware (i.e., 6-cm dishes, 6-well plates).
3. Irradiated or mitomycin-C-treated mouse embryonic fibroblasts (MEFs).
4. Matrigel Basement Membrane Matrix (BD Biosciences; cat. no. 354230):

    Thaw the frozen vial of matrigel on ice over night in the refrigerator. Dilute the thawed matrigel 1:20 in ice-cold DMEM with cooled pipettes. Prepare 5–15 mL aliquots in cooled 15 mL tubes and freeze at −20°C.

5. *Procedure for coating dishes with Matrigel*:

    Thaw the aliquots overnight on ice and dilute the aliquots 1:2 in ice-cold DMEM to yield a final dilution of 1:40. Cover dishes or plates with the diluted Matrigel solution and let them stand for 1 h at 37°C (i.e. 0.2 mL/cm^2). Aspirate the Matrigel solution and rinse dishes once with DMEM prior to plating cells. It is important that the Matrigel solution never reaches room temperature before coating, as this will lead to clumps of gelatinous precipitate on the dishes.

6. Gelatin (Sigma-Aldrich, cat. no. G2500). Dilute to 0.1% in PBS and sterilize by autoclaving. Sterilized solutions can be stored at 4°C for up to 1 year.
7. *Procedure for coating dishes with gelatin*:

    Incubate dishes with 0.1% gelatin solution for 30 min to 24 h at 37°C. Aspirate the gelatin solution and wash once in PBS prior to plating cells.

8. *MEF medium*:

    DMEM (Invitrogen, cat. no. 11965-118)

    10% FBS (Invitrogen, cat. no. 26140-095)

    1% L-glutamine (200 mM, Invitrogen, cat. no. 25030-081)

    1% Penicillin/streptomycin (Invitrogen, cat. no. 15140-122)

9. *hESC medium*:

    DMEM/F12 (Invitrogen, cat. no. 31330)

    20% knockout serum replacement (Invitrogen, cat. no. 10828-028)

    0.5% L-glutamine (200 mM, Invitrogen, cat. no. 25030-081)

0.1% 2-mercaptoethanol (Invitrogen, cat. no. 31350-010)

0.5% penicillin/streptomycin (Invitrogen, cat. no. 15140-122)

1% MEM nonessential amino acids (Invitrogen, cat. no. 11140-035)

10 ng/mL FGF2 (R&D Systems, cat. no. 233-FB)

10. *Preparation of conditioned hESC medium*:
    (a) Plate irradiated or mitomycin-C-treated MEFs at a high density (30–50,000 cells/cm^2) on gelatin-coated dishes in MEF medium and let them attach overnight.
    (b) After 24 h, wash cells once in PBS and add hESC medium (0.3–0.4 mL medium/cm^2).
    (c) Harvest the medium 24 h later and store at 4°C till use. The cells can be used for generating CM for up to 2 weeks with harvest every day.
    (d) Filter the conditioned medium with a 0.22 μm filter and add additional 10 ng/mL of FGF2 before use.

11. *hESC freezing medium*:

    50% hESC medium

    43% FBS (Invitrogen, cat. no. 26140-095)

    7% DMSO

12. Accutase (Invitrogen, cat. no. A11105-01). Aliquot and store at −20°C. Stable for 2 weeks at 4°C.

13. Dispase (STEMCELL Technologies, cat. no. 07913). Aliquot and store at −20°C. Stable for 1 week at 4°C.

14. ROCK inhibitor (Y-27632 from Tocris Bioscience). Resuspend in H$_2$O to 10 mM and store aliquots at −20°C. Aliquots are stable at 4°C for at least 2 weeks.

2.3. Neural Differentiation of hESCs

1. *KSR medium*:

   KnockOut DMEM (Invitrogen, cat. no. 10829-018)

   15% KnockOut Serum Replacement (Invitrogen, cat. no. 10828-028)

   1% L-glutamine (200 mM, Invitrogen, cat. no. 25030-081)

   0.1% 2-mercaptoethanol (Invitrogen, cat. no. 31350-010)

   1% penicillin/streptomycin (Invitrogen, cat. no. 15140-122)

   1% MEM non-essential amino acids (Invitrogen, cat. no. 11140-035)

2. *Modified N2 medium*:

   DMEM:F12 (Invitrogen, cat. no. 32500-035)

   8.6 mM glucose

   24 mM NaHCO$_3$

20 nM progesterone

30 nM sodium selenite

60 μM putrescine

25 μg/mL insulin (Sigma-Aldrich, cat. no. 16634). Dissolve stock in 0.1 M HCl.

0.1 mg/mL apotransferrin

1% penicillin/streptomycin (Invitrogen, cat. no. 15140-122)

3. SB 431542 (Tocris Bioscience cat. no. 1614). Dissolve to 20 mM in DMSO and store at −20°C for up to 3 months. Do not dissolve in EtOH since this will destabilize the compound.

4. Recombinant Human Noggin/Fc Chimera (R&D Systems, cat. no. 3344-NG-050).

5. Recombinant Human BDNF (R&D Systems, cat. no. 248-BD).

6. Ascorbic acid (Sigma-Aldrich, cat. no. A5960). Make stock solutions of 200 mM in H_2O.

7. Matrigel Basement Membrane Matrix (BD Bioscience; cat. no. 354230, see Sect. 2.2.4).

8. Polyornithine (Sigma-Aldrich, cat. no. P3655). Make sterile filtered stock solutions of 1.5 mg/mL in H_2O and store at −20°C. Working solutions (15 μg/mL in H_2O) can be stored at 4°C for up to a week.

9. Laminin (Invitrogen, cat. no. 23017-015). Make stock solutions of 1 mg/mL in PBS and store at −80°C.

10. Fibronectin (Invitrogen, cat. no. 33010-018). Make stock solutions of 1 mg/mL in PBS and store at −20°C. Aliquots can be stored at 4°C for up to 2 weeks.

11. *Procedure for coating dishes with polyornithine, laminin and fibronectin*

    Dilute PO stock solutions 1:100 in H_2O to yield a final concentration of 15 μg/mL. Add the solution to wells and incubate at 37°C overnight (i.e. 0.2 mL/cm^2). Aspirate the PO solution and wash three times in H_2O. Prepare FN+lam solution by adding 1:200 of FN and 1:200 of laminin to PBS to yield a final concentration of 5 μg/mL for each. Add the FN+lam solution to the PO-coated wells at 0.2 mL/cm^2 and incubate at 37°C for 24–72 h. Wash the plates once in PBS prior to plating the cells.

2.4. Monitoring of Reporter Gene Expression

1. Inverted fluorescence microscope.
2. FACS cytometer instrument for flow cytometric analysis of GFP expression.
3. 4% paraformaldehyde solution for fixation.
4. *Blocking buffer*.

PBS

5% serum (use serum from host species of secondary antibodies)

0.1% Triton-X100

5. Primary antibodies for staining:

    (a) Oct-3/4 (Santa Cruz Biotechnology, Inc., cat. no. sc-5279, use 1:200)

    (b) Pax6 (Developmental Studies Hybridoma Bank, use 1:100)

    (c) βIII-tubulin (Promega G7121, use 1:1,000)

6. RNA isolation kit (RNeasy mini kit, QIAGEN, cat. no. 74104).

7. SuperScript III reverse transcriptase (Invitrogen, cat. no. 18080-044).

8. Random primers (Invitrogen, cat. no. 48190-011).

9. Primers for real-time PCR of cDNA from differentiating cells

    (a) Primers for pluripotency-associated genes (i.e., Oct4, Nanog)

    (b) Primers for neural progenitor markers (i.e., Pax6, Sox1, Ngn2)

    (c) Primers for neuronal markers (i.e., MAP2, Tau)

    (d) Primers for virus-specific sequences (i.e., GFP or WPRE)

10. SYBR green master mix (Roche Applied Science, cat. no. 04887352001).

11. Real-time PCR instrument for quantitative PCR (i.e., LightCycler 480).

3. Methods

3.1. Lentiviral Vector Production and Titering

Caution: All plasticware which has been in contact with virus and with virus-transduced cells for up to 7 days after transduction should be thoroughly decontaminated with Virkon before disposal. The following lentiviral vector production protocol is based on a calcium phosphate transfection method using a third-generation lentiviral vector system (14, 16, 17). This protocol can be scaled up as needed.

1. 24 h before transfection, plate 9×10^6 293T cells in growth medium in a 15-cm tissue culture dish.

2. 2 h before transfection, change medium to 22.5 mL IMDM medium.

3. For transfection, mix plasmids: 9 μg pMDG + 12.5 μg pMDL + 6.25 μg pRSV-Rev + 32 μg transfer vector (microRNA-regulated reporter plasmid). Bring the plasmid mix to a final volume of 1,125 μL with 0.1× TE buffer.

4. Add 125 μL 2.5 M $CaCl_2$ solution to the DNA/TE mix and leave at room temperature for 5 min.

5. While vortexing, add 1,250 μL 2× HBS solution dropwise to the DNA/TE/$CaCl_2$ mix to form DNA/calcium phosphate precipitate.

6. Add the total precipitate to the 15-cm dish with 293T cells and rock the dish back and forth to distribute the precipitate evenly across the dish.

7. After 12–14 h, change medium to 16 mL fresh IMDM medium.

8. 30–36 h after changing the medium, harvest the virus-containing supernatant.

9. Spin down the supernatant at $800 \times g$ for 10 min to remove cellular debris and subsequently filter the supernatant with a 0.2-μm filter.

10. Ultracentrifuge the filtered supernatant at 19,500 RPM for 2 h at 4°C.

11. After ultracentrifugation, decant the supernatant and add 100–200 μL PBS to the pellet without pipetting. Leave the pellet in PBS for at least 2 h at 4°C to dissolve the virus particles.

12. Mix and aliquot the viral suspension and store at −80°C.

13. To determine the viral titer, plate 1×10^5 293T cells in a 6-well plate and transduce on the same day with various dilutions of virus suspension (i.e. 0.01–3 μL virus suspension) (see Note 1).

14. 3 days later, dissociate the transduced 293T cells with trypsin and spin down in growth medium.

15. Resuspend the cells in 500 μL growth medium with 2% PFA and leave at room temperature for 15 min before FACS analysis (see Note 2).

16. Calculate the concentration of transducing units (TU) per mL, i.e., if 0.1 μL virus suspension results in 7% GFP+ cells of the total population, then 0.01 μL virus suspension contains $0.07 \times 100{,}000 = 7{,}000$ TU → this corresponds to 7×10^8 TU/mL (see Note 3).

3.2. Culturing and Transduction of hESCs

hESCs are cultured on irradiated or mitomycin-C-treated mouse embryonic fibroblasts (MEFs) in hESC medium. Colonies should be routinely passaged every 7 days with dispase onto new MEFs. MEFs should be plated no later than 48 h and no earlier than 8 h before plating of hESCs. This protocol does not describe routine culturing of hESCs, but focuses on how to generate efficient lentiviral transduction of hESCs. The addition of ROCK inhibitor (Y-27632) will improve survival of the cells after Accutase passage and during transduction.

1. For lentiviral transduction, use hESC cultures of good quality with high density (high hESC/MEF ratio). Before using the cells, remove differentiated colonies from the dish with a vacuum-connected Pasteur pipette.

2. On the day of transduction, coat 6-well plates with matrigel (see Sects. 2.2.4 and 2.2.5, coat 1× well for each virus condition).

3. Wash hESCs once in PBS, and add Accutase to dissociate cells (500 µL to a 6-well). Leave cells with Accutase in the incubator for 5–10 min.

4. Use a 1-mL pipette to triturate cells to single cells, and spin down cells in 10 mL hESC medium at $200 \times g$ for 5 min.

5. Wash cells once again in 10 mL hESC medium and spin down.

6. Resuspend the cells in conditioned hESC medium with 10 µM Y-27632. Count the cells and plate 5×10^5 cells/well in 6-well plates in 2 mL conditioned medium with 10 µM Y-27632.

7. 2 h after plating, add virus to the cells for a final MOI of 20–30 (i.e., $1–1.5 \times 10^7$ transducing units for 5×10^5 cells) (see Note 4).

8. 1, 2, and 3 days after transduction, change medium to conditioned hESC medium + 10 µM Y-27632.

9. 4 days after transduction, passage transduced cells with Accutase: Wash cells once in PBS and add 500 µL accutase/well. Leave for 5–10 min in the incubator until all cells have loosened from the dish. Use a 1 mL pipette to triturate cells to single cells, and spin down cells in 10 mL hESC medium at $200 \times g$ for 5 min.

10. Wash cells once again in 10 mL hESC medium and spin down.

11. Count and plate the transduced hESCs on new MEFs at a concentration of 4×10^5 cells/well in 6-well plates in hESC medium with 10 µM Y-27632. Save an aliquot of the cell suspension in media for FACS analysis (i.e., 300 µL cell suspension with a concentration of $\geq 5 \times 10^5$ cells/mL) (see Notes 2 and 5).

12. Change medium and expand cells with dispase until sufficient number of cells is obtained for freezing and for performing experiments (see Note 6).

3.3. Neural Differentiation of hESCs

The differentiation protocol described here is based on a previously published protocol for neural induction (1). The protocol can be scaled up as needed. If specific subtypes of neural cells are desired, additional factor can be added to the protocol during differentiation. For the GFP-mir292T construct, GFP will start to turn on at day 4 of differentiation, and it will increase up to day 20 (see Fig. 2b).

1. 6–8 days before differentiation, passage hESCs with dispase as normally, and plate colonies onto Matrigel-coated plates in conditioned hESC medium (see Sects. 2.2.4 and 2.2.5). Change medium every day to conditioned hESC medium until the colonies are ready to be passaged again. This passage on Matrigel will eliminate the presence of MEFs before starting differentiation.
2. 1–3 days before the start of differentiation, prepare Matrigel-coated 48-well plates and passage colonies from Sect. 3.3.1 with Accutase to yield single cells (see Sects. 2.2.4, 2.2.5, and 3.2.3). Spin down and wash cells × 2 in hESC medium.
3. Plate single cells onto the Matrigel-coated 48-well plates at a density of 150–200,000 cells/well in conditioned hESC medium + 10 µM Y-27632.
4. When the cells are 80–90% confluent (1–2 days after plating), start differentiation (see Note 7).
5. To initiate differentiation (= d0), aspirate medium and replace with KSR medium + 10 µM SB431542 and 200 ng/mL noggin (see Note 8).
6. Replace this medium on d2 and d4.
7. On d5 of differentiation, replace medium with 75% KSR medium + 25% N2 medium + 10 µM SB431542 and 200 ng/mL noggin.
8. On d7 of differentiation, replace medium with 50% KSR medium + 50% N2 medium + 10 µM SB431542, 200 ng/mL noggin, and 100 ng/mL FGF8a.
9. On d9 of differentiation, replace medium with 25% KSR medium + 75% N2 medium + 100 ng/mL FGF8a, 20 ng/mL BDNF, and 200 µM ascorbic acid.
10. On d10 of differentiation, aspirate differentiation media and add 100 µL Accutase to coat the well. Incubate at 37°C for 15 min or until all cells are rendered to single cells.
11. Triturate the cells in the dish using a 1-mL pipette with additional N2 medium until the cells are in a single cell suspension.

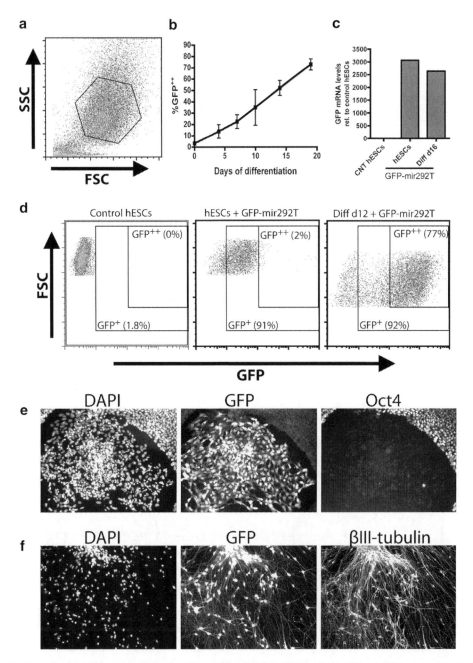

Fig. 2. *Monitoring reporter gene expression in differentiating hESCs.* Analysis of the GFP-mir292T reporter vector in differentiating hESCs by flow cytometry, qRT-PCR analysis, and staining. (**a**) FACS plot showing gating settings for the live cell population. (**b**) Time course of GFP expression analyzed by flow cytometry in neural differentiation of GFP-mir292T hESCs. (**c**) Analysis of vector GFP mRNA expression in hESCs before (hESCs) and after (diff d16) differentiation for 16 days. (**d**) Representative FACS plots of untransduced hESCs (control) and GFP-mir292T hESCs before and after neural differentiation. The middle panel shows the baseline leakiness of the GFP-mir292T construct in undifferentiated hESCs. (**e**) Staining for Oct4 and DAPI at an early timepoint during differentiation shows localization of GFP expression exclusively in the GFP-negative population. (**f**) Staining for βIII-tubulin and DAPI after terminal neuronal differentiation shows bright GFP expression in the neuronal cell population.

12. Wash and centrifuge cells twice in N2 media.
13. Resuspend the cells in N2 media, and determine the cell concentration using a hemocytometer.
14. Adjust the cell concentration to 1×10^7/mL.
15. Plate the cells in droplets onto dried wells coated with polyornithine, fibronectin, and laminin. Spot one or several droplets of 5 µL of the cell suspension in each well and let the cells attach for 15 min in the incubator before adding N2 containing BDNF (20 ng/mL), FGF8a (100 ng/mL), and ascorbic acid (200 µM).
16. Culture cells in this medium for 4–10 days or until mature neuronal cells are visible.

3.4. Monitoring of Reporter Gene Expression

At different timepoints during differentiation, the microRNA-regulated GFP levels can be monitored at the protein level either by fluorescence microscopy or flow cytometry on live cells or by staining of fixed cells. For an accurate and quantitative measure of the time course of GFP protein regulation, flow cytometry is the preferable method. Note that some microRNAs regulate their target vector through degradation of the viral mRNA, whereas others exert translational control of reporter protein synthesis. To analyze whether your target vector is regulated at the mRNA or the protein level, a flow cytometric time course analysis of GFP expression must be complemented by a qRT-PCR analysis of viral mRNA levels. For the GFP-mir292T vector, GFP expression in human cells is exclusively regulated at the protein level. Thus, high levels of vector-derived mRNA can be detected both in the pluripotent and the differentiated state, but GFP fluorescence is only detected after differentiation (see Fig. 2b, c)

1. *Staining of transduced cells during differentiation:*
   (a) Perform differentiations in 48-well plates, and at different timepoints during differentiation, fix the differentiating cells in 4% PFA for 15 min at 37°C.
   (b) Wash cells three times in PBS and incubate cells in blocking buffer at room temperature for 1 h.
   (c) Dilute primary antibodies in blocking buffer and incubate on cells overnight at 4°C or room temperature.
   (d) Wash three times in PBS and add secondary fluorescent antibodies diluted in blocking buffer together with DAPI or another nuclear dye.
   (e) Wash three times in PBS and evaluate colabeling of GFP with pluripotency and differentiation markers. (See Note 9 and Fig. 2e, f).

2. *Flow cytometric analysis on live transduced cells*
   (a) Perform differentiations in 48-well plates, and at different timepoints during differentiation, analyze cells by flow cytometry.
   (b) On the day of analysis, remove media from the plate, and wash twice in PBS to remove any dead or floating cells.
   (c) Add Accutase (100 µL/well) and incubate at 37°C for 10–15 min.
   (d) Triturate cells with a 100 µL pipette until a single cell suspension is achieved. It is important that all cells in the suspension are single cells, since doublets will give biased results on the FACS. If cells are not yet completely dissociated, incubate the cell suspension in Accutase for 10–15 min more.
   (e) Add 400 µL N2 medium to the cells and transfer the cell suspension to a FACS tube for analysis. If cell clumps or stringy fibers of cells are present in the suspension due to DNA leakage, the cell suspension should be filtered through a 40 µm cell strainer before analysis.
   (f) Run the sample through the FACS machine and gate for live cells in the FSC/SSC channels (see Note 2 and Fig. 2a).
   (g) Analyze GFP fluorescence intensity for the live cell population in the FL1 channel (see Note 5 and Fig. 2d).
3. *Quantitative RT-PCR analysis of viral mRNA levels:*
   (a) Perform differentiations in 48-well plates, and at different timepoints during differentiation, lyse cells in 350 µL RLT buffer + β-mercaptoethanol, to extract total RNA following the RNeasy kit manual.
   (b) Perform cDNA synthesis on 1–5 µg of total RNA, using SuperScript III and random primers following the producer's manual.
   (c) Perform real-time quantitative PCR on cDNA using primers for detection of viral mRNA as well as primers for pluripotency and differentiation-associated genes following the Sybr Green manual (see Note 10).

4. Notes

1. With this virus production protocol, it is expected to achieve a total of 1×10^7 to 1×10^8 TU from one 15 cm dish.
2. cytometry on live cells should only be done >7 days after transduction since this involves transfer of transduced cells outside

the viral vector certified cell culture laboratory. If analysis is done less than 7 days after transduction, cells should be fixed in a medium solution containing 2% PFA solution for 15 min before analysis.

3. The calculations of viral titer should be based on transductions resulting in <15% GFP positive cells. Higher transduction efficiencies will underestimate the viral load due to the presence of cells containing more than one integrated vector.

4. If high-titer virus is a limiting factor, transductions can be done in smaller wells and scaled down to fewer cells and fewer viruses. This will only mean that it will take longer to expand the transduced cells in large quantities.

5. Most miR-regulated reporter constructs will have some degree of background leakiness, which can be hard to see in the microscope, but which is clearly visualized by flow cytometry. Thus, for the GFP-mir292T construct, the efficiency of transduction can be visualized in the hESC population by flow cytometry. Ideally, >90% of the cells should show a shift to the right in the FL1 channel if the transduction efficiency is high. If the efficiency of transduction is not satisfactory in the first round, the cells can be transduced once again by taking them back to Step 1 in Sect. 3.2 and if necessary, by adjusting the amount of virus added to the cells in Step 7 in Sect. 3.2.

6. Ideally, a transduced cell population should be stable over time; however, we observe some instability of the transgene expression with repeated passaging. This can be due to either silencing of the transgene or selective growth of nontransduced cells in the culture over time. Thus, we recommend that transduced cell populations are passaged no more than six times. After this, a new round of transduction should be performed.

7. The confluency of the cells at the start of differentiation is crucial for the efficiency of differentiation, so it is important that the differentiation is not started before the cells are 80–90% confluent. For H9 cells, if 180,000 cells/well are plated and the quality and survival of the cells are good, then the cells should be ready to start differentiation on the next day. If you wish to start differentiation after 2 days, 130,000–150,000 cells can be plated in stead. To start 3 days later, 100,000 cells can be plated. Note that the survival of the cells will drop exponentially with lower plating densities. The optimal plating densities will depend on the cell line used.

8. When differentiating the cells, the medium volume in the well should be high since the cell density is extremely high and medium is only changed every other day (i.e., use 600–800 µL medium/well for a 48-well plate)

9. In the neural differentiation paradigm, it is relevant to follow the regulation of GFP in parallel with the loss of pluripotency and the acquisition of neural fate. For this purpose, we recommend staining for Oct-3/4 as a marker of pluripotency, Pax6 as a marker of early neural progenitors and βIII-tubulin as a marker of neuronal cells (see Sect. 2.4.5 for antibodies and dilutions). The EGFP reporter protein maintains fluorescent properties after PFA fixation, so it is not normally necessary to stain for GFP.

10. When performing qRT-PCR on viral sequences from transduced cells, it is important to keep in mind that any contamination with genomic DNA or viral plasmid will give rise to a signal which is indistinguishable from that of the viral vector cDNA. Thus, for transduced cells, we recommend that DNase treatment is always included in the RNA purification protocol and that a parallel control lacking SuperScript enzyme is included in the PCR reaction. This control will reveal whether contamination with plasmids or genomic DNA is present in your samples.

11. The optimal design of the microRNA-regulated vector ultimately depends on the cell type and the experimental paradigm. Examples of variations include the choice of promoter, miRNA-target sequence, and reporter gene. We refer previous papers where it is extensively described how microRNA-regulated lentiviral vectors can be optimized for various purposes (14, 15). It should be noted that most microRNA-regulated vectors can be transferred between species due to the high degree of conservation among microRNAs. The miR-292T vector used this chapter is based on the sequence of the murine mmu-miR-292 but also works efficiently in human cells due to presence of several close homologues, including hsa-miR-371.

References

1. Chambers, S. M., et al. (2009) Highly efficient neural conversion of human ES and iPS cells by dual inhibition of SMAD signaling, *Nat Biotechnol* **27**, 275–280.
2. Reubinoff, B. E., et al. (2001) Neural progenitors from human embryonic stem cells, *Nat Biotechnol* **19**, 1134–1140.
3. Zhang, S. C., et al. (2001) In vitro differentiation of transplantable neural precursors from human embryonic stem cells, *Nat Biotechnol* **19**, 1129–1133.
4. Koch, P., et al. (2009) Emerging concepts in neural stem cell research: autologous repair and cell-based disease modelling, *Lancet Neurol* **8**, 819–829.
5. Aubert, J., et al. (2003) Screening for mammalian neural genes via fluorescence-activated cell sorter purification of neural precursors from Sox1-gfp knock-in mice, *Proc Natl Acad Sci USA* **100 Suppl 1**, 11836–11841.
6. Copeland, N. G., Jenkins, N. A., and Court, D. L. (2001) Recombineering: a powerful new tool for mouse functional genomics, *Nat Rev Genet* **2**, 769–779.
7. Hedlund, E., et al. (2008) Embryonic stem cell-derived Pitx3-enhanced green fluorescent protein midbrain dopamine neurons survive enrichment by fluorescence-activated cell sorting and function in an animal model of Parkinson's disease, *Stem Cells* **26**, 1526–1536.

8. Tomishima, M. J., et al. (2007) Production of green fluorescent protein transgenic embryonic stem cells using the GENSAT bacterial artificial chromosome library, *Stem Cells* **25**, 39–45.
9. Giudice, A., and Trounson, A. (2008) Genetic modification of human embryonic stem cells for derivation of target cells, *Cell Stem Cell* **2**, 422–433.
10. Irion, S., et al. (2007) Identification and targeting of the ROSA26 locus in human embryonic stem cells, *Nat Biotechnol* **25**, 1477–1482.
11. Placantonakis, D. G., et al. (2009) BAC transgenesis in human embryonic stem cells as a novel tool to define the human neural lineage, *Stem Cells* **27**, 521–532.
12. Sachdeva, R., et al. Tracking differentiating neural progenitors in pluripotent cultures using microRNA-regulated lentiviral vectors, *Proc Natl Acad Sci USA* **107**, 11602–11607.
13. Bartel, D. P. (2004) MicroRNAs: genomics, biogenesis, mechanism, and function, *Cell* **116**, 281–297.
14. Brown, B. D., et al. (2007) Endogenous microRNA can be broadly exploited to regulate transgene expression according to tissue, lineage and differentiation state, *Nat Biotechnol* **25**, 1457–1467.
15. Brown, B. D., et al. (2006) Endogenous microRNA regulation suppresses transgene expression in hematopoietic lineages and enables stable gene transfer, *Nat Med* **12**, 585–591.
16. Dull, T., et al. (1998) A third-generation lentivirus vector with a conditional packaging system, *J Virol* **72**, 8463–8471.
17. Zufferey, R., et al. (1997) Multiply attenuated lentiviral vector achieves efficient gene delivery in vivo, *Nat Biotechnol* **15**, 871–875.

INDEX

A

Acrylate group 417, 418
Adhesion promoter 418, 421
Adipocyte
    gene expression 356–357
    marker gene 356
Adipogenesis 351, 352, 355
Adipogenic
    cocktails 355, 356
    differentiation 351–357
    differentiation medium 354, 356
    induction 354–356
Adipose tissue 173–175, 179–182, 188, 189
AFP. *See* Alpha-fetoprotein (AFP)
Aggregate formation 38–40, 48, 148, 398, 400–402
Aggregation 39, 40, 49, 268, 309, 354
Air–liquid interface 407–409
    culture 405–410
Albumin secretion 374, 383–384
Alginate-poly-L-lysin-alginate (APA)
    microcapsule 52–55
Alpha-fetoprotein (AFP) 126, 127, 373, 381, 385,
    388, 389, 392, 393, 398–403, 449, 455, 457–458
ALP staining 122–125
Amnion 249–263
Amniotic membrane 249, 250, 253, 254,
    263, 283, 329
Antigene staining 142, 143
APA. *See* Alginate-poly-L-lysin-alginate (APA)
Attachment culture 356
Autologous feeder 161–171

B

Basic fibroblast growth factor (bFGF) 6, 9, 11, 15,
    16, 23, 60, 62, 64, 73, 74, 97, 98, 116, 126,
    135–137, 141, 154, 156, 162, 164, 170, 177,
    185–187, 189, 236, 251, 252, 268–270, 314,
    316, 352, 361, 375, 379, 430–434, 436, 438,
    440, 441, 446, 450
    human recombinant 15, 62, 97, 116, 135,
    162, 164, 177, 251, 269, 352

BMP. *See* Bone morphogenetic protein (BMP)
Bone marrow (BM) 94, 173–175, 179–181,
    296, 322, 338, 349
Bone morphogenetic protein (BMP) 60, 73, 77,
    280, 296, 339, 344, 345, 360, 362, 364–366,
    452–456, 458, 459

C

Capillary force lithography 414, 416, 417
Cardiomyocyte differentiation 43, 46
CD34$^+$
    cell 99, 114–117, 120–124, 126,
    142, 268, 269, 271, 272, 275, 330, 331, 343
    progenitors 114, 117, 120
Cell bank 89, 309
Cell dissociation 28, 31–33, 39, 43, 46, 177
Cell microencapsulation 52, 55–56
CELLstart 4, 6, 8–9, 11
Chondrocyte 173, 359
Chondrogenic differentiation 359–366
Coating 7–11, 26, 31, 33, 38, 106, 155–157,
    163, 206, 208, 224, 272, 301, 308, 315,
    371, 374, 406, 407, 418, 419, 441, 468, 470
Coculture 126, 159, 206, 277, 296,
    312, 321–334, 398–401, 406
Collagenase IV 43, 46, 74, 117, 137, 158, 162,
    206, 223, 251, 252, 261, 262, 280, 284, 340, 346,
    352, 353, 363, 371, 398, 408, 409, 429–431,
    433–436
Collagen I coating 155, 157, 275, 389, 392
Colony formation assay 326–327
Cord blood, CD133$^+$ 93–110
Cord formation assay 281, 290, 293
Cryoperservation 85–89, 174, 298, 441
Culture of embryonic stem cells 3, 37–49, 450
Cytochrome P450 .... 374, 383, 385,
    388–390, 392, 393, 403

D

Defined culture conditions 26
Dexamethasone 153, 155, 354, 355, 362, 364, 389, 399

DiI-acetylated low-density lipoprotein (DiI-LDL),
   uptake assay 270, 273
Disaggregation 76–77, 82, 278
Dissection 76–77, 234, 253
DMEM. *See* Dulbecco's modified Eagle's medium (DMEM)
Dulbecco's modified Eagle's medium (DMEM) 6, 14, 27, 43, 53, 62, 73, 87, 96, 115, 135, 153, 162, 176, 193, 205, 234, 251, 269, 279, 297, 324, 338, 352, 360, 371, 388, 398, 407, 415, 429, 446, 467

E

EB. *See* Embryoid body (EB)
E-cadherin 25–33, 39, 43, 124, 381
Ectoderm 44, 48, 71, 106, 109, 128, 130, 152, 158, 159, 187, 188, 351, 445
Embryoid body (EB) 117–118, 125–127, 152, 158, 187, 312, 316, 343, 355, 371, 376–377, 431
   differentiation 117–118, 125–127
   formation 152, 158, 187, 312, 316, 371, 376–377
Embryonic germ 79, 95, 109, 152, 242, 445
Embryonic stem cells (ESCs) 3, 13, 25, 37, 60, 85, 105, 152, 161, 173, 191, 204, 231, 267, 277, 295, 311, 321, 351, 359, 370, 388, 397, 407, 413, 429, 445, 465
   culture 3, 8–10, 37–49, 450
   feeder cells 3–12, 14, 18–19, 21–23, 42, 60, 62–66, 68, 275, 277, 302, 309, 322, 328, 431, 451
   medium 61–68, 73–74, 77–79, 81, 95, 97, 98, 105, 109, 116, 117, 122, 125, 126, 129, 137, 163, 164, 169–171, 236, 391
   passaging 14, 18–20, 22, 23, 37, 40, 46, 51, 60, 330
   pluripotent 38, 60, 67, 85–89, 152, 161, 191, 192, 203, 204, 297, 304, 311–319, 321–334, 337–338, 351, 360, 370, 388, 450, 465, 466
   thawing 6, 86–89, 94, 99, 117, 130, 274, 292, 293, 330, 384
Endoderm
   hepatic progenitors 370, 392
Endogenous
   gene expression 220, 466, 470–471, 475–477
   microRNA expression patterns 465–479
Endothelial induction 270, 272, 273, 296
Endothelial stem cells 133–148, 267–275, 277–309, 311–319
ESCs. *See* Embryonic stem cells (ESCs)
Extra embryo 71, 77, 250, 452, 453, 455, 456, 458

F

FACS 43, 47, 102, 121, 141, 174, 300, 301, 304, 307, 308, 313, 316–318, 326, 331–332, 338, 340–342, 345, 348, 360, 399, 448, 460, 468, 470, 472, 473, 475, 477

Feeder cells
   autologous 161–171
   fibroblasts 26, 193, 197–198
   irradiation 87
   mitomycin-C treatment 121, 136, 167, 169–171, 291–292
   preparation 18–19, 62, 64–65, 167
Feeder-free culture medium
   CELL-start-STEMPRO 4, 6, 8–9, 11
   Matrigel-mTeSR1, 4, 6
   vitronectin-mTeSR1 .... 4, 6, 10, 12
Feeder growth medium 15, 19
Fetal livers 322, 325, 327–329, 397–403
Fibroblasts 3, 14, 26, 42, 51, 62, 75, 87, 93, 114, 133, 151, 161, 174, 191–200, 204, 232, 251, 268, 279, 309, 328, 343, 352, 361, 371, 388, 398, 406, 415, 429, 450, 468
Floating culture 278, 282, 293, 354
Flow cytometric analysis 120, 121, 211, 468, 470, 477
Foetal gonad collection 76–77
Freezing medium 44, 87–89, 99, 110, 164, 207, 208, 223, 234–236, 238, 251, 256–258, 262, 280, 284, 292, 469

G

Gametes 71
Gelatin coating 155, 163–164, 272, 418, 419
Gene therapy 233
Germ cells 71–82
Green fluorescent protein (GFP) 103, 104, 110, 115, 119–121, 123, 129, 193, 195, 196, 199, 398, 466–468, 470–472, 474–479

H

HEK293 cells 27
Hemangioblast 268, 295, 331, 338, 341, 343
Hematopoiesis 322–325, 327, 330, 332, 337
Hematopoietic progenitor 94, 105, 268, 325–326, 331–334, 337–349
Hematopoietic stem cells (HSC) 94, 95, 114, 278, 322, 337, 338
Hemogenic endothelium 340, 341, 346, 348, 349
Hepatic differentiation
   maturation 378, 379, 384, 388, 392
   multistage 387–396
   progenitor induction and expansion 392
Hepatocyte
   culture 152–153, 155, 157, 159, 392, 394
   differentiation 370, 378
Heterogeneity
   heterogeneous 458
   sporadic differentiation pluripotency 446, 450, 451, 458

HFF. *See* Human foreskin fibroblasts (HFF)
HiPSC. *See* Human induced pluripotent stem cells (HiPSC)
HSC. *See* Hematopoietic stem cells (HSC)
Human adipose tissues 175–176, 179–182, 188
Human amnion cells 249–263
Human bone marrow 175–176, 179–181
Human embryonic stem cells (hESCs)
  cryopreservation... 85–89
  culture 3, 4, 6, 8, 12, 14, 38, 39, 43, 45, 147, 271, 272, 275, 305, 363, 365, 374, 379, 432, 433, 440, 450, 452, 473
  differentiation 267–275, 295–309, 311–319, 413–421, 427–441
  feeder-free culture..................... 3–12, 277–294, 302, 305
  generation 38
  growth 3–23, 52, 269, 296, 409
  medium.................. 4, 271, 330, 331, 342, 345, 346, 361, 363, 366, 407–409, 433, 468–469, 472–474
  passage 3, 8–10, 22, 26, 32, 38, 40, 42, 54, 88, 105, 271, 278, 282, 284–286, 288, 289, 293, 302, 303, 305, 309, 327, 328, 330, 433–436, 440
Human foreskin fibroblasts (HFF)
  growth 41–15
  irradiation 14–15, 42
Human hair follicle 203–226
Human hepatocyte culture 155, 157
Human hepatocyte reprogramming........................... 157–158
Human induced pluripotent stem cells (HiPSC) 25, 151–159, 161–171, 191–200, 203, 210, 311–319, 322, 337–349, 351–357, 387–396
  characterization 26, 152, 154–155, 208, 217–220
  colony identification 157–158
  cryopreservation... 85–89
  derivation.. 161–171
  differentiation 351–357
  EB formation.................. 106–109, 152, 158, 187, 312, 313, 319, 353–356, 436
  expansion 59–68, 141–142, 146, 158–159, 210, 387, 388, 434–437
  feeder-free culture....... 241, 245, 246, 291, 302, 315–316
  generation 113–130, 133–148, 151–159, 167–170, 176–177, 181–187, 191–200, 249–263, 337–349
  growth 331
  medium............................. 207, 211, 214–216, 222–224
  picking and expanding colonies 105, 129, 141–142, 169, 198, 216, 233, 236, 240–241, 245, 246, 260–261
  reprogramming.......................... 125–127, 129, 133, 134, 139–143, 148, 152, 157–158, 174, 192, 198, 200, 204, 220, 225, 233, 241, 242, 322, 343, 370
Human mesenchymal stem cells.............................. 173–190

Human primordial germ cells....................................... 71–82
Human skin biopsy... 231–246
Human umbilical vein endothelial cells................. 133–148
Hypoxia 13–23, 68, 297
Hypoxic workstation 16, 21, 22

I

IBMX. *See* 3-Isobutyl–1-methylxanthine (IBMX)
Immunoblotting 28–31
Immunocytochemistry..................... 106, 109, 138, 147, 415, 420–421, 438–440, 466
Immunofluorescence
  imaging... 18, 21
  microscopy................... 21, 158, 178–179, 182, 187–188, 393, 421
  staining 16–17, 20–21, 270, 273, 303–304, 306, 421
Immunohistochemistry 40, 44, 45, 47, 109, 142, 143, 394
Immunological tolerance 250
Immunostaining 20, 123–124, 142, 143, 146, 242, 289, 291, 298, 304, 316, 331, 372–373, 380, 381, 410, 420
Implantation 56, 57
Indirect immunofluorescence localization 79
Induced pluripotent stem cells
  culture 87–88, 95, 98, 106, 109, 144, 153–154, 156, 158, 171, 176, 177, 185–187, 193, 197, 198, 200, 207, 210, 211, 220–225, 232, 233, 259, 262, 312, 315, 321–334, 347, 354, 388–389, 391, 392, 394, 431, 432, 434–436, 438
  generation 94, 114, 133–148, 167–170, 174, 176–179, 181–187, 192–194, 197–199, 203–226, 231–246, 249–263
Infection 60, 95, 103–105, 110, 119, 120, 122, 123, 129, 138, 139, 166–170, 185–187, 193, 196–200, 235, 240, 244, 245, 259–260
  fibroblasts 196–199
  lentivirus 166–167
  primary cells 193, 196–197
  virus 60, 122, 139, 185–187, 244, 259–260
Insulin 62, 64, 109, 153, 155, 354, 355, 372, 378, 389, 470
In vitro
  culture.. 52, 54, 73, 302, 323
  differentiation
    hESC.. 297
    iPSC 106–109, 177–178, 187, 188, 428
  models 232, 369, 371
In vivo differentiation 40, 109, 188, 189
Irradiation...................... 14–15, 23, 64, 65, 87, 88, 114, 161, 213, 221, 222, 258, 271, 279, 285, 291, 292
3-Isobutyl–1-methylxanthine (IBMX).................... 354, 355
Isolation of CD34+ cells by MACS 269, 272

K

Karyotyping.................40, 61, 89, 144–145, 216, 217, 242, 300, 302, 303, 305, 309, 370, 388, 409

Keratinocyte 93, 94, 114, 134, 152, 191–200, 205, 206, 208–215, 224

Knockout serum replacement (KSR/KOSR).............. 43, 73, 74, 97, 115, 135, 153, 156, 162, 177, 194, 199, 206, 236, 251, 269, 279, 298, 316, 352, 360, 361, 372, 388, 389, 407, 430–432, 468, 469

L

Lentivirus
    collection 166
    infection...166–167
    titering... 466–468, 471–472
    transduction........................133–148, 167, 170, 471, 473

Lipid accumulation..356, 357

Liver engraftment assay..390, 394

Long-term hypoxia...13–23

Long-term maintenance
    hESC 14, 18–20

Lung epithelium..405–410

M

Magnetic-activated cell sorting 269

Manual microdissection...18

Matrigel
    coating 7, 11, 31, 141, 145, 147, 155, 159, 209, 211, 285, 287, 301, 303, 308, 309, 315, 316, 319, 371, 374–376, 379, 391, 400–402, 474
    matrix 148, 154, 270, 290, 293, 315, 375
    plug assay 281, 290–291

Mechanical passage 60, 62, 66, 67, 407

Medium...............................4, 14, 26, 38, 53, 61, 73, 86, 95, 115, 135, 152, 162, 174, 193, 205, 234, 251, 268, 279, 296, 314, 324, 340, 352, 361, 371, 388, 398, 406, 415, 428, 467

MEF. *See* Mouse embryonic fibroblasts (MEF)

2-mercaptoethanol 6, 9, 11, 28, 87, 97, 98, 116, 162, 164, 177, 207, 236, 269, 270, 274, 280, 284, 352–354, 360, 388, 398, 431, 469

Mesenchymal cells, murine fetal liver...............................398

Mesenchymal stem cells (MSCs)
    culture..174, 175, 179–180
    medium..175–176, 181, 185

Mesendoderm..452–456

Mesoderm 44, 48, 106, 109, 128, 130, 151, 152, 158, 159, 187, 188, 331, 351, 352, 445, 452, 460

Methylcellulose culture....................................... 341–344, 349

Micromass culture 359–366

MicroRNA 232, 465–479

Mitomycin-C
    inactivation 75–76, 161, 167
    treatment 161, 167, 171, 183–184, 291, 402

Mold.. 415–418, 421

Mouse...................................7, 16, 26, 43, 44, 47, 72, 73, 75, 97, 109, 127, 128, 130, 134, 151, 152, 154, 161, 162, 166, 168, 174, 179, 188, 189, 239, 243, 250, 256, 279, 282, 312, 323, 325, 327, 328, 354, 355, 372, 373, 378, 379, 394, 395, 439, 460, 461, 466

Mouse embryonic fibroblasts (MEF) 3, 26, 31, 51, 68, 74, 134, 154, 161, 192, 232, 251, 271, 309, 342, 360, 388, 400, 406, 429, 450, 468
    gamma-irradiation..161
    medium...................................... 156, 235, 236, 241, 251, 256–258, 260, 263, 389, 407, 409, 432, 434, 435, 468, 469
    mitomycin C treatment 75–76
    passaging 75, 139, 148, 284, 342
    preparation................. 193, 197–198, 235, 239–240, 279, 282–285, 363, 391, 400, 407, 429, 434

mTeSR–1 .... 4, 6–8, 10–12, 28, 31, 32, 87, 134, 135, 137, 141, 142, 144, 145, 147, 148, 154, 158, 245, 297, 302, 303, 305, 309, 314–316, 430, 435, 436

Multistage hepatic differentiation 387–396

Murine stromal cells..321–334

N

Nanog............. 17, 20, 22, 87, 89, 94, 95, 106, 109, 124, 140, 142–144, 154, 158, 191, 204, 205, 218, 221, 232, 242, 243, 303–304, 311, 343, 438, 439, 449–452, 455, 457, 458, 471

Nano-scale structure...418

Neural stem cells 93, 152, 192, 438

Neuroepithelial cells 145–147, 352

Neuronal lineage...133–148, 414

Neurospheres formation and expansion...................146–147

Nonessential amino acid (NEAA).............................. 15, 16, 27, 43, 53, 62, 64, 87, 117, 135, 136, 153, 156, 164, 193, 194, 196, 197, 199, 206, 207, 236, 270, 284, 314, 316, 360–362, 372, 378, 388, 389, 398, 407, 415, 429–432, 434, 446, 450, 469
    solution 73, 74, 97, 98, 116, 148, 162, 177, 269, 279, 298, 299, 324, 326, 352, 353, 432

Nonviral integration-free induced pluripotent
    stem cells..203–226

Nucleofection 205, 208, 212–214, 219, 224, 225

O

Oct4 .... 16, 17, 20–22, 40, 43, 44, 54, 87, 89, 93–110, 133, 136, 137, 142, 154, 156, 158, 191, 192, 195, 196, 204, 205, 218, 221, 232–234, 240, 242, 243, 311, 328, 343, 381, 471, 475

Oil Red O staining 356, 357
Oxygen plasma 415, 418, 419

P

Passaging 14, 18–20, 22, 23, 37, 40, 46, 51,
 60, 62–68, 80, 134, 136–137, 139, 142, 144, 147,
 166, 170, 171, 215–216, 222, 225, 250, 309, 316,
 330, 371, 375–376, 391, 407, 434–437, 478
Patient specific iPS cells 94, 231, 232,
 250, 322, 337
Periodic acid-Schiff (PAS) assay 388, 390, 393
Phenotype 72, 93, 147, 216, 217,
 220, 225, 249, 297, 302–304, 331, 387, 391,
 438, 450
Photoinitiator 417
Pioglitazone 354, 355
Placenta 75, 134, 239, 249, 250, 253, 254,
 256, 328, 329
Pluripotency 4, 14, 17, 20–22, 26, 40, 51, 72,
 93–95, 106, 109, 126, 127, 134, 140, 142, 143,
 152, 158, 159, 187, 189, 191, 204, 211, 216,
 220, 221, 225, 232, 242, 250, 295, 297, 302,
 309, 322, 338, 370, 397, 413, 438–440, 450,
 452–456, 458, 471, 476, 477, 479
Pluripotent stem cell
 culture 25–33, 62, 63, 66, 67
 expansion 59–68
 hepatic endoderm differentiation 370, 371
Polybrene 97, 103, 115, 122, 135, 140, 153,
 157, 163, 164, 167, 168, 177, 185, 193, 196, 197,
 236, 240, 251, 260
Polyethylene terephthalate (PET) 415, 417
Polyurethane acrylate (PUA) 415–418, 420, 421
Precursor 71, 268, 296, 338, 417,
 418, 438
Primary cell culture 234, 237–238
Primary hepatocyte 152, 159, 369, 370
Primordial germ 71–82

R

Radiation curable 417
Real-time PCR 17, 21–22, 307, 448, 471
Recombinant E-Cadherin-IgG Fc
 fusion protein 25–33
 production 26–29
 purification 25–33
Replica 415–418
Reprogramming
 CD34⁺ cord blood cells 115–117, 120–124
 somatic cells 93, 137, 174, 191, 204, 232, 233,
 235–236, 240, 250, 322, 370
Retrovirus
 collection 196
 concentration 156, 244

infection 103–105, 110, 120, 123, 168–170,
 235, 240, 244
production 97, 102–103, 153, 156, 167–169,
 184–185, 233–235, 238–239, 244
transduction 97, 103–105, 115, 119–120,
 192, 232, 233
vector 95, 97, 113, 119, 136,
 137, 153, 156, 162, 177, 184, 191–200, 234,
 238, 250
Reverse transcriptase-polymerase chain reaction
 (RT-PCR) 20, 54, 124, 126, 127,
 140, 141, 143, 144, 184, 208, 217–219,
 270–271, 274, 307, 349
Ridge/grove pattern 413–421
RT-PCR. *See* Reverse transcriptase-polymerase chain
 reaction (RT-PCR)

S

Scaffold 52, 414, 427, 428
Scale-up 37–49, 52, 110
Scanning electron microscopy (SEM) 415, 418, 419
SCID mouse
 grafting 127
 operation 127, 128
 tumor removal 127, 128
SDS-PAGE 28–31
Self-renewal 14, 18, 38, 60, 147, 161, 171, 231,
 295, 303, 359, 413, 428
Semi-quantitative real-time PCR 21–22, 357, 382
Silicon master 414, 416, 417
Single-cell
 inoculation 39
 transcript profiling 445–461
Somatic stem cells 13, 173
Southern blotting 204, 219–220
Sox2 .... 87, 93–110, 115, 119, 122, 123,
 133, 136, 137, 143, 144, 153, 156, 162, 168, 174,
 184, 189, 191–193, 195, 196, 204, 205, 216, 221,
 232–234, 240, 242, 243, 250, 259, 260, 311, 343,
 439, 449, 450, 455, 457, 458
Sputter 418
Stemness 52, 125, 322, 450
StemPror hESC SFM 4, 6, 8–9, 11

T

Teratoma
 assay 109, 118, 122, 127, 128, 152
 formation 40, 42, 47, 118, 122,
 127, 128, 158–159, 179, 188, 189, 216, 217, 242,
 312, 322
Thawing 6, 86–89, 95, 96, 99–100, 117,
 130, 165, 207, 220, 223–224, 226, 244, 252,
 257–258, 262–263, 274, 284, 285, 292,
 293, 308, 315, 330, 353, 434

Three dimensional (3D)
  cell culture 38, 51–57, 277, 312, 341
  differentiation 51–57
  matrix 428
  neural differentiation 427–441
Topography 414, 418
Transcript, gene expression profiling 446
Transduction
  lentiviral vector 133–148, 170, 473
  retroviral vector...........................103–105, 115, 119–120, 170, 192, 232, 233
Transfection........................ 97, 103, 104, 115, 119, 136–138, 163, 166, 168, 177, 183, 193, 195–196, 200, 204, 235, 238, 243, 244, 251, 259, 370, 387, 471, 472
Transgene 97, 133, 142–144, 199, 204–206, 210, 211, 217–219, 224, 232, 250, 466, 467, 478
Transgenesis 466
Trypsin 15, 27, 49, 56, 62, 73, 102, 115, 135, 154, 162, 177, 205, 234, 251, 269, 279, 299, 314, 324, 340, 352, 360, 389, 398, 446, 466
Two-dimensional (2D) endothelial
  differentiation 267–275

U

Ultraviolet (UV)-assisted
  curable 417
  light 417, 418, 466
Umbilical cord blood cells 100–101

Undifferentiated human embryonic stem cell.................3–12
Up-scaling 61
Ureagenesis...374, 384

V

Valproic acid (VPA) 177, 185, 233, 236, 240
Vascular endothelial
  cells..277–294
  differentiation 277, 278, 282, 288–289, 292
  growth factor 135, 136, 268, 270, 273, 280, 282, 296, 297, 299, 314, 316, 326, 339, 345
  maturation 288–289
Vectors.........................95, 97, 113, 114, 119, 136–138, 140, 143, 144, 153, 156, 162, 177, 183, 184, 189, 191–200, 203–205, 210, 211, 217–219, 225, 232, 234, 238, 240, 245, 250–251, 259, 466, 467, 471–472, 475, 476, 478, 479
VEGF. *See* Vascular endothelial, growth factor
Viability assessment... 88–89, 100
Viral transfection..370, 387
Visualization of neural differentiation 465–479
Vitronectin 4, 6, 10, 12, 26

X

Xeno-free
  cell culture system..59–68
  medium..61